Introduction to Computational Chemistry

Frank Jensen
Odense University, Odense, Denmark

JOHN WILEY & SONS

Chichester · New York · Weinheim · Brisbane · Singapore · Toronto

Other Wiley Editorial Offices

John Wiley & Sons, Inc., 605 Third Avenue,
New York, NY 10158-0012, USA

WILEY-VCH Verlag GmbH, Pappelallee 3,
D-69469 Weinheim, Germany

Jacaranda Wiley Ltd, 33 Park Road, Milton,
Queensland 4064, Australia

John Wiley & Sons (Asia) Pte Ltd, 2 Clementi Loop #02-01,
Jin Xing Distripark, Singapore 129809

John Wiley & Sons (Canada) Ltd, 22 Worcester Road,
Rexdale, Ontario M9W 1L1, Canada

British Library Cataloguing in Publication Data

A catalogue record for this book is available from the British Library

ISBN 0 471 98055 4; 0 471 98425 6 (pbk.)

Typeset in 10/12 Times by Thomson Press (India) Ltd., New Delhi
Printed and bound in Great Britain by Bookcraft (Bath) Ltd
This book is printed on acid-free paper responsibly manufactured from sustainable forestation,
for which at least two trees are planted for each one used for paper production.

Contents

Preface

Computational chemistry is rapidly emerging as a subfield of theoretical chemistry, where the primary focus is on solving chemically related problems by calculations. For the newcomer to the field there are three main problems:

(1) Deciphering the code. The language of computational chemistry is littered with acronyms, what do these abbreviations stand for in terms of underlying assumptions and approximations?
(2) Technical problems. How does one actually run the program and what does one look for in the output?
(3) Quality assessment. How good is the number that has been calculated?

Point (1) is part of every new field: there is not much you can do about it. If you want to live in another country, you have to learn the language. If you want to use computational chemistry methods, you need to learn the acronyms. I have tried in the present book to include a good fraction of the most commonly used abbreviations and standard procedures.

Point (2) is both hardware and software specific. It is not well suited to a textbook, as the information rapidly becomes out of date. The average lifetime of computer hardware is a few years, the time between new versions of software is even less. Problems of type (2) need to be solved "on location". I have made one exception, however, and have included a short discussion of how to make Z-matrices. A Z-matrix is a convenient way of specifying a molecular geometry in terms of internal coordinates, and it is used by many electronic structure programs. Furthermore, geometry optimizations are often performed in Z-matrix variables, and since optimizations in a good set of internal coordinates are significantly faster than in Cartesian, it is important to have a reasonable understanding of Z-matrix construction.

As computer programs evolve they become easier to use. Modern programs often communicate with the user in terms of a graphical interface, and many methods have become essential "black box" procedures: if you can draw the molecule, you can also do the calculation. This effectively means that you no longer have to be a highly trained theoretician to run even quite sophisticated calculations.

The ease by which calculations can be performed means that point (3) has become the central theme in computational chemistry. It is quite easy to run a series of calculations which produce results that are absolutely meaningless. The program will not tell you whether the chosen method is valid for the problem you are studying. Quality assessment is thus an absolute requirement. This, however, requires much more experience and insight than just running the program. Basic understanding of the theory behind the method is needed, and knowledge of the performance of the method for other systems. If you are breaking new ground, where there is no previous experience, you need a way of calibrating the results.

The lack of quality assessment is probably one of the reasons why computational chemistry has (had) a somewhat bleak reputation. "If five different computational methods give five widely different results, what has computational chemistry contributed? You just pick the number closest to experiments and claim that you can reproduce experimental data accurately." One commonly see statements of the type "The theoretical results for property X are in disagreement. Calculation at the CCSD(T)/6-31G(d,p) level predicts that ..., while the MINDO/3 method gives opposing results. There is thus no clear consent from theory." This is clearly a lack of understanding of the quality of the calculations. If the results disagree, there is a very high probability that the CCSD(T) results are basicly correct, and the MINDO/3 results are wrong. If you want to make predictions, and not merely reproduce known results, you need to be able to judge the quality of your results. This is by far the most difficult task in computational chemistry. I hope the present book will give some idea of the limitations of different methods.

Computers don't solve problems, people do. Computers just generate numbers. Although computational chemistry has evolved to the stage where it often can be competitive with experimental methods for generating a value for a given property of a given molecule, the number of possible molecules (there are an estimated 10^{200} molecules with a molecular weight less than 850) and their associated properties is so huge that only a very tiny fraction will ever be amenable to calculation (or experiment). Furthermore, with the constant increase in computational power, a calculation which can barely be done today will be possible on medium sized machines in 5–10 years. Prediction of properties using methods that do not provide converged results (with respect to theoretical level) will typically only have a lifetime of a few years before being surpassed by prediction using more accurate calculations.

The real strength of computational chemistry is the ability to generate data (for example by analysing the wave function) from which a human may gain *insight*, and thereby rationalize the behaviour of a large class of molecules. Such insights and rationalizations are much more likely to be useful over a longer period of time than the raw results themselves. A good example is the concept used by organic chemists with molecules composed of functional groups, and representing reactions by "pushing electrons". This may not be particularly accurate from a quantum mechanical point of view, but it is very effective in rationalizing a large body of experimental results, and has good predictive power.

Just as computers do not solve problems, mathematics by itself does not provide insight. It merely provides formulas, a framework for organizing thoughts. It is in this spirit that I have tried to write this book. Only the necessary mathematical background (obviously a subjective criterion) has been provided, the aim being that the reader should be able to understand the premises and limitations of different methods, and follow the main steps in running a calculation. This means that in many cases I have omitted to tell the reader of some of the finer details, which may annoy the purists. However, I believe the large overview is necessary before embarking on a more stringent and detailed derivation of the mathematics. The goal of this book is to provide an overview of commonly used methods, giving enough theoretical background to understand why for example the AMBER force field is used for modelling proteins but MM2 is used for small organic molecules, or why Coupled Cluster inherently is an iterative method, while Perturbation Theory and Configuration Interaction inherently are non-iterative methods, although the CI problem in practice is solved by iterative techniques.

The prime focus of this book is on calculating molecular structures and (relative) energies, and less on molecular properties or dynamical aspects. In my experience, predicting structures and energetics is the main use of computational chemistry today, although this may well change in the coming years. I have tried to include most methods that are already extensively used, together with some that I expect to become generally available in the near future. The amount of detailing in the description of the methods depends partly on how practical and commonly used the methods are (both in terms of computational resources and software), and partly reflects my own limitations in terms of understanding. Although simulations (e.g. molecular dynamics) are becoming increasingly powerful tools, only a very rudimentary introduction is provided in Chapter 16. This area is outside my expertise, and several excellent textbooks are already available.

Computational chemistry contains a strong practical element. Theoretical methods must be translated into working computer programs in order to produce results. Different algorithms, however, may display different types of behaviour in practice, and it becomes necessary to be able to evaluate whether a certain type of calculation can be carried out with the available computers. The book thus contains some guidelines for evaluating the type of resources necessary for carrying out a given calculation.

The present book grew out of a series of lecture notes that I have used for teaching a course in computational chemistry at Odense University, and the style of the book reflects it origin. It is difficult to master all disciplines in the vast field of computational chemistry. A special thanks to H. J. Aa. Jensen, K. V. Mikkelsen, T. Saue, S. P. A. Sauer, M. Schmidt, P. M. W. Gill, P.-O. Norrby, D. L. Cooper, T. U. Helgaker and H. G. Petersen for having read various parts of the book and provided input. Remaining errors are of course my sole responsibility. A special thanks to M. Robb for providing the figure upon which the cover is based. A good part of the final transformation from a set of lecture notes to the present book was done during sabbatical leave spent with Prof. L. Radom at the Research School of Chemistry, Australia National University, Canberra, Australia. A special thanks to him for his hospitality during the stay.

A few comments on the layout of the book. Definitions or common phrases are marked in *italic*; these can be found in the index. <u>Underline</u> is used for emphasizing important points. Operators, vectors and matrices are denoted in **bold**, scalars in normal text. Although I have tried to keep the notation as consistent as possible, different branches in computational chemistry often use different symbols for the same quantity. In order to comply with common usage, I have elected sometimes to switch notation between chapters. The second derivative of the energy, for example, is called the force constant k in force field theory, the corresponding matrix is denoted $\mathbf{F}$ when discussing vibrations, and called the Hessian $\mathbf{H}$ for optimization purposes.

I have assumed that the reader has no prior knowledge of concepts specific to computational chemistry, but has a working understanding of introductory quantum mechanics and elementary mathematics, especially linear algebra, vector, differential and integral calculus. The following features specific to chemistry are used in the present book without further introduction. Adequate descriptions may be found in a number of quantum chemistry textbooks (J. P. Lowe, *Quantum Chemistry*, Academic Press, 1993; I. N. Levine, *Quantum Chemistry*, Prentice Hall, 1992; P. W. Atkins, *Molecular Quantum Mechanics*, Oxford University Press, 1983).

(1) The Schrödinger equation, with the consequences of quantized solutions and quantum numbers.
(2) The interpretation of the square of the wave function as a probability distribution, the Heisenberg uncertainty principle and the possibility of tunnelling.
(3) The solutions for the hydrogen atom, atomic orbitals.
(4) The solutions for the harmonic oscillator and rigid rotor.
(5) The molecular orbitals for the H_2 molecule generated as a linear combination of two *s*-functions, one on each nuclear centre.
(6) Point group symmetry, notation and representations, and the group theoretical condition for when an integral is zero.

I have elected to include a discussion of the variational principle and perturbational methods, although these are often covered in courses in elementary quantum mechanics. The properties of angular momentum coupling are used at the level of knowing the difference between a singlet and a triplet state. I do not believe that it is necessary to understand the details of vector coupling to understand the implications.

Although I have tried to keep each chapter as self-contained as possible, there are unavoidable dependencies. The part in Chapter 3 describing HF methods is prequisite for understanding Chapter 4. Both these chapters use terms and concepts for basis sets which are treated in Chapter 5. Chapter 5 in turn, relies on concepts in Chapters 3 and 4, i.e. these three chapters form the core for understanding modern electronic structure calculations. Many of the concepts in Chapters 3 and 4 are also used in Chapters 6, 7, 9, 11 and 15 without further introduction, although these five Chapters probably can be read with some benefit without a detailed understanding of Chapters 3 and 4. Chapter 8 and to a certain extent also Chapter 10 are fairly advanced for an introductory textbook, such as this one, and can be skipped. They do, however, represent areas that are probably going to be more and more important in the coming years. Function optimization, which is described separately in Chapter 14, is part of many areas, but a detailed understanding is not required for following the arguments in the other chapters. Chapters 12 and 13 are fairly self-contained, and form some of the background for the methods in the other chapters. In my own course I normally take Chapters 12, 13, and 14 fairly early in the course, as they provide background for Chapters 3, 4 and 5.

If you would like to make comments, advise me of possible errors, make clarifications, add references etc., or view the current list of misprints and corrections, please visit the author's web site (URL:http://bogense.chem.ou.dk/~icc).

1 Introduction

Chemistry is the science dealing with construction, transformation and properties of molecules. Theoretical chemistry is the subfield where mathematical methods are combined with fundamental laws of physics to study processes of chemical relevance (some books in the same area are given in reference 1).

Molecules are traditionally considered as being "composed" of atoms or, in a more general sense, as a collection of charged particles, positive nuclei and negative electrons. The only important physical force for chemical phenomena is the Coulombic interaction between these charged particles. Molecules differ because they contain different nuclei and number of electrons, or the nuclear centres may be in different geometrical positions. Those in the latter categories may be "chemically different" molecules like ethanol and dimethyl ether, or different "conformations" of, for example, butane. Given a set of nuclei and electrons, theoretical chemistry can attempt to calculate things such as:

(1) The geometrical arrangements of the nuclei that correspond to stable molecules.
(2) Their relative energies.
(3) Their properties (dipole moment, polarizability, NMR coupling constants etc.).
(4) The rate by which one stable molecule can transform into another.
(5) The time dependence of molecular structures and properties.

The only systems that can be solved exactly are those composed of only one or two particles. The latter can be separated into two pseudo one-particle problems by introducing a "centre-of-mass" coordinate system. Numerical solutions to a given accuracy (which may be so high that the solutions are essentially "exact") can in many cases be generated for many-body systems, but only by performing a very large number of mathematical operations. Prior to the advent of electronic computers (i.e. before 1950), the number of systems that could be treated with a high accuracy was thus very limited. During the sixties and seventies electronic computers evolved from a few very expensive, difficult to use, machines to become generally available for researchers all over the world. The performance for a given price has been steadily increasing and the use of computers is now widespread in many branches of science. This has spawned a new field in chemistry, computational chemistry, where the computer is used as an "experimental" tool, much like, for example, an NMR spectrometer.

Computational chemistry is focused on obtaining results relevant to chemical problems, not directly on developing new theoretical methods. There is of course a strong interplay between traditional theoretical chemistry and computational chemistry. Developing new theoretical models may enable new problems to be studied, and results from calculations may reveal limitations and suggest improvements in the underlying theory. Depending on the accuracy wanted, and the nature of the system at hand, one can today obtain useful information for systems containing up to several thousand particles. One of the main problems in computational chemistry is selecting a suitable level of theory for a given problem, and being able to evaluate the quality of the obtained results. This book will try to put the variety of modern computational methods into perspective, hopefully giving the reader a chance of estimating which types of problem can benefit from calculations.

1.1 Background

A molecule may be considered as a number of electrons surrounding a set of positively charged nuclei. The Coulombic attraction between these two types of particle forms the basis for atoms and molecules. The potential between two particles with charges q_i and q_j separated by a distance r_{ij} is (in suitable units) given by

$$V_{ij} = V(r_{ij}) = \frac{q_i q_j}{r_{ij}} \tag{1.1}$$

Besides the interaction potential, an equation is also needed for describing the dynamics of the system, i.e. how the system evolves in time. In classical mechanics this is Newton's second law (**F** is the force, **a** is the acceleration, **r** is the position vector and m the particle mass).

$$\mathbf{F} = m\mathbf{a}$$
$$-\frac{dV}{d\mathbf{r}} = m\frac{d^2\mathbf{r}}{dt^2} \tag{1.2}$$

Electrons are very light particles and cannot be described by classical mechanics. They display both wave and particle characteristics, and must be described in terms of a wave function, Ψ. The quantum mechanical equation corresponding to Newtons second law is the time-dependent Schrödinger equation ($\hbar$ is Plancks constant divided by 2π).

$$\mathbf{H}\Psi = i\hbar\frac{\partial\Psi}{\partial t} \tag{1.3}$$

If the Hamilton operator, **H**, is independent of time, the time dependence of the wave function can be separated out as a simple phase factor.

$$\mathbf{H}(\mathbf{r}, t) = \mathbf{H}(\mathbf{r})$$
$$\Psi(\mathbf{r}, t) = \Psi(\mathbf{r})e^{-iEt/\hbar} \tag{1.4}$$
$$\mathbf{H}(\mathbf{r})\Psi(\mathbf{r}) = E\Psi(\mathbf{r})$$

The time-independent Schrödinger equation describes the particle–wave duality, the square of the wave function giving the probability of finding the particle at a given position.

For a general *N*-particle system the Hamilton operator contains kinetic (**T**) and potential (**V**) energy for all particles.

$$\mathbf{H} = \mathbf{T} + \mathbf{V}$$

$$\mathbf{T} = \sum_{i=1}^{N} \mathbf{T}_i = -\sum_{i=1}^{N} \frac{\hbar^2}{2m_i} \mathbf{V}_i^2$$

$$\mathbf{V}_i^2 = \left(\frac{\partial^2}{\partial x_i^2} + \frac{\partial^2}{\partial y_i^2} + \frac{\partial^2}{\partial z_i^2} \right) \tag{1.5}$$

$$\mathbf{V} = \sum_{i=1}^{N} \sum_{j>1}^{N} V_{ij}$$

where the potential energy operator is the Coulomb potential (eq. (1.1)). As nuclei are much heavier than electrons, their velocities are much smaller. The Schrödinger equation can therefore to a good approximation be separated into one part which describes the electronic wave function for a fixed nuclear geometry, and another part which describes the nuclear wave function, where the energy from the electronic wave function plays the role of a potential energy. This separation is called the *Born–Oppenheimer* (BO) approximation. The electronic wave function depends parametrically on the nuclear coordinates, it depends only on the position of the nuclei, not their momenta. The picture is that the nuclei move on *Potential Energy Surfaces* (PES), which are solutions to the electronic Schrödinger equation. Denoting nuclear coordinates with **R** and subscript n, and electron coordinates with **r** and e, this can be expressed as follows.

$$\mathbf{H}_{tot} \Psi_{tot}(\mathbf{R}, \mathbf{r}) = E_{tot} \Psi_{tot}(\mathbf{R}, \mathbf{r})$$

$$\mathbf{H}_{tot} = \mathbf{H}_e + \mathbf{T}_n$$

$$\mathbf{H}_e = \mathbf{T}_e + \mathbf{V}_{ne} + \mathbf{V}_{ee} + \mathbf{V}_{nn}$$

$$\Psi_{tot}(\mathbf{R}, \mathbf{r}) = \Psi_n(\mathbf{R}) \Psi_e(\mathbf{R}, \mathbf{r}) \tag{1.6}$$

$$\mathbf{H}_e \Psi_e(\mathbf{R}, \mathbf{r}) = E_e(\mathbf{R}) \Psi_e(\mathbf{R}, \mathbf{r})$$

$$(\mathbf{T}_n + E_e(\mathbf{R})) \Psi_n(\mathbf{R}) = E_{tot} \Psi_n(\mathbf{R})$$

The Born–Oppenheimer approximation is usually very good. For the hydrogen molecule the error is of the order of 10^{-4}, and for systems with heavier nuclei, the approximation becomes better. As we shall see later, it is only possible in a few cases to solve the electronic part of the Schrödinger equation to an accuracy of 10^{-4}, i.e. neglect of the nuclear–electron coupling is usually only a minor approximation compared with other errors.

Once the electronic Schrödinger equation has been solved for a large number of nuclear geometries (and possibly also for several electronic states), the PES is known. This can then be used for solving the nuclear part of the Schrödinger equation. If there are *N* nuclei, there are 3*N* coordinates that define the geometry. Of these coordinates, three describe the overall translation of the molecule, and three describe the overall rotation of the molecule with respect to three axes. For a linear molecule, only two coordinates are necessary for describing the rotation. This leaves 3N–6(5) coordinates to describe the internal movement of the nuclei, the vibrations, often chosen to be "vibrational normal coordinates". It should be noted that nuclei are heavy enough for quantum effects to be almost negligible, they behave to a good approximation as classical particles. Indeed, if nuclei showed significant quantum aspects, the concept of molecular structure (i.e. different configurations and conformations) would not have any meaning, the nuclei would simply tunnel through barriers and end up in the global minimum. Furthermore, it would not be possible to speak of a molecular geometry, since the Heisenberg uncertainty principle would not permit a measure of nuclear positions to an accuracy much smaller than the molecular dimension.

Methods aimed at solving the electronic Schrödinger equation are broadly referred to as "electronic structure calculations". An accurate determination of the electronic wave function is very demanding. Constructing a complete PES for molecules containing more than 3–4 atoms is virtually impossible. Consider for example mapping the PES by calculating E_e for every 0.1 Å over say a 1 Å range (a very coarse mapping). With three atoms there are three internal coordinates, giving 10^3 points to be calculated. Already four atoms produces six internal coordinates, giving 10^6 points, which only can be done by a very determined effort. Larger systems are out of reach. Constructing global PESs for all but the smallest molecules is thus impossible. However, by restricting the calculations to the "chemically interesting" part of the PES it is possible to obtain useful information. The interesting parts of a PES are usually nuclear arrangements which have low energies. For example, nuclear movements near a minimum on the PES, which corresponds to a stable molecule, are molecular vibrations. Chemical reactions correspond to larger movements, and may in the simplest approximation be described by locating the lowest energy path leading from one minimum on the PES to another. These considerations lead to the following definition:

Chemistry is knowing the energy as a function of the nuclear coordinates.

The large majority of what is commonly referred to as molecular properties may similarly be defined as:

Properties are knowing how the energy changes upon adding a perturbation.

In the following we will look at some aspects of solving the electronic Schrödinger equation, how to deal with the movement of nuclei on the PES, and various technical points of commonly used methods. A word of caution here. Although it is the nuclei that move, and the electrons follow "instantly" (according to the Born–Oppenheimer approximation), it is common also to speak of "atoms" moving. An isolated atom consists of a nucleus and some electrons, but in a molecule the concept of an atom is not well defined. Analogously to the isolated atom, an atom in a molecule should consist of a nucleus and some electrons. But how does one partition the total electron distribution in a molecule so that a given portion belongs to a given nucleus? Nevertheless, the words nucleus and atom are often used interchangeably.

Much of the following will concentrate on describing individual molecules. Experiments are rarely done on single molecules, rather they are performed on macroscopic samples with perhaps 10^{23} molecules. The link between the properties of a single molecule, or a small collection of molecules, and the macroscopic observable, is statistical mechanics. Briefly, macroscopic properties, such as temperature, heat capacities, entropies etc., are the net effect of a very large number of molecules having a certain distribution of energies. If all the possible energy states can be determined for an individual molecule or a small collection of molecules, statistical mechanics can be used for calculating macroscopic properties.

Reference

1. A. R. Leach, *Molecular Modelling. Principles and Applications*, Longman, 1996; A. Hinchcliffe *Modelling Molecular Structure*, Wiley, 1996; D. M. Hirst, *A Computational Approach to Chemistry*, Blackwell, 1990; T. Clark, *A Handbook of Computational Chemistry*, Wiley, 1985.

2 Force Field Methods

2.1 Introduction

As mentioned in Chapter 1, one of the major problems is calculating the electronic energy for a given nuclear configuration. In *Force Field* (FF) methods this step is bypassed by writing E_e as a parametric function of the nuclear coordinates. The parameters that enter the function are fitted to experimental or higher level computational data. The molecules are modelled as atoms held together by bonds. The atoms may have different sizes and "softness", and the bonds may be more or less "stiff", i.e. the molecule is described by a "ball and spring" model.[1] Force field methods are therefore also referred to as *Molecular Mechanics* (MM) methods. Many different force fields exist, in this chapter we will use Allingers MM2 and MM3 (Molecular Mechanics versions 2 and 3) to illustrate specific details.[2]

In addition to bypassing the solution of the electronic Schrödinger equation, the quantum aspects of the nuclear motion are also neglected. This means that the dynamics of the atoms is treated by classical mechanics, i.e. Newton's second law. For time-independent phenomena, the problem reduces to calculating the energy at a given geometry. Often the interest is in finding geometries of stable molecules and/or different conformations, and possibly also interconversions between conformations. The problem is then reduced to finding energy minima (and possibly also some first-order saddle points) on the potential energy surface.

The foundation of force field methods is the observation that molecules tend to be composed of units which are structurally similar in different molecules. For example, all C–H bond lengths are roughly constant in all molecules, between 1.06 and 1.10 Å. The C–H stretch vibrations are also similar, between 2900 and 3300 cm^{-1}, implying that the C–H force constants are also comparable. If the C–H bonds are further divided into groups, like those attached to single, double or triple bonded carbon, the variation within each of these groups becomes even less. The same grouping holds for other features as well, e.g. all C=O bonds are approximately 1.21 Å long and have vibrational frequencies $\sim 1700\,\text{cm}^{-1}$, all doubly bonded carbons are essentially planar etc. The transferability also holds for energetic features. If for example the heat of formation of linear alkanes, $CH_3(CH_2)_nCH_3$, is plotted as a function of n, a straight line is observed. This shows that each CH_2 group contributes essentially the same amount of energy (for a general discussion of estimating heat of formation from group additivities see Ref. 3.

The picture of molecules being composed of structural units, "functional groups", which behave similarly in different molecules forms the very basis of organic chemistry. The drawing of molecular structures where alphabetic letters represent atoms and lines represent bonds is used universally. Organic chemists often build ball and stick, or CPK space-filling, models of their molecules to examine their shapes. Force field methods are in a sense a generalization of these models, with the added feature that the atoms and bonds are not fixed at one size and length. Furthermore, force field calculations enable predictions of relative energies and barriers for interconversion of different conformations.

Table 2.1 MM2(91) atom types

Type	Symbol	Description	Type	Symbol	Description
1	C	sp^3-carbon	28	H	enol or amide
2	C	sp^2-carbon, alkene	48	H	ammonium
3	C	sp^2-carbon, carbonyl, imine	36	D	deuterium
4	C	sp-carbon	20	lp	lone pair
22	C	cyclopropane	15	S	sulfide (R_2S)
29	C·	radical	16	S+	sulfonium (R_3S^+)
30	C+	carbocation	17	S	sulfoxide (R_2SO)
38	C	sp^2-carbon, cyclopropene	18	S	sulfone (R_2SO_2)
50	C	sp^2-carbon, aromatic	42	S	sp^2-sulfur, thiophene
56	C	sp^3-carbon, cyclobutane	11	F	fluoride
57	C	sp^2-carbon, cyclobutene	12	Cl	chloride
58	C	carbonyl, cyclobutanone	13	Br	bromide
67	C	carbonyl, cyclopropanone	14	I	iodide
68	C	carbonyl, ketene	26	B	boron, trigonal
71	C	ketonium carbon	27	B	boron, tetrahedral
8	N	sp^3-nitrogen	19	Si	silane
9	N	sp^2-nitrogen, amide	25	P	phosphine (R_3P)
10	N	sp-nitrogen	60	P	phosphor, pentavalent
37	N	azo or pyridine (−N=)	51	He	helium
39	N+	sp^3-nitrogen, ammonium (R_4N^+)	52	Ne	neon
40	N	sp^2-nitrogen, pyrrole	53	Ar	argon
43	N	azoxy (−N=N−O)	54	Kr	krypton
45	N	azide, central atom	55	Xe	xenon
46	N	nitro (−NO$_2$)	31	Ge	germanium
72	N	imine, oxime (=N−)	32	Sn	tin
6	O	sp^3-oxygen	33	Pb	lead (R_4Pb)
7	O	sp^2-oxygen, carbonyl	34	Se	selenium
41	O	sp^2-oxygen, furan	35	Te	tellurium
47	O⁻	carboxylate	59	Mg	magnesium
49	O	epoxy	61	Fe	iron(II)
69	O	amine oxide	62	Fe	iron(III)
70	O	ketonium oxygen	63	Ni	nickel(II)
5	H	hydrogen, except on N or O	64	Ni	nickel(III)
21	H	alcohol (OH)	65	Co	cobalt (II)
23	H	amine (NH)	66	Co	cobalt (III)
24	H	carboxyl (COOH)			

Note that special atom types are defined for carbon atoms involved in small rings, like cyclopropane and cyclobutane. The reason for this will be discussed in Section 2.2.2.

The idea of molecules being composed of atoms, which are structurally similar in different molecules, is implemented in force field models as atom *types*. The atom type depends on the atomic number and the type of chemical bonding it is involved in. The type may be denoted either by a number or by a simple letter code. In MM2, for example, there are 71 different atom types (type 44 is missing). Type 1 is an sp^3-hybridized carbon; an sp^2-hybridized carbon may be type 2, 3, or 50, depending on the neighbour atom(s). Type 2 is used if the bonding is to another sp^2-carbon (simple double bond), type 3 is used if the carbon is bonded to an oxygen (carbonyl group) and type 50 is used if the carbon is part of an aromatic ring with delocalized bonds. Table 2.1 gives a complete list of the MM2(91) atom types, where (91) indicates the year in which the parameter set was released. The atom type numbers roughly reflect the order in which the corresponding functional groups were parameterized.

2.2 The Force Field Energy

The force field energy is written as a sum of terms, each describing the energy required for distorting a molecule in a specific fashion.

$$E_{FF} = E_{str} + E_{bend} + E_{tors} + E_{vdw} + E_{el} + E_{cross} \tag{2.1}$$

E_{str} is the energy function for stretching a bond between two atoms, E_{bend} represents the energy required for bending an angle, E_{tors} is the torsional energy for rotation around a bond, E_{vdw} and E_{el} describe the non-bonded atom–atom interactions, and finally E_{cross} describes coupling between the first three terms. Given such an energy function of the nuclear coordinates, geometries and relative energies can then be calculated. Stable molecules correspond to minima on the potential energy surface, and they can be located by minimizing E_{FF} as a function of the nuclear coordinates. Exactly how such a multidimensional function optimization may be carried out is described in Chapter 14.

2.2.1 The Stretch Energy

E_{str} is the energy function for stretching a bond between two atom types A and B. In its simplest form, it is written as a Taylor expansion around a "natural", or "equilibrium", bond length R_0. Terminating the expansion at second order gives the expression

$$E_{str}(R^{AB} - R_0^{AB}) = E(0) + \frac{dE}{dR}(R^{AB} - R_0^{AB}) + \frac{1}{2}\frac{d^2E}{dR^2}(R^{AB} - R_0^{AB})^2 \tag{2.2}$$

The derivatives are evaluated at $R = R_0$ and the $E(0)$ term is normally set to zero; this is just the zero point for the energy scale. The second term is zero as the expansion is around the equilibrium value. In its simplest form the stretch energy can thus be written as

$$E_{str}(R^{AB} - R_0^{AB}) = k^{AB}(R^{AB} - R_0^{AB})^2 = k^{AB}(\Delta R^{AB})^2 \tag{2.3}$$

where k^{AB} is the "force constant" for the A–B bond. This is the form of a harmonic oscillator, the potential is quadratic in the displacement from the minimum.

The harmonic form is the simplest possible, and is in fact sufficient for determination of most equilibrium geometries. There are certain strained and crowded systems where the results from a harmonic approximation are significantly different from experimental values, and if the force field should be able to reproduce features such as vibrational frequencies, the functional form for E_{str} must be improved. The straightforward approach is to include more terms in the Taylor expansion:

$$E_{str}(\Delta R^{AB}) = k_2^{AB}(\Delta R^{AB})^2 + k_3^{AB}(\Delta R^{AB})^3 + k_4^{AB}(\Delta R^{AB})^4 + \cdots \qquad (2.4)$$

This of course has a price: more parameters have to be assigned.

Polynomial expansions of the stretch energy do not have the correct limiting behaviour. The cubic anharmonicity constant k_3 is normally negative, and if the Taylor expansion is terminated at third order, the energy will go toward $-\infty$ for long bond lengths. Minimization of the energy with such an expression can cause the molecule to fly apart if a poor starting geometry has been chosen. The quartic constant k_4 is normally positive and the energy will go toward $+\infty$ for long bond lengths if the Taylor series is terminated at fourth order. The correct limiting behaviour for a bond stretched to infinity is that the energy converges towards the dissociation energy. A simple function which satisfies this criteria is the Morse potential.[4]

$$E_{Morse}(\Delta R) = D[1 - e^{\alpha \Delta R}]^2 \qquad (2.5)$$

Here D is the dissociation energy and α is related to the force constant ($\alpha = \sqrt{k/2D}$). This function reproduces the actual behaviour quite accurately over a wide range of distances, as seen in Figure 2.1. There are, however, some difficulties with the Morse potential in actual applications. For long bond lengths the restoring force is quite small. Distorted structures, which may either be a poor starting geometry or one that develops during a simulation, will therefore display a slow convergence towards the equilibrium bond length. For minimization purposes and simulations at ambient temperatures (e.g. 300 K) it is sufficient that the potential is reasonably accurate to ~ 10 kcal/mol above the minimum (the average kinetic energy per atom is 0.89 kcal/mol at 300 K). Most force fields therefore employ a simple polynomium for the stretch energy. The number of parameters is often reduced by taking the cubic, quartic etc. constants as a constant fraction of the harmonic force constant. A popular method is to require that the nth order derivative at R_0 matches the corresponding derivative of the Morse potential. For a fourth-order expansion this leads to the following expression.

$$E_{str}(\Delta R^{AB}) = k^{AB}(\Delta R^{AB})^2[1 - \alpha(\Delta R^{AB}) + \tfrac{7}{12}\alpha^2(\Delta R^{AB})^2] \qquad (2.6)$$

The α constant is the same as that appearing in the Morse function, but is usually taken as a fitting parameter.

Figure 2.1 compares the performance of various functional forms for the stretch energy in CH_4. The "exact" form is taken from electronic structure calculations ([8,8]-CASSCF/6-311++G(2df,2pd)). The simple harmonic approximation (P2) is seen to be accurate to about ± 0.1 Å from the equilibrium geometry and the quartic approximation (P4) up to ± 0.3 Å. The Morse potential reproduces the real curve quite accurately up to an elongation of 0.8 Å, and of course it becomes exact again in the dissociation limit. For the large majority of systems, including simulations, the only important chemical region is up to ~ 10 kcal/mol above the bottom of the curve. In this region a fourth-order polynomial is essentially indistinguishable from either a Morse or the exact curve, as shown in Figure 2.2. Even a simple harmonic approximation does a quite good job.

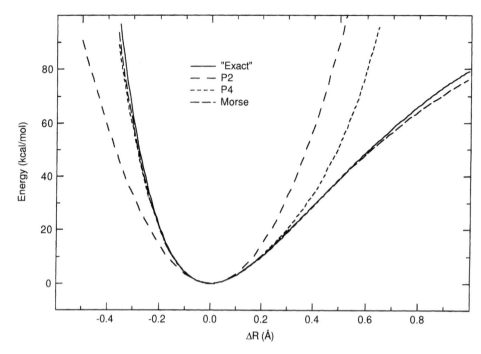

Figure 2.1   The stretch energy for CH_4

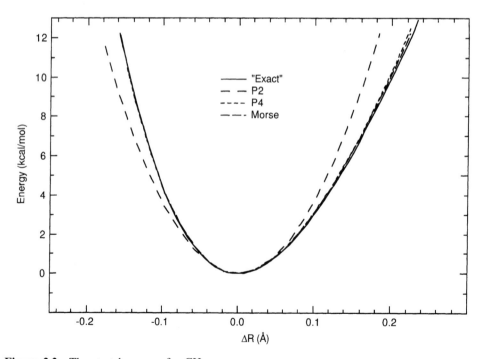

Figure 2.2   The stretch energy for CH_4

So far we have used two different words for the R_0 parameter, the "natural" or "equilibrium" bond length. The latter is slightly misleading. The R_0 parameter is not the equilibrium bond length for any molecule! Instead it is the parameter which, when used to calculate the minimum energy structure of a molecule, will produce a geometry having the experimental equilibrium bond length. If there was only one stretch energy in the whole force field energy expression (i.e. a diatomic molecule), R_0 would be the equilibrium bond length. However, in a polyatomic molecule the other terms in the force field energy will usually produce a minimum energy structure with bond lengths slightly longer than R_0. R_0 is the hypothetical bond length if no other terms are included, and the word "natural" bond length is a better description of this parameter than "equilibrium" bond length. Essentially all molecules have bond lengths which deviate very little from their "natural" values, typically less than 0.03 Å. For this reason a simple harmonic, or possibly a cubic expansion, is usually sufficient for reproducing experimental geometries.

For each bond type, i.e. a bond between two atom types A and B, there are at least two parameters to be determined, k^{AB} and R_0^{AB}. The higher-order expansions and the Morse potential have one additional parameter (α or D) to be determined.

2.2.2 The Bending Energy

E_{bend} is the energy required for bending an angle formed by three atoms A–B–C, where there is a bond between A and B, and between B and C. Similarly to E_{str}, E_{bend} is usually expanded as a Taylor series around a "natural" bond angle and terminated at second order, giving the harmonic approximation.

$$E_{bend}(\theta^{ABC} - \theta_0^{ABC}) = k^{ABC}(\theta^{ABC} - \theta_0^{ABC})^2 \tag{2.7}$$

While the simple harmonic expansion is quite adequate for most applications, there may be cases where higher accuracy is required. The next improvement is to include a third-order term, analogous to E_{str}. This can in general give a very good description over a large range of angle, as illustrated in Figure 2.3 for CH_4. The "exact" form is again taken from electronic structure calculations (MP2/6-311++G(3df,3pd)). The simple harmonic approximation (P2) is seen to be accurate to about $\pm 30°$ from the equilibrium geometry and the cubic approximation (P3) up to $\pm 70°$. In order to reproduce also for example vibrational frequencies, higher-order terms are often included. Analogous to E_{str}, the higher-order force constants are often taken as a fixed fraction of the harmonic constant. The constants beyond third order can rarely be assigned values with high confidence due to insufficient experimental information. Fixing the higher-order constants in terms of the harmonic constant of course reduces the quality of the fit. While a third-order polynomial is capable of reproducing the actual curve very accurately if the cubic constant is fitted independently, the assumption that it is a fixed fraction (independent of the atom type) of the harmonic constant deteriorates the fit, but it still represents an improvement relative to a simple harmonic approximation.

In the chemically important region below ~ 10 kcal/mol above the bottom, a second-order expansion is analogous to E_{str} normally sufficient, as illustrated in Figure 2.4.

Angles where the central atom is di- or trivalent (ethers, alcohols, sulfides, amines and enamines) represent a special problem. In these cases an angle of 180° corresponds to an

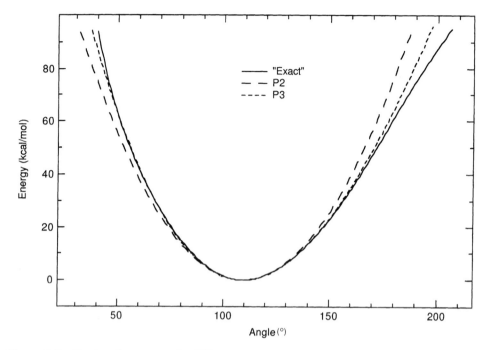

Figure 2.3   The bending energy for CH_4

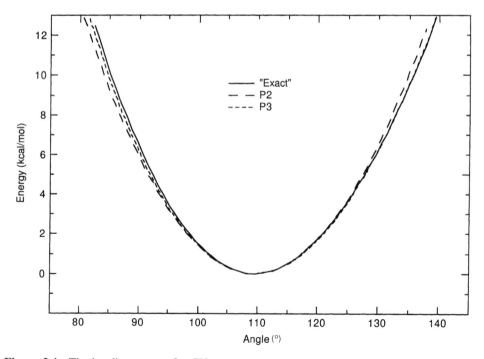

Figure 2.4   The bending energy for CH_4

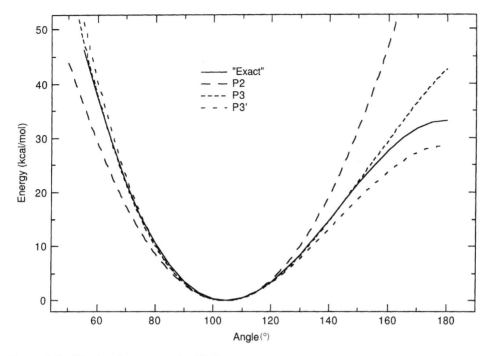

Figure 2.5   The bending energy for H_2O

energy maximum, i.e. the derivative of the energy with respect to the angle should be zero and the second derivative should be negative. This may be enforced by suitable boundary conditions on Taylor expansions of at least order three. A third-order polynomial fixes the barrier for linearity in terms of the harmonic force constant and the equilibrium angle $(\Delta E^{\neq} = k(\theta - \theta_0)^2/6)$. A fourth-order polynomial enables an independent fit of the barrier to linearity. Such constrained polynomial fittings are rarely done. Instead the bending function is taken to be identical for all atom types, for example a fourth-order polynomial with cubic and quartic constants as a fixed fraction of the harmonic constant.

These features are illustrated for H_2O in Figure 2.5, where the "exact" form is taken from a parametric fit to a large number of spectroscopic data.[5] The simple harmonic approximation (P2) is seen to be accurate to about $\pm 20°$ from the equilibrium geometry and the cubic approximation (P3) up to $\pm 40°$. Enforcing the cubic polynomial to have a zero derivative at $180°$ (P3') gives a qualitative correct behaviour, but reduces the overall fit, although it still is better than a simple harmonic approximation.

Although such refinements over a simple harmonic potential clearly improve the overall performance, they have little advantage in the chemically important region up to ~ 10 kcal/mol above the minimum. As for the stretch energy term, the energy cost for bending is so large that most molecules only deviate a few degrees from their natural bond angles. This again indicates that including only the harmonic term is adequate for most applications.

As noted above, special atom types are often defined for small rings. This is due to the very different equilibrium angles for such rings. Consider for example cyclopropane. The carbons are formally sp^3-hybridized, but have an equilibrium CCC angle of 60° in contrast to 110° in an acyclic system. With a low-order polynomial for the bend energy, the energy cost for such a deformation is large. For cyclobutane, for example, E_{bend} will dominate the total energy and cause the calculated structure to be planar, in contrast to the puckered geometry found experimentally.

For each combination of three atom types, A, B and C, there are at least two bending parameters to be determined, k^{ABC} and θ_0^{ABC}.

2.2.3 The Out-of-plane Bending Energy

If the central atom B in the angle is sp^2-hybridized, there is a significant energy penalty associated with making the centre pyramidal; the four atoms prefer to be located in a plane. If the four atoms are exactly in a plane, the sum of the three angles with B as the central atom should be exactly 360°. However, a quite large pyramidalization may be achieved without seriously distorting any of these three angles. Taking the bond distances to 1.5 Å, and moving the central atom 0.2 Å out of the plane, only reduces the angle sum to 354.8° (i.e. only a 1.7° decrease per angle). The corresponding out-of-plane angle, χ, is 7.7° for this case. If the ABC, ABD and CBD angle distortions should reflect the energy cost associated with the pyramidalization, very large force constants must be used. This would have as a consequence that the in-plane angle deformations for a planar structure become unrealistically stiff. For this reason, a special *out-of-plane energy bend* term (E_{oop}) is usually added, while the in-plane angles (ABC, ABD, CBD) are treated as in the general case above. E_{oop} may be written as a harmonic term in the angle χ (the equilibrium angle for a planar structure is zero) or as a quadratic function in the distance d, as shown in Figure 2.6 and eq. (2.8).

$$E_{oop}(\chi^B) = k^B(\chi^B)^2 \quad \text{or} \quad E_{oop}(d) = k^B d^2 \qquad (2.8)$$

Such terms may also be used for increasing the inversion barrier in sp^3-hybridized atoms (i.e. an extra energy penalty for being planar), and E_{oop} is also sometimes called E_{inv}. Inversion barriers are in most cases (e.g. in amines, NR$_3$) adequately modelled without an explicit E_{inv} term, the barrier arising naturally from the increase in bond angles upon inversion. The energy cost for non-planarity of sp^2-hybridized atoms may also be accounted for by an "improper" torsional energy, as described in Section 2.2.4.

For each sp^2-hybridized atom there is one additional out-of-plane force constant to be determined, k^B.

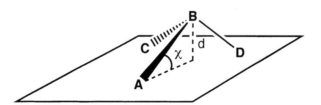

Figure 2.6 Out-of-plane variable definitions

2.2.4 The Torsional Energy

E_{tors} is the energy change associated with rotation around a B–C bond in a four atom sequence A–B–C–D, where A–B, B–C and C–D are bonded. Looking down the B–C bond, the torsional angle is defined as the angle formed by the A–B and C–D bonds as shown in Figure 2.7. The angle ω may be taken to be in the range $[0°, 360°]$ or $[-180°, 180°]$.

The torsional energy is fundamentally different from E_{str} and E_{bend} in two aspects. First, the energy function must be periodic in the angle ω: if the bond is rotated 360° the energy should return to the same value. Second, the cost in energy for distorting a molecule by rotation around a bond is often low, i.e. large deviations from the minimum energy structure may occur. A Taylor expansion in ω is therefore not a good idea. To encompass the periodicity, E_{tors} is written as a Fourier series.

$$E_{tors}(\omega) = \sum_{n=1} V_n \cos(n\omega) \tag{2.9}$$

The $n = 1$ term describes a rotation which is periodic by 360°, the $n = 2$ term is periodic by 180°, the $n = 3$ term is periodic by 120° and so on. The V_n constants determine the size of the barrier for rotation around the B–C bond. Depending on the situation, some of these V_n constants may be zero. Consider for example ethane. The most stable conformation is one where the hydrogens are staggered relative to each other, the eclipsed conformation represents an energy maximum. As the three hydrogens at each end are identical, it is clear that there are three energetically equivalent staggered, and three equivalent eclipsed, conformations. The rotational energy profile must therefore have three minima and three maxima. In the Fourier series only those terms which have $n = 3, 6, 9$ etc. can therefore have V_n constants different from zero.

For rotation around single bonds in substituted systems other terms may be necessary. In the butane molecule, for example, there are still three minima, but the two *gauche* (torsional angle $\sim \pm 60°$) and *anti* (torsional angle $= 180°$) conformations now have different energies. The barriers separating the two *gauche* and the *gauche* and *anti* conformations are also of different height. This may be introduced by adding a term corresponding to $n = 1$.

For the ethylene molecule the rotation around the C=C bond must be periodic by 180°, and thus only $n = 2, 4$ etc. terms can enter. The energy cost for rotation around a double bond is of course much higher than that for rotation around a single bond in ethane, which would be reflected in a larger value of the V_2 constant. For rotation around the C=C bond in a molecule like 2-butene, there would again be a large V_2 constant, analogously to ethylene, but in addition there are now two different orientations of the two methyl groups relative to each other, *cis* and *trans*. The full

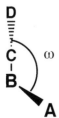

Figure 2.7 Torsional angle definition

rotation is periodic with a period of 360°, with deep energy minima at 0° and 180°, but with slightly different energies for these two minima. This energy difference would show up as an V_1 constant, i.e. the V_2 constant essentially determines the barrier and location of the minima for rotation around the C=C bond, and the V_1 constant determines the energy difference between the *cis* and *trans* isomers.

Molecules that are composed of atoms having a maximum valency of 4 (as essentially all organic molecules) are with a few exceptions found to have rotational profiles showing at most three minima. The first three terms in the Fourier series eq. (2.9) are sufficient for qualitatively reproducing such profiles. Force fields which are aimed at large systems often limit the Fourier series to only one term, depending on the bond type (e.g. single bonds only have cos (3ω) and double bonds only cos (2ω)).

Systems with bulky substituents on sp^3-hybridized atoms are often found to have four minima, the *anti* conformation being split into two minima with torsional angles of approximately $\pm 170°$. Other systems, notably polyfluoroalkanes, also split the *gauche* minima into two, often called *gauche* (angle of approximately $\pm 50°$) and *ortho* (angle of approximately $\pm 90°$) confomations, creating a rotational profile with six minima.[6] Rotations around a bond connecting sp^3- and sp^2-hybridized atoms (like CH_3NO_2) also display profiles with six minima.[7] These exceptions from the regular three minima rotational profile around single bonds are caused by repulsive and attractive van der Waals interactions, and can still be modelled by having only terms up to cos(3ω) in the torsional energy expression. Higher-order terms may be included to modify the detailed shape of the profile, and a few force fields employ terms with $n = 4$ and 6. Cases where higher-order terms probably are necessary are those with rotation around bonds to octahedral coordinated metals, such as Ru(pyridine)$_6$ or a dinuclear complex such as $Cl_4Mo-MoCl_4$. Here the rotation is periodic by 90° and thus require a cos(4ω) term.

It is customary to shift the zero point of the potential by adding a factor of 1 to each term. Most rotational profiles resemble either the ethane or ethylene examples above, and a popular expression for the torsional energy is

$$\begin{aligned}
E_{\text{tors}}(\omega^{\text{ABCD}}) = &\tfrac{1}{2} V_1^{\text{ABCD}}[1 + \cos{(\omega^{\text{ABCD}})}] \\
&+ \tfrac{1}{2} V_2^{\text{ABCD}}[1 - \cos{(2\omega^{\text{ABCD}})}] \\
&+ \tfrac{1}{2} V_3^{\text{ABCD}}[1 + \cos{(3\omega^{\text{ABCD}})}]
\end{aligned} \tag{2.10}$$

The $+$ and $-$ signs are chosen so that the onefold rotational term has a minimum for an angle of 180°, the twofold rotational term minima for angles of 0° and 180°, and the threefold rotational term minima for angles of 60°, 180° and 300° $(-60°)$ as illustrated in Figure 2.8. The factor 1/2 is included so that the V_i parameters directly give the height of the barrier if only one term is present. The V_i parameters may also be negative, which corresponds to changing the minima on the rotational energy profile to maxima, and vice versa.

As mentioned in Section 2.2.3, the out-of-plane energy may also be described by an "improper" torsional angle. For the example shown in Figure 2.6, a torsional angle ABCD may be defined, even though there is no bond between C and D. The out-of-plane E_{oop} may then be described by an angle ω^{ABCD}, for example as a harmonic function $(\omega - \omega_0)^2$. Note that the definition of such improper torsional angles is not unique, the angle ω^{ABDC} (for example) is equally good. In practice there is little difference between describing E_{oop} as in eq. (2.8) or as an improper torsional angle.

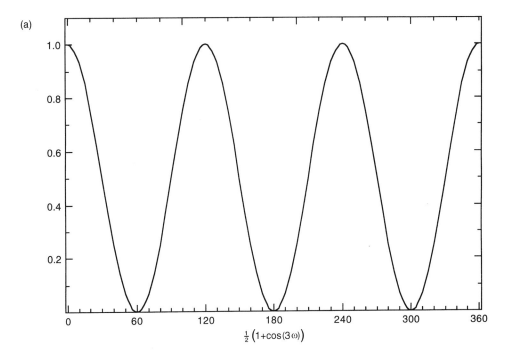

$\frac{1}{2}\left(1+\cos(3\omega)\right)$

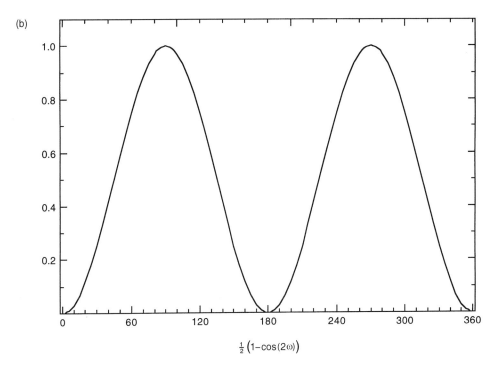

$\frac{1}{2}\left(1-\cos(2\omega)\right)$

Figure 2.8   Torsional energy functions

(c)

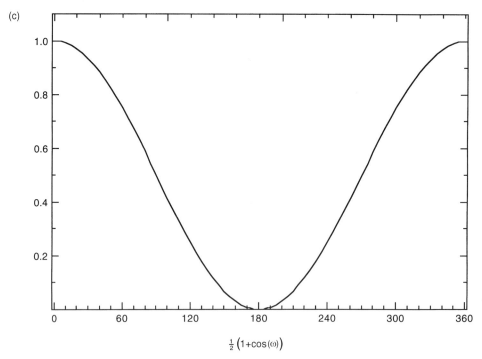

$$\tfrac{1}{2}\left(1+\cos(\omega)\right)$$

Figure 2.8 (*contd.*)

For each combination of four atom types, A, B, C and D, there are generally three torsional parameters to be determined, V_1^{ABCD}, V_2^{ABCD} and V_3^{ABCD}.

2.2.5 The van der Waals Energy

E_{vdw} is the van der Waals energy describing the repulsion or attraction between atoms that are not directly bonded. Together with the electrostatic term E_{el} (Section 2.2.6) they describe the non-bonded energy. E_{vdw} may be interpreted as the part of the interaction which is not related to electrostatic energy due to (atomic) charges. This may for example be the interaction between two methane molecules, or two methyl groups at different ends of the same molecule. At large interatomic distances E_{vdw} is zero, and for small distances it becomes very repulsive. In quantum mechanical terms this is due to the overlap of the electron clouds of the two atoms; the electrons repel each other because they are negatively charged. At intermediate distances, however, there is a slight attraction between two such electron clouds. This is due to the phenomenon known as electron correlation (discussed in Chapter 4). Pictorially, the attraction is due to induced dipole–dipole interactions. Even if the molecule (or part of a molecule) has no permanent dipole moment, the motion of the electrons may create a slightly uneven distribution at a given time. This dipole moment will induce a charge polarization in the neighbouring molecule (or another part of the same molecule), creating an attraction. It can be derived theoretically that this attraction varies as the inverse sixth power of the distance between the two fragments. Actually the induced dipole–dipole interaction is only one of such terms: there are also contributions from induced quadrupole–dipole, quadrupole–quadrupole etc. interactions. These vary as R^{-8}, R^{-10} etc.; the R^{-6} dependence is only the asymptotic behaviour at long distances. The force associated with this potential is often referred to as a "dispersion" or "London" force.[8] The van der Waals term is the only interaction between rare gas atoms (and thus the reason why say argon can become a liquid and a solid) and it is the main interaction between non-polar molecules such as alkanes.

E_{vdw} is very positive at small distances, has a minimum which is slightly negative at a distance corresponding to the two atoms just "touching" each other, and goes towards zero as the distance becomes large. A general functional form which fits these considerations is

$$E_{vdw}(R^{AB}) = E_{repulsive}(R^{AB}) - \frac{C^{AB}}{(R^{AB})^6} \tag{2.11}$$

It is not possible to derive theoretically the functional form of the repulsive interaction, it is only required that it goes towards zero as R goes to infinity, and it should approach zero faster than the R^{-6} term as the energy goes towards zero from below.

A popular potential which obeys these general requirements is the *Lennard-Jones* (LJ) potential[9] where the repulsive part is given by an R^{-12} dependence.

$$E_{LJ}(R) = \frac{C_1}{R^{12}} - \frac{C_2}{R^6} \tag{2.12}$$

Here C_1 and C_2 are suitable constants. The Lennard-Jones potential can also be written as

$$E_{LJ}(R) = \varepsilon \left[\left(\frac{R_0}{R} \right)^{12} - 2 \left(\frac{R_0}{R} \right)^6 \right] \tag{2.13}$$

where R_0 is the minimum energy distance and ε the dept of the minimum. There are no theoretical arguments for choosing the exponent in the repulsive part to be 12, this is purely a computational convenience. In fact there is some evidence that an exponent of 9 or 10 gives better results.

From electronic structure theory it is known that the repulsion is due to overlap of the electronic wave functions, and furthermore that the electron density falls off approximately exponentially with the distance from the nucleus (the exact wave function for the hydrogen atom *is* an exponential function). There is therefore some justification for choosing the repulsive part as an exponential function. The general form of the "Exponential $- R^{-6}$" E_{vdw} function, also known as a "*Buckingham*" or "*Hill*" type potential[10] is

$$E_{vdw}(R) = Ae^{-BR} - \frac{C}{R^6} \tag{2.14}$$

where A, B and C are suitable constants. It is sometimes written in a little more convoluted form as

$$E_{vdw}(R) = \varepsilon \left[\frac{6}{\alpha - 6} e^{\alpha(1 - R/R_0)} - \frac{\alpha}{\alpha - 6} \left(\frac{R_0}{R} \right)^6 \right] \tag{2.15}$$

where R_0 and ε have been defined in eq. (2.13), and α is a free parameter. Choosing an α value of 12 gives long-range behaviour identical to the Lennard-Jones potential, while a value of 13.772 reproduces the Lennard-Jones force constant at the equilibrium distance. The α parameter may also be taken as a fitting constant. The Exp.-6 potential has a problem for short interatomic distances where it "turns over". As R goes toward zero, the exponential becomes a constant (A) while the R^{-6} term goes toward $-\infty$. Minimizing the energy of a structure which accidentally has a very short distance between two atoms will thus result in nuclear fusion! Special precautions have to be taken to avoid this when using Exp.-6 van der Waals potentials.

A third functional form, which has an exponential dependence and the correct general shape, is the Morse potential, eq. (2.5). It does not have the R^{-6} dependence at long range, but as mentioned above, in reality there are also R^{-8}, R^{-10} etc. terms. The D and α parameters of a Morse function describing E_{vdw} will of course be much smaller than for E_{str}, and R_0 will be longer.

For small systems, where accurate interaction energy profiles are available, it has been shown that the Morse function actually gives a slightly better description than an Exp.-6, which again performs significantly better than a Lennard-Jones 12–6 potential.[11] This is illustrated for the H_2–He interaction in Figure 2.9.

The main difference between the three functions is in the repulsive part at short distances: the Lennard-Jones potential is much too hard, and the Exp.-6 also tends to overestimate the repulsion. It furthermore has the problem of "inverting" at short distances. For chemical purposes these "problems" are irrelevant, energies in excess of 100 kcal/mol are sufficient to break most bonds, and will never be sampled in actual calculations. The behaviour in the attractive part of the potential, which is essential for intermolecular interactions, is very similar for the three functions, as shown in

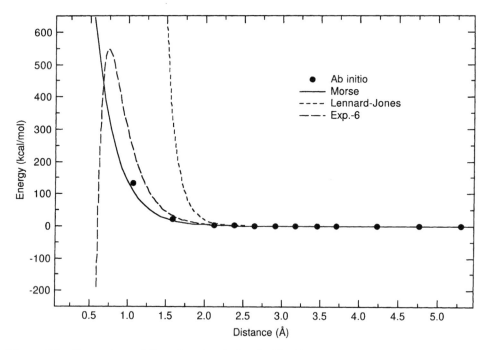

Figure 2.9 Comparison of E_{vdw} functionals for the H_2–He potential

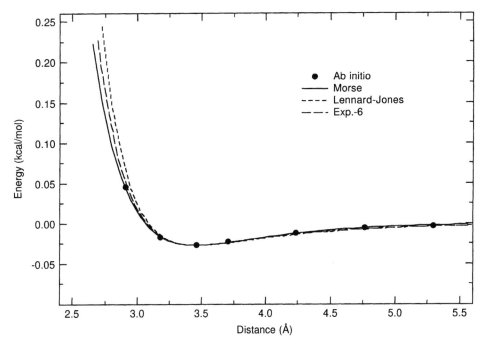

Figure 2.10   Comparison of E_{vdw} functionals for the attractive part of the H_2–He potential

Figure 2.10. Part of the better description of the Morse and Exp.-6 potentials may be due to the fact that they have three parameters, while the Lennard-Jones potential only employs two. Since the equilibrium distance and the well depth fix two constants, there is no additional flexibility in the Lennard-Jones function to fit the form of the repulsive interaction.

Most force fields employ the Lennard-Jones potential, despite the known inferiority to an exponential type function. Let us examine the reason for this in a little more detail.

Essentially all force field calculations use Cartesian coordinates of the atoms as the variables in the energy expression. To obtain the distance between two atoms one need to calculate

$$R_{ij} = \sqrt{(x_i - x_j)^2 + (y_i - y_j)^2 + (z_i - z_j)^2} \qquad (2.16)$$

In the exponential type potentials, the distance is multiplied by a constant and used as the argument for the exponential. Computationally it takes significantly longer time (typical factor of ~ 5) to perform mathematical operations like taking the square root and calculating exponential functions, than to do simple multiply and add. The Lennard-Jones potential has the advantage that the distance itself is not needed, only R raised to even powers. Using square root and exponential functions is thus avoided. The power of 12 in the repulsive part is chosen as it is simply the square of the power of 6 in the attractive part. Calculating E_{vdw} for an exponential type potential is computationally more demanding than for the Lennard-Jones potential. For large molecules, the calculation of the non-bonded energy in the force field energy expression is by far the one which takes most time, as will be demonstrated in Section 2.5. The difference between the above functional forms is in the repulsive part of E_{vdw} which usually is not very important. In actual calculations the Lennard-Jones potential gives results comparable to the more accurate functions, and it is computationally more efficient.

The van der Waals distance, R_0^{AB}, and softness parameters, ε^{AB}, depend on both atom types. These parameters are in all force fields written in terms of parameters for the individual atom types. There are several ways of combining atomic parameters to diatomic parameters, some of them being quite complicated.[12] A commonly used method is to take the van der Waals minimum distance as the sum of two van der Waals radii, and the interaction parameter as the geometrical mean of atomic "softness" constants.

$$R_0^{AB} = R_0^A + R_0^B$$
$$\varepsilon^{AB} = \sqrt{\varepsilon^A \varepsilon^B} \tag{2.17}$$

In some force fields, especially those using the Lennard-Jones form in eq. (2.12), the R_0^{AB} parameter is defined as the geometrical mean of atomic radii, implicitly via the geometrical mean rule used for the C_1 and C_2 constants.

For each atom type there are two parameters to be determined, the van der Waals radius and the atom softness, R_0^A and, ε^A. It should be noted that since the van der Waals energy is calculated between pairs of atoms, but parameterized against experimental data, the derived parameters represent an <u>effective</u> pair potential, which at least partly includes many-body contributions.

The van der Waals energy is the interaction between the electron clouds surrounding the nuclei. In the above treatment the atoms are assumed to be spherical. There are two instances where this may not be a good approximation. The first is when one (or both) of the atoms is hydrogen. Hydrogen has only one electron, which is always involved in bonding to the neighbouring atom. For this reason the electron distribution around the hydrogen nucleus is not spherical, rather the electron distribution is displaced towards the other atom. One way of modelling this anisotropy is to displace the position which is used in calculating E_{vdw} inwards along the bond. MM2 and MM3 use this approach with a scale factor of ~ 0.92, i.e. the distance which enter E_{vdw} is calculated between points located 0.92 times the X–H bond distance, as shown in Figure 2.11.

The electron density around the hydrogen will also be significantly dependent on the nature of the X atom, e.g. electronegative atoms such as oxygen or nitrogen will lead to a smaller effective van der Waals radius for the hydrogen, than if it is bonded to carbon. Many force fields therefore have several different types of hydrogen, depending on whether it is bonded to carbon, nitrogen, oxygen etc., and the hydrogen type may

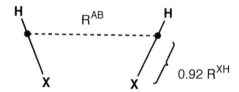

Figure 2.11 Illustration of the distance reduction which can be used for E_{vdw} involving hydrogens

further depend on the type of neighbour (e.g. alcohol or acid oxygen), see for example Table 2.1.

The other case where the spherical approximation may be less than optimal is for atoms having lone pairs, like oxygen and nitrogen. The lone pair electrons are more diffuse than the electrons involved in bonding, and the atom is thus "larger" in the lone pair direction. Some force fields choose to model lone pairs by assigning pseudo atoms at the lone pair positions. These pseudo atoms (type 20 in Table 2.1) behave as any other atom type with bond distances and angles, and have their own van der Waals parameters. They are significantly smaller than normal hydrogen atoms, and thus make the oxygen or nitrogen atom "bulge" in the lone pair direction. In some cases sulfur is also assigned lone pairs, although it has been argued that the second row atoms are more spherical due to the increased number of electrons, and therefore should not need lone pairs.

It is unclear at present whether it is necessary to include these effects to achieve good models. The effects are small, and it may be that the error introduced by assuming spherical atoms can be absorbed in the other parameters. Introducing lone pair pseudo atoms and making special treatment for hydrogens again make the time-consuming part of the calculation, the non-bonded energy, even more demanding. Most force fields thus neglect these effects.

Hydrogen bonds require special attention. Such bonds are formed between hydrogens attached to electronegative atoms like oxygen and nitrogen, and lone pairs, especially on oxygen and nitrogen. They have a bond strength of typically 2–5 kcal/mol, while normal single bonds are 60–110 kcal/mol and van der Waals interactions are 0.1–0.2 kcal/mol. The main part of the hydrogen bond energy normally comes from electrostatic attraction between the positively charged hydrogen and negatively charged heteroatom (see Section 2.2.6). Additional stabilization may be modelled by assigning special deep and short van der Waals interactions (via large ε and small R_0 parameters). This does not mean that the van der Waals radius for say a hydrogen bonded to an oxygen is especially short, since this would affect all van der Waals interactions involving this atom type. Instead only those pairs of interactions which are capable of forming hydrogen bonds are identified (by their atoms types) and the normal E_{vdw} parameters are replaced by special "hydrogen bonding" parameters. The functional form of E_{vdw} may also be different. One commonly used function is a modified Lennard-Jones potential of the form

$$E_{\text{H-bond}}(R) = \varepsilon \left[5 \left(\frac{R_0}{R} \right)^{12} - 6 \left(\frac{R_0}{R} \right)^{10} \right] \tag{2.18}$$

In some cases $E_{\text{H-bond}}$ also includes a directional term such as $(1 - \cos \theta^{\text{XHY}})$ or $(\cos \theta^{\text{XHY}})^4$ multiplied onto the distance dependent part in eq. (2.18). The current trend seem to be that force fields are moving away from such specialized parameters and/or functional forms, and instead accounting for hydrogen bonding purely in terms of electrostatic interactions.

2.2.6   The Electrostatic Energy

The other part of the non-bonded interaction is due to internal distribution of the electrons, creating positive and negative parts of the molecule. A carbonyl group, for example, has a negative oxygen and a positive carbon. At the lowest approximation this may be modelled by assigning charges to each atom. Alternatively, the bond may be assigned a bond dipole moment. These two descriptions give similar, but not identical, results. Only in the long-distance limit of interaction between such molecules do the two descriptions give identical results.

The interaction between point charges is given by the Coulomb potential.

$$E_{el}(R^{AB}) = \frac{Q^A Q^B}{\varepsilon R^{AB}} \quad (2.19)$$

where ε is a dielectric constant. The atomic charges may be treated as fitting parameters, analogous to the van der Waals constants. It is more common, however, to assign them on the basis of fitting to the electrostatic potential calculated by electronic structure methods, as described in Section 9.2. Since hydrogen bonding to a large extent is due to attraction between the electron deficient hydrogen and an electronegative atom such as oxygen or nitrogen, a proper choice of partial charges may adequately model this interaction.

The MM2 and MM3 force fields use a bond dipole description for E_{el}. The interaction between two dipoles is given by

$$E_{el}(R^{AB}) = \frac{\mu^A \mu^B}{\varepsilon (R^{AB})^3} (\cos \chi - 3\cos \alpha_A \cos \alpha_B) \quad (2.20)$$

where the angles are defined as shown in Figure 2.12.

Properly parameterized there is little difference in the performance of the two ways of representing E_{el}. There are exceptions where two strong bond dipoles are immediate neighbours (for example α-halogen ketones). The dipole model will lead here to a stabilizing electrostatic interaction for a transoid configuration (torsional angle of 180°), while the atomic charge model will be purely repulsive for all torsional angles (since all 1,3-interactions are neglected). In either case, however, a proper rotational profile may be obtained by suitable choices of the constants in E_{tors}. The atomic charge model is easier to parameterize by fitting to an electronic wave function, and is preferred by almost all force fields.

The "effective" dielectric constant ε can be included to model the effect of surrounding molecules (solvent) and the fact that interactions between distant sites may be "through" part of the same molecule, i.e. a polarization effect. A value of 1 for ε

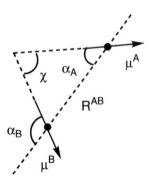

Figure 2.12   Variable definition for a dipole–dipole interaction

corresponds to a vacuum, while a large ε reduces the importance of long range charge–charge interactions. Typically a value between 1 and 4 is used, although there is little theoretical justification for any specific value. In some applications the dielectric constant is made distance dependent (e.g. $\varepsilon = \varepsilon_0 R^{AB}$, changing the Coulomb interaction to $Q_A Q_B / \varepsilon_0 (R^{AB})^2$) to model the "screening" by solvent molecules. There is no theoretical justification for this, but it increases the efficiency of the calculation as a square root operation is avoided (discussed in Section 2.2.5), and it seems to provide reasonable results.

As for the van der Waals energy, the standard electrostatic term only contains two-body contributions. For polar species the three-body contribution is quite significant, perhaps 10–20% of the two-body term.[13] The three-body effect may be considered as the interaction between two atomic charges being modified because a third atom polarizes the charges. Such "many-body" effects may be modelled by including an atom polarization.[14] The electrostatic interaction is then given by an "intrinsic" contribution due to atomic charges plus a dipolar term arising from the electric field created by the other atomic charges times the polarizability tensor. As each of the atoms contributes to the field, the final set of atomic dipoles is solved self-consistently by iterative methods. Addition of such improvements significantly increases the computational time, and has only seen limited use. The neglect of polarization is probably one of the main limitations of modern force fields, at least for polar systems. It should be noted that the average polarization is included implicitly in the parameterization, since atomic charges are often selected to give a dipole moment which is larger than the observed value for an isolated molecule (i.e. the effective dipole moment for H_2O in the solid state is 2.5 D, compared to 1.8 D in the gas phase[15]).

How far apart (in terms of number of bonds between them) should two atoms be before a non-bonded energy term contributes to E_{FF}? It is clear that two atoms directly bonded should not have an E_{vdw} or E_{el} term; their interaction is described by E_{str}. It is also clear that the interaction between two hydrogens at each end of say $CH_3(CH_2)_{50}CH_3$ is identical to the interaction between two hydrogens belonging to two different molecules, and they therefore should have an E_{vdw}/E_{el} term. But where should the dividing line be? Most force fields include E_{vdw}/E_{el} for atom pairs which are separated by three bonds or more, although interactions which are 1,4 in many cases are scaled down by a factor of between 1 and 2. This means that the rotational profile for an A–B–C–D sequence is determined both by E_{tors} and E_{vdw}/E_{el} terms for the A–D pair. In a sense E_{tors} may be considered as a correction necessary for obtaining the correct rotational profile once the non-bonded contribution has been accounted for. Some force fields choose also to include E_{vdw} for atoms which are 1,3 with respect to each other, these are called *Urey-Bradley* force fields. In this case the energy required to bend a three atom sequence is a mixture of E_{bend} and E_{vdw}. Most modern force fields calculate E_{str} between all atoms pairs which are 1,2 with respect to each other in terms of bonding, E_{bend} for all pairs which are 1,3, E_{tors} between all pairs which are 1,4, and E_{vdw}/E_{el} between all pairs which are 1,4 or higher.

2.2.7   Cross Terms

The first five terms in the general energy expression, eq. (2.1), are common to all force fields. The last term, E_{cross}, covers coupling between these fundamental, or diagonal, terms. Consider for example a molecule like H_2O. It has an equilibrium angle of $104.5°$ and a O–H distance of 0.958 Å. If the angle is compressed to say $90°$, and the optimal bond length is determined by electronic structure calculations, one finds that the distance now is 0.968 Å, i.e. slightly longer. Similarly if the angle is widened, the lowest energy bond length is shorter than the equilibrium bond length. This may qualitatively be understood by noting that the hydrogens come closer together if the angle is reduced. This leads to an increased repulsion between the hydrogens, which is partly alleviated by making the bonds longer. If only the first five terms in the force field energy are included, this coupling between bond distance and angle cannot be modelled. It may be taken into account by including a term which depends on both bond length and angle. E_{cross} may in general include a whole series of terms which couple two (or more) of the bonded terms.

The components in E_{cross} are usually written as products of Taylor-like expansions in the individual coordinates. The most important of these is the stretch/bend term which for an A–B–C sequence may be written as

$$E_{str/bend} = k^{ABC}(\theta^{ABC} - \theta_0^{ABC})[(R^{AB} - R_0^{AB}) + (R^{BC} - R_0^{BC})] \qquad (2.21)$$

Other examples of such cross terms are

$$E_{str/str} = k^{ABC}(R^{AB} - R_0^{AB})(R^{BC} - R_0^{BC})$$
$$E_{bend/bend} = k^{ABCD}(\theta^{ABC} - \theta_0^{ABC})(\theta^{BCD} - \theta_0^{BCD})$$
$$E_{str/tors} = k^{ABCD}(R^{AB} - R_0^{AB})\cos(n\omega^{ABCD}) \qquad (2.22)$$
$$E_{bend/tors} = k^{ABCD}(\theta^{ABC} - \theta_0^{ABC})\cos(n\omega^{ABCD})$$
$$E_{bend/tors/bend} = k^{ABCD}(\theta^{ABC} - \theta_0^{ABC})(\theta^{BCD} - \theta_0^{BCD})\cos(n\omega^{ABCD})$$

Usually the constants involved in these cross terms are not taken to depend on all the atom types involved in the sequence. For example the stretch/bend constant in principle depends on all three atoms, A, B and C. However, it is usually taken to depend only on the central atom, i.e. $k^{ABC} = k^B$, or chosen as a universal constant independent of atom type. It should be noted that cross terms of the above type are inherently unstable if the geometry is far from equilibrium. Stretching a bond to infinity, for example, will make $E_{str/bend}$ go towards $-\infty$ if θ is less than θ_0. If the bond stretch energy itself is harmonic (or quartic) this is not a problem as it approaches $+\infty$ faster, however, if a Morse type potential is used, special precautions will have to be made to avoid long bonds in geometry optimizations and simulations.

Another type of correction, which is related to cross terms, is the modification of parameters based on atoms not directly involved in the interaction described by the parameter. Carbon–carbon bond lengths, for example, become shorter if there are electronegative atoms present at either end. Such electronegativity effects may be modelled by adding a correction to the natural bond length R_0^{AB} based on the atoms which are attached to the A–B bond.[16]

$$R_0^{AB-C} = R_0^{AB} + \Delta R_0^C \qquad (2.23)$$

Although cross terms between the bonded potentials are part of all force fields designed to achieve high accuracy, the coupling between the geometry and the atomic charges is rarely addressed. From electronic structure calculations it is known that the optimum set of atomic charges to some extent depends on the molecular conformation. This could be modelled by having cross terms between the electrostatic and bonded terms, or by allowing the atomic charges to polarize each other.[17] This significantly complicates the force field, reducing the computational efficiency, and is not (yet) in common use.

2.2.8 Small Rings and Conjugated Systems

We have already mentioned that small rings present a problem as their equilibrium angles are very different from those of their acyclic cousins. One way of alleviating this problem is to assign new atom types. If a sufficient number of cross terms is included, however, the necessary number of atom types can actually be reduced. Some force fields have only one sp^3-carbon atom type, covering bonding situations from cyclopropane to linear alkanes with the same set of parameters. The necessary flexibility in the parameter space is here transferred from the atom types to the parameters in the cross terms, i.e. the cross terms modify the diagonal terms so that a more realistic behaviour is obtained for large deviations from the natural value.

One additional class of bonding that requires special consideration in force fields is conjugated systems. Consider for example 1,3-butadiene. According to the MM2 type convention (Table 2.1), all carbon atoms are of type 2. This means that the same set of parameters are used for the terminal and central C–C bonds. Experimentally the bond lengths are 1.35 Å and 1.47 Å, very different, which is due to the partial delocalization of the π-electrons in the conjugated system.[18] The outer C–C bond is slightly reduced in double bond character (and thus has a slightly longer bond length than in ethylene) while the central bond is roughly halfway between a single and a double bond. Similarly, without special precautions, the barriers for rotation around the terminal and central bonds are calculated to be the same, and assume a value characteristic of a localized double bond, ~ 55 kcal/mol. Experimentally, however, the rotational barrier for the central bond is only ~ 6 kcal/mol.[19]

There are two main approaches for dealing with conjugated systems. One is to identify certain bonding combinations and use special parameters for these cases, analogously to the treatment of hydrogen bonds in E_{vdw}. If for example four type 2 carbons are located in a linear sequence, they constitute a butadiene unit and special stretch and torsional parameters should be used for the central and terminal bonds. Similarly, if six type 2 carbons are in a ring, they constitute an aromatic ring and a set of special aromatic parameters is used. Or the atom type 2 may be changed to a type 50, identifying from the start that these carbons should be treated with a different parameter set. The main problem with this approach is that there are many such "special" cases which require separate parameters. Three conjugated double bonds for example may either be linearly or cross-conjugated (1,3,5-hexatriene and 2-vinyl-1,3-butadiene), each requiring a set of special parameters different from those used for 1,3-butadiene. The central bond in biphenyl will be different from the central bond in 1,3-butadiene. Modelling the bond alterations in fused aromatics such as naphthalene or phenanthrene requires complicated bookkeeping to keep track of all the different bond lengths etc.

The other approach, which is somewhat more general, is to perform a simple electronic structure calculation to determine the degree of delocalization within the π-system. This approach is used in the MM2 and MM3 force fields, often denoted MMP2 and MMP3.[20] The electronic structure calculation is of the Pariser–Pople–Parr (PPP) type (Section 3.9.3), which is only slightly more advanced than a simple Hückel calculation. From the calculated π-molecular orbitals, the π-bond order ρ for each bond can be calculated. There is now a connection between the π-bond order and the length (and force constant) of the bond. Also the energy required for rotation around the bond correlates with the bond order. The connections used in MM2 are (n_i is the number of electrons in the ith MO, and β_{BC} is a resonance parameter)

$$\rho_{AB} = \sum_{i}^{MO} n_i c_{Ai} c_{Bi}$$
$$R_0^{AB} = 1.503 - 0.166\rho_{AB} \tag{2.24}$$
$$k^{AB} = 5.0 + 4.6\rho_{AB}$$
$$V_2^{ABCD} = 15.0\rho_{BC}\beta_{BC}$$

The natural bond length varies between 1.503 Å and 1.337 Å for bond orders between 0 and 1, these are the values for pure single and double bonds between two sp^2-carbons. Similarly the force constant varies between the values used for isolated single and double bonds. The rotational barrier for an isolated double bond is 60 kcal/mol, since there are four torsional contributions for a double bond.

This approach, however, requires addition of a second level of iterations in a geometry optimization. At the initial geometry, a PPP calculation is performed, the π-bond orders are calculated and suitable bond parameters (R_0^{AB}, k^{AB} and V_2^{ABCD}) are assigned. These parameters are then used for optimizing the geometry. The optimized geometry will usually differ from the initial geometry, thus the parameters used in the optimization are no longer valid. At the "optimized" geometry a new PPP calculation is performed and a new set of parameters derived. The structure is reoptimized, and a new PPP calculation is carried out etc. This is continued until the geometry change between two macro-iterations is negligible.

For commonly encountered conjugated systems like butadiene and benzene, the *ad hoc* assignment of new parameters is usually preferred as it is simpler than the computationaly more demanding PPP method. For less common conjugated systems the PPP approach is more elegant and has the definite advantage that the common user does not need to worry about assigning new parameters. If the system of interest contains

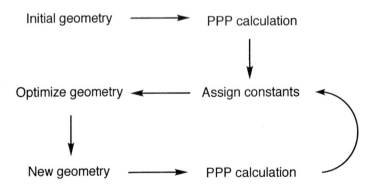

Figure 2.13   Illustration of the two-level optimization involved in a MMP2 calculation

conjugation, and a force field which uses the parameter replacement method is chosen, the user should check that proper bond lengths and reasonable rotational barriers are calculated (i.e. that the force field has identified the conjugated moiety and contains suitable substitution parameters). Otherwise very misleading results may be obtained without any indication from the force field of problems.

2.2.9 Comparing Energies of Structurally Different Molecules

The zero point of the energy in each term has so far been chosen for convenience. This is inconsequential for comparing energies of different conformations of the same molecule, i.e. where the atom types and bonding are the same. It is also possible to obtain meaningful results from comparing structurally different molecules if they contain the same number and types of structural units. For comparing relative stabilities of chemically different molecules such as dimethyl ether and ethyl alcohol, or for comparing experimental heats of formation, the zero point of the energy scale must be the same. The force field energy, E_{FF}, is often called the *steric energy* as in some sense it is the excess energy relative to a hypothetical molecule with non-interacting fragments. The numerical value of E_{FF} by itself has no meaning!

To convert the steric energy to heat of formation, terms can be added depending on the number and types of bond present in the molecule. This again rests on the assumption of transferability, e.g. all C–H bonds have a dissociation energy close to 100 kcal/mol. A "heat of formation" parameter can be assigned to each bond type, and the numerical value again determined by fitting to experimental data. This value will roughly be the bond dissociation energy relative to the heat of formation of the atoms involved, i.e. for a C–H bond it will be $\Delta H_{CH} = D_{C-H} - \Delta H_f(H) - \Delta H_f(C)$. To achieve a better fit, parameters may also be assigned to larger units than the individual bond, such as methyl groups or an OCH_2O unit. The latter type corrections are usually fairly small.

$$\Delta H_f = E_{FF} + \sum^{bonds} \Delta H_{AB} + \sum^{groups} \Delta H_G \qquad (2.25)$$

These heat of formation parameters may be considered as shifting the zero point of E_{FF} to a common origin. Since corrections from larger moieties are small, it follows that energy differences between systems having the same groups (for example methyl-cyclohexane and ethyl-cyclopentane) can be calculated directly from differences in steric energy.

If the heat of formation parameters are derived on the basis of fitting to a large variety of compounds, a specific set of parameters is obtained. A slightly different set of parameters may be obtained if only certain "strainless" molecules are included in the parameterization. Typically molecules like straight chain alkanes and cyclohexane are defined as strainless. Using these strainless heat of formation parameters, a strain energy may be calculated as illustrated in Figure 2.14.

Deriving such heat of formation parameters requires a large body of experimental ΔH_f values. For many classes of compound there are not sufficient data available. Only a few force fields, notably MM2 and MM3, attempt to parameterize also heats of formation. Most force fields are only concerned with reproducing geometries and possibly conformational relative energies, for which the steric energy is sufficient.

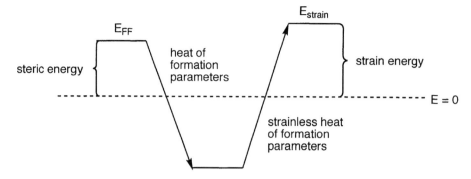

Figure 2.14   Illustrating the difference between steric energy and heat of formation

2.3   Force Field Parameterization

Having settled on the functional description and a suitable number of cross terms, the problem of assigning numerical values to the parameters arises. This is by no means trivial. Consider for example MM2(91) with 71 atom types. Not all of these can form stable bonds with each other, hydrogens and halogens can only have one bond etc. For the sake of argument, however, assume that the effective number of atom types capable of forming bonds between each other is 30.

- Each of the 71 atom types has two van der Waals parameters, R_0^A and ε^A, giving 142 parameters.
- There are $30 \times 30/2 = 450$ possible different E_{str} terms, each requiring at least two parameters, k^{AB} and R_0^{AB}, for a total of at least 900 parameters.
- There are $30 \times 30 \times 30/2 = 13\,500$ possible different E_{bend} terms, each requiring at least two parameters, k^{ABC} and θ_0^{ABC}, for a total of at least 27 000 parameters.
- There are $30 \times 30 \times 30 \times 30/2 = 405\,000$ possible different E_{tors} terms, each requiring at least three parameters, V_1^{ABCD}, V_2^{ABCD} and V_3^{ABCD}, for a total of at least 1 215 000 parameters.
- Cross terms may further add some million possible parameters.

To achieve just a rudimentary assignment of the value of one parameter, at least 3–4 independent data should be available. To parameterize MM2 for all molecules described by the 71 atom types would thus require of the order of 10^7 independent experimental data, not counting cross terms. This is clearly impossible. Furthermore, the parameters that are the most numerous, the torsional constants, are also the ones that are the hardest to obtain experimental data for. Experimental techniques normally probe a molecule near its equilibrium geometry. Getting energetical information about the whole rotational profile is very demanding and has only been done for a handful of small molecules. In recent years it has therefore become common to rely on data from electronic structure calculations to derive force field parameters. Calculating for example rotational energy profiles is computationally fairly easy. The so-called "Class II" force fields rely heavily on data from electronic structure calculations to derive force field parameters, especially the bonded parameters (stretch, bend and torsional).

While the non-bonded terms are relatively unimportant for the "local" structure, they are the only contributors to intermolecular interactions, and the major factor in determining the global structure of a large molecule, like protein folding. The electrostatic part of the interaction may be assigned on the basis of fitting parameters to the electrostatic potential derived from an electronic wave function, as discussed in Section 9.2. van der Waals interactions, however, are difficult to calculate reliably by electronic structure methods, requiring a combination of electron correlation and very large basis sets. The van der Waals parameters are therefore usually assigned on the basis of fitting to experimental data for either the solid or liquid state.[21] Since the parameterization implicitly takes many-body effects into account, two (slightly) different sets of van der Waals parameters may be obtained, depending on where the fitting is done to reproduce the crystal or solution phase data. Furthermore, it is possible that several combinations of van der Waals parameters for different atoms may be able to reproduce the properties of a liquid, i.e. the parameters are not unique.

The above considerations illustrate the inherent contradiction in designing highly accurate force fields. To get high accuracy for a wide variety of molecules, and a range of properties, many and functional complex terms must be included in the force field expression. For each additional parameter introduced in an energy term, the potential number of new parameters to be derived grows as the number of atom types to a power between 1 and 4. The higher accuracy that is needed, the more finely the fundamental units must be separated, i.e. more atoms types must be used. In the extreme limit, each atom which is not symmetry related is a new atom type in each new molecules. In this limit each molecule will have its own set of parameters to be used just for this one molecule. To derive these parameters, the molecule must be subjected to many different experiments, or a large number of electronic structure calculations. This is the approach used in "inverting" spectroscopic data to produce a potential energy surface. From a force field point of view the resulting function is essentially worthless, it just reproduces known results. To be useful a force field should be able to predict unknown properties of molecules from known data on other molecules, i.e. a sophisticated form of inter- or extrapolation. If the force field becomes very complicated, the amount of work required to derive the parameters may be larger than the work required for measuring the property of interest for a given molecule.

The fundamental assumption of force fields is that structural units are transferable between different molecules. A compromise between accuracy and generality must thus be made. In MM2(91) for example the actual number of parameters compared to the theoretical estimated possible (based on the 30 effective atom types above) is shown in Table 2.2.

Table 2.2 Comparison of possible and actual number of MM2(91) parameters

Term	Estimated number of parameters	Actual number of parameters
E_{vdw}	142	142
E_{str}	900	290
E_{bend}	27000	824
E_{tors}	1215000	2466

As seen from Table 2.2 there are a large number of possible compounds for which there are no parameters, and for which it is then impossible to perform force field calculations on (a good listing of available force field parameters can be found in ref. 22. Actually the situation is not as bad as it would appear from Table 2.2. Although only $\sim 0.2\%$ of the possible combinations for the torsional constants have been parameterized, these encompass the majority of the chemically interesting compounds. It has been estimated that $\sim 20\%$ of the ~ 15 million known compounds can be modelled by the parameters in MM2, the majority with a good accuracy. However, the problem of lacking parameters is very real, and anyone who has used a force field for all but the most rudimentary problems have encountered the problem. How does one progress if there are insufficient parameters for the molecule of interest?

There are two possible routes. The first is to estimate the missing parameters by comparison to force field parameters for similar systems. If for example there are no torsional parameters for rotation around a H–X–Y–O bond in your molecule, but parameters exist for H–X–Y–C, then it is probably a good approximation to use the same values. In other cases it may be less obvious what to chose. What if your system has an O–X–Y–O torsion, and parameters exist for O–X–Y–C and C–X–Y–O, but they are very different. What to choose then? One or the other, or the average? After a choice has been made, the results should ideally be evaluated to determine how sensitive they are to the exact value of the guessed parameters. If for example the guessed parameters can be varied by $\pm 50\%$ without seriously affecting the final results, the property of interest is insensitive to the guessed parameters, and can be trusted to the usual degree of the force field. If, on the other hand, the final results vary by a factor of 2 if the guessed parameters are changed by 10%, a better estimate of the critical parameters should be sought from external sources. If many parameters are missing from the force field, such an evaluation of the sensitivity to parameter changes becomes impractical, and one should consider either the second route described below, or abandon force field methods altogether.

The second route to missing parameters is to use external information, experimental or electronic structure calculations. If for example the missing parameters are the bond length and force constant for a specific bond type, it is possible that an experimental bond distance may be obtained from an X-ray structure and the force constant estimated from measured vibrational frequencies. Or missing torsional parameters may be obtained from a rotational energy profile calculated by electronic structure calculations. Again, if many parameters are missing, this approach is very time-consuming, and may not give final results as good as one may have expected from the "rigorous" way of deriving the parameters. The reason for this is discussed below.

Assume now that the functional form of the force field has been settled. The next task is to select a set of reference data, for the sake of argument let us assume that they are derived from experiments, but they could also be taken from electronic structure calculations. The problem is then to assign numerical values to all the parameters such that the results from force field calculations match as close as possible the reference data set. The reference data may be of very different types and accuracy. It may for example contain bond distances, bond angles, relative energies, vibrational frequencies, dipole moments etc. These data of course have different units. How should they be weighted? How much weight should be put on reproducing a bond length of 1.532 Å relative to an energy difference of 2.5 kcal/mol? Should the same weight be used for all bond distances, if for example one distance is determined to ± 0.001 Å while another is known only to ± 0.07 Å? The selection is further complicated by the fact that different experimental methods may give slightly different answers for say the bond distance, even in the limit of no experimental uncertainty. The reason for this is that different experimental methods do not measure the same property. X-ray diffraction, for example, determines the electron distribution, while microwave spectroscopy primarily depends on the nuclear position. The maximum in the electronic distribution may not be exactly identical to the nuclear position, and these two techniques will therefore give slightly different bond lengths.

Once the question of assigning weights for the reference data has been decided, the fitting process can begin. It may be formulated in terms of an error function.[23]

$$\text{ErrF (parameters)} = \sum_{\text{data}} \text{weight} \cdot (\text{reference value} - \text{calculated value})^2 \qquad (2.26)$$

The problem is now to find the minimum of ErrF with the parameters as variables. From an initial set of guess parameters, force field calculations are performed for the whole set of reference molecules and the results compared with the reference data. The deviation is calculated and a new improved set of parameters can be derived. This is continued until a minimum has been found for the ErrF function. To find the best set of force field parameters corresponds to finding the global minimum for the multidimensional ErrF function. The simplest optimization procedure performs a cyclic minimization, reducing the ErrF value by varying one parameter at a time. More advanced methods rely on the ability to calculate the gradient (and possibly also the second derivative) of the ErrF with respect to the parameters. Such information may be used in connection with an optimization procedures as described in Chapter 14.

The parameterization process may be done sequentially or in a combined fashion. In the sequential method a certain class of compound, such as hydrocarbons, is parameterized first. These parameters are held fixed, and a new class of compound, for example alcohols and ethers, is then parameterized. This method is in line with the basic assumption of force fields: parameters are transferable. The advantage is that only a fairly small number of parameters are fitted at a time. The ErrF is therefore a relatively low-dimensional function, and one can be reasonably certain that a "good" minimum has been found (although it may not be the global minimum). The disadvantage is that the final set of parameters necessarily provides a poorer fit (as defined from the value of the ErrF) than if all the parameters are fitted simultaneously.

The combined approach tries to fit all the constants in a single parameterization step. Considering that the number of force field parameters may be many thousands, it is clear that the ErrF function will have a very large number of local minima. To find the global minimum of such a multivariable function is very difficult. It is thus likely that the final set of force field parameters derived by this procedure will in some sense be less than optimal, although it may still be "better" than that derived by the sequential procedure. Furthermore, many of the parameter sets which give low ErrF values (including the global minimum) may be "unphysical", e.g. force constants for similar bonds may be very different. Owing to the large dimensionality of the problem, such combined optimizations require the ability to calculate the gradient of the ErrF with respect to the parameters. Writing such programs is not trivial. There is also a more fundamental problem when new classes of compounds are introduced at a later time than the original

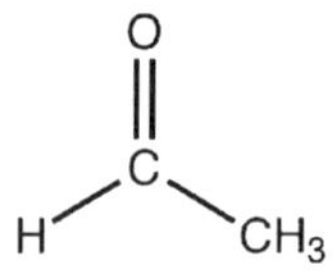

Figure 2.15   The structure of acetaldehyde

parameterization. To be consistent, the whole set of parameters should be reoptimized. This has the consequence that (all) parameters change every time a new class of compounds is introduced, or whenever more data are included in the reference set. Such "time-dependent" force fields are clearly not desirable. Most parameterization procedures therefore employ a sequential technique, although the number of compound types parameterized in each step varies.

There is one additional point to be mentioned in the parameterization process, which is also important for understanding why the addition of missing parameters by comparison with existing data or from external sources is somewhat problematic. This is the question of redundant variables. Consider for example acetaldehyde. In the bend energy expression there will be four angle terms describing the geometry around the carbonyl carbon, an HCC, an HCO, a CCO, and an out-of-plane bend. Assuming the latter to be zero at the moment, it is clear that the former three angles are not independent. If the HCO and CCO angle are given, the HCC angle must be 360° –HCO–CCO. Nevertheless, there will be three natural angle parameters, and three force constants associated with these angles. The force field parameters, as defined by the E_{FF} expression, are not independent. The implicit assumption in force field parameterization is that given sufficient amounts of data, this redundancy will cancel out. In the above case, additional data for other aldehydes and ketones may be used for (at least partly) removing this ambiguity in assigning angle bend parameters. This clearly illustrates that force field parameters are just that, parameters. They do not necessarily have any direct connection with experimental force constants. Experimental vibrational frequencies can be related to a unique set of force constants, but only in the context of a non-redundant set of coordinates. It is also clear that errors in the force field due to inadequacies in the functional forms used for each of the energy terms to some extent will be absorbed by the parameter redundancy. Adding new parameters from external sources, or estimating missing parameters by comparison with those for "similar" fragments, may partly destroy this cancellation of errors. This is also the reason why parameters are not transferable between different force fields, the values of the parameters are strongly dependent on the functional form of the energy terms.

The parameter redundancy is also the reason that care should be exercised when trying to decompose energy differences into individual terms. Although it may be possible to rationalize the preference of one conformation over another by for example increased steric repulsion between certain atom pairs, this is intimately related to the chosen functional form for the non-bonded energy, and the balance between this and the angle bend/torsional terms. The rotational barrier in ethane, for example, may be reproduced solely by an HCCH torsional energy term, solely by an H–H van der Waals repulsion or solely by H–H electrostatic repulsion. Different force fields will have (slightly) different balances of these terms, and while one force field may contribute a conformational difference primarily to steric interactions, another may have the torsional energy as the major determining factor, and a third may "reveal" that it is all due to electrostatic interactions.

2.3.1   Parameter Reductions in Force Fields

The overwhelming problem in developing force fields is the lack of enough high quality reference data. As illustrated above, there are literally millions of possible parameters in even quite simple force fields. The most numerous of these are the torsional parameters, followed by the bending constants. As force fields are designed for predicting properties of unknown molecules, it is invariable that the problem of lacking parameters will be encountered frequently. Furthermore, many of the existing parameters may be based on very few reference data, and therefore associated with substantial uncertainty.

Many modern force field programs are commercial. Having the program tell the user that his or her favourite molecule cannot be calculated owing to lack of parameters is not good for business. Making the user derive new parameters, and getting the program to accept them, may require more knowledge than the average user, who is just interested in the answer, has. Many force fields thus have "generic" parameters. This is just a fancy word for the program making more or less educated guesses for the missing parameters.

One way of reducing the number of parameters is to reduce the dependence on atom types. Torsional parameters, for example, can be taken to depend only on the types of the two central atoms. All C–C single bonds would then have the same set of torsional parameters. This does not mean that the rotational barriers for all C–C bonds are identical, since van der Waals and/or electrostatic terms also contribute. Such a reduction replaces all tetra-atomic parameters with diatomic constants, i.e. $V^{ABCD} \rightarrow V^{BC}$. Similarly, the triatomic bending parameters may be reduced to atomic constants by assuming that the bending parameters only depend on the central atom type ($k^{ABC} \rightarrow k^B$, $\theta_0^{ABC} \rightarrow \theta_0^B$). Generic constants are often taken from such reduced parameter sets. In the case of missing torsional parameters, they may also simply be omitted, i.e. setting the constants to zero. A good force field program informs the user of the quality of the parameters used in the calculation, especially if such generic parameters are used. This is useful for evaluating the quality of the results. Some programs unfortunately use the necessary number of generic parameters to carry out the calculations without notifying the user. In extreme cases, one may perform calculations on molecules for which essentially no "good" parameters exist, and get totally useless results out. The ability to perform a calculation is no guarantee that the results can be trusted!

The quality of force field parameters is essential for judging how much faith can be put in the results. If the molecule at hand only uses parameters which are based on many good quality experimental results, the computational results can be trusted to be almost of experimental quality. If, on the other hand, the employed parameters are based only on a few experimental data, and/or many generic parameters are used, the results should be treated with care. Using low quality parameters for describing an "uninteresting" part of the molecule, like a substituted aromatic ring in a distant side chain, is not problematic. In some cases such uninteresting parts may simply be substituted by other simpler groups (for example a methyl group). However, if the low quality parameters directly influence the property of interest, the results may potentially be misleading.

2.3.2   *Force Fields for Metal Coordination Compounds*

One area which is especially plagued with the problems of assigning suitable functions for describing the individual energy terms and deriving good parameters is coordination chemistry.[24] The bonding around metals is much more varied than for organic molecules, not just two to four bonds as with organic compounds. Furthermore, for a given number of ligands, more than one geometrical arrangement is usually possible. A 4-coordinated metal, for example, may either be tetrahedral or square planar, and a 5-coordinated metal may either have a square pyramidal or trigonal bipyramidal structure. This is in contrast to 4-coordinated atoms like carbon or sulfur which always are very close to tetrahedral. The increased number of ligands combined with the multitude of possible geometries significantly increases the problems of assigning suitable functional forms for each of the energy terms. Consider for example a "simple" compound like $Fe(CO)_5$, which has a trigonal bipyramidal structure as shown in Figure 2.16.

$$
\begin{array}{c}
\text{CO} \\
| \\
\text{OC} - \text{Fe} \cdots \text{CO} \\
| \quad \searrow \text{CO} \\
\text{CO}
\end{array}
$$

Figure 2.16   The structure of iron pentacarbonyl

It is immediately clear that a C–Fe–C angle bend must have three energy minima corresponding to 90°, 120° and 180°, indicating that a simple Taylor expansion around a (single) natural value is not suitable. Furthermore, the energy cost for a geometrical distortion (bond stretching and bending) is usually much smaller around a metal atom than for a carbon atom. This has the consequence that coordination compounds are much more dynamic, displaying phenomena such as pseudo rotations, ligand exchange and large geometrical variations for changes in the ligands. In iron pentacarbonyl there exist a whole series of equivalent trigonal bipyramidal structures which readily interconvert, i.e. the energy cost for changing the C–Fe–C angle from 90° to 120° and to 180° is small. Deviations up to 30° from the "natural" angle by introducing bulky substituents on the ligands are not uncommon. Furthermore, the distance of a given metal-ligand bond is often sensitive to the nature of the other ligands. An example here is the *trans* effect, where a metal-ligand bond distance can vary by perhaps 0.2 Å depending on the nature of the ligand on the opposite side.

Another problem encountered in metal systems is the lack of well-defined bonds. Consider for example an olefin coordinated to a metal: should this be considered as a single bond between the metal and the center of the C–C bond, or as a metalocyclopropane with two M–C bonds? A cyclopentadiene ligand may similarly be modelled either with a single bond to the center of the ring, or with five M–C bonds. In reality these represent limiting behaviours; the structures on the left in Figure 2.17 correspond to a weak interaction while those on the right involve strong electron donation from the ligand to the metal (and vice versa). A whole range of intermediate cases are found in coordination chemistry. The description with bonds between the metal and all the ligand atoms suffers from the lack of (free) rotation of the ligand. The

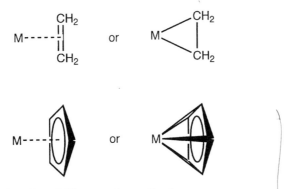

Figure 2.17   The ambiguity of modelling metal coordination

coordination to the "centre" of the ligand may be modelled by placing a *pseudo atom* at that position, and relating the ligand atoms to the pseudo atom (a pseudo atom is just a point in space, also sometimes called a dummy atom, see Appendix E). Alternatively the coordination may be described entirely by non-bonded interactions (van der Waals and electrostatic).

One possible, although not very elegant, solution to these problems is to assign different atom types for each bonding situation. In the $Fe(CO)_5$ example this would mean distinguishing between equatorial and axial CO units. There would then be three different C–Fe–C bending terms, C_{eq}–Fe–C_{eq}, C_{eq}–Fe–C_{ax} and C_{ax}–Fe–C_{ax}, with natural angles of 120°, 90° and 180°, respectively. This approach sacrifices the dynamics of the problem, interchanging an equatorial and an axial CO no longer produces energetically equivalent structures. Similarly, the same metal atom in two different geometries (such as tetrahedral and square planar) would be assigned two different types, or in general a new type for each metal in a specific oxidation and spin state, and with a specific number of ligands. This approach encounters the parameter "explosion", as discussed above. It also biases the results in the direction of the user's expectations; if a metal atom is assigned a square planar atom type, the structure will end up close to square planar, even though the real geometry may be tetrahedral. The object of a computational study, however, is often a series of compounds which have similar bonding around the metal atom. In such cases the specific parameterization may be quite useful, but the limitations should of course be kept in mind. Most force field modelling of coordination compounds to date has employed this approach, tailoring an existing method to also reproduce properties (most notably geometries) of a small set of reference systems.

Part of the problem may be solved by using more flexible functional forms for the individual energy terms, most notably the stretching and bending energies. The stretch energy may be chosen as a Morse potential (eq. (2.5)), allowing for quite large distortions away from the natural distance, and also able to account for dissociation. However, phenomena like the *trans* effect are inherently electronic in nature (similar to the delocalization in conjugated systems) and are not easily accounted for in a force field description.

The multiple minima nature of the bending energy and the low barriers for interconversion resemble the torsional energy for organic molecules. An expansion of E_{bend} in terms of cosine or sine functions of the angle is therefore more natural than a

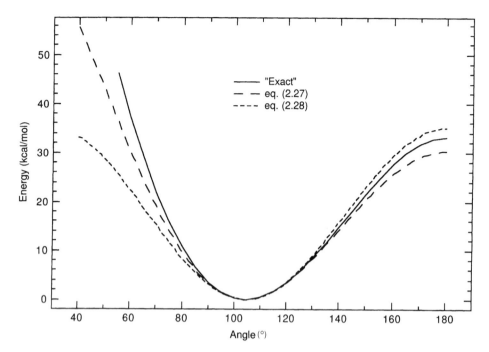

Figure 2.18 Comparing different E_{bend} functionals for the H_2O potential

simple Taylor expansion in the angle. Furthermore, bending around main-group atoms often has an energy maximum for an angle of 180°, with a low barrier. The following examples have a zero derivative for a linear angle, and are reasonable for describing bond bending like that encountered in the H_2O example, Figure 2.5.

$$E_{\text{bend}}(\theta) = k\left(\sin^2\frac{\theta}{2} - \sin^2\frac{\theta_0}{2}\right)^2 \qquad (2.27)$$

$$E_{\text{bend}}(\theta) = k(1 + \cos(n\theta + \psi_0)) \qquad (2.28)$$

The latter functional form contains a constant n which determines the periodicity of the potential (ψ_0 is a phase factor), and allows bending energies with multiple minima, analogous to the torsional energy. It does, however, have problems of unwanted oscillations if an energy minimum with a natural angle close to 180° is desired (this requires n to be large, creating many additional minima). It is also unable to describe situations where the minima are not regular spaced, such as the $Fe(CO)_5$ system (minima for angles of 90°, 120° and 180°). The performance of eqs. (2.27) and (2.28) for the H_2O case is given in Figure 2.18, which can be compared to Figure 2.5.

In both potentials in eqs. (2.27) and (2.28) the barrier towards linearity is given implicitly by the force constant. A more general expression which allows even quite complicated energy functionals to be fitted is a general Fourier expansion.

$$E_{\text{bend}}(\theta) = \sum_n k_n \cos(n\theta) \qquad (2.29)$$

An alternative approach consists of neglecting L–M–L bending terms, and instead including non-bonded 1,3-interactions. The geometry around the metal is then defined exclusively by the van der Waals and electrostatic contributions (i.e. placing the ligands as far apart as possible), and the model is known as *Points-On-a-Sphere* (POS).[25] It is basicly equivalent to the *VSEPR* (Valence Shell Electron-Pair Repulsion) model,[26] with VSEPR focusing on the electron pairs which make up a bond, and POS focusing on the atoms and their size. For alkali, alkaline earth and rare earth metals, where the bonding is mainly electrostatic, POS gives quite reasonable results, but it is unable to model systems where the d-orbitals have a preferred bonding arrangement. Tetra-coordinate metal atoms, for example, will in such models always end up being tetrahedral, although d^8-metals normally are square planar.

The final problem encountered in designing force fields for metal complexes is the lack of sufficient numbers of experimental data. Geometrical data for metal compounds are much more scarce than for organic structures, and the soft deformation potentials mean that vibrational frequencies often are difficult to assign to specific modes. Deriving parameters from electron structure calculations is troublesome because the presence of multiple ligands means that the number of atoms is quite large, and the metal atom itself contains many electrons. Furthermore, there are often many different low-lying electronic states due to partly occupied d-orbitals, indicating that single reference methods (i.e. HF type calculations) are insufficient for even a qualitatively correct wave function. Finally, relativistic effects become important for some of the metals in the lower part of the periodic system. These effects have the consequence that electronic structure calculations of sufficient quality are computationally expensive to carry out.

2.3.3 Universal Force Fields

The combination of many atom types and lack of a sufficient number of reference data have spawned developments of force fields with reduced parameters sets, such as the *Universal Force Field* (UFF).[27] The idea is to derive di-, tri- and tetra-atomic parameters (E_{str}, E_{bend}, E_{tors}) from atomic constants (such as atom radii, ionization potentials, electronegativities, polarizabilities, etc.). Such force fields are in principle capable of covering molecules composed of elements from the whole periodic table, these have been labelled as "all elements" in Table 2.3. They give less accurate results compared to conventional force fields, but geometries are often calculated qualitatively correctly. Relative energies, however, are much more difficult to obtain accurately, and conformational energies for organic molecules are generally quite poor. Another approach is to use simple valence bonding arguments (e.g. hybridization) to derive the functional form for the force field, as employed in the VALBOND approach.[28] It should be noted, however, that such "universal" force fields have only been proposed in recent years, and it is likely that improvements will be forthcoming.

2.4 Differences in Force Fields

There are many different force fields in use. They differ in three main aspects:

(1) The functional form of each energy term.
(2) The number of cross terms included.
(3) The type of information used for fitting the parameters.

Table 2.3 Comparison of functional forms used in common force fields. The torsional energy, E_{tors}, is in all cases given as a Fourier series in the torsional angle

Force Field	Types	E_{str}	E_{bend}	E_{oop}	E_{vdw}	E_{el}	E_{cross}	Molecules
EAS	2	P2	P3	none	Exp.–6	none	none	alkanes
EFF	2	P4	P3	none	Exp.–6.	none	ss,bb,sb, st,btb	alkanes
MM2	71	P3	P2+6	P2	Exp.–6	dipole	sb	general
MM3	153	P4	P6	P2	Exp.–6	dipole or charge	sb,bb,st	general (all elements)
MM4	3	P6	P6	imp.	Exp.–6	charge	ss,bb,sb, tt,st,tb,btb	hydrocarbons
CVFF	53	P2 or Morse	P2	P2	6–12	charge	ss,bb,sb, btb	general
CFF 91/93/95	48	P4	P4	P2	6–9	charge	ss,bb,st, sb,bt,btb	general
TRIPOS	31	P2	P2	P2	6–12	charge	none	general
MMFF	99	P4	P3	P2	7–14	charge	sb	general
COSMIC	25	P2	P2		Morse	charge	none	general
DREIDING	37	P2 or Morse	P2(cos)	P2(cos)	6–12 or Exp.–6	charge	none	general
AMBER	41	P2	P2	imp.	6–12 10–12	charge	none	proteins, nucleic acids, carbohydrates
OPLS	41	P2	P2	imp.	6–12	charge	none	proteins, nucleic acids, carbohydrates
CHARMM	29	P2	P2	imp.	6–12	charge	none	proteins
GROMOS		P2	P2	P2(imp.)	6–12	charge	none	proteins, nucleic acids, carbohydrates
ECEPP		fixed	fixed	fixed	6–12 10–12	charge	none	proteins
MOMEC		P2	P2	P2	Exp.–6	none	none	metal coordination
SHAPES		P2	$\cos(n\theta)$	imp.	6–12	charge	none	metal coordination
ESFF	97	Morse	P2(cos)	P2	6–9	charge	none	all elements
UFF	126	P2 or Morse	$\cos(n\theta)$	imp.	6–12	charge	none	all elements

Notation: Pn: Polynomial of order n; Pn(cos): polynomial of order n in cosine to the angle; $\cos(n\theta)$: Fourier term(s) in cosine to the angle; Exp.–6: exponential $+ R^{-6}$; n–m: $R^{-n} + R^{-m}$; fixed: not a variable; imp.: improper torsional angle; ss: stretch–stretch; bb: bend–bend; sb: stretch–bend; st: stretch–torsional; bt: bend–torsional; tt: torsional–torsional; btb: bend–torsional–bend.

(continued)

There are two general trends. If the force field is designed primarily to treat large systems, such as proteins or DNA, the functional forms are kept as simple as possible. This means that only harmonic functions are used for E_{str} and E_{bend} (or these terms are omitted, forcing all bond lengths and angles to be constant), no cross terms are included, and the Lennard-Jones potential is used for E_{vdw}. Such force fields are often called "harmonic" or "Class I". The other branch concentrates on reproducing small to medium size molecules to a high degree of accuracy. These force fields will include a number of cross terms, use at least cubic or quartic expansions of E_{str} and E_{bend}, and possibly an exponential type potential for E_{vdw}. The current efforts in developing small molecule force fields go in the direction of striving to reproduce not only geometries and relative energies, but also vibrational frequencies. Such force fields are often called "Class II" force fields.

Further simplification may be achieved in force fields designed for treating macromolecules by not considering hydrogens explicitly, the so-called *united atom* approach (an option present in for example the AMBER, CHARMM, GROMOS and DREIDING force fields). Instead of modelling a CH_2 group as a carbon and two hydrogens, a single "CH_2 atom" may be assigned. This united atom will have a larger van der Waals radius to account for the hydrogens. The use of such united atoms effectively reduces the number of variables by a factor of $\sim 2-3$, thereby allowing correspondingly larger systems to be treated. Of course the coarser the atomic description, the less detailed the final results. Which description, and thus which type of force field to use, depends on what type of information is sought. If the interest is in geometries and relative energies of different conformations of say hexose, then an elaborate force field is necessary. However, if the interest is in studying the dynamics of

EAS: E. M. Engler, J. D. Andose and P. v. R. Schleyer, *J. Am. Chem. Soc.*, **95** (1973), 8005; EFF: J. L. M. Dillen and *J. Comput. Chem.*, **16** (1995), 595, 610; MM2: N. L. Allinger, *J. Am. Chem. Soc.*, **99** (1977), 8127; MM3: N. L. Allinger, Y. H. Yuh and J. H. Lii, *J. Am. Chem. Soc.*, **111** (1989), 8551; J. H. Lii and N. L. Allinger, *J. Am. Chem. Soc.*, **111** (1989), 8566, 8576; "all elements" MM3: N. L. Allinger, X. Zhou and J. Bergsma, *J. Mol. Struct. Theochem.*, **312** (1994), 69; MM4: N. L. Allinger, K. Chen and J.-H. Lii, *J. Comput. Chem.*, **17** (1996), 642; N. Nevins, K. Chen and N. L. Allinger, *J. Comput. Chem.*, **17** (1996), 669; N. Nevins, J.-H. Lii and N. L. Allinger, *J. Comput. Chem.*, **17** (1996), 695; N. L. Allinger, K. Chen, J. A. Katzenellenbogen, S. R. Wilson and G. M. Anstead, *J. Comput. Chem.*, **17** (1996,) 747; CVFF: S. Lifson, A. T. Hagler and P. Dauber, *J. Am. Chem. Soc.*, **101** (1979), 5111, 5122, 5131; CFF91/93/95: M. J. Hwang, J. P. Stockfisch and A. T. Hagler, *J. Am. Chem. Soc.*, **116** (1994), 2515; TRIPOS: M. Clark, R. D. Cramer III and N. van Opdenbosch, *J. Comput. Chem.*, **10** (1989), 982; J. R. Maple, M.-J. Hwang, T. P. Stockfisch, U. Dinur, M. Waldman, C. S. Ewig and A. T. Hagler, *J. Comput. Chem.*, **15** (1994), 162; MMFF: T. A. Halgren, *J. Comput. Chem.*, **17** (1996), 490; COSMIC: S. D. Morley, R. J. Abraham, I. S. Haworth, D. E. Jackson, M. R. Saunders and J. G. Vinter, *J. Computu.-Aided Mol. Des.*, **5** (1991), 475; DREIDING: S. L. Mayo, B. D. Olafson and W. A. Goddard III, *J. Phys. Chem.*, **94** (1990), 8897; AMBER: W. D. Cornell, P. Cieplak, C. I. Bayly, I. R. Gould, K. M. Merz Jr, D. M. Ferguson, D. C. Spellmeyer, T. Fox, J. W. Caldwell and P. A. Kollman, *J. Am. Chem. Soc.*, **117** (1995), 5179; OPLS: W. Damm, A. Frontera, J. Tirado-Rives and W. L. Jorgensen, *J. Comput Chem.*, **18**, (1997), 1995; CHARMM: R. Brooks, R. E. Bruccoleri, B. D. Olafson, D. J. States, S. Swaminathan and M. Karplus, *J. Comput. Chem.* **4** (1983), 187; GROMOS: W. F. Van Gunsterenm and H. J. C. Berendsen, Groningen Molecular Simulation (GROMOS) library manual; ECEPP: G. Nemethy, K. D. Gibsen, K. A. Palmer, C. N. Yoon, G. Paterlini, A. Zagari, S. Rumsey and H. A. Sheraga, *J. Phys. Chem.*, **96** (1992) 6472; MOMEC: P. Comba and T. W. Hambley, *Molecular Modeling of Inorganic Compounds*, VCH, 1995; SHAPES: V. S. Allured, C. M. Kelly and C. R. Landis, *J. Am. Chem. Soc.*, **113** (1991), 1; ESFF: S. Barlow, A. L. Rohl, S. Shi, C. M. Freeman and D. O'Hare, *J. Am. Chem. Soc.*, **118** (1996), 7578; UFF: A. K. Rappé, C. J. Casewit, K. S. Colwell, W. A. Goddard III and W. M. Skiff, *J. Am. Chem. Soc.*, **114** (1992), 10024; C. J. Casewit, K. S. Colwell and A. K. Rappé, *J. Am. Chem. Soc.*, **114** (1992), 10035, 10046.

a protein consisting of hundreds of amino acids, a crude model may be the only one possible, considering the sheer size of the problem.

Table 2.3 gives a description of the functional form used in some of the common force fields. The torsional energy is written as a Fourier series, typically of order three, in all cases. Many of the force fields undergo developments, and the number of atom types increases as more and more systems become parameterized; thus Table 2.3 may be considered as a "snapshot" of the situation when the data were collected. The "universal" type force fields, described in Section 2.3.3, are in principle capable of covering molecules composed of elements from the whole periodic table, these have been labelled as "all elements".

2.5 Computational Considerations

It has already been stated that the evaluation of the non-bonded energy is by far the most time-consuming. Consider a series of calculations of linear alkanes $CH_3(CH_2)_{n-2}CH_3$. The number of individual contributions to each energy term is given in Table 2.4.

As can be seen from Table 2.4, the number of bonded contributions, E_{str}, E_{bend} and E_{tors}, grows <u>linearly</u> with the system size. The non-bonded contributions, E_{vdw} (and E_{el}), grow as the <u>square</u> of the system size. This is fairly obvious, for a large molecule most of the atom pairs are not bonded, or bonded to a common atom, and thus contribute with an E_{vdw} term. Already for $CH_3(CH_2)_{98}CH_3$, which contains a mere 302 atoms, the non-bonded terms account for $\sim 96\%$ of the computational effort. For a 1000 atom system the percentage is 98.8%, and for 10 000 atoms it is 99.88%. <u>In the limit of large molecules, the computational time for calculating the force field energy grows approximately as the square of the number of atoms.</u> The majority of these non-bonded energy contributions are numerically very small, as the distance between the atom pairs is large. A considerable saving in computational time can be achieved by truncating the van der Waals potential at some distance, say 10 Å. If the distance is larger than this cutoff, the contribution is neglected. This is not quite as clean as it may sound at first. Although it is true that the contribution from a pair of atoms is very small if they are separated by 10 Å, there may be a large number of such atom pairs. The <u>individual</u> contribution falls off quickly, but the <u>number</u> of contributions also rises. Many force fields use cutoff distances around 10 Å, but it has been shown that the total van der Waals energy only is converged if the cut-off distance is of the order of 20 Å. However, using a cut-off of 20 Å may significantly increase the computational time (by a factor of perhaps 5–10) relative to a cut-off of 10 Å.

The introduction of a cut-off distance does not by itself lead to a significant computational saving, since all the distances must be computed prior to the decision of

Table 2.4 Number of terms for each energy contribution in $CH_3(CH_2)_{n-2}CH_3$

n	N_{atoms}	E_{str}	E_{bend}	E_{tors}	E_{vdw}
10	32	31 (5%)	30 (10%)	81 (14%)	405 (70%)
20	62	61 (3%)	60 (6%)	171 (8%)	1710 (83%)
50	152	151 (1%)	300 (3%)	441 (4%)	11025 (93%)
100	302	301 (1%)	600 (1%)	891 (2%)	44550 (96%)
	N	$(N-1)$	$2(N-2)$	$3(N-5)$	$N(N-1)/2 - 3N + 5$

whether to include the contribution. A substantial increase in computational efficiency, however, can be obtained by keeping a *non-bonded* or *neighbour list* over atom pairs. From a given starting geometry a list is prepared over atom pairs which are within the cut-off distance (or slightly larger). During a minimization or simulation, only the contributions from the atom pairs on the list are evaluated, which avoids the calculation of distances between all pairs of atoms. Since the geometry changes during the minimization or simulation, the non-bonded list must be updated at suitable intervals, for example every 10 or 20 steps.

The use of a cut-off distance reduces the formal scaling in the large system limit from N_{atom}^2 to N_{atom} since the non-bonded contributions now only are evaluated within the local "sphere" determined by the cut-off radius. However, a cut-off distance of $\sim 10\,\text{Å}$ is so large that the large system limit is not achieved in practical calculations. The actual scaling is thus more like N_{atom}^n, where n is perhaps 1.5–1.8. In static applications, however, it is not the energy of a single geometry that is of interest, but that of an optimized geometry. The larger the molecule, the more degrees of freedom, and the more complicated the geometry optimization is. The gain by introducing a non-bonded cut-off is partly offset by the increase in computational effort in the geometry optimization. Thus as a rough guideline the increase in computational time upon changing the size of the molecule can be taken as being proportional to N_{atom}^2.

The introduction of a cut-off distance, beyond which E_{vdw} is set to zero, is quite reasonable as the neglected contributions are small. This is not true for the other part of the non-bonded energy, the Coulomb interaction. Contrary to the van der Waals energy, which falls of as R^{-6}, the charge–charge interaction varies as R^{-1}. This is actually true only for the interaction between molecules carrying a net charge. The charge distribution in neutral molecules or fragments makes the long-range interaction behave as a dipole–dipole interaction. Consider for example the interaction between two carbonyl groups. The carbons carry a positive and the oxygens a negative charge. Seen from a distance, however, this looks like a bond dipole moment, not two net charges. The interaction between two dipoles behaves like R^{-3}, not R^{-1} (the van der Waals interaction is between two induced dipoles, making this interaction $(R^{-3})^2 = R^{-6}$). Nevertheless, an R^{-3} interaction requires a larger cut-off than the van der Waals R^{-6}.

Table 2.5 shows the interaction energy between two carbonyl groups in terms of the MM3 E_{vdw} and E_{el}, the latter described either by an atom point charge or a bond dipole model. The bond dipole moment is 1.86 debye, corresponding to atomic charges of ± 0.32 separated by a bond length of $1.208\,\text{Å}$. For comparison, the interaction between two net charges of 0.32 is also given.

From Table 2.5 it is clearly seen that E_{vdw} becomes small (less than ~ 0.001 kcal/mol) beyond a distance of $\sim 10\,\text{Å}$. The electrostatic interaction reaches the same level of importance at a distance of $\sim 30\,\text{Å}$. The Table also shows that the interaction between point charges behaves much like a dipole–dipole interaction, i.e. an R^{-3} dependence. However, the interaction between net charges is very long range; even at $100\,\text{Å}$ separation, there is a 0.34 kcal/mol energy contribution. The "cut-off" distance corresponding to a contribution of 0.001 kcal/mol is of the order of $3000\,\text{Å}$!

There are different ways of implementing the cut-off approximation. The simplest is to neglect all contributions if the distance is larger than the cut-off. This is in general not a very good method as the energy function becomes discontinuous. Derivatives of the energy function also become discontinuous, which causes problems in optimization

Table 2.5 Comparing the distance behaviour of non-bonded energy contributions. (Distances are in Å, energies in kcal/mol)

Distance	E_{vdw}	$E_{dipole-dipole}$	$E_{point charges}$	$E_{net charges}$
5	-0.219	0.398	0.382	6.822
10	-0.00143	0.0498	0.0492	3.411
15	-0.000128	0.01475	0.01468	2.274
20	-2.28×10^{-5}	0.00622	0.00620	1.705
30	-2.01×10^{-6}	0.00184	0.00184	1.137
50	-9.40×10^{-8}	0.000398	0.000398	0.682
100	-1.46×10^{-9}	4.98×10^{-5}	4.98×10^{-5}	0.341

procedures and for performing simulations. A better method is to use two cut-off distances between which a switching function connects the correct E_{vdw} or E_{el} smoothly with zero. Such interpolations solve the mathematical problems associated with optimization and simulation, but the chemical significance of the cut-off of course still remains. This is especially troublesome in simulation studies where the distribution of solvent molecules can be very dependent on the use of cut-offs. The modern approaches for evaluating the electrostatic contribution involve the use of fast multipole or Ewald sum methods (see Section 16.2.1), both of which are able to calculate the electrostatic energy exactly (to within a specified numerical precision) with an effort which scales less than quadratic with the number of particles (linear for fast multipole and $N^{3/2}$ for Ewald sum methods).

Obtaining a good description of the electrostatic interaction between molecules (or between different parts of the same molecule) is one of the big problems in force field work. Many commercial applications of force field methods are trying to design molecules which interact in a specific fashion. Such interactions are usually pure non-bonded. For polar molecules, such as for example amino acids, the electrostatic interactions are very important. Unfortunately, atom centred point charge models, or bond dipoles, are fairly crude approximations of the real interaction. An improved description can be obtained by including charges which are not centred on atoms,[29] or include dipole, quadrupole etc. moments at atomic positions, but this significantly increases the computational problem. We will return to this in Section 9.2. The approximation of the electrostatic interaction by nuclear point charges and the neglect of polarization are probably the largest sources of errors in modern force fields.

2.6 Validation of Force Fields

The quality of a force field calculation depends on two things: how appropriate is the mathematical form of the energy expression, and how accurate are the parameters. If elaborate forms for the individual interaction terms have been chosen, and a large number of experimental data is available for assigning the parameters, the results of a calculation may well be as good as those obtained from experiment, but at a fraction of the cost. This is the case for simple systems such as hydrocarbons. Even a force field with complicated functional forms for each of the energy contributions contains only a handful of parameters when carbon and hydrogen are the only atom types, and experimental data exist for hundreds of such compounds. The parameters can therefore be assigned with a high degree of confidence. Other well-known compound types, such as ethers and alcohol, can achieve almost as good results. For less common species, like sulfones, or polyfunctional molecules, much less experimental information is available, and the parameters are less well defined.

Force field methods are primarily geared to predict two properties: geometries and relative energies. Structural features are in general much easier to predict than relative energies. Each geometrical feature usually only depends on a few parameters, e.g. bond distances are essentially determined by R_0 and the corresponding force constant, bond angles by θ_0, and conformational minima by V_1, V_2 and V_3. It is therefore relatively easy to assign parameters which reproduce a given geometry. Relative energies of different conformations, however, are much more troublesome, since they are a consequence of many small contributions, i.e. the exact shape of the individual energy terms and the balance between them. The largest contributions to conformational energy differences are the non-bonded and torsional terms, and it is therefore important to have good representations of for example the whole torsional energy profile. Even though a given force field may be parameterized to reproduce rotational energy profiles for ethane and ethanol, and contains a good description of hydrogen bonding between for example two ethanol molecules, there is no guarantee that it will be successful in reproducing the relative energies of different conformations of say 1,2-dihydroxyethane.[30] For large systems it is inevitable that small inaccuracies in the functional forms for the energy terms and parameters will influence the shape of the whole energy surface to the point where minima may disappear or become saddle points. Essentially all force fields, no matter how elaborate the functional forms and parameterization, will have artificial minima, and fail to predict real minima, even for quite small systems. For cyclododecane (which is one of the largest molecules to have subjected to an exhaustive search), the MM2 force field predicts 122 different conformations, but the MM3 surface contains only 98 minima.[31] Given that cyclododecane belongs to a class of well-parameterized molecules, the saturated hydrocarbons, and that MM2 and MM3 are among the most accurate force fields, this clearly illustrates the point.

Validation of a force field is typically done by showing how accurately it reproduces reference data, which may or may not have been used in the actual parameterization. Since different force fields employ different sets of reference data, it is difficult to compare their accuracy directly. Indeed there is no single "best" force field, each has its advantages and disadvantages. They perform best for the type of compounds used in the parameterization, but may give questionable results for other systems. Table 2.6 gives some typical accuracies for ΔH_f that can be obtained with the MM2 force field.

Table 2.6 Average errors in heat of formations (kcal/mol) by MM2

Compound type	Average error in ΔH_f
Hydrocarbons	0.42
Ethers and Alcohols	0.50
Carbonyl compounds	0.81
Aliphatic amines	0.46
Aromatic amines	2.90
Silanes	1.08

Table 2.6 is sourced from ref. 32.

The average error is the difference between the calculated and experimental ΔH_f. In this connection it should be noted that the average error in the experimental data for the hydrocarbons is 0.40 kcal/mol, i.e. MM2 essentially reproduces the experiments to within the experimental uncertainty.

There is one final thing that needs to be mentioned in connection with the validation of a force field, namely the reproducibility. The results of a calculation are determined by the mathematical expressions for the energy terms and the parameter set (assuming that the computer program is working correctly). A new force field is usually parameterized for a fairly small set of functional groups initially, and may then evolve by addition of parameters for a larger diversity later. This sometimes has the consequence that some of the initial parameters must be modified to give an acceptable fit. Furthermore, new experimental data may warrant changes in existing parameters. In some cases different sets of parameters are derived by different research groups for the same type of functional group. The result is that the parameter set for a given force field is not constant in time, and sometimes not in geographical location either. There may also be differences in the implementation details of the energy terms, the E_{oop} in MM2, for example, is defined as a harmonic term in the bending angle (Figure 2.6), but may be substituted by an improper torsional angle in some computer programs. The consequence is that there often are several different "flavours" of a given force field, depending on the exact implementation, the original parameter set (which may not be the most recent), and any local additions to the parameters. A vivid example is the MM2 force field; which exist in several different implementations which do not give exactly the same results but nevertheless are denoted as "MM2" results.

2.7 Practical Considerations

It should be clear that force field methods are models of the real quantum mechanical systems. The total neglect of electrons as individual particles forces the user to define explicitly the bonding present in the molecule prior to any calculations. The user must decide how to describe a given molecule in terms of the selected force field. The input to a calculation consists of three sets of information.

(1) The types of atom present.
(2) How they are connected, i.e. which atoms are bonded to each other.
(3) A start guess of the geometry.

The first two sets of information determine the functional form of E_{FF}, i.e. they enable the calculation of the potential energy surface for the molecule, as long as the atomic connectivity is conserved. Normally the molecule will then be optimized by minimizing E_{FF}, which requires a start guess of the geometry. The information necessary for the program to perform the calculation is read in via a file on the computer. In older programs the input file had to be prepared manually by the user. It should be noted, however, that all the above three sets of information can be uniquely defined from a (three-dimensional) drawing of a molecule. Modern programs therefore usually have a graphical interface which allows the molecule simply to be drawn on the screen. The interface then automatically assigns suitable atom types based on the selected atomic symbols and the connectivity, and converts the drawing to Cartesian coordinates.

2.8 Advantages and Limitations of Force Field Methods

The main advantage of force field methods is the speed at which calculations can be performed. This enables large systems to be treated. Even with modest size computers (such as a workstation or a large PC), molecules with several thousand atoms can be optimized. This makes applications viable for modelling biomolecular macromolecules, such as proteins and DNA, and molecular modelling is now used by many pharmaceutical companies. The ability to treat a large number of particles also makes force field methods the only realistic method for performing simulations where solvent effects or crystal packing can be studied (Chapter 16).

For systems where good parameters are available, it is possible to make very good predictions of geometries and relative energies of a large number of molecules in a short time. It is also possible to determine barriers for interconversion between different conformations. One of the main problems is of course the lack of good parameters. If the molecule is slightly out of the ordinary, it is very likely that only poor quality parameters, or none at all, exist. Obtaining suitable values for these missing parameters can be a frustrating experience. Force field methods are good for predicting properties for classes of molecules where a lot of information already exists. For unusual molecules, their use is very limited.

Finally, force field methods are "zero-dimensional". It is not possible to asses the probable error of a given result within the method. The quality of the result can only be judged by comparison with other calculations on similar types of molecules, for which relevant experimental data exist.

2.9 Transition Structure Modelling

Structural changes can be divided into two general types: those of a conformational nature and those involving bond breaking/forming. There are intermediate cases, such as bonds involving metal coordination, but since metal coordination is difficult to model anyway, we will neglect such systems at present. The bottleneck for structural changes is the highest energy point along the reaction path, called the Transition State or Transition Structure (TS) (Chapter 12). Conformational TSs have the same atom types and bonding for both the reactant and product, and can be located on the force field energy surface by standard optimization algorithms. Since conformational changes are often localized to rotation around a single bond, simply locating the maximum energy structure for rotation (so-called "torsional angle driving", see Section 14.5.9) around this bond represents a quite good approximation to the real TS.

Modelling TSs for reactions involving bond breaking/forming within a force field methodology is much more difficult. In this case the reactant and product are not described by the same set of atom types and/or bonding. There may even be a different number of atoms at each end of the reaction (for example lone pairs disappearing). This means that there are two different force field energy functions for the reactant and product, i.e. the energy as a function of the reactant coordinate is not continuous. Nevertheless, methods have been developed for modelling differences in activation energies between similar reactions by means of force field techniques. We will here describe two approaches.

2.9.1   *Modelling the TS as a Minimum Energy Structure*

One of the early applications of TS modelling was the work on steric effects in S_N2 reactions by DeTar and co-workers, and it has more recently been advocated by Houk and co-workers.[33] The approach consists in first locating the TS for a typical example of the reaction with electronic structure methods, often at the *ab initio* HF level. The force field function is then modified so that an energy minimum is created with a geometry which matches the TS geometry found by the electronic structure method. The modification defines new parameters for all the energy terms involving the partly formed/broken bonds. The stretch energy terms have natural bond lengths taken from the electronic structure calculation, and force constants which typically are half the strength of normal bonds. Similarly, bond angle terms are modified with respect to equilibrium values and force constants, the former taken from the electronic structure data and the latter usually estimated. These modifications often necessitate definition of new "transition state" atom types. Once the force field parameters have been defined, the structure is minimized as usual. Sometimes a few cycles of parameter adjustments and reoptimizations are necessary for obtaining a set of parameters capable of reproducing the desired TS geometry. When the modified force field is capable of reproduce the *ab initio* TS geometry, it can be used for predicting TS geometries and relative energies of reactions related to the model system. As long as the differences between the systems are purely "steric", it can be hoped that relative energy differences (energy differences between the reactant and the TS model) will correlate with relative activation energies. Purely electronic effects, like Hammett type effects due to para-substitution in aromatic systems, can of course not be modelled by force field techniques.

2.9.2   *Modelling the TS as a Minimum Energy Structure on the Reactant/Product Energy Seam*

There are two principal problems with the above modelling technique. First, the TS is modelled as a minimum on the energy surface, while it should be a first-order saddle point. This has the consequence that changes in the TS position along the reaction coordinate due to differences in the reaction energy will be in the wrong direction (Section 15.6). In many cases this is probably not important. For reactions having a reasonable barrier, the TS geometry appears to be relatively constant, which may be rationalized in terms of the Marcus equation (Section 15.5). The second problem is the more or less *ad hoc* assignment of parameters. Even for quite simple reactions many new parameters must be added. Inventing perhaps 40 new parameters to reproduce maybe five relative activation energies raises the nagging question of whether TS modelling is just a fancy way of describing five data points by 40 variables.

Both of these problems are eliminated in the intersecting potential energy surface modelling technique.[34] The force field TS is here modelled as the lowest point on the seam of the reactant and product energy functions, as shown in Figure 2.19. Locating the minimum energy structure on the seam is an example of a constrained optimization, the energy should be minimized subject to the constraint that the reactant and product energies are identical. Although this is computationally somewhat more complicated

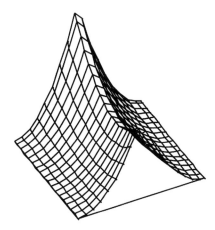

Figure 2.19 Modelling a transition structure as a minimum on the intersection of two potential energy surfaces

than the simple minimization required in the Houk approach, it can be handled in a quite efficient manner.

In the seam minimization approach only the force field parameters for describing the reactant and products are necessary, alleviating the problem of assigning parameters specific to the TS. Furthermore, differences in reactivity due to differences in reaction energy are automatically included. The remaining question is how accurately the lowest energy point on the seam resembles the actual TS. This is difficult to evaluate rigorously as it is intimately connected with the accuracy of the force field used for describing the reactant and product structures. It is clear that the TS will have bond distances and angles significantly different from these of equilibrium structures. This method of TS modelling therefore requires a force field which is accurate over a much wider range of geometries than normal. Especially important is the stretch energy, which must be able to describe bond breaking. A polynomial expansion is therefore not suitable, and for example a Morse function is necessary. Furthermore, many of the commonly employed cross terms (Section 2.2.7) become unstable at long bonds lengths, and must be modified. When such modifications are incorporated, however, the intersecting energy surface model appears to give surprisingly good results.

There are of course also disadvantages in this approach: these are essentially the same as the advantages! The seam method automatically includes the effect of different reaction energies, since a more exothermic reaction will move the TS toward the reactant and produce a lower activation energy (Section 15.5). This, however, requires that the force field is able to calculate relative energies of the reactant and product, i.e. the ability to convert steric energies to heat of formation. As mentioned in Section 2.2.9, there are only a few force fields which have been parameterized for this. In practice this is not a major problem. Usually the reaction energy for a prototypical example of the reaction of interest can be obtained from experimental data or estimated. Using the normal force field assumption of transferability of heat of formation parameters, the difference in reaction energy is thus equal to the difference in steric energy. Only the reaction energy for a single reaction of the given type therefore needs to be estimated, and relative activation energies are not sensitive to the exact value used.

If the minimum energy seam structure does not accurately represent the actual TS (compared for example to that obtained from electronic structure calculations) the lack of specific TS parameters becomes a disadvantage. In the Houk approach it is fairly easy to adjust the relevant TS parameters to reproduce the desired TS geometry. In the intersecting energy surface method, the TS geometry is a complicated result of the force field parameters for the reactant and product, and the force field energy functions. Modifying the force field parameters, or the functional form of some of the energy terms, in order to achieve the desired TS geometry without destroying the description of the reactant/product, is far from trivial.

Since the TS is given in terms of the diabatic energy surfaces for the reactant and product, it is also clear that activation energies will be too high. For evaluating relative activation energies of similar reaction this is not a major problem since the important aspect is the relative energies. More recently the overestimation of the activation energy has been improved by adding a "resonance" term to the force field, evaluated entirely from force field parameters. This removes the discontinuous derivatives at the seam and produces a smooth surface connecting the reactant and product, thereby allowing the TS to be located analogously to conformational TSs.

2.10   Hybrid Force Field–Electronic Structure Methods

Force field methods are inherently unable to describe the details of bond breaking/ forming reactions, since there is an extensive rearrangement of the electrons, which is neglected in the classical model. If the system of interest is too large to treat entirely by electronic structure methods, there are two possible approximative methods that can be used. In some cases the system can be "pruned" to a size that can be treated, by replacing "unimportant" parts of the molecule by smaller model groups, e.g. substitution of a hydrogen or methyl group for a phenyl ring. For studying enzymes, however, it is usually assumed that the whole system is important for holding the active size in the proper arrangement, and the "backbone" conformation may change during the reaction. Hybrid methods have been designed for modelling such cases, where the active size is calculated by electronic structure methods (usually semi-empirical, low-level *ab initio* or DFT methods), while the backbone is calculated by a force field method.[35] Such methods are often denoted *Quantum Mechanics–Molecular Mechanics* (QM-MM).

The main problem with QM–MM schemes is deciding how the two parts should be connected. Partial charges on the MM atoms can be incorporated into the electronic HF equations, analogously to nuclear charges (i.e. adding V_{ne}-like terms to the one-electron matrix elements in eq. (3.55)), and the QM atoms thus feel the electric potential due to all the MM atoms. van der Waals potentials from the MM atoms are also added to the QM part in order to prevent QM and MM atoms from bumping into each other. In many cases the MM and QM parts belong to the same molecule, and the connection between the two parts must be made by cutting a molecular bond. The QM part is terminated by adding "link" atoms such as hydrogens to each of the dangling bonds. The forces from the force field backbone are added to the atoms treated by the electronic structure method, and vice versa. It should be noted that there is no unique way of deciding which part should be treated by a force field and which by quantum mechanics.

The concept has been generalized in the *ONIOM* method[36] to include several layers, for example using high level *ab initio* (e.g. CCSD(T)) in the central part, lower-level electronic structure theory (e.g. MP2) in an intermediate layer and a force field to treat the outer layer.

References

1. U. Dinur and A. T. Hagler, *Rev. Comput. Chem.*, **2** (1991), 99; U. Burkert and N. L. Allinger, *Molecular Mechanics*, ACS Monograph, 1982; A. K. Rappe and C. J. Casewit, *Molecular Mechanics Across Chemistry*, University Science Books, 1997.
2. N. L. Allinger, *J. Am. Chem. Soc.*, **99** (1977), 8127; N. L. Allinger, Y. H. Yuh and J. H. Lii, *J. Am. Chem. Soc.*, **111** (1989), 8551.
3. S. W. Benzon, *Thermochemical Kinetics*, Wiley, 1976.
4. P. M. Morse, *Phys. Rev.*, **34** (1929), 57.
5. P. Jensen, *J. Mol. Spectosc.*, **133** (1989), 438.
6. F. Neumann, H. Teramae, J. W. Downing and J. Michl, *J. Am. Chem. Soc.*, **120** (1998), 573.
7. P. C. Chen, *Int. J. Quantum Chem.*, **62** (1997), 213.
8. F. London, *Z. Physik*, **63** (1930), 245.
9. J. E. Lennard-Jones, *Proc. R. Soc. London, Ser. A*, **106** (1924), 463.
10. T. L. Hill, *J. Chem. Phys.*, **16** (1948), 399.
11. J. R. Hart and A. K. Rappé, *J. Chem. Phys.*, **97** (1992), 1109.
12. T. A. Halgren, *J. Am. Chem. Soc.*, **114** (1992), 7827.
13. M. P. Hodges, A. J. Stone and S. S. Xantheas, *J. Phys. Chem.*, A **101** (1997), 9163.
14. K. Ramnarayan, B. G. Rao and U. C. Singh, *J. Chem. Phys.*, **92** (1990), 7057.
15. E. Whalley, *Chem. Phys. Lett.*, **53** (1978), 449.
16. H. D. Thomas, K. Chen and N. L. Allinger, *J. Am. Chem.* Soc., **116** (1994), 5887.
17. A. K. Rappé and W. A. Goddard III, *J. Phys. Chem.*, **95** (1991), 3358; U. Dinur and A. T. Hagler, *J. Comput. Chem.*, **16** (1995), 154.
18. K. Kveseth, R. Seip and D. A. Kohl, *Acta Chem. Scand.*, **A34** (1980), 31.
19. R. Engeln, D. Consalvo and J. Reuss, *Chem. Phys.*, **160** (1992), 427.
20. J. T. Sprague, J. C. Tai, Y. Yuh and N. L. Allinger, *J. Comput. Chem.*, **8** (1987), 581.
21. C. D. Berweger, W. F. van Gunsteren and F. Muller-Plathe, *Chem. Phys. Lett.*, **232** (1995), 429.
22. E. Osawa and K. B. Lipkowitz, *Rev. Comp. Chem.*, **6** (1995), 355.
23. P.-O. Norrby and T. Liljefors, *J. Comput. Chem.*, **19** (1998), 1146.
24. C. R. Landis, D. M. Root and T. Cleveland, *Rev. Comput. Chem.*, **6** (1995), 73; P. Comba, *Coord. Chem. Rev.*, **123** (1993), 1; B. P. Hay, *Coord. Chem. Rev.*, **126** (1993), 177.
25. B. P. Hay, *Coord. Chem. Rev.*, **126** (1993), 177.
26. R. J. Gillespie and I. Hargittai: *The VSEPR Model of Molecular Geometry*, Allyn and Bacon, 1991.
27. A. K. Rappé, C. J. Casewit, K. S. Colwell, W. A. Goddard III and W. M. Skiff, *J. Am. Chem. Soc.*, **114** (1992), 10024.
28. D. M. Root, C. R. Landis and T. Cleveland, *J. Am. Chem. Soc.*, **115** (1993), 4201.
29. R. W. Dixon and P. A. Kollman, *J. Comput. Chem.*, **18** (1997), 1632.
30. S. Reiling, J. Brickmann, M. Schlenkrich and P. A. Bopp, *J. Comput. Chem.*, **17** (1996), 133.
31. I. Kolossvary and W. G. Guida, J. Am. Chem. Soc., **118** (1996), 5011.
32. N. L. Allinger, S. H.-M. Chang, D. H. Glaser and H. Hönig, *Isr. J. Chem.*, **20** (1980), 51; J. P. Bowen, A. Pathiaseril, S. Profeta Jr and N. L. Allinger, *J. Org. Chem.*, **52** (1987), 5162; S. Profeta Jr and N. L. Allinger, *J. Am. Chem. Soc.*, **107** (1985), 1907; J. C. Tai and N. L. Allinger,

J. Am. Chem. Soc., **110** (1988), 2050; M. R. Frierson, M. R. Imam, V. B. Zalkow and N. L. Allinger, *J. Org. Chem.*, **53** (1988), 5248.

33. J. E. Eksterowicz and K. N. Houk, *Chem. Rev.*, **93** (1993), 2439.
34. F. Jensen, *J. Comput. Chem.*, **15** (1994), 1199.
35. M. J. Field, P. A. Bash and M. J. Karplus, *J. Comput. Chem.*, **11** (1990), 700.
36. M. Svensson, S. Humbel, R. D. J. Froese, T. Matsubara, S. Sieber and K. Morokuma, *J. Phys. Chem.*, **100** (1996), 19357.

3 Electronic Structure Methods

If we are interested in describing the electron distribution in detail, there is no substitute for quantum mechanics. Electrons are very light particles, and they cannot be described even qualitatively correctly by classical mechanics. We will in this and subsequent chapters concentrate on solving the time-independent Schrödinger equation, which in short-hand operator form is given as

$$\mathbf{H}\Psi = E\Psi \qquad (3.1)$$

If solutions are generated without reference to experimental data, the methods are usually called *ab initio* (latin: "from the beginning"), in contrast to semi-empirical models, which are described in Section 3.9.

A word of caution before we start. A rigorous approach to many of the derivations requires keeping track of several different indices and validating why certain transformations are possible. The derivations will be performed less rigorously, trying to illustrate the flow of arguments, rather than focus on mathematical details.

3.1 The Adiabatic and Born–Oppenheimer Approximations

Let us first review the Born-Oppenheimer approximation in a bit more detail.[1] The total Hamilton operator can be written as the kinetic and potential energies of the nuclei and electrons.

$$\mathbf{H}_{tot} = \mathbf{T}_n + \mathbf{T}_e + \mathbf{V}_{ne} + \mathbf{V}_{ee} + \mathbf{V}_{nn} \qquad (3.2)$$

The Hamilton operator is first transformed to the centre of mass system, where it may be written as (using atomic units, see Appendix D):

$$\mathbf{H}_{tot} = \mathbf{T}_n + \mathbf{H}_e + \mathbf{H}_{mp}$$
$$\mathbf{H}_e = \mathbf{T}_e + \mathbf{V}_{ne} + \mathbf{V}_{ee} + \mathbf{V}_{nn}$$
$$\mathbf{H}_{mp} = -\frac{1}{2M_{tot}}\left(\sum_{i=1}^{N}\mathbf{\nabla}_i\right)^2$$

(3.3)

Here $\mathbf{H}_e$ is the *electronic Hamilton operator* and $\mathbf{H}_{mp}$ is called the *mass-polarization* (M_{tot} is the total mass of all the nuclei and the sum is over all electrons). We note that $\mathbf{H}_e$ depends only on the nuclear <u>positions</u> (via $\mathbf{V}_{ne}$ and $\mathbf{V}_{nn}$, see eq. (3.23)) and not on their <u>momenta</u>.

Assume for the moment that the full set of solutions to the electronic Schrödinger equation is available, where $\mathbf{R}$ denotes nuclear positions and $\mathbf{r}$ electronic coordinates.

$$\mathbf{H}_e(\mathbf{R})\Psi_i(\mathbf{R},\mathbf{r}) = E_i(\mathbf{R})\Psi_i(\mathbf{R},\mathbf{r}), \quad i = 1,2\ldots\infty \qquad (3.4)$$

Since the Hamilton operator is *hermitic* ($\int\Psi_i^*\mathbf{H}\Psi_j d\mathbf{r} = \int\Psi_j\mathbf{H}^*\Psi_i^* d\mathbf{r}$), the solutions can be chosen to be orthogonal and normalized (*orthonormal*).

$$\int\Psi_i^*(\mathbf{R},\mathbf{r})\Psi_j(\mathbf{R},\mathbf{r})d\mathbf{r} = \delta_{ij}$$
$$\delta_{ij} = 1, \quad i = j$$
$$\delta_{ij} = 0, \quad i \neq j$$

(3.5)

Without introducing any approximations, the total (exact) wave function can be written as an expansion in the complete set of electronic functions, with the expansion coefficients being functions of the nuclear coordinates.

$$\Psi_{tot}(\mathbf{R},\mathbf{r}) = \sum_{i=1}^{\infty}\Psi_{ni}(\mathbf{R})\Psi_i(\mathbf{R},\mathbf{r}) \qquad (3.6)$$

Inserting eq. (3.6) into the Schrödinger equation (3.1) gives

$$\sum_{i=1}^{\infty}(\mathbf{T}_n + \mathbf{H}_e + \mathbf{H}_{mp})\Psi_{ni}(\mathbf{R})\Psi_i(\mathbf{R},\mathbf{r}) = E_{tot}\sum_{i=1}^{\infty}\Psi_{ni}(\mathbf{R})\Psi_i(\mathbf{R},\mathbf{r}) \qquad (3.7)$$

The nuclear kinetic energy is essentially a differential operator, and we may write it as:

$$\mathbf{T}_n = \sum_a -\frac{1}{2M_a}\mathbf{\nabla}_a^2 = \mathbf{\nabla}_n^2$$
$$\mathbf{\nabla}_a = \left(\frac{\partial}{\partial X_a},\frac{\partial}{\partial Y_a},\frac{\partial}{\partial Z_a}\right)$$
$$\mathbf{\nabla}_a^2 = \left(\frac{\partial^2}{\partial X_a^2}+\frac{\partial^2}{\partial Y_a^2}+\frac{\partial^2}{\partial Z_a^2}\right)$$

(3.8)

where the mass dependence, sign and summation is implicitly included in the $\mathbf{\nabla}_n^2$

symbol. Expanding (3.7) gives

$$\sum_{i=1}^{\infty} (\mathbf{V}_n^2 + \mathbf{H}_e + \mathbf{H}_{mp}) \Psi_{ni} \Psi_i = E_{tot} \sum_{i=1}^{\infty} \Psi_{ni} \Psi_i$$

$$\sum_{i=1}^{\infty} \{ \mathbf{V}_n^2 (\Psi_{ni} \Psi_i) + \mathbf{H}_e \Psi_{ni} \Psi_i + \mathbf{H}_{mp} \Psi_{ni} \Psi_i \} = E_{tot} \sum_{i=1}^{\infty} \Psi_{ni} \Psi_i$$

$$\sum_{i=1}^{\infty} \{ \mathbf{V}_n [(\Psi_i \mathbf{V}_n \Psi_{ni}) + (\Psi_{ni} \mathbf{V}_n \Psi_i)] + \Psi_{ni} \mathbf{H}_e \Psi_i + \Psi_{ni} \mathbf{H}_{mp} \Psi_i \} = E_{tot} \sum_{i=1}^{\infty} \Psi_{ni} \Psi_i$$

$$\sum_{i=1}^{\infty} \{ \Psi_i (\mathbf{V}_n^2 \Psi_{ni}) + 2(\mathbf{V}_n \Psi_i)(\mathbf{V}_n \Psi_{ni}) + \Psi_{ni} (\mathbf{V}_n^2 \Psi_i) + \Psi_{ni} E_i \Psi_i + \Psi_{ni} \mathbf{H}_{mp} \Psi_i \}$$

$$= E_{tot} \sum_{i=1}^{\infty} \Psi_{ni} \Psi_i \tag{3.9}$$

where we have used the fact that $\mathbf{H}_e$ and $\mathbf{H}_{mp}$ only act on the electronic wave function, and Ψ_i is an exact solution to the electronic Schrödinger equation (eq. (3.4)). We will now use the orthonormality of the Ψ_i by multiplying from the left by a specific electronic wave function Ψ_j^* and integrating over the electron coordinates.

We will take the opportunity at this point to introduce the *bra-ket* notation.

$$\int \Psi^* \mathbf{H} \Psi dv = \langle \Psi | \mathbf{H} | \Psi \rangle$$

$$\int \Psi^* \Psi dv = \langle \Psi | \Psi \rangle \tag{3.10}$$

The *bra* $\langle n |$ denotes a complex conjugate wave function with quantum number n standing to the left of the operator, while the *ket* $| m \rangle$, denotes a wave function with quantum number m standing to the right of the operator, and the combined *bracket* denotes that the whole expression should be integrated over all coordinates. Such a bracket is often referred to as a *matrix element*. The orthonormality condition eq. (3.5) can then be written as.

$$\langle \Psi_i | \Psi_j \rangle = \delta_{ij} \tag{3.11}$$

With this change in notation, eq. (3.9) becomes after integration

$$\mathbf{V}_n^2 \Psi_{ni} + E_j \Psi_{ni} + \sum_{i=1}^{\infty} \{ 2 \langle \Psi_j | \mathbf{V}_n | \Psi_i \rangle (\mathbf{V}_n \Psi_{ni}) + \langle \Psi_j | \mathbf{V}_n^2 | \Psi_i \rangle \Psi_{ni}$$

$$+ \langle \Psi_j | \mathbf{H}_{mp} | \Psi_i \rangle \Psi_{ni} \} = E_{tot} \Psi_{nj} \tag{3.12}$$

The electronic wave function has now been removed from the first two terms while the curly bracket contains terms which couple different electronic states. The first two of these are the first- and second-order *non-adiabatic coupling elements*, respectively, while the last is the mass polarization. The non-adiabatic coupling elements are important for systems involving more than one electronic surface, such as photochemical reactions.

In the *adiabatic* approximation the form of the total wave function is restricted to one electronic surface, i.e. all coupling elements in eq. (3.12) are neglected (only the terms with $i = j$ survive). Except for spatially degenerate wave functions, the diagonal first-order non-adiabatic coupling element is zero.

$$(\mathbf{V}_n^2 + E_j + \langle \Psi_j | \mathbf{V}_n^2 | \Psi_j \rangle + \langle \Psi_j | \mathbf{H}_{mp} | \Psi_j \rangle) \Psi_{nj} = E_{tot} \Psi_{nj} \tag{3.13}$$

Neglecting the mass polarization and reintroducing the kinetic energy operator gives

$$(\mathbf{T}_n + E_j + \langle \Psi_j | \mathbf{V}_n^2 | \Psi_j \rangle) \Psi_{nj} = E_{tot} \Psi_{nj} \tag{3.14}$$

or more explicitly

$$(\mathbf{T}_n + E_j(\mathbf{R}) + U(\mathbf{R})) \Psi_{nj}(\mathbf{R}) = E_{tot} \Psi_{nj}(\mathbf{R}) \tag{3.15}$$

The $U(\mathbf{R})$ term is known as the *diagonal correction*, and is smaller than $E_j(\mathbf{R})$ by a factor roughly equal to the ratio of the electronic and nuclear masses (eq. (3.8)). It is usually a slowly varying function of $\mathbf{R}$, and the shape of the energy surface is therefore determined almost exclusively by $E_j(\mathbf{R})$.[2] In the *Born–Oppenheimer* (BO) approximation the diagonal correction is neglected, and the resulting equation takes on the usual Schrödinger form, where the electronic energy plays the role of a potential energy.

$$\begin{aligned} (\mathbf{T}_n + E_j(\mathbf{R})) \Psi_{nj}(\mathbf{R}) = E_{tot} \Psi_{nj}(\mathbf{R}) \\ (\mathbf{T}_n + V_j(\mathbf{R})) \Psi_{nj}(\mathbf{R}) = E_{tot} \Psi_{nj}(\mathbf{R}) \end{aligned} \tag{3.16}$$

In the Born–Oppenheimer picture the nuclei move on a *potential energy surface* (PES) which is a solution to the <u>electronic</u> Schrödinger equation. The PES is independent of the nuclear masses (i.e. it is the same for isotopic molecules), this is not the case when working in the adiabatic approximation since the diagonal correction (and mass polarization) depends on the nuclear masses. Solution of (3.16) for the nuclear wave function leads to energy levels for molecular vibrations (Section 13.1) and rotations, which in turn are the fundamentals for many forms of spectroscopy, such as IR, Raman, microwave etc.

The Born–Oppenheimer (and adiabatic) approximation is usually a good approximation, but breaks down when two (or more) solutions to the electronic Schrödinger equation come close together energetically. Consider for example stretching the bond in the LiF molecule. Near the equilibrium distance the molecule is very polarized, i.e. it can be described essentially by an ionic wave function, Li^+F^-. The molecule, however, dissociates into neutral atoms (all bonds break homolytically in the gas phase), i.e. the wave function at long distance is of a covalent type, Li·F·. At the equilibrium distance the covalent wave function is higher in energy than the ionic, and the situation reverses as the bond distance increases. At some point they must "cross", however, as they have the same symmetry, they do not actually cross. Instead they make an *avoided crossing*. In the region of the avoided crossing, the wave function changes from being mainly ionic to covalent over a short distance, and the adiabatic, and therefore also the Born–Oppenheimer, approximation, breaks down.

For the majority of systems the Born–Oppenheimer approximation introduces only very small errors. Once the Born–Oppenheimer approximation is made, the

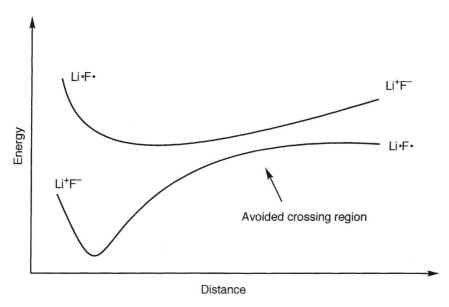

Figure 3.1 Avoided crossing of potential energy surfaces for LiF

problem is reduced to solving the electronic Schrödinger equation for a set of nuclear geometries.

The electronic Hamilton operator is normally written as

$$\mathbf{H}_e = \mathbf{T}_e + \mathbf{V}_{ne} + \mathbf{V}_{ee} + \mathbf{V}_{nn} \tag{3.17}$$

In doing this we have implicitly neglected relativistic effects. These are normally negligible for the first 3 rows in the periodic table (i.e. $Z < 36$), but become important for the fourth and fifth rows, and for transition metals. A more detailed discussion can be found in Chapter 8. Besides the neglect of relativistic effects, and the introduction of the Born–Oppenheimer approximation, a number of operators which in most cases give only small contributions are also neglected. These may for example be external fields, or operators describing spin–orbit or spin–spin coupling. Spin-dependent terms are relativistic in origin, but can be introduced in an *ad hoc* fashion in non-relativistic theory, and calculated as corrections (for example by means of perturbation theory) after the electronic Schrödinger equation has been solved. This will be discussed in more detail in Chapter 10.

3.2 Self-consistent Field Theory

Having stated the limitations (non-relativistic Hamilton operator and the Born–Oppenheimer approximation), we are ready to consider the electronic Schrödinger equation. It can only be solved exactly for the H_2^+ molecule, and similar one-electron systems. In the general case we have to rely on approximate (numerical) methods. By neglecting relativistic effects, we also have to introduce electron spin as an *ad hoc* quantum effect. Each electron has a spin quantum number of 1/2. In the presence of an external magnetic field there are two possible states, corresponding to alignment <u>along</u> or <u>opposite</u> to the field. The corresponding spin functions are denoted α and β, and obey the following orthonormality conditions.

$$\langle\alpha|\alpha\rangle = \langle\beta|\beta\rangle = 1$$
$$\langle\alpha|\beta\rangle = \langle\beta|\alpha\rangle = 0 \tag{3.18}$$

To generate approximate solutions we will employ the variational principle, which states that any approximate wave function has an energy above or equal to the exact energy (see Appendix B for a proof). The equality holds only if the wave function is the exact function. By making a trial wave function containing a number of parameters, we can generate the "best" trial function of the given form by minimizing the energy as a function of these parameters.

The energy of an approximate wave function can be calculated as the expectation value of the Hamilton operator, divided by the norm of the wave function.

$$E_e = \frac{\langle\Psi|\mathbf{H}_e|\Psi\rangle}{\langle\Psi|\Psi\rangle} \tag{3.19}$$

For a normalized wave function the denominator is 1, and therefore $E_e = \langle\Psi|\mathbf{H}_e|\Psi\rangle$. The total electronic wave function must be antisymmetric (change sign) with respect to the interchange of any two electron coordinates (since electrons are fermions, having a spin of 1/2). The *Pauli principle*, which states that two electrons cannot have all quantum numbers equal, is a direct consequence of this antisymmetry requirement. The antisymmetry of the wave function can be achieved by building it from *Slater Determinants* (SDs). The columns in a Slater determinant are single electron wave functions, *orbitals*, while the electron coordinates are along the rows. Let us in the following assume that we are interested in solving the electronic Schrödinger equation for a molecule. The one-electron functions are thus *Molecular Orbitals* (MO), which are given as a product of a spatial orbital times a spin function (α or β), also known as *spinorbitals*, and may be taken to be orthonormal. For the general case of N electrons and N spinorbitals, a Slater determinant is given as

$$\Phi_{SD} = \frac{1}{\sqrt{N!}}\begin{vmatrix} \phi_1(1) & \phi_2(1) & \dots & \phi_N(1) \\ \phi_1(2) & \phi_2(2) & \dots & \phi_N(2) \\ \dots & \dots & \dots & \dots \\ \phi_1(N) & \phi_2(N) & \dots & \phi_N(N) \end{vmatrix}, \quad \langle\phi_i|\phi_j\rangle = \delta_{ij} \tag{3.20}$$

We now make one further approximation, taking the trial wave function to consist of a <u>single</u> Slater determinant. As will be seen later, this implies that electron correlation is neglected, or equivalently, the electron–electron repulsion is only included as an average effect. Having selected a single determinant trial wave function the variational principle can be used to derive the *Hartree–Fock* (HF) equations. The HF model is a kind of branching point, either additional approximations may be invoked, leading to semi-empirical methods, or it can be improved by adding additional determinants, generating solutions which can be made to converge towards the exact solution of the electronic Schrödinger equation.[3]

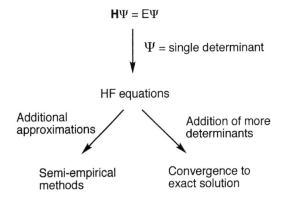

$$\mathbf{H}\Psi = E\Psi$$

Ψ = single determinant

HF equations

Additional approximations

Addition of more determinants

Semi-empirical methods

Convergence to exact solution

Figure 3.2 The HF model as a starting point for either more approximate or more accurate treatments

3.3 The Energy of a Slater Determinant

We will first evaluate the energy of a single Slater determinant. It is convenient to write it as a sum of permutations over the "diagonal" of the determinant. We will denoted the diagonal product by Π, and use the symbol Φ to represent the determinant wave function.

$$\Phi = \mathbf{A}[\phi_1(1)\phi_2(2)\ldots\phi_N(N)] = \mathbf{A}\Pi$$

$$\mathbf{A} = \frac{1}{\sqrt{N!}}\sum_{p=0}^{N-1}(-1)^p\mathbf{P} = \frac{1}{\sqrt{N!}}\left[\mathbf{1} - \sum_{ij}\mathbf{P}_{ij} + \sum_{ijk}\mathbf{P}_{ijk} - \ldots\right] \tag{3.21}$$

The $\mathbf{1}$ operator is the identity, while $\mathbf{P}_{ij}$ generates all possible permutations of two electron coordinates, $\mathbf{P}_{ijk}$ all possible permutations of three electron coordinates etc. It may be shown that the *antisymmetrizing operator* $\mathbf{A}$ commutes with $\mathbf{H}$, and that $\mathbf{A}$ acting twice gives the same as $\mathbf{A}$ acting once, multiplied by the square root of N factorial.

$$\mathbf{AH} = \mathbf{HA}$$

$$\mathbf{AA} = \sqrt{N!}\mathbf{A} \tag{3.22}$$

Consider now the Hamilton operator. The nuclear–nuclear repulsion does not depend on electron coordinates and is a constant for a given nuclear geometry. The nuclear–electron attraction is a sum of terms, each depending only on one electron coordinate. The same holds for the electron kinetic energy. The electron–electron repulsion, however, depends on two electron coordinates.

$$\mathbf{H}_e = \mathbf{T}_e + \mathbf{V}_{ne} + \mathbf{V}_{ee} + \mathbf{V}_{nn}$$

$$\mathbf{T}_e = -\sum_i^N \frac{1}{2}\mathbf{V}_i^2$$

$$\mathbf{V}_{ne} = -\sum_i^N \sum_a \frac{Z_a}{|\mathbf{R}_a - \mathbf{r}_i|} \tag{3.23}$$

$$\mathbf{V}_{ee} = \sum_i^N \sum_{j>i}^N \frac{1}{|\mathbf{r}_i - \mathbf{r}_j|}$$

$$\mathbf{V}_{nn} = \sum_a \sum_{b>a} \frac{Z_a Z_b}{|\mathbf{R}_a - \mathbf{R}_b|}$$

The operators may be collected according to the number of electron indices.

$$\mathbf{h}_i = -\frac{1}{2}\mathbf{\nabla}_i^2 - \sum_a \frac{Z_a}{|\mathbf{R}_a - \mathbf{r}_i|}$$

$$\mathbf{g}_{ij} = \frac{1}{|\mathbf{r}_i - \mathbf{r}_j|} \tag{3.24}$$

$$\mathbf{H}_e = \sum_{i=1}^{N} \mathbf{h}_i + \sum_{i=1}^{N}\sum_{j>i}^{N} \mathbf{g}_{ij} + \mathbf{V}_{nn}$$

The one electron operator $\mathbf{h}_i$ describes the motion of electron i in the field of all the nuclei, and $\mathbf{g}_{ij}$ is a two electron operator giving the electron–electron repulsion. We note that the zero point of the energy corresponds to the particles being at rest ($\mathbf{T}_e = 0$) and infinitely removed from each other ($\mathbf{V}_{ne} = \mathbf{V}_{ee} = \mathbf{V}_{nn} = 0$).

The energy may be written in terms of the permutation operator as (using eqs. (3.21)–(3.22))

$$\begin{aligned} E &= \langle\Phi|\mathbf{H}|\Phi\rangle \\ &= \langle\mathbf{A}\Pi|\mathbf{H}|\mathbf{A}\Pi\rangle \\ &= \sqrt{N!}\langle\Pi|\mathbf{H}|\mathbf{A}\Pi\rangle \\ &= \sum_p (-1)^p \langle\Pi|\mathbf{H}|\mathbf{P}\Pi\rangle \end{aligned} \tag{3.25}$$

The nuclear repulsion operator does not depend on electron coordinates and can immediately be integrated to yield a constant.

$$\langle\Phi|\mathbf{V}_{nn}|\Phi\rangle = V_{nn}\langle\Phi|\Phi\rangle = V_{nn} \tag{3.26}$$

For the one-electron operator only the identity operator can give a non-zero contribution. For coordinate 1 this yields

$$\begin{aligned} \langle\Pi|\mathbf{h}_1|\Pi\rangle &= \langle\phi_1(1)\phi_2(2)\ldots\phi_N(N)|\mathbf{h}_1|\phi_1(1)\phi_2(2)\ldots\phi_N(N)\rangle \\ &= \langle\phi_1(1)|\mathbf{h}_1|\phi_1(1)\rangle\langle\phi_2(2)|\phi_2(2)\rangle\ldots\langle\phi_N(N)|\phi_N(N)\rangle \\ &= \langle\phi_1(1)|\mathbf{h}_1|\phi_1(1)\rangle = h_1 \end{aligned} \tag{3.27}$$

as all the MOs ϕ_i are normalized. All matrix elements involving a permutation operator give zero. Consider for example

$$\begin{aligned} \langle\Pi|h_1|\mathbf{P}_{12}\Pi\rangle &= \langle\phi_1(1)\phi_2(2)\ldots\phi_N(N)|\mathbf{h}_1|\phi_2(1)\phi_1(2)\ldots\phi_N(N)\rangle \\ &= \langle\phi_1(1)|\mathbf{h}_1|\phi_2(1)\rangle\langle\phi_2(2)|\phi_1(2)\rangle\ldots\langle\phi_N(N)|\phi_N(N)\rangle \end{aligned} \tag{3.28}$$

This is zero as the integral over electron 2 is an overlap of two different MOs, which are orthogonal (eq. (3.20)).

For the two electron operator, only the identity and $\mathbf{P}_{ij}$ operators can give a non-zero contribution. A three electron permutation will again give at least one overlap integral between two different MOs, which will be zero. The term arising from the identity operator is

$$\langle \Pi | \mathbf{g}_{12} | \Pi \rangle = \langle \phi_1(1)\phi_2(2)\ldots\phi_N(N) | \mathbf{g}_{12} | \phi_1(1)\phi_2(2)\ldots\phi_N(N) \rangle$$
$$= \langle \phi_1(1)\phi_2(2) | \mathbf{g}_{12} | \phi_1(1)\phi_2(2) \rangle \ldots \langle \phi_N(N) | \phi_N(N) \rangle \qquad (3.29)$$
$$= \langle \phi_1(1)\phi_2(2) | \mathbf{g}_{12} | \phi_1(1)\phi_2(2) \rangle = J_{12}$$

and is called a *Coulomb* integral. It represents a classical repulsion between two charge distributions described by $\phi_1^2(1)$ and $\phi_2^2(2)$. The term arising from the $\mathbf{P}_{ij}$ operator is

$$\langle \Pi | \mathbf{g}_{12} | \mathbf{P}_{12}\Pi \rangle = \langle \phi_1(1)\phi_2(2)\ldots\phi_N(N) | \mathbf{g}_{12} | \phi_2(1)\phi_1(2)\ldots\phi_N(N) \rangle$$
$$= \langle \phi_1(1)\phi_2(2) | \mathbf{g}_{12} | \phi_2(1)\phi_1(2) \rangle \ldots \langle \phi_N(N) | \phi_N(N) \rangle \qquad (3.30)$$
$$= \langle \phi_1(1)\phi_2(2) | \mathbf{g}_{12} | \phi_2(1)\phi_1(2) \rangle = K_{12}$$

and is called an *exchange* integral, which has no classical analogy. Note that the order of the MOs in the J and K matrix elements is according to the electron indices. The energy can thus be written as

$$E = \sum_{i=1}^{N} h_i + \sum_{i=1}^{N} \sum_{j>i}^{N} (J_{ij} - K_{ij}) + V_{nn} \qquad (3.31)$$

where the minus sign for the exchange term comes from the factor of $(-1)^p$ in the antisymmetrizing operator, eq. (3.21). The energy may also be written in a more symmetrical form as

$$E = \sum_{i=1}^{N} h_i + \frac{1}{2} \sum_{i=1}^{N} \sum_{j=1}^{N} (J_{ij} - K_{ij}) + V_{nn} \qquad (3.32)$$

where the factor of 1/2 allows the double sum to run over all electrons (it is easily seen from eqs. (3.29) and (3.30) that the Coulomb "self-interaction" J_{ii} is exactly canceled by the corresponding "exchange" element K_{ii}).

For the purpose of deriving the variation of the energy, it is convenient to express the energy in terms of Coulomb and Exchange operators.

$$E = \sum_{i}^{N} \langle \phi_i | \mathbf{h}_i | \phi_i \rangle + \frac{1}{2} \sum_{ij}^{N} (\langle \phi_j | \mathbf{J}_i | \phi_j \rangle - \langle \phi_j | \mathbf{K}_i | \phi_j \rangle) + V_{nn}$$
$$\mathbf{J}_i | \phi_j(2) \rangle = \langle \phi_i(1) | \mathbf{g}_{12} | \phi_i(1) \rangle | \phi_j(2) \rangle \qquad (3.33)$$
$$\mathbf{K}_i | \phi_j(2) \rangle = \langle \phi_i(1) | \mathbf{g}_{12} | \phi_j(1) \rangle | \phi_i(2) \rangle$$

Note that the $\mathbf{J}$ operator involves "multiplication" by a matrix element with the same orbital on both sides, while the $\mathbf{K}$ operator "exchanges" the two functions on the right-hand side of the $\mathbf{g}_{12}$ operator.

The objective is now to determine a set of MOs which makes the energy a minimum, or at least stationary with respect to a change in the orbitals. The variation, however, must be carried out in such a way that the MOs remain orthogonal and normalized. This is a constrained optimization, and can be handled by means of *Lagrange multipliers* (see Section 14.6). The condition is that a small change in the orbitals should not change the Lagrange function, i.e. the Lagrange function is stationary with respect to an orbital variation.

$$L = E - \sum_{ij}^{N} \lambda_{ij}(\langle\phi_i|\phi_j\rangle - \delta_{ij})$$

$$\delta L = \delta E - \sum_{ij}^{N} \lambda_{ij}(\langle\delta\phi_i|\phi_j\rangle + \langle\phi_i|\delta\phi_j\rangle) = 0$$

(3.34)

The variation of the energy is given by

$$\delta E = \sum_{i}^{N} (\langle\delta\phi_i|\mathbf{h}_i|\phi_i\rangle + \langle\phi_i|\mathbf{h}_i|\delta\phi_i\rangle)$$

$$+ \frac{1}{2}\sum_{ij}^{N} (\langle\delta\phi_i|\mathbf{J}_i - \mathbf{K}_j|\phi_i\rangle + \langle\phi_i|\mathbf{J}_j - \mathbf{K}_j|\delta\phi_i\rangle$$

$$+ \langle\delta\phi_j|\mathbf{J}_i - \mathbf{K}_i|\phi_j\rangle + \langle\phi_j|\mathbf{J}_i - \mathbf{K}_i|\delta\phi_j\rangle)$$

(3.35)

The third and fifth terms are identical (since the summation is over all i and j), as are the fourth and sixth terms. They may be collected to cancel the factor of 1/2, and the variation can be written in terms of a *Fock operator*, $\mathbf{F}_i$.

$$\delta E = \sum_{i}^{N} \langle\delta\phi_i|\mathbf{h}_i|\phi_i\rangle + \langle\phi_i|\mathbf{h}_i|\delta\phi_i\rangle + \sum_{ij}^{N} (\langle\delta\phi_i|\mathbf{J}_j - \mathbf{K}_j|\phi_i\rangle + \langle\phi_i|\mathbf{J}_j - \mathbf{K}_j|\delta\phi_i\rangle)$$

$$\delta E = \sum_{i}^{N} (\langle\delta\phi_i|\mathbf{F}_i|\phi_i\rangle + \langle\phi_i|\mathbf{F}_i|\delta\phi_i\rangle)$$

$$\mathbf{F}_i = \mathbf{h}_i + \sum_{j}^{N} (\mathbf{J}_j - \mathbf{K}_j)$$

(3.36)

The Fock operator is an effective one-electron energy operator, describing the kinetic energy of an electron, the attraction to all the nuclei and the repulsion to all the other electrons (via the $\mathbf{J}$ and $\mathbf{K}$ operators). Note that the Fock operator is associated with the variation of the total energy, not the energy itself. The Hamilton operator (3.23) is not a sum of Fock operators.

The variation of the Lagrange function (3.34) now becomes

$$\delta L = \sum_{i}^{N} (\langle\delta\phi_i|\mathbf{F}_i|\phi_i\rangle + \langle\phi_i|\mathbf{F}_i|\delta\phi_i\rangle) - \sum_{ij}^{N} \lambda_{ij}(\langle\delta\phi_i|\phi_j\rangle + \langle\phi_i|\delta\phi_j\rangle)$$

(3.37)

The variational principle states that the desired orbitals are those that make $\delta L = 0$. Making use of the fact that $\langle\phi|\delta\phi\rangle = \langle\delta\phi|\phi\rangle^*$ and $\langle\phi|\mathbf{F}|\delta\phi\rangle = \langle\delta\phi|\mathbf{F}|\phi\rangle^*$ we get

$$\delta L = \sum_{i}^{N} \langle\delta\phi_i|\mathbf{F}_i|\phi_i\rangle - \sum_{ij}^{N} \lambda_{ij}\langle\delta\phi_i|\phi_j\rangle$$

$$+ \sum_{i}^{N} \langle\delta\phi_i|\mathbf{F}_i|\phi_i\rangle^* - \sum_{ij}^{N} \lambda_{ij}\langle\delta\phi_j|\phi_i\rangle^* = 0$$

(3.38)

The variation of <u>either</u> $\langle\delta\phi|$ <u>or</u> $\langle\delta\phi|^*$ should make $\delta L = 0$. The first two terms in (3.38) should thus be zero, and the last two terms should be zero. Taking the complex conjugate of the last two terms and subtracting them from the first two gives

$$\sum_{ij}^{N} (\lambda_{ij} - \lambda_{ji}^*)\langle\delta\phi_i|\phi_j\rangle = 0 \tag{3.39}$$

which means that the Lagrange multipliers are elements of a Hermitian matrix $(\lambda_{ij} = \lambda_{ji}^*)$. The final set of *Hartree–Fock equations* may be written as

$$\mathbf{F}_i\phi_i = \sum_{j}^{N} \lambda_{ij}\phi_j \tag{3.40}$$

The equations may be simplified by choosing a unitary transformation (Chapter 13) which makes the matrix of Lagrange multipliers diagonal, i.e. $\lambda_{ij} \to 0$ and $\lambda_{ii} \to \varepsilon_i$. This special set of molecular orbitals (ϕ') are called *canonical* MOs, and they transform eq. (3.40) into a set of pseudo-eigenvalue equations.

$$\mathbf{F}_i\phi_i' = \varepsilon_i\phi_i' \tag{3.41}$$

The Lagrange multipliers can be interpreted as MO energies, i.e. they are the expectation value of the Fock operator in the MO basis (multiply eq. (3.41) by $\phi_i'^*$ from the left and integrate).

$$\varepsilon_i = \langle\phi_i'|\mathbf{F}_i|\phi_i'\rangle \tag{3.42}$$

The Hartree–Fock equations form a set of pseudo-eigenvalue equations, as the Fock operator depends on all the occupied MOs (via the Coulomb and Exchange operators, eqs. (3.36) and (3.33)). A specific Fock orbital can only be determined if all the other occupied orbitals are known, and iterative methods must therefore be employed for determining the orbitals. A set of functions which is a solution to eq. (3.41) are called *Self-Consistent Field* (SCF) orbitals.

The canonical MOs may be considered as a convenient set of orbitals for carrying out the variational calculation. The total energy, however, depends only on the total wave function, which is a Slater determinant written in terms of the occupied MOs, eq. (3.20). The total wave function is unchanged by a unitary transformation of the occupied MOs among themselves (rows and columns in a determinant can be added and subtracted without affecting the determinant itself). After having determined the canonical MOs, other sets of MOs may be generated by forming linear combinations, such as localized MOs, or MOs displaying hybridization, to be discussed in more detail in Chapter 9.

The orbital energies can be considered as matrix elements of the Fock operator with the MOs (dropping the prime notation and letting ϕ be the canonical orbitals). The total energy can be written either as eq. (3.32) or in terms of MO energies (using the definition of $\mathbf{F}$ in eqs. (3.36) and (3.42)).

$$E = \sum_{i}^{N} \varepsilon_i - \frac{1}{2}\sum_{ij}^{N} (J_{ij} - K_{ij}) + V_{nn}$$

$$\varepsilon_i = \langle\phi_i|\mathbf{F}_i|\phi_i\rangle = h_i + \sum_{j}^{N} (J_{ij} - K_{ij}) \tag{3.43}$$

The total energy is <u>not</u> simply a sum of MO orbital energies. The Fock operator contains terms describing the repulsion to all other electrons (**J** and **K** operators), and the sum over MO energies therefore counts the electron–electron repulsion twice, which must be corrected for. It is also clear that the total energy cannot be exact, as it describes the repulsion between an electron and all the other electrons, assuming that their spatial distribution is described by a set of orbitals. The electron–electron repulsion is only accounted for in an average fashion, and the HF method is therefore also referred to as a *Mean Field* approximation. As mentioned previously, this is due to the approximation of a single Slater determinant as the trial wave function.

3.4 Koopmans' Theorem

The canonical MOs are convenient for the physical interpretation of the Lagrange multipliers. Consider the energy of a system with one electron removed from orbital number k, and assume that the MOs are identical for the two systems (eq. (3.32)).

$$
\begin{aligned}
E_N &= \sum_{i=1}^{N} h_i + \frac{1}{2} \sum_{i=1}^{N} \sum_{j=1}^{N} (J_{ij} - K_{ij}) + V_{nn} \\
E_{N-1}^{k} &= \sum_{i=1}^{N-1} h_i + \frac{1}{2} \sum_{i=1}^{N-1} \sum_{j=1}^{N-1} (J_{ij} - K_{ij}) + V_{nn}
\end{aligned}
\tag{3.44}
$$

Subtracting the two total energies gives

$$
E_N - E_{N-1}^{k} = h_k + \frac{1}{2} \sum_{i=1}^{N} (J_{ik} - K_{ik}) + \frac{1}{2} \sum_{j=1}^{N} (J_{kj} - K_{kj})
\tag{3.45}
$$

The last two sums are identical and the energy difference becomes

$$
E_N - E_{N-1}^{k} = h_k + \sum_{i=1}^{N} (J_{ki} - K_{ki}) = \varepsilon_k
\tag{3.46}
$$

which is seen (eq. (3.43)) to be exactly the orbital energy ε_k. The ionization energy within the "frozen MO" approximation is given simply as the orbital energy, a result known as *Koopmans' theorem*.[4] Similarly, the electron affinity of a neutral molecule is given as the orbital energy of the corresponding anion, or, since the MOs are assumed constant, as the energy of the kth unoccupied orbital energy in the neutral species.

$$
E_{N+1}^{k} - E_N = \varepsilon_k
\tag{3.47}
$$

Computationally, however, there is a significant difference between the eigenvalue of an occupied orbital for the anion and the eigenvalue corresponding to an unoccupied orbital in the neutral species when the orbitals are expanded in a set of basis functions (Section 3.5). Eigenvalues corresponding to occupied orbitals are well defined and they converge to a specific value as the size of the basis set is increased. In contrast, unoccupied orbitals in a sense are only the "left over" functions in a given basis set, and their number increases as the basis set is made larger. The lowest unoccupied eigenvalue usually converges to zero, corresponding to a solution for a free electron, described by a linear combination of the most diffuse basis functions. Equating ionization potentials to occupied orbital energies is therefore justified based on the frozen MO approximation, but taking unoccupied orbital energies as electron affinities is somewhat questionable, since continuum solutions are mixed in.

3.5   The Basis Set Approximation

For small highly symmetric systems, like atoms and diatomic molecules, the Hartree–Fock equations may be solved by mapping the orbitals on a set of grid points. These are referred to as *numerical Hartree–Fock* methods.[5] However, essentially all calculations use a basis set expansion to express the unknown MOs in terms of a set of known functions. Any type of basis function may in principle be used: exponential, Gaussian, polynomial, cube, plane wave etc. There are two guidelines for choosing the basis functions. One is that they should have a behaviour which agrees with the physics of the problem, this ensures that the convergence as more basis functions are added is reasonably rapid. For bound atomic and molecular systems this means that the functions should go towards zero as the distance between the nucleus and the electron becomes large. The second guideline is practical: the chosen functions should make it easy to calculate all the required integrals. The first criterion suggests the use of exponential functions located on the nuclei, such functions are known to be exact solutions for the hydrogen atom. Unfortunately exponential functions turn out to be computationally difficult. Gaussian functions are computationally much easier to handle, and although they are poorer at describing the electronic structure on a one-to-one basis, the computational advantages more than make up for this. We will return to the precise description of basis sets in Chapter 5, but for now simply assume that a set of M basis functions located on the nuclei has been chosen.

Each MO is expanded in terms of the basis functions, conventionally called *atomic orbitals* (MO=LCAO, *Linear Combination of Atomic Orbitals*), although they are generally not solutions to the atomic HF problem.

$$\phi_i = \sum_{\alpha}^{M} c_{\alpha i} \chi_\alpha \tag{3.48}$$

The Hartree–Fock equations (3.41) may be written as:

$$\mathbf{F}_i \sum_{\alpha}^{M} c_{\alpha i} \chi_\alpha = \varepsilon_i \sum_{\alpha}^{M} c_{\alpha i} \chi_\alpha \tag{3.49}$$

Multiplying from the left by a specific basis function and integrating yields the *Roothaan–Hall* equations (for a closed shell system).[6] These are the Fock equations in the <u>atomic orbital basis</u>, and all the M equations may be collected in a matrix notation.

$$\mathbf{FC} = \mathbf{SC\varepsilon}$$
$$F_{\alpha\beta} = \langle \chi_\alpha | \mathbf{F} | \chi_\beta \rangle \tag{3.50}$$
$$S_{\alpha\beta} = \langle \chi_\alpha | \chi_\beta \rangle$$

The **S** matrix contains the overlap elements between basis functions, and the **F** matrix contains the Fock matrix elements. Each $F_{\alpha\beta}$ element contains two parts from the Fock operator (eq. (3.36)), integrals involving the one-electron operators, and a sum over occupied MOs of coefficients multiplied with two-electron integrals involving the electron–electron repulsion operator. The latter is often written as a product of a *density matrix* and two-electron integrals:

$$\langle\chi_\alpha|\mathbf{F}|\chi_\beta\rangle = \langle\chi_\alpha|\mathbf{h}|\chi_\beta\rangle + \sum_j^{\text{occ. MO}} \langle\chi_\alpha|\mathbf{J}_j - \mathbf{K}_j|\chi_\beta\rangle$$

$$= \langle\chi_\alpha|\mathbf{h}|\chi_\beta\rangle + \sum_j^{\text{occ. MO}} (\langle\chi_\alpha\phi_j|\mathbf{g}|\chi_\beta\phi_j\rangle - \langle\chi_\alpha\phi_j|\mathbf{g}|\phi_j\chi_\beta\rangle)$$

$$= \langle\chi_\alpha|\mathbf{h}|\chi_\beta\rangle + \sum_j^{\text{occ. MO}}\sum_\gamma^{\text{AO}}\sum_\delta^{\text{AO}} c_{\gamma j}c_{\delta j}(\langle\chi_\alpha\chi_\gamma|\mathbf{g}|\chi_\beta\chi_\delta\rangle - \langle\chi_\alpha\chi_\gamma|\mathbf{g}|\chi_\delta\chi_\beta\rangle)$$

$$= \langle\chi_\alpha|\mathbf{h}|\chi_\beta\rangle + \sum_\gamma^{\text{AO}}\sum_\delta^{\text{AO}} D_{\gamma\delta}(\langle\chi_\alpha\chi_\gamma|\mathbf{g}|\chi_\beta\chi_\delta\rangle - \langle\chi_\alpha\chi_\gamma|\mathbf{g}|\chi_\delta\chi_\beta\rangle)$$

$$D_{\gamma\delta} = \sum_j^{\text{occ. MO}} c_{\gamma j}c_{\delta j} \tag{3.51}$$

For use in Section 3.8, it can also be written in a more compact notation

$$F_{\alpha\beta} = h_{\alpha\beta} + \sum_{\gamma\delta} G_{\alpha\beta\gamma\delta}D_{\gamma\delta}$$

$$\mathbf{F} = \mathbf{h} + \mathbf{G}\cdot\mathbf{D} \tag{3.52}$$

where $\mathbf{G}\cdot\mathbf{D}$ denotes the contraction of the $\mathbf{D}$ matrix with the four-dimensional $\mathbf{G}$ tensor.

The total energy (3.32) in term of integrals over basis functions is given as

$$E = \sum_i^N \langle\phi_i|\mathbf{h}_i|\phi_i\rangle + \frac{1}{2}\sum_{ij}^N (\langle\phi_i\phi_j|\mathbf{g}|\phi_i\phi_j\rangle - \langle\phi_i\phi_j|\mathbf{g}|\phi_j\phi_i\rangle) + V_{nn}$$

$$E = \sum_i^N\sum_{\alpha\beta}^M c_{\alpha i}c_{\beta i}\langle\chi_\alpha|\mathbf{h}_i|\chi_\beta\rangle + \frac{1}{2}\sum_{ij}^N\sum_{\alpha\beta\gamma\delta}^M c_{\alpha i}c_{\gamma j}c_{\beta i}c_{\delta j}(\langle\chi_\alpha\chi_\gamma|\mathbf{g}|\chi_\beta\chi_\delta\rangle$$
$$- \langle\chi_\alpha\chi_\gamma|\mathbf{g}|\chi_\delta\chi_\beta\rangle) + V_{nn}$$

$$E = \sum_{\alpha\beta}^M D_{\alpha\beta}h_{\alpha\beta} + \frac{1}{2}\sum_{\alpha\beta\gamma\delta}^M D_{\alpha\beta}D_{\gamma\delta}(\langle\chi_\alpha\chi_\gamma|\mathbf{g}|\chi_\beta\chi_\delta\rangle - \langle\chi_\alpha\chi_\gamma|\mathbf{g}|\chi_\delta\chi_\beta\rangle) + V_{nn}$$

$$\tag{3.53}$$

The latter expression may also be written as

$$E = \sum_{\alpha\beta}^M D_{\alpha\beta}h_{\alpha\beta} + \frac{1}{2}\sum_{\alpha\beta\gamma\delta}^M (D_{\alpha\beta}D_{\gamma\delta} - D_{\alpha\delta}D_{\delta\beta})\langle\chi_\alpha\chi_\gamma|\mathbf{g}|\chi_\beta\chi_\delta\rangle + V_{nn} \tag{3.54}$$

The one- and two-electron integrals in the atomic basis are given as (eq. (3.24))

$$\langle \chi_\alpha | \mathbf{h} | \chi_\beta \rangle = \int \chi_\alpha(1) \left(-\frac{1}{2} \nabla^2 \right) \chi_\beta(1) d\mathbf{r}_1 + \sum_a \int \chi_\alpha(1) \frac{Z_a}{|\mathbf{R}_a - \mathbf{r}_1|} \chi_\beta(1) d\mathbf{r}_1$$

$$\langle \chi_\alpha \chi_\gamma | \mathbf{g} | \chi_\beta \chi_\delta \rangle = \int \chi_\alpha(1) \chi_\gamma(2) \frac{1}{|\mathbf{r}_1 - \mathbf{r}_2|} \chi_\beta(1) \chi_\delta(2) d\mathbf{r}_1 d\mathbf{r}_2 \qquad (3.55)$$

The two-electron integrals are often written in a notation without the **g** operator present.

$$\int \chi_\alpha(1) \chi_\gamma(2) \frac{1}{|\mathbf{r}_1 - \mathbf{r}_2|} \chi_\beta(1) \chi_\delta(2) d\mathbf{r}_1 d\mathbf{r}_2 = \langle \chi_\alpha \chi_\gamma | \mathbf{g} | \chi_\beta \chi_\delta \rangle$$

$$= \langle \chi_\alpha \chi_\gamma | \chi_\beta \chi_\delta \rangle \qquad (3.56)$$

This is known as the *physicist's* notation, where the ordering of the functions is given by the electron indices. They may also be written in an alternative order with both functions depending on electron 1 on the left, and the functions depending on electron 2 on the right; this is known as the *Mulliken* or *chemist's* notation.

$$\int \chi_\alpha(1) \chi_\beta(1) \frac{1}{|\mathbf{r}_1 - \mathbf{r}_2|} \chi_\gamma(2) \chi_\delta(2) d\mathbf{r}_1 d\mathbf{r}_2 = (\chi_\alpha \chi_\beta | \chi_\gamma \chi_\delta) \qquad (3.57)$$

The braket notation has the electron indices $\langle 12|12 \rangle$ while the parenthesis notation has the order $(11|22)$. In many cases the integrals are written with only the indices given, i.e. $\langle \chi_\alpha \chi_\gamma | \chi_\beta \chi_\delta \rangle = \langle \alpha \gamma | \beta \delta \rangle$. Since Coulomb and exchange integrals often are used as their difference, the following double bar notations are also used frequently.

$$\langle \chi_\alpha \chi_\beta \| \chi_\gamma \chi_\delta \rangle = \langle \chi_\alpha \chi_\beta | \chi_\gamma \chi_\delta \rangle - \langle \chi_\alpha \chi_\beta | \chi_\delta \chi_\gamma \rangle$$

$$(\chi_\alpha \chi_\beta \| \chi_\gamma \chi_\delta) = (\chi_\alpha \chi_\beta | \chi_\gamma \chi_\delta) - (\chi_\alpha \chi_\gamma | \chi_\beta \chi_\delta) \qquad (3.58)$$

The Roothaan–Hall equation (3.50) is a determination of the eigenvalues of the Fock matrix, see Chapter 13 for details. To determine the unknown MO coefficients $c_{\alpha i}$, the Fock matrix must be diagonalized. However, the Fock matrix is only known if all the MO coefficients are known, eq. (3.51). The procedure therefore starts off by some guess of the coefficients, forms the F matrix, and diagonalizes it. The new set of coefficients is then used for calculating a new Fock matrix etc., as illustrated in Figure 3.3. This is continued until the set of coefficients used for constructing the Fock matrix is equal to those resulting from the diagonalization (to within a certain threshold). This set of coefficients determines an SCF solution. The potential (or field) generated by the SCF electron density is identical to that produced by solving for the electron distribution. The Fock matrix, and therefore the total energy, depends only on the occupied MO. Solving the Roothaan–Hall equations produces a total of M (= number of basis functions) MOs, i.e. there are N occupied and $M - N$ unoccupied, or *virtual*, MOs. The virtual orbitals are orthogonal to all the occupied orbitals, but have no direct physical interpretation, except as electron affinities (via Koopmans' theorem).

To construct the Fock matrix, eq. (3.51), integrals over all pairs of basis functions and the one-electron operator **h** are needed. For M basis functions there are of the order of M^2 of such *one-electron integrals*. These one-integrals are also known as *core* integrals, they describe the interaction of an electron with the whole frame of bare nuclei. The second part of the Fock matrix involves integrals over four basis functions and the **g** two-electron operator. There are of the order of M^4 of these *two-electron integrals*. In conventional HF methods the two-electron integrals are calculated and saved before the

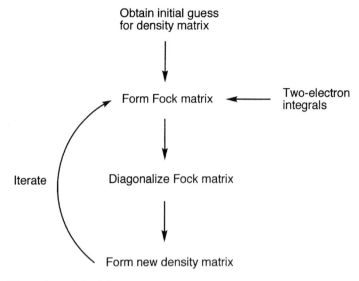

Figure 3.3 Illustration of the SCF procedure

SCF procedure is begun, and then used in each SCF iteration. Formally, in the large basis set limit the SCF procedure involves a computational effort which increases as the number of basis functions to the fourth power. Below it will be shown that the scaling may be substantially smaller in actual calculations.

For the two-electron integrals, the four basis functions may be located on 1, 2, 3 or 4 different atomic centers. It has already been mentioned that exponential type basis functions ($\chi \propto \exp(-\alpha r)$) fundamentally are better suited for electronic structure calculations, however, it turns out that the calculation of especially 3- and 4-centre two-electron integrals is very time-consuming for exponential functions. Gaussian functions ($\chi \propto \exp(-\alpha r^2)$) are much easier for calculating two-electron integrals. This is because the product of two Gaussians located at two different positions ($\mathbf{R}_A$ and $\mathbf{R}_B$) with different exponents (α and β) can be written as a single Gaussian located intermediately between the two original positions. This allows compact formulas for all types of one- and two-electron integrals to be derived.

$$G_A(\mathbf{r}) = \left(\frac{2\alpha}{\pi}\right)^{3/4} e^{-\alpha(\mathbf{r}-\mathbf{R}_A)^2}$$

$$G_B(\mathbf{r}) = \left(\frac{2\beta}{\pi}\right)^{3/4} e^{-\beta(\mathbf{r}-\mathbf{R}_B)^2}$$

$$G_A(\mathbf{r})G_B(\mathbf{r}) = K e^{-\gamma(\mathbf{r}-\mathbf{R}_c)^2}$$

$$\gamma = \alpha + \beta$$

$$\mathbf{R}_c = \frac{\alpha\mathbf{R}_A + \beta\mathbf{R}_B}{\alpha + \beta}$$

$$K = \left(\frac{2}{\pi}\right)^{3/2} (\alpha\beta)^{3/4} e^{-\frac{\alpha\beta}{\alpha+\beta}(\mathbf{R}_A-\mathbf{R}_B)^2} \tag{3.59}$$

As the number of basis functions increases, the accuracy of the MOs becomes better. In the limit of a complete basis set (infinite number of basis functions), the results are identical to those obtained by a numerical HF method, this is known as the *Hartree–Fock limit*. This is <u>not</u> the exact solution to the Schrödinger equation, only the best single determinant wave function that can be obtained. In practical calculations the HF limit is never reached, and the term Hartree–Fock is normally used to cover also SCF solutions within an incomplete basis set. *Ab initio* HF methods, where all the necessary integrals are calculated from a given basis set, are one-dimensional. As the size of the basis set is increased, the variational principle ensures that the results become better (at least in an energetic sense). The quality of a results can therefore be assessed by running calculations with increasingly larger basis sets.

3.6   Alternative Formulation of the Variational Problem

The objective is to minimize the total energy as a function of the molecular orbitals, subject to the orthonormality constraint. In the above formulation this is handled by means of Lagrange multipliers. The final Fock matrix in the MO basis is diagonal, with the diagonal elements being the orbital energies. During the iterative sequence, i.e. before the orbitals have converged to an SCF solution, the Fock matrix is not diagonal. Starting from an initial set of molecular orbitals, the problem may also be formulated as a rotation of the orbitals (unitary transformation) in order to make the operator diagonal.[7] Since the operator depends on the orbitals, the procedure again becomes iterative.

The orbital rotation is given by a unitary matrix $\mathbf{U}$, which can be written as an exponential transformation.

$$\phi' = \mathbf{U}\phi = e^{\mathbf{X}}\phi \tag{3.60}$$

The $\mathbf{X}$ matrix contains the parameters describing the unitary transformation of the M orbitals, being of the size of $M \times M$. The orthogonality is incorporated by requiring that the $\mathbf{X}$ matrix is antisymmetric, $x_{ij} = -x_{ji}$, i.e.

$$\mathbf{U}^\dagger\mathbf{U} = (e^{\mathbf{X}})^\dagger(e^{\mathbf{X}}) = (e^{\mathbf{X}^\dagger})(e^{\mathbf{X}}) = (e^{-\mathbf{X}})(e^{\mathbf{X}}) = e^{-\mathbf{X}+\mathbf{X}} = \mathbf{1} \tag{3.61}$$

Normally the orbitals are real, and the unitary transformation becomes an orthogonal transformation. In the case of only two orbitals, the $\mathbf{X}$ matrix contains the rotation angle α, and the $\mathbf{U}$ matrix describes a 2 by 2 rotation. The connection between $\mathbf{X}$ and $\mathbf{U}$ is illustrated in Chapter 13 (Figure 13.2) and involves diagonalization of $\mathbf{X}$ (to give eigenvalues of $\pm i\alpha$), exponentiation (to give complex exponentials which may be written as $\cos\alpha \pm i\sin\alpha$), follow by backtransformation.

$$\mathbf{X} = \begin{pmatrix} 0 & \alpha \\ -\alpha & 0 \end{pmatrix}$$
$$\mathbf{U} = e^{\mathbf{X}} = \begin{pmatrix} \cos\alpha & \sin\alpha \\ -\sin\alpha & \cos\alpha \end{pmatrix} \tag{3.62}$$

In the general case the $\mathbf{X}$ matrix contains rotational angles for rotating all pairs of orbitals.

It should be noted that the unoccupied orbitals do not enter the energy expression (3.32), and a rotation between the virtual orbitals can therefore not change the energy. A rotation between the occupied orbitals corresponds to making linear combinations of these, but this does not change the total wave function. The occupied–occupied and virtual–virtual blocks of the **X** matrix can therefore be chosen as zero. The variational parameters are the elements in the **X** matrix which describe the mixing of the occupied and virtual orbitals, i.e. there are a total of $N_{occ} \times (M - N_{occ})$ parameters. The goal of the iterations is to make the off-diagonal elements in the occupied–virtual block of the Fock matrix zero. Alternatively stated, the off-diagonal elements are the gradients of the energy with respect to the orbitals, and the stationarity condition is that this gradient vanish.

3.7 Restricted and Unrestricted Hartree–Fock

So far there have not been any restrictions on the MOs used to build the determinantal trial wave function. The Slater determinant has been written in terms of spinorbitals, eq. (3.20), being products of a spatial orbital times a spin function (α or β). If there are no restrictions on the form of the spatial orbitals, the trial function is an *Unrestricted Hartree–Fock* (UHF) wave function.[8] The term *Different Orbitals for Different Spins* (DODS) is also sometimes used. If the interest is in systems with an even number of electrons and a singlet type of wave function (a *closed shell* system), the restriction that each spatial orbital should have two electrons, one with α and one with β spin, is normally made. Such wave functions are known as *Restricted Hartree–Fock* (RHF). Open-shell systems may also be described by restricted type wave functions, where the spatial part of the doubly occupied orbitals is forced to be the same; this is known as *Restricted Open-shell Hartree–Fock* (ROHF). For open-shell species a UHF treatment leads to well-defined orbital energies, which may be interpreted as ionization potentials, Section 3.4. For an ROHF wave function it is not possible to chose a unitary transformation which makes the matrix of Lagrange multipliers in eq. (3.40) diagonal, and orbital energies from an ROHF wave function are consequently not uniquely defined, and cannot be equated to ionization potentials by a Koopman type argument.

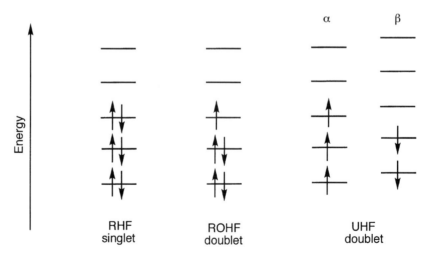

Figure 3.4 Illustrating an RHF singlet, and ROHF and UHF doublet states

The UHF wave function allows different spatial orbitals for the two electrons in an orbital. As restricted type wave functions put constraints on the variation parameters, the energy of a UHF wave function is always lower than or equal to a corresponding R(O)HF type wave function. For singlet states near the equilibrium geometry, it is usually not possible to lower the energy by allowing the α and β MOs to be different. For an open-shell system like a doublet, however, it is clear that forcing the α and β MOs to be identical is a restriction. If the unpaired electron has α spin, it will interact differently with the other α electrons than with the β electrons, and consequently the optimum α and β orbitals will be different. The UHF description, however, has the disadvantage that the wave function is normally not an eigenfunction of the $\mathbf{S}^2$ operator (unless it is equal to the RHF solution), where the $\mathbf{S}^2$ operator evaluates the value of the total electron spin squared. This means that a "singlet" UHF wave function may also contain contributions from higher-lying triplet, quintet etc. states. Similarly, a "doublet" UHF wave function will contain spurious (unphysical) contributions from higher-lying quartet, sextet etc. states. This will be discussed in more detail in Chapter 4.

Semi-empirical methods (Section 3.9) sometimes employ the so-called *half-electron* (HE) method for describing open-shell systems, such as doublets and triplets. In this model a doublet state is described by putting two half-electrons in the same orbitals with opposite spins, i.e. constructing an RHF type wave function where all electron spins are paired. A triplet state may similarly be modelled as having two orbitals, each occupied by two half-electrons with opposite spins. The main motivation behind this artificial construct is that open- and closed-shell systems (such as a triplet and a singlet state) will have different amounts of electron correlation. Since semi-empirical methods perform the parameterization based on single determinant wave functions, the HE method basicly cancels the difference in electron correlations, and allows open- and closed-shell systems to be treated on an equal footing in terms of energy. It has the disadvantage that the open-shell nature is no longer present in the wave function, it is for example not possible to calculation spin densities (i.e. determine where the unpaired electron(s) is(are) most likely to be).

3.8 SCF Techniques

As seen in Section 3.5, the Roothaan–Hall (or Pople–Nesbet for the UHF case) equations must be solved iteratively since the Fock matrix depends on its own solutions. The procedure illustrated in Figure 3.3 involves the following steps.

(1) Calculate all one- and two-electron integrals.
(2) Generate a suitable start guess for the MO coefficients.
(3) Form the initial density matrix
(4) Form the Fock matrix as the core (one-electron) integrals + the density matrix times the two-electron integrals.
(5) Diagonalize the Fock matrix (see Chapter 13 for details). The eigenvectors contain the new MO coefficients.
(6) Form the new density matrix. If it is sufficiently close to the previous density matrix, we are done, otherwise go to step (4).

There are several points hidden in this scheme. Will the procedure actually converge at all? Will the SCF solution correspond to the desired energy minimum (and not a maximum or saddle point)? Can the number of iterations necessary for convergence be reduced? Does the most efficient method depend on the type of computer and/or the size of the problem?

Let us look at some of the SCF techniques used in practice.

3.8.1 SCF Convergence

There is <u>no</u> guarantee that the above iterative scheme will converge. For geometries near equilibrium and using small size basis sets, the straightforward SCF procedure often converges. Distorted geometries (such as transition structures) and large basis sets containing diffuse functions, however, rarely converge, and metal complexes, where several states with similar energies are possible, are even more troublesome. There are different tricks that can be tried to help convergence.[9]

(1) *Extrapolation*. This is a method for trying to make the convergence faster by extrapolating previous Fock matrices to generate a (hopefully) better Fock matrix than the one calculated directly from the current density matrix. Typically the last three matrices are used in the extrapolation.

(2) *Damping*. Often the reason for divergence, or very slow convergence, is due to oscillations. A given density matrix $\mathbf{D}_n$ gives a Fock matrix $\mathbf{F}_n$ which upon diagonalization gives a density matrix $\mathbf{D}_{n+1}$. The Fock matrix $\mathbf{F}_{n+1}$ from $\mathbf{D}_{n+1}$ gives a density matrix $\mathbf{D}_{n+2}$ which is close to $\mathbf{D}_n$, but $\mathbf{D}_n$ and $\mathbf{D}_{n+1}$ are very different. The damping procedure tries to solve this by replacing the current density matrix with a weighted average, $\mathbf{D}_{n+1} = \alpha \mathbf{D}_n + (1 - \alpha)\mathbf{D}_{n+1}$. The weighting factor α may be chosen as a constant or changed dynamically during the SCF procedure.

(3) *Level Shifting*. This technique[10] is perhaps best understood in the formulation of a rotation of the MOs which form the basis for the Fock operator, Section 3.6. At convergence the Fock matrix elements in the MO basis between occupied and virtual orbitals are zero. The iterative procedure involves mixing (making linear

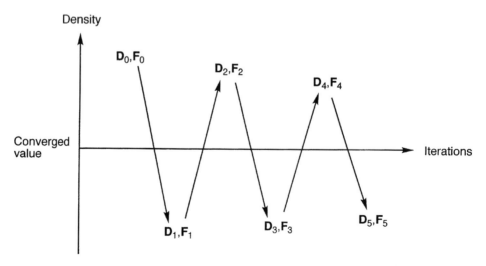

Figure 3.5   An oscillating SCF procedure

combinations of) occupied and virtual MOs. During the iterative procedure these mixings may be large, causing oscillations or making the total energy increase. The degree of mixing may be reduced by artificially increasing the energy of the virtual orbitals. If a sufficiently large constant is added to the virtual orbital energies, it can be shown that the total energy is guaranteed to decrease, thereby forcing convergence. The more the virtual orbitals are raised in energy, the more stable is the convergence, but the <u>rate</u> of convergence also decreases with level shifting. For large enough shifts, convergence is guaranteed, but it is likely to occur very slowly.

(4) *Direct Inversion in the Iterative Subspace* (DIIS). This procedure was developed by Pulay and is an extrapolation procedure.[11] It has proved to be very efficient in forcing convergence, and in reducing the number of iterations at the same time. It is now one of the most commonly used methods for helping SCF convergence. The idea is as follows. As the iterative procedure runs, a sequence of Fock and density matrices ($\mathbf{F}_0, \mathbf{F}_1, \mathbf{F}_2, \ldots$ and $\mathbf{D}_0, \mathbf{D}_1, \mathbf{D}_2, \ldots$) is produced. At each iteration it is also assumed that an estimate of the "error" ($\mathbf{E}_0, \mathbf{E}_1, \mathbf{E}_2, \ldots$) is available, i.e. how far the current Fock/density matrix is from the converged solution. The converged solution has an error of zero, and the DIIS method forms a linear combination of the error indicators which in a least squares sense is a minimum (as close to zero as possible). In the function space generated by the previous iterations we try to find the point with lowest error, which is not necessarily one of the points actually calculated. It is common to use the trace (sum of diagonal elements) of the matrix product of the error matrix with itself as a scalar indicator of the error.

$$\text{ErrF}(\mathbf{c}) = \text{trace}\,(\mathbf{E}_{n+1} \cdot \mathbf{E}_{n+1})$$

$$\mathbf{E}_{n+1} = \sum_{i=0}^{n} c_i \mathbf{E}_i \tag{3.63}$$

$$\sum_{i=0}^{n} c_i = 1$$

Minimization of the ErrF subject to the normalization constraint is handled by the Lagrange method (Chapter 14), and leads to the following set of linear equations, where λ is the multiplier associated with the normalization.

$$\begin{pmatrix} a_{11} & a_{12} & \ldots & a_{1n} & -1 \\ a_{21} & a_{22} & \ldots & a_{2n} & -1 \\ \ldots & \ldots & \ldots & \ldots & \ldots \\ a_{n1} & a_{n2} & \ldots & a_{nn} & -1 \\ -1 & -1 & -1 & -1 & 0 \end{pmatrix} \begin{pmatrix} c_1 \\ c_2 \\ \ldots \\ c_n \\ -\lambda \end{pmatrix} = \begin{pmatrix} 0 \\ 0 \\ \ldots \\ 0 \\ -1 \end{pmatrix} \tag{3.64}$$

$$a_{ij} = \text{trace}\,(\mathbf{E}_i \cdot \mathbf{E}_j)$$

$$\mathbf{Ac} = \mathbf{B}$$

In iteration n the $\mathbf{A}$ matrix has dimension $n \times n$, where n usually is less than 20. The coefficients $\mathbf{c}$ can be obtained by directly inverting the $\mathbf{A}$ matrix and multiplying it onto the $\mathbf{B}$ vector ($\mathbf{c} = \mathbf{A}^{-1}\mathbf{B}$), i.e. in the "subspace" of the "iterations" the linear equations are solved by "direct inversion", thus the name DIIS. Having obtained the coefficients which minimize the error function at iteration n, the same set of coefficients is used for generating an extrapolated Fock matrix $(\mathbf{F}^*)$ at iteration n, which is used in place of $\mathbf{F}_n$ for generating the new density matrix.

$$\mathbf{F}_n^* = \sum_{i=0}^{n} c_i \mathbf{F}_i \tag{3.65}$$

The only remaining question is the nature of the error function. Pulay suggested the difference **FDS** − **SDF** ($\mathbf{S}$ is the overlap matrix), which is related to the gradient of the SCF energy with respect to the MO coefficients. This has been found to work well in practice.

(5) *"Quadratically Convergent" or "Second-Order" SCF.* As mentioned in Section 3.6, the variational procedure can be formulated in terms of an exponential transformation of the MOs, with the (independent) variational parameters contained in an $\mathbf{X}$ matrix. Note that the $\mathbf{X}$ variables are preferred over the MO coefficients in eq. (3.48) for optimization, since the latter are not independent (the MOs must be orthonormal). The exponential may be written as a series expansion, and the energy expanded in terms of the $\mathbf{X}$ variables describing the occupied–virtual mixing of the orbitals.[12]

$$e^{\mathbf{X}} = 1 + \mathbf{X} + \tfrac{1}{2}\mathbf{XX} + \dots$$
$$E(\mathbf{X}) = E(\mathbf{0}) + E'(\mathbf{0})\mathbf{X} + \tfrac{1}{2}\mathbf{X}E''(\mathbf{0})\mathbf{X} + \dots \tag{3.66}$$

The first and second derivatives of the energy with respect to the $\mathbf{X}$ variables ($E'(\mathbf{0})$ and $E''(\mathbf{0})$) can be written in term of Fock matrix elements and two-electron integrals in the MO basis.[13] For an RHF type wave function these are given as

$$\frac{\partial E}{\partial x_{ia}} = 4\langle \phi_i | \mathbf{F} | \phi_a \rangle$$
$$\frac{\partial^2 E}{\partial x_{ia}\partial x_{jb}} = 4\delta_{ij}\langle \phi_a | \mathbf{F} | \phi_b \rangle - 4\delta_{ab}\langle \phi_i | \mathbf{F} | \phi_j \rangle \tag{3.67}$$
$$+ 4(\langle \phi_i \phi_b | \phi_a \phi_j \rangle - \langle \phi_i \phi_j | \phi_b \phi_a \rangle - \langle \phi_i \phi_a | \phi_j \phi_b \rangle)$$

The gradient of the energy is an off-diagonal element of the molecular Fock matrix, which is easily calculated from the atomic Fock matrix. The second derivative, however, involves two-electron integrals which require an AO to MO transformation (see Section 4.2.1), and is therefore computationally expensive.

In this formulation the problem is no different from other types of non-linear optimizations, and the same types of techniques, such as steepest descent, conjugated gradient or Newton–Raphson methods can be used, see Chapter 14 for details. The Newton–Raphson method has the advantage of being quadratically convergent, i.e. sufficiently near the minimum that it converges very fast. The main problem in using Newton–Raphson methods for wave function optimization is computational efficiency. The exact calculation of the second derivative matrix is somewhat demanding, and each iteration in a Newton–Raphson optimization therefore takes longer than the simple Roothaan–Hall iterative scheme. Owing to the fast convergence near the minimum, a Newton–Raphson approach normally takes fewer iterations than for example DIIS, but the overall computational time is still a factor of ~2 longer. Alternative schemes, where an approximation to the second derivative matrix is used (pseudo-Newton–Raphson), have also been developed, and they are often competitive with DIIS.[14] It should be kept in mind that the simple Newton–Raphson is unstable, and require some form of stabilization, for example by using the augmented Hessian techniques discussed in Chapter 14. Direct minimizations (steepest descent) and Newton–Raphson schemes are primarily used as a "last resort" for forcing convergence in difficult cases.

3.8.2 Use of Symmetry

From group theory it may be shown that an integral can only be non-zero if the integrand belongs to the totally symmetric representation. Furthermore, the product of two functions can only be totally symmetric if they belong to the same irreducible representation. As both the Hamilton and Fock operators are totally symmetric, integrals of the following type can only be non-zero if the basis functions involving the same electron coordinate belong to the same representation.

$$\int \chi_\alpha(1)\chi_\beta(1)d\mathbf{r}_1, \quad \int \chi_\alpha(1)\mathbf{F}\chi_\beta(1)d\mathbf{r}_1, \quad \int \chi_\alpha(1)\mathbf{H}\chi_\beta(1)d\mathbf{r}_1 \tag{3.68}$$

Similar considerations hold for the two-electron integrals.

By forming suitable linear combinations of basis functions (*symmetry adapted functions*), many one- and two-electron integrals need not be calculated as they are known to be exactly zero due to symmetry. Furthermore, the Fock (in an HF calculation) or Hamilton matrix (in a CI calculation) will become block diagonal, as only matrix elements between functions having the same symmetry can be non-zero. The savings depend on the specific system, but as a guideline the computational time is reduced roughly by a factor corresponding to the order of the point group (number of symmetry operations). Although the large majority of molecules do not have any symmetry, a sizable portion of the small molecules for which *ab initio* electronic structure calculations are possible, are symmetric. Almost all *ab initio* programs employ symmetry as a tool for reducing the computational effort.

3.8.3 Ensuring that the HF Energy is a Minimum

The standard iterative procedure produces a solution where the variation of the HF energy is stationary with respect to all orbital variations, i.e. the first derivatives of the energy with respect to the MO coefficients are zero. To be sure this corresponds to an energy minimum, the second derivatives should also be calculated.[15] This is a matrix the size of the number of occupied MOs times the number of virtual MOs, and the eigenvalues of this matrix should all be positive for it to be an energy minimum. A negative eigenvalue means that it is possible to get to a lower energy state by "exciting" an electron from an occupied to an unoccupied orbital, i.e. the solution is unstable. In practice the stability is rarely checked, it is assumed that the iterative procedure has converged to a minimum. It should be noted that a positive definite second-order matrix only ensures that the solution is a <u>local</u> minimum, there may be other minima with lower energies. The second derivative matrix is identical to that arising in quadratic convergent SCF methods (Section 3.8.1).

The question of whether the energy is a minimum is closely related to the concept of wave function stability. If a lower energy RHF solution can be found, the wave function is said to posses a *singlet instability*. It is possible that an RHF type wave function is a minimum in the coefficient space, but is a saddle point if the constraint of double occupancy of each MO is relaxed. This indicates that a lower-energy wave function of the UHF type can be constructed, and is called a *triplet instability*. It should be noted that in order to generate such UHF wave functions for a singlet state, an initial guess of the SCF coefficients must be specified which has the spatial parts of at least one set of α and β MOs different. There are other types of such instabilities, such as relaxing the constraint that MOs should be real (allowing complex orbitals), or the constraint that a MO should only have a single spin function. Relaxing the latter produces the "*general*" HF method, where each MO is written as a spatial part having α spin plus another spatial part having β spin.[16] Such wave functions are no longer eigenfunctions of the $\mathbf{S}_z$ operator, and not commonly used.

Another aspect of wave function instability concerns *symmetry breaking*, i.e. the wave function has a lower symmetry than the nuclear framework.[17] It occurs for example for the allyl radical with an ROHF type wave function. The nuclear geometry has C_{2v} symmetry, but the C_{2v} symmetric wave function corresponds to a (first-order) saddle point. The lowest energy ROHF solution has only C_s symmetry, and corresponds to a localized double bond and a localized electron (radical). Relaxing the double occupancy constraint, and allowing the wave function to become UHF, re-establish the correct C_{2v} symmetry. Such symmetry breaking phenomena usually indicate that the type of wave function used is not flexible enough for even a qualitatively correct description.

3.8.4   Initial Guess Orbitals

The quality of the initial guess orbitals influences the number of iterations necessary for achieving convergence. As each iteration involves a computational effort proportional to M^4, it is of course desirable to generate as good a guess as possible. In some cases different start orbitals may also result in convergence to different SCF solutions, or make the difference between convergence and divergence. One possible way of generating a set of start orbitals is to diagonalize the Fock matrix consisting only of the one-electron contributions, the "core" matrix. This corresponds to initializing the density matrix as a zero matrix, totally neglecting the electron–electron repulsion in the first step. This is generally a poor guess, but is available for all types of basis sets and easily implemented. Essentially all programs therefore have it as an option.

More sophisticated procedures involve taking the start MO coefficients from a semi-empirical calculation, such as Extended Hückel Theory (EHT) or Intermediate Neglect of Differential Overlap (INDO) (Sections 3.12 and 3.9). The EHT method has the advantage that it is readily parameterized for all elements, and it can provide start orbitals for systems involving elements from essentially the whole periodic table. An INDO calculation normally provides better start orbitals, but at a price. The INDO calculation itself is iterative, and it may suffer from convergence problems, just as the *ab initio* SCF does itself.

Many systems of interest are symmetric. The MOs will transform as one of the irreducible representations in the point group, and most programs use this to speed up calculations. The initial guess for the start orbitals involves selecting how many MOs of each symmetry should be occupied, i.e. the electron configuration. Different start configurations produce different final SCF solutions. Many programs automatically select the start configuration based on the orbital energies of the starting MOs, which may be "wrong" in the sense that it does not produce the desired solution. Of course a given solution may be checked to see if it actually corresponds to an energy minimum, but as stated above, this is rarely done. Furthermore, there may be several (local) minima, thus the verification that the found solution is an energy minimum is no guarantee that it is the global minimum. The reason different initial electron configurations may generate different final solutions is due to the fact that matrix elements between orbitals belonging to different representations are exactly zero, thus only orbitals belonging to the same representation can mix. Forcing the program to run the calculation without symmetry usually does not help. Although turning the symmetry off will make the program actually calculate all matrix elements, those between MOs of different symmetry will still be zero (except for numerical inaccuracies). In many cases it is necessary to manually specify which orbitals should be occupied initially to generate the desired solution.

3.8.5 Direct SCF

The number of two-electron integrals formally grows as the fourth power of the size of the basis set. Owing to permutational symmetry (the following integrals are identical $\langle \chi_1\chi_2|\chi_3\chi_4\rangle = \langle \chi_3\chi_2|\chi_1\chi_4\rangle = \langle \chi_1\chi_4|\chi_3\chi_2\rangle = \langle \chi_3\chi_4|\chi_1\chi_2\rangle = \langle \chi_2\chi_1|\chi_4\chi_3\rangle = \langle \chi_4\chi_1|\chi_2\chi_3\rangle = \langle \chi_2\chi_3|\chi_4\chi_1\rangle = \langle \chi_4\chi_3|\chi_2\chi_1\rangle$) the total number is approximately $1/8 \, M^4$. Each integral is a floating point number associated with four indices indicating which basis functions are involved in the integral. Storing a floating point number in double precision (which is necessary for calculating the energy with an accuracy of ~ 14 digits) requires 64 bits = 8 bytes. A basis set with 100 functions thus generates $\sim 12.5 \times 10^6$ integrals, requiring ~ 100 Mbyte of disk space or memory. The disk space required for storing the integrals increases rapidly, thus a basis set with 500 functions requires ~ 60 Gbyte of disk space (or memory). This is out of reach for most computers. In practice the storage requirement is somewhat less, since many of the integrals are small, and can be ignored. Typically a cut-off around $\sim 10^{-10}$ is employed; if the integral is less than this value it is not stored, and consequently makes a zero contribution to the construction of the Fock matrix in the iterative procedure. However, the disk space requirement effectively limits conventional HF methods to basis sets smaller than ~ 300 functions.

Older computers had only very limited amounts of memory, and disk storage of the integrals was the only option. Modern machines often have quite significant amounts of memory, a few Gbytes is not uncommon. For small and medium sized systems it may be possible to store all the integrals in the memory instead of on disk. Such "in-core" methods are very efficient for performing an HF calculation. The integrals are only calculated once, and each SCF iteration is just a multiplication of the integral tensor with a density matrix to form the Fock matrix, eq. (3.52). Essentially all machines have optimized routines for doing matrix multiplication efficiently. The only limitation is the cubic (M^3) growth of the memory requirement with basis set size, which in practice restricts such in-core methods to basis sets with less than ~ 150 functions.

The disk space (or memory) requirement can be reduced dramatically by performing the SCF in a *direct* fashion.[18] In the direct SCF method the integrals are calculated from scratch in each iteration. At first this would appear to involve a computational effort which is larger than a conventional HF calculation by a factor close to the number of iterations. There are, however, a number of considerations which often make direct SCF methods computationally quite competitive or even advantageous.

In disk based methods all the integrals are first calculated and written to disk. To reduce the disk space requirement, the four indices associated with each integral are "packed" into a single number, and written to disk. In each iteration the whole set of integrals must be read, and the indices "unpacked" before the integrals are multiplied with the proper density matrix elements and added to the Fock matrix. Typically half the time in an SCF procedure is spent calculating the integrals and writing them to disk, the other half is spent reading, unpacking and forming the Fock matrix maybe 20 times. In a direct approach there are no overheads due to packing and unpacking of indices, and writing/reading of integrals.

In disk based methods only integrals larger than a certain cut-off are saved. In direct methods it is possible to ignore additional integrals. The contribution to a Fock matrix element is a product of density matrix elements and two-electron integrals. In disk based methods, the density matrix is not known when the integrals are calculated, and all integrals above the cut-off must be saved and processed in each iteration. In direct methods the density matrix is known at the time when the integrals are calculated. Thus if the product of the density matrix elements and the integral is less than the cut-off, the integral can be ignored. Of course this is only a saving if an estimate of the size of the integral is available before it is actually calculated. One such estimate is the *Schwarz inequality*.

$$|\langle \alpha\gamma | \beta\delta \rangle| \leq \sqrt{\langle \alpha\alpha | \beta\beta \rangle} \cdot \sqrt{\langle \gamma\gamma | \delta\delta \rangle} \tag{3.69}$$

The number of two-centre integrals on the right-hand side is quite small (of the order of M^2) and can easily be calculated beforehand. Thus if the product of the density matrix elements and the upper limit of the integral is less than the cut-off, the integral does not need to be calculated. In practice integrals are calculated in *batches*, where a batch is a collection of integrals having the same exponent. For a $\langle pp | pp \rangle$ type batch there are thus 81 individual integrals. The integral screening is normally done at the batch level, i.e. if the largest term is smaller than a given cut-off, the whole batch can be neglected.

The above *integral screening* is even more advantageous if the Fock matrix is formed incrementally. Consider two sequential density and Fock matrices in the iterative procedure (eq. (3.52)).

$$\mathbf{F}_n = \mathbf{h} + \mathbf{G} \cdot \mathbf{D}_n$$
$$\mathbf{F}_{n+1} = \mathbf{h} + \mathbf{G} \cdot \mathbf{D}_{n+1} \tag{3.70}$$
$$\mathbf{F}_{n+1} = \mathbf{F}_n + \mathbf{G} \cdot (\mathbf{D}_{n+1} - \mathbf{D}_n)$$

The change in the Fock matrix depends on the change in the density matrix. Combined with the above screening procedure, it is thus only necessary to calculate those integrals to be multiplied with density matrix elements which have changed significantly since last iteration. As the SCF converges there are fewer and fewer integrals that need to be calculated.

The formal scaling of HF methods is M^4, M being the number of basis functions. This assessment is based on the fact that the total number of two-electron integrals increases as M^4. However, as we have just seen, we do not need to calculate all the two-electron integrals, many can be neglected without affecting the final results. The observed scaling is therefore less than the quartic dependence, but the exact power depends on how the size of the problem is increased. If the number of atoms is increased for a fixed basis set per atom, the scaling depends on the dimensionality of the atomic arrangement and the size of the atomic basis. The most favourable case is a small compact basis set (like a minimum basis) and an essentially one-dimensional system, like polyacetylene, $H-(C \equiv C)_n-H$, or linear alkanes. In this case the scaling is close to M^2 once the number of functions exceeds ~ 100. A two-dimensional arrangement of atoms (like a slab of graphite) has a slightly larger exponential dependence, while a three-dimensional system (like a diamond structure) has a power dependence close to $M^{2.3}$.[19] It should be noted that most molecular systems have a dimensionality between 2 and 3; the presence of "holes" in the structure reduces the effective dimensionality below 3. With a larger basis set, especially if diffuse functions are present, the screening of integrals becomes much less efficient, or equivalently, the molecular system must be significantly larger to achieve the limiting scaling. In practice, however, the increase in the total number of basis functions is often not due to an enlargement of the molecular system, but rather to the use of an increasingly larger basis set per atom for a fixed sized molecule. For such cases the observed scaling is often worse than the theoretical M^4 dependence, since the integral screening becomes less and less efficient.

The combination of these effects means that the increase in computational time of a direct SCF calculation over a disk based method is less than initially expected. For a medium size SCF calculation which requires say 20 iterations, the increase in CPU time may only be a factor of 2 or 3. Owing to the more efficient screening, however, the direct method actually becomes more and more advantageous relative to disk based methods as the size of the system increases. At some point direct methods will therefore require less CPU time than a conventional method. Exactly where the crossover point occurs depends on the way the number of basis functions is increased, the machine type and the efficiency of the integral code. Small compact basis sets in general experience the crossover point quite early (perhaps around 100 functions) while it occurs later for large extended basis sets. Since conventional disk based methods are limited to 200–300 basis functions, direct methods are normally the only choice for large calculations. Direct methods are essentially only limited by the available CPU time, and calculations involving up to several thousand basis functions have been reported.

Although direct methods for small and medium size systems require more CPU time than disk based methods, this is in many cases irrelevant. For the user the determining factor is the time from submitting the calculation to the results being available. Over the years the speed of CPUs has increased much faster than the speed of data transfer to and from disk. Many modern machines have quite slow data transfer to disk compared to CPU speed. Measured by the elapsed wall clock time, disk based HF methods are often the slowest in delivering the results, despite the fact that they require the least CPU time.

Simply speaking, the CPU may be spending most of its time waiting for data to be transferred from disk. Direct methods, on the other hand, use the CPU with a near 100% efficiency. For machines without fast disk transfer (like workstation type machines) the crossover point for direct vs. conventional methods in terms of wall clock time may be so low that direct methods are always preferred.

Finally it should be mentioned that there is a strong research effort towards designing computational chemistry programs to run on parallel computers. These types of machines have more than one CPU, typically in the range 10–1000. Making direct SCF calculations run efficiently in a parallel fashion is fairly easy, each processor is given the task of calculating a certain batch of integrals and the total Fock matrix is simply the sum of contributions from each individual CPU.

3.8.6   Linear Scaling Techniques

The computational bottleneck in HF methods is the calculation of the two-electron Coulomb and exchange terms arising from the electron–electron repulsion. In non-metallic systems the exchange term is quite short-ranged, while the Coulomb interaction is long-ranged. In the large system limit, the Coulomb integrals thus dominate the computational cost. By use of screening techniques, as described in the previous section, the scaling in the large system limit will eventually be reduced from M^4 to M^2. Similar considerations hold for DFT methods, see Chapter 6 for details.

The *Fast Multipole Moment* (FMM) method (Section 16.2.1) was originally developed for calculating interactions between point charges. A direct calculation involves a summation over all pairs, i.e. a computational effort which increases as M^2. The idea in FMM is to split the total interaction into near- and far-fields. The near-field is evaluated directly, while the far-field is calculated by dividing the physical space into boxes, and the interaction between all the charges in one box and all the charges in another is approximated as interactions between multipoles located at the centre of the boxes. The further away from each other two boxes are, the larger the boxes can be for a given accuracy, thereby reducing the formal M^2 behaviour to linear scaling, i.e. proportional to M.

The original FMM has been refined by adjusting the accuracy of the multipole expansion as a function of the distance between boxes, producing the *very Fast Multipole Moment* (vFMM) method.[20] Both of these have been generalized to continuous charge distributions, as is required for calculating the Coulomb interaction between electrons in a quantum description.[21] The use of FMM methods in electronic structure calculations enables the Coulomb part of the electron–electron interaction to be calculated with a computational effort which depends linearly on the number of basis functions, once the system becomes sufficiently large.

Instead of dividing the physical space into a near- and a far-field, the Coulomb operator itself may be partitioned into a short- and long-ranged part.[22] The short-ranged operator is evaluated exactly, while the long-ranged part is evaluated for example by means of a Fourier transformation. The net effect is again that the total Coulomb interaction can be calculated with a computational effort which only scales linearly with system size.

Although the exchange term in principle is short-ranged, and thus should benefit significantly from integral screening, this is normally not observed in practical calculations. This has been attributed to basis set incompleteness,[23] and the analysis allowed a formulation of a more aggressive screening technique which enables also the exchange part of the electron–electron interaction to be reduced to an order M method.

With the advent of methods which enables the construction of the Fock matrix to be done with a computational effort which scales linearly with system size, the diagonalization step for solving the HF equations eventually becomes the computational bottleneck, since matrix diagonalization depends on the third power of the problem size. It is, however, possible to reformulate the SCF problem in terms of a minimization of an energy functional which depends directly on the density matrix elements.[24] This functional can then be minimized for example by conjugate gradient methods (Section 14.2), taking advantage of the fact that the density matrix becomes sparse for large systems. The HF method therefore appears to have reached the "holy grail" of quantum chemistry, linear scaling with system size.

3.9 Semi-empirical Methods

The cost of performing an HF calculation scales formally as the fourth power of the number of basis functions. This arises from the number of two-electron integrals necessary for constructing the Fock matrix. Semi-empirical methods reduce the computational cost by reducing the number of these integrals.[25] Although linear scaling methods can reduce the scaling of *ab initio* HF methods to $\sim M^1$, this is only the limiting behavior in the large basis set limit, and *ab initio* methods will still require a significantly larger computational effort than semi-empirical methods.

The first step in reducing the computational problem is to consider only the valence electrons explicitly, the core electrons are accounted for by reducing the nuclear charge or introducing functions to model the combined repulsion due to the nuclei and core electrons. Furthermore, only a minimum basis set (the minimum number of functions necessary for accommodating the electrons in the neutral atom) is used for the valence electrons. Hydrogen thus has one basis function, and all atoms in the second and third rows of the periodic table have four basis functions (one s- and one set of p-orbitals, p_x, p_y and p_z). The large majority of semi-empirical methods to date use only s- and p-functions, and the basis functions are taken to be Slater type orbitals (see Chapter 5), i.e. exponential functions.

The central assumption of semi-empirical methods is the *Zero Differential Overlap* (ZDO) approximation, which neglects all products of basis functions depending on the same electron coordinates when located on different atoms. Denoting an atomic orbital on centre A as μ_A (it is customary to denote basis functions with: μ, ν, λ and σ in semi-empirical theory, while we are using χ_α, χ_β, χ_γ, and χ_δ for *ab initio* methods), the ZDO approximation corresponds to: $\mu_A(i) \cdot \nu_B(i) = 0$. Note that it is the <u>product</u> of functions on different atoms that is set equal to zero, not the <u>integral</u> over such a product. This has the following consequences (eqs. (3.50 and 3.55)):

(1) The overlap matrix **S** is reduced to a unit matrix.
(2) One-electron integrals involving three centres (two from the basis functions and one from the operator) are set to zero.
(3) All three- and four-center two-electron integrals, which are by far the most numerous of the two-electron integrals, are neglected.

To compensate for these approximations, the remaining integrals are made into parameters, and their values are assigned on the basis of calculations or experimental data. Exactly how many integrals are neglected, and how the parameterization is done, defines the various semi-empirical methods.

Rewriting eq. (3.51) with semi-empirical labels gives the following expression for a Fock matrix element, where a two-electron integral is abbreviated as $\langle \mu\nu|\lambda\sigma\rangle$, (eq. (3.56)).

$$F_{\mu\nu} = h_{\mu\nu} + \sum_{\lambda\sigma}^{\text{AO}} D_{\lambda\sigma}[\langle \mu\nu|\lambda\sigma\rangle - \langle \mu\lambda|\nu\sigma\rangle]$$

$$h_{\mu\nu} = \langle \mu|\mathbf{h}|\nu\rangle$$

$$(3.71)$$

Approximations are made for the one- and two-electron parts as follows.

3.9.1 Neglect of Diatomic Differential Overlap Approximation (NDDO)

In the *Neglect of Diatomic Differential Overlap* (NDDO) approximation there are no further approximations than those mentioned above. Using μ and ν to denote either an s- or p-type (p_x, p_y or p_z) orbital, the NDDO approximation is defined by the following equations.

Overlap integrals (eq. (3.50)):

$$S_{\mu\nu} = \langle \mu_A|\nu_B\rangle = \delta_{\mu\nu}\delta_{AB}$$

$$(3.72)$$

One-electron operator (eq. (3.24)):

$$\mathbf{h} = -\tfrac{1}{2}\nabla^2 - \sum_a \frac{Z_a'}{|\mathbf{R}_a - \mathbf{r}|} = -\tfrac{1}{2}\nabla^2 - \sum_a \mathbf{V}_a$$

$$(3.73)$$

Here Z_a' denotes that the nuclear charge has been reduced by the number of core electrons.

One-electron integrals (eq. (3.55)):

$$\langle \mu_A|\mathbf{h}|\nu_A\rangle = \langle \mu_A| -\tfrac{1}{2}\nabla^2 - \mathbf{V}_A|\nu_A\rangle - \sum_{a\neq A}\langle \mu_A|\mathbf{V}_a|\nu_A\rangle$$

$$\langle \mu_A|\mathbf{h}|\nu_B\rangle = \langle \mu_A| -\tfrac{1}{2}\nabla^2 - \mathbf{V}_A - \mathbf{V}_B|\nu_B\rangle$$

$$\langle \mu_A|\mathbf{V}_c|\nu_B\rangle = 0$$

$$(3.74)$$

Owing to the orthogonality of the atomic orbitals, the first one-centre matrix element in eq. (3.74) is zero unless the two functions are identical.

$$\langle \mu_A| -\tfrac{1}{2}\nabla^2 - \mathbf{V}_A|\nu_A\rangle = \delta_{\mu\nu}\langle \mu_A| -\tfrac{1}{2}\nabla^2 - \mathbf{V}_A|\mu_A\rangle$$

$$(3.75)$$

Two-electron integrals (eqs. (3.55) and (3.56)):

$$\langle \mu_A\nu_B|\lambda_C\sigma_D\rangle = \delta_{AC}\delta_{BD}\langle \mu_A\nu_B|\lambda_A\sigma_B\rangle$$

$$(3.76)$$

3.9.2 *Intermediate Neglect of Differential Overlap Approximation (INDO)*

In addition to the NDDO approximations, the *Intermediate Neglect of Differential Overlap* (INDO) approximation neglects all two-centre two-electron integrals which are not of the Coulomb type. Furthermore, in order to preserve rotational invariance, i.e. the total energy should be independent of a rotation of the coordinate system, some of the integrals must be made independent of the orbital type (i.e. an integral involving a p-orbital must be the same as with an s-orbital). This has as a consequence that one-electron integrals involving two different functions on the same atom and a $\mathbf{V}_a$ operator from another atom disappear. The INDO method involves the following additional approximations, beside those for NDDO.

One-electron integrals (eq. (3.74)):

$$\langle \mu_A | \mathbf{h} | \nu_A \rangle = -\delta_{\mu\nu} \sum_a \langle \mu_A | \mathbf{V}_a | \mu_A \rangle$$

$\langle \mu_A | \mathbf{V}_a | \mu_A \rangle$ is independent of orbital type (s or p)

$$(3.77)$$

Two-electron integrals (eq. (3.76)):

$$\langle \mu_A \nu_B | \lambda_C \sigma_D \rangle = \delta_{\mu_A \lambda_C} \delta_{\nu_B \sigma_D} \langle \mu_A \nu_B | \mu_A \nu_B \rangle$$

$\langle \mu_A \nu_B | \mu_A \nu_B \rangle$ is independent of orbital type (s or p)

$$(3.78)$$

The surviving integrals are commonly denoted by γ.

$$\langle \mu_A \nu_A | \mu_A \nu_A \rangle = \langle \mu_A \mu_A | \mu_A \mu_A \rangle = \gamma_{AA}$$

$$\langle \mu_A \nu_B | \mu_A \nu_B \rangle = \gamma_{AB}$$

$$(3.79)$$

The INDO method is intermediate between the NDDO and CNDO methods in terms of approximations.

3.9.3 *Complete Neglect of Differential Overlap Approximation (CNDO)*

In the *Complete Neglect of Differential Overlap* (CNDO) approximation only the Coulomb one-centre and two-centre two-electron integrals remain (eq. (3.78)).

$$\langle \mu_A \nu_B | \lambda_C \sigma_D \rangle = \delta_{AC} \delta_{BD} \delta_{\mu\lambda} \delta_{\nu\sigma} \langle \mu_A \nu_B | \mu_A \nu_B \rangle$$

$\langle \mu_A \nu_B | \mu_A \nu_B \rangle$ is independent of orbital type (s or p)

$$(3.80)$$

The integrals are again parameterized as in eq. (3.79). The approximations for the one-electron integrals in CNDO are the same as for INDO. The *Pariser–Pople–Parr* (PPP) method can be considered as a CNDO approximation where only π-electrons are treated.

The main difference between CNDO, INDO and NDDO is the treatment of the two-electron integrals. While CNDO and INDO reduce these to just two parameters (γ_{AA} and γ_{AB}), all the one- and two-center integrals are kept in the NDDO approximation. Within an sp-basis, however, there are only 27 different types of one- and two-center integrals, while the number rises to over 500 for a basis containing s-, p- and d-functions.

3.10   Parameterization

An *ab initio* HF calculation with a minimum basis set is rarely able to give more than a qualitative picture of the MOs, it is of very limited value for predicting quantitative features. Introduction of the ZDO approximation decreases the quality of the (already poor) wave function, i.e. a direct employment of the above NDDO/INDO/CNDO schemes is not useful. To "repair" the deficiencies due to the approximations, parameters are introduced in place of some or all of the integrals.

There are three methods that can be used for transforming the NDDO/INDO/CNDO approximations into working computational models.

(1) The remaining integrals can be calculated from the functional form of the atomic orbitals.
(2) The remaining integrals can be made into parameters, which are assigned values based on a few (usually atomic) experimental data.
(3) The remaining integrals can be made into parameters, which are assigned values based on <u>fitting</u> to many (usually molecular) experimental data.

Method (2) derives specific atomic properties, such as ionization potentials and excitation energies, in terms of the parameters, and assigns their values accordingly. Method (3) takes the parameters as fitting constants, and assigns their values based on a least squares fit to a large set of experimental data, analogously to the fitting of force field parameters (Section 2.3).

The CNDO, INDO and NDDO <u>methods</u> use a combination of (1) and (2) for assigning parameters.[26] Some of the non-zero integrals are calculated from the atomic orbitals, others are assigned values based on atomic ionization potentials and electron affinities. Many different versions exist; they differ in the exact way in which the parameters have been derived. Some of the names associated with these methods are CNDO/1, CNDO/2, CNDO/S, CNDO/FK, CNDO/BW, INDO/1, INDO/2, INDO/S and SINDO1. These methods are rarely used in modern computational chemistry, mainly because the "modified" methods described below usually perform better. Exceptions are INDO based methods, such as *SINDO1*[27] and *INDO/S*.[28] SINDO (*Symmetric orthogonalized INDO*) employs the INDO approximations described above, but not the ZDO approximation for the overlap matrix. The INDO/S method (INDO parameterized for *Spectroscopy*) is especially designed for calculating electronic spectra of large molecules or systems involving heavy atoms.

The group centred around M. J. S. Dewar has used a combination of (2) and (3) for assigning parameter values, resulting in a class of commonly used methods. The molecular data used for parameterization are geometries, heats of formation, dipole moments and ionization potentials. These methods are denoted "modified" as their parameters have been obtained by fitting.

3.10.1   *Modified Intermediate Neglect of Differential Overlap (MINDO)*

Three versions of *Modified Intermediate Neglect of Differential Overlap* (*MINDO*) models exist, MINDO/1, MINDO/2 and MINDO/3. The first two attempts at parameterizing INDO gave quite poor results, but MINDO/3, introduced in 1975,[29] produced the first general purpose quantum chemical method which could successfully predict molecular properties at a relatively low computational cost. The parameterization of MINDO contains diatomic variables in the two-centre one-electron term, thus parameters β_{AB} must be derived for all pairs of bonded atoms. The parameters I_μ are ionization potentials.

$$\begin{aligned}
\langle \mu_A | \mathbf{h} | \nu_B \rangle &= \langle \mu_A | - \tfrac{1}{2} \mathbf{V}^2 - \mathbf{V}_A - \mathbf{V}_B | \nu_B \rangle \\
&= S_{\mu\nu} \beta_{AB} (I_\mu + I_\nu) \\
S_{\mu\nu} &= \langle \mu_A | \nu_B \rangle
\end{aligned} \tag{3.81}$$

MINDO/3 has been parameterized for H, B, C, N, O, F, Si, P, S and Cl, although certain combinations of these elements have been omitted. MINDO/3 is rarely used in modern computational chemistry, having been succeeded in accuracy by the NDDO methods below. Since there are parameters in MINDO which depend on two atoms, the number of parameters increases as the square of the number of elements. It is unlikely that MINDO will be parameterized beyond the above-mentioned in the future.

3.10.2 Modified NDDO Models

The MNDO, AM1 and PM3 methods[30] are parameterizations of the NDDO model, where the parameterization is in terms of atomic variables, i.e. referring only to the nature of a single atom. MNDO, AM1 and PM3 are derived from the same basic approximations (NDDO), and differ only in the way the core–core repulsion is treated, and how the parameters are assigned. Each method considers only the valence s- and p-functions, which are taken as Slater type orbitals with corresponding exponents, ζ_s and ζ_p.

The *one-center one-electron integrals* have a value corresponding to the energy of a single electron experiencing the full nuclear charge (U_s or U_p) plus terms from the potential due to all the other nuclei in the system (eq. (3.74)). The latter is parameterized in terms of the (reduced) nuclear charges and a two-electron integral.

$$\begin{aligned}
h_{\mu\nu} &= \langle \mu_A | \mathbf{h} | \nu_A \rangle = \delta_{\mu\nu} U_\mu - \sum_{a \neq A} Z'_a \langle \mu_A s_a | \nu_A s_a \rangle \\
U_\mu &= \langle \mu_A | - \tfrac{1}{2} \mathbf{V}^2 - \mathbf{V}_A | \mu_A \rangle
\end{aligned} \tag{3.82}$$

Here μ_A and ν_A are either s- or p-type functions on atom A.

The *two-center one-electron integrals* given by the second equation in (3.74) are written as a product of the corresponding overlap integral times the average of two atomic "resonance" parameters, β.

$$\begin{aligned}
\langle \mu_A | \mathbf{h} | \nu_B \rangle &= \langle \mu_A | - \tfrac{1}{2} \mathbf{V}^2 - \mathbf{V}_A - \mathbf{V}_B | \nu_B \rangle \\
&= S_{\mu\nu} \tfrac{1}{2} (\beta_\mu + \beta_\nu) \\
S_{\mu\nu} &= \langle \mu_A | \nu_B \rangle
\end{aligned} \tag{3.83}$$

Here μ and ν again indicate either s- or p-type functions and the overlap is calculated explicitly (note that this is not consistent with the ZDO approximation, and the inclusion is the origin of the "Modified" label).

There are only five types of *one-centre two-electron integral* surviving the NDDO approximation within an sp-basis (eq. (3.76)).

$$\langle ss|ss \rangle = G_{ss}$$
$$\langle sp|sp \rangle = G_{sp}$$
$$\langle ss|pp \rangle = H_{sp} \qquad\qquad (3.84)$$
$$\langle pp|pp \rangle = G_{pp}$$
$$\langle pp'|pp' \rangle = G_{p2}$$

The G-type parameters are Coulomb terms, while the H parameter is an exchange integral. The G_{p2} integral involves two different types of p-functions (i.e. p_x, p_y or p_z).

There are a total of 22 different *two-centre two-electron integrals* arising from an sp-basis; these are modelled as interactions between multipoles. Electron 1 in an $\langle s\mu|s\mu \rangle$ type integral for example, is modelled as a monopole, in a $\langle s\mu|p\mu \rangle$ type integral as a dipole and in a $\langle p\mu|p\mu \rangle$ type integral as a quadrupole. The dipole and quadrupole moments are generated as fractional charges located at specific points away from the nuclei, where the distance is determined by the orbital exponents ζ_s and ζ_p. The main reason for adopting a multipole expansion of these integrals was the limited computational resources available when these methods were developed initially. In the limit of the two nuclei being placed on top of each other, a two-centre two-electron integral becomes a one-center two-electron integral, which puts boundary conditions on the functional form of the multipole interaction. The bottom line is that all two-center two-electron integrals are written in terms of the orbital exponents and the one-center two-electron parameters given in eq. (3.84).

The *core–core repulsion* is the repulsion between nuclear charges, properly reduced by the number of core-electrons. The "exact" expression for this term is simply the product of the charges divided by the distance, $Z'_A Z'_B / R_{AB}$. Due to the inherent approximations in the NDDO method, however, this term is not cancelled by electron–electron terms at long distances, resulting in a net repulsion between uncharged molecules or atoms even when their wave functions do not overlap. Consequently the core–core term must be modified to generate the proper limiting behaviour, which means that two-electron integrals must be involved. The specific functional form depends on the exact method, and is given below.

Each of the MNDO, AM1 and PM3 methods involves at least 12 parameters per atom: orbital exponents, $\zeta_{s/p}$; one-electron terms, $U_{s/p}$ and $\beta_{s/p}$; two-electron terms, G_{ss}, G_{sp}, G_{pp}, G_{p2}, H_{sp}; parameters used in the core–core repulsion, α; and for the AM1 and PM3 methods also a, b and c constants, as described below.

3.10.3 *Modified Neglect of Diatomic Overlap (MNDO)*

The core–core repulsion of the *Modified Neglect of Diatomic Overlap* (MNDO) model[31] has the form:

$$V_{nn}^{MNDO}(A, B) = Z'_A Z'_B \langle s_A s_B | s_A s_B \rangle (1 + e^{-\alpha_A R_{AB}} + e^{-\alpha_B R_{AB}}) \qquad (3.85)$$

where the α exponents are taken as fitting parameters.

Interactions involving O–H and N–H bonds are treated differently

$$V_{nn}(A, H) = Z'_A Z_H \langle s_A s_H | s_A s_H \rangle \left(1 + \frac{e^{-\alpha_A R_{AH}}}{R_{AH}} + e^{-\alpha_H R_{AH}} \right) \qquad (3.86)$$

In addition, MNDO uses the approximation, $\zeta_s = \zeta_p$ for some of the lighter elements. MNDO has been parameterized for the elements: H, B, C, N, O, F, Al, Si, P, S, Cl, Zn, Ge, Br, Sn, I, Hg and Pb. The G_{ss}, G_{sp}, G_{pp}, G_{p2}, H_{sp} parameters are taken from atomic spectra, while the others are fitted to molecular data. Although MNDO has been succeeded by the AM1 and PM3 methods, it is still used for some types of calculation where MNDO is known to give better results.

Some known limitations of the MNDO model are.

(1) Sterically crowded molecules, like neopentane, are too unstable.
(2) Four membered rings are too stable.
(3) Weak interactions are unreliable, for example it does not predict hydrogen bonds.
(4) Hypervalent molecules, like sulfoxides and sulfones, are too unstable.
(5) Activation energies for bond breaking/forming reactions are too high.
(6) Non-classical structures are predicted to be unstable relative to classical structures (for example ethyl cation).
(7) Oxygenated substituents on aromatic rings are out-of-plane (for example nitrobenzene).
(8) Peroxide bonds are too short by ~ 0.17 Å
(9) The C–X–C angle in ethers and sulfides is too large by $\sim 9°$.

MNDOC[32] has the same functional form as MNDO, however, electron correlation is explicitly calculated by second-order perturbation theory. The derivation of the MNDOC parameters is done by fitting the correlated MNDOC results to experimental data. Electron correlation in MNDO is only included implicitly via the parameters, from fitting to experimental results. Since the training set only includes ground-state stable molecules, MNDO has problems treating systems where the importance of electron correlation is substantially different from "normal" molecules. MNDOC consequently performs significantly better for systems where this is not the case, such as transition structures and excited states.

3.10.4 Austin Model 1 (AM1)

After some experience with MNDO, it became clear that there were certain systematic errors. For example the repulsion between two atoms which are 2–3 Å apart is too high. This has as a consequence that activation energies in general are too large. The source was traced to too repulsive an interaction in the core–core potential. To remedy this, the core–core function was modified by adding Gaussian functions, and the whole model was reparameterized. The result was called *Austin Model 1* (AM1)[33] in honour of Dewar's move to the University of Austin. The core–core repulsion of AM1 has the form:

$$V_{nn}(A, B) = V_{nn}^{MINDO}(A, B) + \frac{Z'_A Z'_B}{R_{AB}}$$

$$\times \left(\sum_k a_{kA} e^{-b_{kA}(R_{AB} - c_{kA})^2} + \sum_k a_{kB} e^{-b_{kB}(R_{AB} - c_{kB})^2} \right) \qquad (3.87)$$

where k is between 2 and 4 depending on the atom. It should be noted that the Gaussian functions more or less were added as patches onto the underlying parameters, which explains why different number of Gaussians are used for each atom. As with MNDO, the G_{ss}, G_{sp}, G_{pp}, G_{p2}, H_{sp} parameters are taken from atomic spectra, while the rest, including the a_k, b_k and c_k constants, are fitted to molecular data. AM1 has been parameterized for the elements: H, B, C, N, O, F, Al, Si, P, S, Cl, Zn, Ge, Br, I and Hg. Some known improvements and limitations of the AM1 model are

(1) AM1 does predict hydrogen bonds with a strength approximately correct, but the geometry is often wrong.
(2) Activation energies are much improved over MNDO.
(3) Hypervalent molecules are improved over MNDO, but still have significantly larger errors than other types of compound.
(4) Alkyl groups are systematically too stable by ~ 2 kcal/mol per CH_2 group.
(5) Nitro compounds are systematically too unstable.
(6) Peroxide bonds are too short by ~ 0.17 Å.
(7) Phosphor compounds have problems when atoms are ~ 3 Å apart, producing incorrect geometries. P_4O_6 for example is predicted to have P–P bonds differing by 0.4 Å, although experimentally they are identical.
(8) The *gauche* conformation in ethanol is predicted to be more stable than the *trans*.

3.10.5 Modified Neglect of Diatomic Overlap, Parametric Method Number 3 (MNDO-PM3)

The parameterization of MNDO and AM1 had been done essentially by hand, taking the G_{ss}, G_{sp}, G_{pp}, G_{p2}, H_{sp} parameters from atomic data and varying the rest until a satisfactory fit had been obtained. Since the optimization was done by hand, only relatively few reference compounds could be included. Stewart[34] made the optimization process automatic, by deriving and implementing formulas for the derivative of a suitable error function with respect to the parameters. All parameters could then be optimized simultaneously, including the two-electron terms, and a significantly larger training set with several hundred data could be employed. In this reparameterization, the AM1 expression for the core–core repulsion (eq. (3.87)) was kept, except that only 2 Gaussians were assigned to each atom. These Gaussian parameters were included as an integral part of the model, and allowed to vary freely. The resulting method was denoted *Modified Neglect of Diatomic Overlap, Parametric Method Number 3* (MNDO-PM3 or PM3 for short), and is essentially AM1 with all the parameters fully optimized. In a sense it has the best set of parameters (or at least a good local minimum) for the given set of experimental data. The optimization process, however, still requires some human intervention, in selecting the experimental data and assigning appropriate weight factors to each set of data. PM3 has been parameterized for the elements: H, Li, C, N, O, F, Mg, Al, Si, P, S, Cl, Zn, Ga, Ge, As, Se, Br, Cd, In, Sn, Sb, Te, I, Hg, Tl, Pb, Bi, Po and At. Parameters for many of the (additional) transition metals are also being developed under the name PM3(tm), which includes d-orbitals.

Some known limitations of the PM3 model are:

(1) Almost all sp^3-nitrogens are predicted to be pyramidal, contrary to experimental observation.

(2) Hydrogen bonds are too short by ~ 0.1 Å.

(3) The *gauche* conformation in ethanol is predicted to be more stable than the *trans*.

(4) Bonds between Si and Cl, Br and I are underestimated, the Si-I bond in H_3SiI, for example, is too short by ~ 0.4 Å.

(5) H_2NNH_2 is predicted to have a C_{2h} structure, while the experimental is C_2, and ClF_3 is predicted to have a D_{3h} structure, while the experimental is C_{2v}.

(6) The charge on nitrogen atoms is often of "incorrect" sign and "unrealistic" magnitude.

Some common limitations to MNDO, AM1 and PM3 are:

(1) Rotational barriers for bonds which have partly double bond character are significantly too low. This is especially a problem for the rotation around the C–N bond in amides, where values of 5–10 kcal/mol are obtained. A purely *ad hoc* fix has been made for amides by adding a force field rotational term to the C–N bond which raises the value to 20–25 kcal/mol, and brings it in line with experimental data. Similarly, the barrier for rotation around the central bond in butadiene is calculated to be only 0.5–2.0 kcal/mol, in contrast to the experimental value of 5.9 kcal/mol.[35]

(2) Weak interactions, such as van der Waals complexes or hydrogen bonds, are poorly predicted. Either the interaction is too weak, or the minimum energy geometry is wrong.

(3) The bond length to nitrosyl groups is underestimated, the N–N bond in N_2O_3, for example, is ~ 0.7 Å too short.

(4) Although MNDO, AM1 and PM3 have parameters for some metals, these are often based on only a few experimental data. Calculations involving metals should thus be treated with care. The PM3(tm) set of parameters are determined exclusively from geometrical data (X-ray), since there are very few reliable energetic data available for transition metal compounds.

3.10.6 The MNDO/d method

With only s- and p-functions included, the MNDO/AM1/PM3 methods are unable to treat a large part of the periodic table. Furthermore, from *ab initio* calculations it is known that d-orbitals significantly improve the results for compounds involving second row elements, especially hypervalent species. The main problem in extending the NDDO formalism to include d-orbitals is the significant increase in distinct two-electron integrals which ultimately must be assigned suitable values. For an sp-basis there are only five one-centre two-electron integrals, while there are 17 in an spd-basis. Similarly, the number of two-centre two-electron integrals raises from 22 to 491 when d-functions are included.

Recently Thiel and Voityuk[36] have constructed a workable NDDO model which also includes d-orbitals for use in connection with MNDO, called *MNDO/d*. With reference to the above description for MNDO/AM1/PM3, it is clear that there are immediately three new parameters: ζ_d, U_d and β_d (eqs. (3.82) and (3.83)). Of the 12 new one-centre two-electron integrals only one (G_{dd}) is taken as a freely varied parameter. The other 11 are calculated analytically based on pseudo-orbital exponents, which are assigned so that the analytical formulas regenerate G_{ss}, G_{pp} and G_{dd}.

With only s- and p-functions present, the two-centre two-electron integrals can be modelled by multipoles up to order 2 (quadrupoles), however, with d-functions present multipoles up to order 4 must be included. In MNDO/d all multipoles beyond order 2 are neglected. The resulting MNDO/d method typically employs 15 parameters per atom, and it currently contains parameters for the following elements (beyond those already present in MNDO): Na, Mg, Al, Si, P, S, Cl, Br, I, Zn, Cd and Hg.

3.10.7 Semi-ab initio Method 1

The philosophy in the *Semi-ab initio Method 1* (SAM1 and SAM1D) model[37] is slightly different from the other "modified" methods. It is again based on the NDDO approximation, but instead of replacing all integrals by parameters, the one- and two-centre two-electron integrals are calculated directly from the atomic orbitals. These integrals are then scaled by a function containing adjustable parameters to fit experimental data (R_{AB} being the interatomic distance).

$$\langle \mu_A \nu_B | \mu_A \nu_B \rangle = f(R_{AB}) \langle \mu_A \nu_B | \mu_A \nu_B \rangle \tag{3.88}$$

The details of the functional form and parameterization have not yet been published. The advantage is that basis sets involving d-orbitals are readily included (defining the SAM1D method), making it possible to perform calculations on a larger fraction of the periodic table. The SAM1 method explicitly uses the minimum STO-3G basis set, but it is in principle also possible to use extended basis sets with this model. The actual calculation of the integrals makes the SAM1 method somewhat slower than the MNDO/ AM1/PM3, but only by a factor of ~ 2. The SAM1/SAM1D methods have been parameterized for the elements: H, Li, C, N, O, F, Si, P, S, Cl, Fe, Cu, Br and I.

3.11 Performance of Semi-empirical Methods

The electronic energy (including the core–core repulsion) calculated by MINDO, MNDO, MNDO/d, AM1 and PM3 is, in analogy with that calculated by *ab initio* methods, the total energy relative to a situation where the nuclei (with their core electrons) and the valence electrons are infinitely separated. The electronic energy is normally converted to a heat of formation by subtracting the electronic energy of the isolated atoms which make up the system, and adding the experimental atomic heat of formation. It should be noted that thermodynamical corrections (e.g. zero-point energies, see Chapter 12) should not be added to the ΔH_f values, as these are included implicitly by the parameterization.

$$\Delta H_f(\text{molecule}) = E_{elec}(\text{molecule}) - \sum_{}^{\text{atoms}} E_{elec}(\text{atoms}) + \sum_{}^{\text{atoms}} \Delta H_f(\text{atoms}) \tag{3.89}$$

Some typical errors in heat of formation for the MNDO, AM1 and PM3 methods are given in Table 3.1.[38] The exact numbers of course depend on which, and how many, compounds have been selected for comparison, thus the numbers should only be taken as a guideline for the accuracy expected. Some typical errors in bond distances are given in Table 3.2.

Angles are typically predicted with an accuracy of a few degrees, the average errors for MNDO, AM1 and PM3 are: 4.3°, 3.3° and 3.9°. Ionization potentials are typically

Table 3.1 Average heat of formation error in kcal/mol (number of compounds)

Compounds:	MNDO	AM1	PM3
H, C, N, O (276)	18.5	10.5	7.9
F (133)	84.2	49.5	11.2
Si (78)	22.9	20.8	14.2
All normal valent (607)	24.3	14.8	11.2
Hypervalent (106)	104.5	62.3	17.3
All (713)	46.2	27.6	11.6

Table 3.2 Average errors in bond distances (Å)

Bonds to:	MNDO	AM1	PM3
H	0.015	0.006	0.005
C	0.002	0.002	0.002
N	0.015	0.014	0.012
O	0.017	0.011	0.006
F	0.023	0.017	0.011
Si	0.030	0.019	0.045

accurate to 0.5–1.0 eV, average errors for MNDO, AM1 and PM3 are: 0.78 eV, 0.61 eV and 0.57 eV. Average errors for dipole moments are 0.45 D, 0.35 D and 0.38 D, respectively.

Since AM1 contains more adjustable parameters than MNDO, and since PM3 can be considered as a version of AM1 with all the parameters fully optimized, it is expected that the error decreases in the order MNDO > AM1 > PM3. This is indeed what is observed in the above tables. It should be noted, however, that the data in the tables refer to averages, thus for specific compounds or classes of compounds the ordering may be different. Bonds between silicon and iodine with PM3 give an example of specific compound being poorly described, although the average description for all compounds is better. It is clear that the PM3 method will perform better than AM1 in an average sense since the two-electron integrals are optimized to give a better fit to the given molecular data set. This does not mean, however, that PM3 necessarily will perform better than AM1 (or MNDO) for properties not included in the training set. Indeed it has been argued that the AM1 method tends to give more "realistic" values for atomic charges than PM3, especially for compounds involving nitrogen. An often quoted example is formamide for which the Mulliken population analysis by different methods is given in Table 3.3. The negative charge on nitrogen produced by PM3 is significantly smaller than that produced by the other methods, but it should be noted that atomic charges are not well-defined quantities, as discussed in Chapter 9. Nevertheless, it may indicate that the electrostatic potential generated by a PM3 wave function is of lower quality than one generated by the AM1 method.

Table 3.4 shows a comparison of some of the elements for which the MNDO, MNDO/ d, AM1, PM3, SAM1 and SAM1d methods have been parameterized.

Table 3.3 Mulliken charges in formamide with different methods

	MNDO	AM1	PM3	HF/6-31G(d,p)	MP2/6-31G(d,p)
C	0.37	0.26	0.16	0.56	0.40
O	− 0.39	− 0.40	− 0.38	− 0.56	− 0.43
N	− 0.49	− 0.62	− 0.13	− 0.73	− 0.63

Table 3.4 Average heat of formation error in kcal/mol (number of compounds)

Compounds	MNDO	AM1	PM3	MNDO/d	SAM1	SAM1d
Al (29)	22.1	10.5	16.4	4.9		
Si (84)	12.0	8.5	6.0	6.3	8.0	11.2
P (43)	38.7	14.5	17.1	7.6	14.4	15.0
S (99)	48.4	10.3	7.5	5.6	8.3	7.9
Cl (85)	39.4	29.1	10.4	3.9	11.1	4.7
Br (51)	16.2	15.2	8.1	3.4	8.7	5.2
I (42)	25.4	21.7	13.4	4.0	6.6	6.6
Zn (18)	21.0	16.9	14.7	4.9		
Hg (37)	13.7	9.0	7.7	2.2		
Al, Si, P, S, Cl, Br, I, Zn, Hg (488)	29.2	15.3	10.0	4.9		
Si, P, S, Cl, Br, I (404)	31.4	16.1	9.5	5.1	9.3	8.2

Considering that the parameters for the MNDO/d method for all first row elements (which are present in most of the training set of compounds) are identical to MNDO, the improvement by addition of *d*-functions is quite impressive. It should also be noted that MNDO/d only contains 15 parameters, compared to 18 for PM3, and that some of the 15 parameters are taken from atomic data (analogously to the MNDO/AM1 parameterization), and not used in the molecular data fitting as in PM3.

The apparent accuracy of 5–10 kcal/mol for calculating heats of formation with semi-empirical methods is slightly misleading. Normally the interest is in relative energies of different species, and since the heat of formation errors are essentially random, relative energies may not be predicted as well (two random errors of 10 kcal/mol may add up to an error of 20 kcal/mol). This is in contrast to *ab initio* methods, which usually are better at predicting relative rather than absolute energies, since errors using these methods tend to be systematic and at least partly cancel out when comparing similar systems.

3.12 Extended Hückel Theory

The Hückel methods perform the parameterization on the Fock matrix elements (eqs. (3.50) and (3.51)), and not at the integral level as do NDDO/INDO/CNDO. This means that Hückel methods are non-iterative, they only require a single diagonalization of the Fock (Hückel) matrix. The *Extended Hückel Theory* (EHT) or *Method* (EHM), developed primarily by Hoffmann[39] again only considers the valence electrons. It makes use of Koopmans' theorem (eq. (3.46)) and assigns the diagonal elements in the **F** matrix to be atomic ionization potentials. The off-diagonal elements are parameterized as averages of the diagonal elements, weighted by an overlap integral. The overlap integrals are actually calculated, i.e. the ZDO approximation is not invoked. The basis functions are taken as Slater type orbitals, with the exponents assigned according to the Slater rules.[40]

$$F_{\mu\mu} = -I_\mu$$
$$F_{\mu\nu} = -K\left(\frac{I_\mu + I_\nu}{2}\right) S_{\mu\nu} \tag{3.89}$$

The K constant is usually taken as 1.75; this value reproduces the rotational barrier in ethane.

Since the diagonal elements depend only on the nature of the atom (i.e. the nuclear charge), this means that for example all carbon atoms have the same ability to attract electrons. After having performed a Hückel calculation, the actual number of electrons associated with atom A, ρ_A, can be calculated according to eq. (3.90) (see section 9.1, eqs. (9.5) and (9.4)).

$$\rho_A = \sum_i^{MO} n_i \left(\sum_{\alpha \in A}^{AO} \sum_\beta^{AO} c_{\alpha i} c_{\beta i} S_{\alpha\beta} \right) \tag{3.90}$$

Subtracting the (reduced) nuclear charge gives the effective (net) atomic charge Q_A.

$$Q_A = Z'_A - \rho_A \tag{3.91}$$

In general it is unlikely that all carbon atoms will have the exact same charge, i.e. owing to the different environments their ability to attract electrons is no longer equal. This may be argued to be inconsistent with the initial assumption of all carbons having the same diagonal elements in the Hückel matrix. In order to achieve "self-consistency", a diagonal element $F_{\mu\mu}$ belonging to atom A may be modified by the calculated atomic charge.

$$F_{\mu\mu} = -I_\mu + \omega Q_A \tag{3.92}$$

The ω parameter determines the weight of the charge on the diagonal elements. Since Q_A is calculated from the results (MO coefficients, eq. (3.90)), but enters the Hückel matrix which produces the results (by diagonalization), such schemes become iterative. Methods where the matrix elements are modified by the calculated charge are often called *charge iteration* or *self-consistent* (Hückel) methods.

The main advantage of the extended Hückel theory is that only atomic ionization potentials are required, and it is easily parameterized to the whole periodic table. Extended Hückel theory can be used for large systems involving transition metals, where it often is the only possible computational model. The very approximate method of extended Hückel theory makes it unsuitable for geometry optimizations without additional modifications[41] or calculations of energetic features at any reasonable level of accuracy. It is primarily used for obtaining qualitative correct MOs, which for example can be used as an initial guess of the density matrix for *ab initio* SCF calculations, or for use in connection with qualitative theories, as discussed in Chapter 15. Orbital energies (and thereby the total energy), however, in many cases show the correct trend for geometry perturbations corresponding to bending or torsional changes, thus qualitative features regarding molecular shapes may often be predicted/rationalized from EHT calculations.

3.12.1 Simple Hückel Theory

In the simple Hückel model the approximations are taken to the limit.[42] Only planar conjugated systems are considered. The σ-orbitals, which are symmetric with respect to a reflection in the molecular plane, are neglected. Only the π-electrons (antisymmetric with respect to the molecular mirror plane) are considered. The overlap matrix is taken as a unit matrix and the diagonal elements of the **F** matrix are assigned a value of α (depending on the atom type). Off-diagonal elements are taken either as β (depending on the two atom types) or zero, depending on whether the two atoms are "neighbours" (connected by a σ-bond) or not.

$$
\begin{aligned}
F_{\mu_A \mu_A} &= \alpha_A \\
F_{\mu_A \nu_B} &= \beta_{AB} \text{ (A and B are neighbours)} \\
F_{\mu_A \nu_B} &= 0 \text{ (A and B are not neighbours)}
\end{aligned}
\tag{3.93}
$$

Atoms are assigned "types", much as in force field methods, i.e. the parameters depend on the nuclear charge and the bonding situation. The α_A and β_{AB} parameters for atom types A and B are related to the corresponding parameters for sp^2-hybridized carbon by means of dimensionless constants h_A and k_{AB}.

$$
\begin{aligned}
\alpha_A &= \alpha_C + h_A \beta_{CC} \\
\beta_{AB} &= k_{AB} \beta_{CC}
\end{aligned}
\tag{3.94}
$$

The carbon parameters α_C and β_{CC} are normally just denoted α and β, and are rarely assigned numerical values. Simple Hückel theory thus only considers the connectivity of the π-atoms, there is no information about the molecular geometry entering the calculation (e.g. whether some bonds are shorter or longer than others, or differences in bond angles).

In analogy to extended Hückel theory, there are also charge iterative methods for simple Hückel theory. The equivalent of eq. (3.90) is

$$
\rho_A = \sum_i^{MO} n_i c_{Ai}^2
\tag{3.95}
$$

and eq. (3.92) becomes

$$
\alpha_A' = \alpha_A + \omega(n_A - \rho_A)\beta
\tag{3.96}
$$

where n_A is the number of π-electrons involved from atom A.

The Hückel method is essentially only used for educational purposes or for very qualitative orbital considerations. It has the ability to produce qualitatively correct MOs, involving a computational effort which is within reach of doing by hand.

3.13 Limitations and Advantages of Semi-empirical Methods

The neglect of all three- and four-centre two-electron integrals reduces the construction of the Fock matrix from a formal order of M^4 to M^2. However, the time required for diagonalization of the **F** matrix grows as the cube of the matrix size, thus semi-empirical methods formally scale as the cube of the number of basis functions in the limit of large molecules. Diagonalization of a matrix becomes significant when the size exceeds $\sim 10\,000 \times 10\,000$. Several iterations are required for solving the SCF equations, and usually the geometry is also optimized, requiring several calculations at different geometries. This places the current limit of semi-empirical methods at around 1000 atoms. It should be noted that the conventional method of solving the HF equations by diagonalizing the Fock matrix rapidly becomes the rate limiting step in semi-empirical methods. Recent developments have therefore concentrated on formulating alternative methods for obtaining the SCF orbitals without the need for diagonalization.[43] Such methods display linear scaling with the number of atoms, allowing calculations to be performed for systems containing several thousand atoms.

The parameterization of MNDO/AM1/PM3 is performed by adjusting the constants involved in the different methods so that the results of HF calculations fit experimental data as closely as possible. This is in a sense wrong. We know that the HF method cannot give the correct result, even in the limit of an infinite basis set and without approximations. The HF results lack electron correlation, as will be discussed in Chapter 4, but the experimental data of course include such effects. This may be viewed as an advantage, the electron correlation effects are implicitly taken into account in the parameterization, and we need not perform complicated calculations to improve deficiencies in the HF procedure. However, it becomes problematic when the HF wave function cannot describe the system even qualitatively correctly, as for example with biradicals and excited states. Additional flexibility can be introduced in the trial wave function by adding more Slater determinants, for example by means of a CI procedure (see Chapter 4 for details). But electron correlation is then taken into account twice, once in the parameterization at the HF level, and once explicitly by the CI calculation.

Semi-empirical methods share the advantage/disadvantage of force field methods, they perform best for systems where much experimental information is already available, but they are unable to predict totally unknown compound types. The dependence on experimental data is not as severe as for the force field method, owing to the more complex functional form of the model. The NDDO methods require only atomic parameters, not di-, tri- and tetra-atomic parameters as do force field methods. Once a given atom has been parameterized, all possible compound types involving this element can be calculated. The smaller number of parameters and the more complex functional form have the disadvantage compared to force field methods that it is very difficult to "repair" a specific problem by reparameterization. The lack of a reasonable rotational barrier in amides, for example, cannot be attributed to an "improper" value for a single (or a few) parameter(s). Too low a rotational barrier in a force field model can easily be fixed by increasing the values of the corresponding torsional parameters. The clear advantage of semi-empirical methods over force field techniques is their ability to describe bond breaking and forming reactions.

Semi-empirical methods are zero-dimensional, just as force field methods are. There is no way of assessing the reliability of a given result within the method. This is due to the selection of a fixed (minimum) basis set. The only way of judging results is by comparing the accuracy of other calculations on similar systems with experimental data.

Semi-empirical models provide a method for calculating the electronic wave function, which may be used for predicting a variety of properties. There is nothing to hinder the calculation of say the polarizability of a molecule (the second derivative of the energy with respect to an external electric field), although it is known from *ab initio* calculations that good results require a large polarized basis set including diffuse functions, and inclusion of electron correlation. Semi-empirical methods like AM1 or PM3 have only a minimum basis (lacking polarization and diffuse functions), electron correlation is only included implicitly by the parameters, and no polarizability data have been used for deriving the parameters. Whether such calculations can produce reasonable results, as compared to experimental data, is questionable, and they certainly require careful calibration. Again it should be emphasized: the ability to perform a calculation is no guarantee that the results can be trusted!

References

1. W. Kolos and L. Wolniewicz, *J. Chem. Phys.*, **41** (1964), 3663; B. T. Sutcliffe, *Adv. Quantum. Chem.*, **28** (1997), 65.
2. N. C. Handy and A. M. Lee, *Chem. Phys. Lett.*, **252** (1996), 425.
3. A. Szabo and N. S. Ostlund *Modern Quantum Chemistry*, McGraw-Hill, 1982; R. McWeeny, *Methods of Molecular Quantum Mechanics*, Academic Press, 1992; W. J. Hehre, L. Radom, J. A. Pople and P. v. R. Schleyer *Ab Initio Molecular Orbital Theory*, Wiley, 1986; J. Simons, *J. Phys. Chem.*, **95** (1991), 1017; J. Simons and J. Nichols, *Quantum Mechanics in Chemistry*, Oxford University Press, 1997.
4. T. A. Koopmans, *Physica*, **1** (1933), 104.
5. J. Kobus, *Adv. Quantum. Chem.*, **28** (1997), 1.
6. C. C. J. Roothaan, *Rev. Mod. Phys.*, **23** (1951), 69; G. G. Hall, *Proc. R. Soc. (London)*, **A205** (1951), 541.
7. M. Head-Gordon and J. A. Pople, *J. Phys. Chem.*, **92** (1988), 3063.
8. J. A. Pople and R. K. Nesbet, *J. Chem. Phys.*, **22** (1954), 571.
9. C. Kollmar, *Int. J. Quantum. Chem.*, **62** (1997), 617.
10. V. R. Saunders and I. H. Hillier, *Mol. Phys.*, **28** (1974), 819.
11. P. Pulay, *J. Comput. Chem.*, **3** (1982), 556.
12. M. Head-Gordon and J. A. Pople, *J. Phys. Chem.*, **92** (1988), 3063.
13. J. Douady, Y. Ellinger, R. Subra and B. Levy, *J. Chem. Phys.*, **72** (1980), 1452.
14. G. Chaban, M. W. Schmidt and M. S. Gordon, *Theor. Chim. Acta.*, **97** (1997), 88.
15. R. Seeger and J. A. Pople, *J. Chem. Phys.*, **66** (1977), 3045; R. Bauernschmitt and R. Aldrichs, *J. Chem. Phys.*, **104** (1996), 9047.
16. P.-O. Löwdin and I. Mayer, *Adv. Quantum. Chem.*, **24** (1992), 79.
17. E. R. Davidson and W. T. Borden, *J. Phys. Chem.*, **87** (1983), 4783.
18. J. Almlöf, K. Faegri Jr and K. Korsell, *J. Comput. Chem.*, **3** (1982), 385; J. Almlöf, *Modern Electronic Structure Theory*, Part I, ed. D. Yarkony, World Scientific, 1995, pp. 110–151.
19. D. L. Strout and G. E. Scuseria, *J. Chem. Phys.*, **102** (1995), 8448.
20. H. G. Petersen, D. Soelvason, J. W. Perram and E. R. Smith, *J. Chem. Phys.*, **101** (1994), 8870.
21. C. A. White and M. Head-Gordon, *J. Chem. Phys.*, **104** (1996), 2620; M. C. Strain, G. E. Scuseria and M. J. Frisch, *Science*, **271** (1996), 51.
22. A. M. Lee, S. W. Taylor, J. P. Dombroski and P. M. W. Gill, *Phys. Rev. A.*, **55** (1997), 3233.
23. E. Schwegler and M. Challacombe, *J. Chem. Phys.*, **105** (1996), 2726.
24. J. M. Millan and G. E. Scuseria, *J. Chem. Phys.*, **105** (1996), 5569.
25. J. Sadley, *Semi-Empirical methods of Quantum Chemistry*, Wiley, 1985; M. C. Zerner, *Rev. Comput. Chem.*, **2** (1991), 313.
26. W. P. Anderson, T. R. Cundari and M. C. Zerner, *Int. J. Quantum. Chem.*, **39** (1991), 31.
27. J. Li, P. C. de Mello and K. Jug, *J. Comput. Chem.*, **13** (1992), 85.

28. M. Kotzian, N. Rösch and M. C. Zerner, *Theot. Chim. Acta* **81** (1992), 201.

29. R. C. Bingham, M. J. S. Dewar and D. H. Lo, *J. Am. Chem. Soc.*, **97** (1975), 1294.

30. J. J. P. Stewart, *Rev. Comput. Chem.*, **1** (1990), 45.

31. M. J. S. Dewar and W. Thiel, *J. Am. Chem. Soc.*, **99** (1977), 4899.

32. W. Thiel, *J. Am. Chem. Soc.*, **103** (1981), 1413, 1421; A. Schweig and W. Thiel, *J. Am. Chem. Soc.*, **103** (1981), 1425.

33. M. J. S. Dewar, E. G. Zoebisch, E. F. Healy and J. J. P. Stewart, *J. Am. Chem. Soc.*, **107** (1985), 3902.

34. J. J. P. Stewart, *J.Comput.Chem.*, **10** (1989), 209, 221.

35. R. Engeln, D. Consalvo and J. Reuss, *Chem. Phys.*, **160** (1992), 427.

36. W. Thiel and A. A. Voityuk, *J. Phys. Chem.*, **100** (1996), 616; W. Thiel, *Adv. Chem. Phys.*, **93** (1996), 703.

37. M. J. S. Dewar, C. Jie and J. Yu, *Tetrahedron*, **49** (1993), 5003.

38. J. J. P. Stewart, *J. Comput.-Aided Mol. Design*, **4** (1990), 1.

39. R. Hoffmann, *J. Chem. Phys.*, **39** (1963), 1397.

40. J. C. Slater, *Phys. Rev.*, **36** (1930), 57.

41. S. L. Dixon and P. C. Jurs, *J. Comput. Chem.*, **15** (1994), 733.

42. K. Yates, *Hückel Molecular Orbital Theory*, Academic Press, 1978.

43. J. J. P. Stewart, *Int. J. Quantum. Chem.*, **58** (1996), 133; A. D. Daniels, J. M. Millam and G. E. Scuseria, *J. Chem. Phys.*, **107** (1997), 425.

4 Electron Correlation Methods

The Hartree–Fock method generates solutions to the Schrödinger equation where the real electron–electron interaction is replaced by an average interaction (Chapter 3). In a sufficiently large basis, the HF wave function is able to account for $\sim 99\%$ of the total energy, but the remaining $\sim 1\%$ is often very important for describing chemical phenomena. The difference in energy between the HF and the lowest possible energy in a given basis set is called the *Electron Correlation* (EC) energy.[1] Physically it corresponds to the motion of the electrons being correlated, on average they are further apart than described by the HF wave function. As shown below, an UHF type of wave function is to a certain extent able to include electron correlation. The proper reference for discussing electron correlation is therefore a restricted (RHF or ROHF) wave function, although many authors use a UHF wave function for open-shell species. In the RHF case all the electrons are paired in molecular orbitals. The two electrons in a MO occupy the same physical space, and differ only in the spin function. The spatial overlap between the orbitals of two such "pair" -electrons is (exactly) one, while the overlap between two electrons belonging to different pairs is (exactly) zero, owing to the orthonormality of the MOs. This not the same as saying that there is no repulsion between electrons in different MOs, since the electron–electron repulsion integrals involve products of MOs ($\langle \phi_i | \phi_j \rangle = 0$ for $i \neq j$, but $\langle \phi_i \phi_j | \mathbf{g} | \phi_i \phi_j \rangle$ and $\langle \phi_i \phi_j | \mathbf{g} | \phi_j \phi_i \rangle$ are not necessarily zero).

Naively it may be expected that the correlation between pairs of electrons belonging to the same spatial MO would be the major part of the electron correlation. However, as the size of the molecule increases, the number of electron pairs belonging to different spatial MOs grows faster than those belonging to the same MO. Consider for example the valence orbitals for CH_4. There are four intraorbital electron pairs of opposite spin, but there are 12 interorbital pairs of opposite spin, and 12 interorbital pairs of the same spin. A typical value for the intraorbital pair correlation of a single bond is ~ 20 kcal/mol, while that of an interorbital pair (where the two MO are spatially close, as in CH_4) is ~ 1 kcal/mol. The interpair correlation is therefore often comparable to the intrapair contribution.

Since the correlation between opposite spins has both intra- and inter-orbital contributions, it will be larger than the correlation between electrons having the same spin. The Pauli principle (or equivalently the antisymmetry of the wave function) has the consequence that there is no intraorbital correlation from electron pairs with the same spin. The opposite spin correlation is sometimes called the *Coulomb correlation*, while the same spin correlation is called the *Fermi correlation*, i.e. the Coulomb correlation is the largest contribution. Another way of looking at electron correlation is in terms of the electron density. In the immediate vicinity of an electron, here is a reduced probability of finding another electron. For electrons of opposite spin, this is often referred to as the *Coulomb hole*, the corresponding phenomenon for electrons of the same spin is the *Fermi hole*.

The HF method determines the best one-determinant trial wave function (within the given basis set). It is therefore clear that in order to improve on HF results, the starting point must be a trial wave function which contains more than one Slater Determinant (SD) Φ. This also means that the mental picture of electrons residing in orbitals has to be abandoned, and the more fundamental property, the electron density, should be considered. As the HF solution usually gives $\sim 99\%$ of the correct answer, electron correlation methods normally use the HF wave function as a starting point for improvements.

A generic multi-determinant trial wave function can be written as

$$\Psi = a_0 \Phi_{\text{HF}} + \sum_{i=1} a_i \Phi_i \tag{4.1}$$

where a_0 usually is close to 1. Electron correlation methods differ in how they calculate the coefficients in front of the other determinants, a_0 being determined by the normalization condition.

As mentioned in Chapter 5, one can think of the expansion of an unknown MO in terms of basis functions as describing the MO "function" in the "coordinate system" of the basis functions. The multi-determinant wave function (4.1) can similarly be considered as describing the total wave function in a "coordinate" system of Slater determinants. The basis set determines the size of the one-electron basis (and thus limits the description of the one-electron functions, the MOs), while the number of determinants included determines the size of the many-electron basis (and thus limits the description of electron correlation).

4.1 Excited Slater Determinants

How are the additional determinants beyond the HF constructed? With N electrons and M basis functions, solution of the Roothaan–Hall equations for the RHF case will yield $N/2$ occupied MOs and $M - N/2$ unoccupied (virtual) MOs. Except for a minimum basis, there will always be more virtual than occupied MOs. A Slater determinant is determined by $N/2$ spatial MOs multiplied by two spin functions to yield N spinorbitals. By replacing MOs which are occupied in the HF determinant by MOs which are unoccupied, a whole series of determinants may be generated. These can be denoted according to how many occupied HF MOs have been replaced by unoccupied MOs, i.e. Slater determinants which are *singly, doubly, triply, quadruply* etc. *excited* relative to the HF determinant, up to a maximum of N excited electrons. These

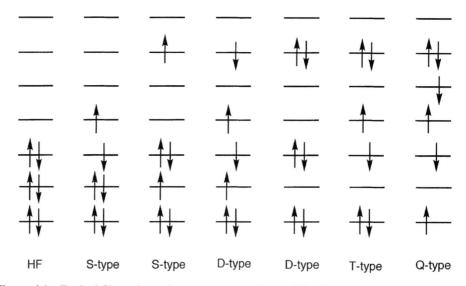

HF	S-type	S-type	D-type	D-type	T-type	Q-type

Figure 4.1 Excited Slater determinants generated from a HF reference

determinants are often referred to as *Singles* (S), *Doubles* (D), *Triples* (T), *Quadruples* (Q) etc.

The total number of determinants that can be generated depends on the size of the basis set, the larger the basis, the more virtual MOs, and the more excited determinants can be constructed. If all possible determinants in a given basis set are included, all the electron correlation (in the given basis) is (or can be) recovered. For an infinite basis the Schrödinger equation is then solved exactly. Note that "exact" is this context is not the same as the experimental value, as the nuclei are assumed to have infinite masses (Born–Oppenheimer approximation) and relativistic effects are neglected. Methods which include electron correlation are thus two-dimensional, the larger the one-electron expansion (basis set size) and the larger the many-electron expansion (number of determinants), the better are the results. This is illustrated in Figure 4.2.

In order to calculate total energies with a "chemical accuracy" of ~1 kcal/mol, it is necessary to use sophisticated methods for including electron correlation and large basis sets, which is only computationally feasible for small systems. Instead the focus is usually on calculating relative energies, trying to make the errors as constant as possible.

EC \ Basic	Minimum	DZ	DZP	· · · · · ·	Infinite
HF (0%)					HF limit
10%					
· · · · · ·					
100%					"Exact"

Figure 4.2 Convergence to the exact solution

The important chemical changes take place in the valence orbitals; the core orbitals are almost constant. In many cases the interest is therefore only in calculating the correlation energy associated with the valence electrons. Limiting the number of determinants to only those which can be generated by exciting the valence electrons is known as the *frozen core* approximation. In some cases the highest virtual orbitals corresponding to the anti-bonding combinations of the core orbitals are also removed from the correlation treatment (*frozen virtuals*). The frozen core approximation is not justified in terms of total energy; the correlation of the core electrons gives a substantial energy contribution. However, it is essentially a constant factor, which drops out when calculating relative energies. Furthermore, if we really want to calculate the core electron correlation, the standard basis sets are insufficient. In order to represent the angular correlation, higher angular moment functions with the same radial size as the filled orbitals are needed, e.g. p- and d-functions with large exponents for correlating the 1s-electrons, as discussed in Section 5.4.5. Just allowing excitations of the core electrons in a standard basis set does not "correlate" the core electrons.

There are three main methods for calculating electron correlation: *Configuration Interaction* (CI), *Many Body Perturbation Theory* (MBPT) *and Coupled Cluster* (CC). A word of caution before we describe these methods in more details. The Slater determinants are composed of spin-MOs, but since the Hamilton operator is independent of spin, the spin dependence can be factored out. Furthermore, to facilitate notation, it is often assumed that the HF determinant is of the RHF type. Finally, many of the expressions below involve double summations over identical sets of functions. To ensure only the unique terms are included, one of the summation indices must be restricted. Alternatively, both indices can be allowed to run over all values, and the overcounting corrected by a factor of $1/2$. Various combinations of these assumptions result in final expressions which differ by factors of $1/2$, $1/4$ etc. from those given here. In the present book the MOs are always spin-MOs, and conversion of a restricted summation to an unrestricted is always noted explicitly.

Finally a comment on notation. The quality of a calculation is given by the level of theory (i.e. how much electron correlation is included) and the size of the basis set. In a commonly used /-notation, introduced by J. A. Pople, this is denoted as "level/basis". If nothing further is specified, this implies that the geometry is optimized at this level of theory. As discussed in Section 5.5, the geometry is usually much less sensitive to the theoretical level than relative energies, and high-level calculations are therefore often carried out using geometries optimized at a lower level. This is denoted as "level2/basis2//level1/basis1", where the notation after the // indicates the level at which the geometry is optimized.

4.2 Configuration Interaction

This is perhaps the easiest method to understand. It is based on the variational principle (Appendix B), analogous to the HF method. The trial wave function is written as a linear combination of determinants with the expansion coefficients determined by requiring that the energy should be a minimum (or at least stationary), a procedure known as *Configuration Interaction* (CI). The MOs used for building the excited Slater determinants are taken from a Hartree–Fock calculation and held fixed. Subscripts S, D, T etc. indicate determinants which are singly, doubly, triply etc. excited relative to the

HF configuration.

$$\Psi_{CI} = a_0 \Phi_{SCF} + \sum_S a_S \Phi_S + \sum_D a_D \Phi_D + \sum_T a_T \Phi_T \ldots = \sum_{i=0} a_i \Phi_i \qquad (4.2)$$

This is an example of a constrained optimization, the energy should be minimized under the constraint that the total CI wave function is normalized. Introducing a Lagrange multiplier (Section 14.6), this can be written as

$$L = \langle \Psi_{CI} | \mathbf{H} | \Psi_{CI} \rangle - \lambda [\langle \Psi_{CI} | \Psi_{CI} \rangle - 1] \qquad (4.3)$$

The first bracket is the energy of the CI wave function, the second bracket is the norm of the wave function. In terms of determinants (eq. (4.2)), these can be written as

$$\langle \Psi_{CI} | \mathbf{H} | \Psi_{CI} \rangle = \sum_{i=0} \sum_{j=0} a_i a_j \langle \Phi_i | \mathbf{H} | \Phi_j \rangle = \sum_{i=0} a_i^2 E_i + \sum_{i=0} \sum_{j \neq i} a_i a_j \langle \Phi_i | \mathbf{H} | \Phi_j \rangle$$

$$\langle \Psi_{CI} | \Psi_{CI} \rangle = \sum_{i=0} \sum_{j=0} a_i a_j \langle \Phi_i | \Phi_j \rangle = \sum_{i=0} a_i^2 \langle \Phi_i | \Phi_i \rangle = \sum_{i=0} a_i^2 \qquad (4.4)$$

The diagonal elements in the sum involving the Hamilton operator are energies of the corresponding determinants. The overlap elements between different determinants are zero as they are built from orthogonal MOs (eq. (3.20)). The variational procedure corresponds to setting all the derivatives of the Lagrange function (4.3) with respect to the a_i expansion coefficients equal to zero.

$$\frac{\partial L}{\partial a_i} = 2 \sum_j a_j \langle \Phi_i | \mathbf{H} | \Phi_j \rangle - 2 \lambda a_i = 0$$

$$a_i (\langle \Phi_i | \mathbf{H} | \Phi_i \rangle - \lambda) + \sum_{j \neq i} a_j \langle \Phi_i | \mathbf{H} | \Phi_j \rangle = 0 \qquad (4.5)$$

$$a_i (E_i - \lambda) + \sum_{j \neq i} a_j \langle \Phi_i | \mathbf{H} | \Phi_j \rangle = 0$$

If there is only one determinant in the expansion ($a_0 = 1$), the last equation shows that the Lagrange multiplier is the (CI) energy, $\lambda = E$.

As there is one equation (4.5) for each i, the variational problem is transformed into solving a set of CI *secular equations*. Introducing the notation $H_{ij} = \langle \Phi_i | \mathbf{H} | \Phi_j \rangle$ the matrix equation becomes

$$\begin{pmatrix} H_{00} - E & H_{01} & \ldots & H_{0j} & \ldots \\ H_{10} & H_{11} - E & \ldots & H_{1j} & \ldots \\ \ldots & \ldots & \ldots & \ldots & \ldots \\ H_{j0} & \ldots & \ldots & H_{jj} - E & \ldots \\ \ldots & \ldots & \ldots & \ldots & \end{pmatrix} \begin{pmatrix} a_0 \\ a_1 \\ .. \\ a_j \\ .. \end{pmatrix} = \begin{pmatrix} 0 \\ 0 \\ .. \\ 0 \\ .. \end{pmatrix} \qquad (4.6)$$

which in shorthand notation may be written as $(\mathbf{H} - E\mathbf{I}) \, \mathbf{a} = \mathbf{0}$ or as $\mathbf{H}\mathbf{a} = E\mathbf{a}$. Solving the secular equations is equivalent to diagonalizing the CI matrix, see Chapter 13. The CI energy is obtained as the lowest eigenvalue of the CI matrix, and the corresponding eigenvector contains the a_i coefficients in front of the determinants in eq. (4.2). The second lowest eigenvalue corresponds to the first excited state etc.

4.2.1 CI Matrix Elements

The CI matrix elements H_{ij} can be evaluated by the strategy employed for calculating the energy of a single determinant used for deriving the Hartree–Fock equations, Section 3.3. This involves expanding the determinants in a sum of products of MOs, thereby making it possible to express the CI matrix elements in terms of MO integrals. There are, however, some general features which make many of the CI matrix elements equal to zero.

The Hamilton operator (eq. (3.23)) does not contain spin, thus if two determinants have different total spin the corresponding matrix element is zero. This situation occurs if an electron is excited from an α spin-MO to a β spin-MO, such as the second S-type determinant in Figure 4.1. When the HF wave function is a singlet, this excited determinant is (part of) a triplet. The corresponding CI matrix element can be written in terms of integrals over MOs, and the spin dependence can be separated out. If there are different numbers of α and β spin-MOs, there will always be at least one integral $\langle \alpha | \beta \rangle = 0$. That matrix elements between different spin states are zero may be fairly obvious. If we are interested in a singlet wave function, only singlet determinants can enter the expansion with non-zero coefficients. However, if the Hamilton operator includes for example the spin–orbit operator, matrix elements between singlet and triplet determinants are not necessarily zero, and the resulting CI wave function will be a mixture of singlet and triplet states.

Consider now the case where an electron with α spin is moved from orbital i to orbital a. The first S-type determinant in Figure 4.1 is of this type. Alternatively, the electron with β spin could be moved from orbital i to orbital a. Both of these excited determinants will have an S_z value of 0, but neither are eigenfunctions of the $\mathbf{S}^2$ operator. The difference and sum of these two determinants describe a singlet state and the $S_z = 0$ component of a triplet (which depends on the exact definition of the determinants).

Such linear combinations of determinants, which are proper spin eigenfunctions, are called *Spin-Adapted Configurations* (SAC) or *Configurational State Functions* (CSF). For higher excited states construction of proper CSFs may involve several determinants. The first D-type excitation in Figure 4.1, for example, must be combined with five other determinants corresponding to rearrangement of the electron spins to make a singlet CSF (actually there are two linearly independent CSFs that can be made). The second D-

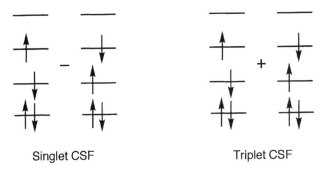

Singlet CSF   Triplet CSF

Figure 4.3 Forming configurational state functions from Slater determinants

type determinant in Figure 4.1 is already a proper CSF. By making suitable linear combinations of determinants the number of non-zero CI matrix elements can therefore be reduced.

If the system contains symmetry, there are additional CI matrix elements which become zero. The symmetry of a determinant is given as the direct product of the symmetries of the MOs. The Hamilton operator always belongs to the totally symmetric representation, thus if two determinants belong to different irreducible representations, the CI matrix element is zero. This is again fairly obvious: if the interest is in a state of a specific symmetry, only those determinants which have the correct symmetry can contribute.

The Hamilton operator consists of a sum of one-electron and two-electron operators, eq. (3.24). If two determinants differ by more than two (spatial) MOs there will always be an overlap integral between two different MOs which is zero (same argument as in eq. (3.28)). CI matrix elements can therefore only be non-zero if the two determinants differ by 0, 1, or 2 MOs, and they may be expressed in terms of integrals of one- and two-electron operators over MOs. These connections are known as the *Slater–Condon* rules. If the two determinants are identical, the matrix element is simply the energy of a single determinant wave function, as given by eq. (3.32). For matrix elements between determinants differing by 1 (exciting an electron from orbital i to a) or 2 (exciting two electrons from orbitals i and j to orbitals a and b) MOs, the results can be shown to be (compare with eq. (3.32), where the **g** operator is implicit in the notation for the two-electron integrals (eq. (3.56)):

$$\langle \Phi_0 | \mathbf{H} | \Phi_i^a \rangle = \langle \phi_i | \mathbf{h} | \phi_a \rangle + \sum_j (\langle \phi_i \phi_j | \phi_a \phi_j \rangle - \langle \phi_i \phi_j | \phi_j \phi_a \rangle)$$

$$\langle \Phi_0 | \mathbf{H} | \Phi_{ij}^{ab} \rangle = \langle \phi_i \phi_j | \phi_a \phi_b \rangle - \langle \phi_i \phi_j | \phi_b \phi_a \rangle$$

(4.7)

The matrix element between the HF and a singly excited determinant is a matrix element of the Fock operator between two different MOs (eq. (3.36)).

$$\langle \phi_i | \mathbf{h} | \phi_a \rangle + \sum_j (\langle \phi_i \phi_j | \phi_a \phi_j \rangle - \langle \phi_i \phi_j | \phi_j \phi_a \rangle) = \langle \phi_i | \mathbf{F} | \phi_a \rangle$$

(4.8)

This is an occupied-virtual off-diagonal element of the Fock matrix in the MO basis, and is identical to the gradient of the energy with respect to an occupied-virtual mixing parameter (except for a factor of 4), see eq. (3.67). If the determinants are constructed from optimized canonical HF MOs, the gradient is zero, and the matrix element is zero. This may also be realized by noting that the MOs are eigenfunctions of the Fock operator, eq. (3.41).

$$\mathbf{F}\phi_a = \varepsilon_a \phi_a$$

$$\langle \phi_i | \mathbf{F} | \phi_a \rangle = \varepsilon_a \langle \phi_i | \phi_a \rangle = \varepsilon_a \delta_{ia}$$

(4.9)

The disappearance of matrix elements between the HF reference and singly excited states is known as *Brillouins theorem*. The HF reference state therefore only has non-zero matrix elements with doubly excited determinants, and the full CI matrix acquires a block diagonal structure.

In order to evaluate the CI matrix elements one- and two-electron integrals over MOs are needed. These can be expressed in terms of the corresponding AO integrals and the

CI matrix	Φ_{HF}	Φ_S	Φ_D	Φ_T	Φ_Q	Φ_Q	...
Φ_{HF}	E_{HF}	0		0	0	0	0
Φ_S	0				0	0	0
Φ_D						0	0
Φ_T	0						0
Φ_Q	0	0					
Φ_Q	0	0	0				
...	0	0	0	0			

Figure 4.4 Structure of the CI matrix

MO coefficients.

$$\langle \phi_i | \mathbf{h} | \phi_j \rangle = \sum_{\alpha}^{M} \sum_{\beta}^{M} c_{\alpha i} c_{\beta j} \langle \chi_\alpha | \mathbf{h} | \chi_\beta \rangle$$

$$\langle \phi_i \phi_j | \phi_k \phi_l \rangle = \sum_{\alpha}^{M} \sum_{\beta}^{M} \sum_{\gamma}^{M} \sum_{\delta}^{M} c_{\alpha i} c_{\beta j} c_{\gamma k} c_{\delta l} \langle \chi_\alpha \chi_\beta | \chi_\gamma \chi_\delta \rangle$$

(4.10)

Such MO integrals are required for all electron correlation methods. The two-electron AO integrals are the most numerous and the above equation appears to involve a computational effect proportional to M^8 (M^4 AO integrals each multiplied by four sets of M MO coefficients). However, by performing the transformation one index at a time, the computational effort can be reduced to M^5.

$$\langle \phi_i \phi_j | \phi_k \phi_l \rangle = \sum_{\alpha} c_{\alpha i} \left(\sum_{\beta} c_{\beta j} \left(\sum_{\gamma} c_{\gamma k} \left(\sum_{\delta} c_{\delta l} \langle \chi_\alpha \chi_\beta | \chi_\gamma \chi_\delta \rangle \right) \right) \right) \quad (4.11)$$

Each step now only involves multiplication of M^4 integrals with M coefficients, i.e. the M^8 dependence is reduced to four M^5 operations. In the large basis set limit, all electron correlation methods formally scale as at least M^5, since this is the scaling for the AO to MO integral transformation. The transformation is an example of a "rotation" of the "coordinate" system consisting of the AOs, to one where the Fock operator is diagonal, the MOs, see Chapter 13. The diagonal system allows a much more compact representation of the matrix elements needed for the electron correlation treatment. The coordinate change is also known as a *four index transformation*, since it involves four indices associated with the basis functions.

4.2.2 Size of the CI Matrix

What is a typical size of the CI matrix? Consider a small system, H_2O with a 6-31G(d) basis. For the purpose of illustration, let us for a moment return to the spinorbital description. There are 10 electrons and 38 spin-MOs, of which 10 are occupied and 28 are empty. There are $K_{10,n}$ possible ways of selecting n electrons out of the 10 occupied orbitals, and $K_{28,n}$ ways of distributing them in the 28 empty orbitals. The number of excited states for a given excitation level is thus $K_{10,n} \cdot K_{28,n}$, and the total number of

Table 4.1 Number of singlet CSFs as a function of excitation level for H_2O with a 6-31G(d) basis

Excitation level n	Number of nth excited CSFs	Total number of CSFs
1	71	71
2	2 485	2 556
3	40 040	42 596
4	348 530	391 126
5	1 723 540	2 114 666
6	5 033 210	7 147 876
7	8 688 680	15 836 556
8	8 653 645	24 490 201
9	4 554 550	29 044 751
10	1 002 001	30 046 752

excited determinants will be a sum over 10 such terms. This is also equivalent to $K_{38,10}$, the total number of ways 10 electrons can be distributed in 38 orbitals.

$$\text{Number of SDs} = \sum_{n=0}^{10} K_{10,n} \cdot K_{28,n} = K_{38,10} = \frac{38!}{10! \cdot (38 - 10)!} \tag{4.12}$$

The number of excited determinants thus grows <u>factorially</u> with the size of the basis set. Many of these excited determinants will of course have different spin multiplicity (triplet, quintet etc. states for a singlet HF determinant), and can therefore be left out in the calculation. Generating only the singlet CSFs, the number of configurations at each excitation level is shown in Table 4.1.

The number of determinants (or CSFs) that can be generated grows wildly with the excitation level! Even if the C_{2v} symmetry of H_2O is employed, there are still a total of 7 536 400 singlet CSFs with A_1 symmetry. If all possible determinants are included we have a *full CI* wave function, there is no truncation in the many-electron expansion besides that generated by the finite one-electron expansion (size of the basis set). This is the best possible wave function within the limitations of the basis set, i.e. it recovers 100% of the electron correlation in the given basis. In this case it corresponds to diagonalizing a matrix of size 30 046 752 × 30 046 752. This is impossible. However, normally the interest is only in the lowest (or a few of the lowest) eigenvalue(s) and -vector(s), and there are special iterative methods (Section 4.2.4) for determining one (or a few) eigenvector(s) of a large matrix.

In the general case of N electrons and M basis functions the total number of singlet CSFs that can be generated is given by

$$\text{Number of CSFs} = \frac{M!(M + 1)!}{\left(\frac{N}{2}\right)! \left(\frac{N}{2} + 1\right)! \left(M - \frac{N}{2}\right)! \left(M - \frac{N}{2} + 1\right)!} \tag{4.13}$$

For H_2O with the above 6-31G(d) basis there are $\sim 30 \times 10^6$ CSFs ($N = 10$, $M = 19$); with the larger 6-311G(2d,2p) basis there are $\sim 106 \times 10^9$ CSFs ($N = 10$, $M = 41$). For $H_2C{=}CH_2$ with the 6-31G(d) basis there are $\sim 334 \times 10^{12}$ CSFs ($N = 16$, $M = 38$).

One of the recent large scale full CI calculations considered H_2O in a DZP type basis with 24 functions.[2] Allowing all possible excitations of the 10 electrons generates 451 681 246 determinants. The variational wave function thus contains roughly half a billion parameters, i.e. the formal size of the CI matrix is of the order of half a billion squared. Although a determination of the lowest eigenvalue of such a problem can be done in a matter of hours on a modern computer, the result is a single number, the ground-state energy of the H_2O molecule. Owing to basis set limitations, however, it is still some 0.2 a.u. (~ 125 kcal/mol) larger than the experimental value. The computational effort for extracting a single eigenvalue and -vector scales essentially linearly with the number of CSFs, and it is possible to handle systems with up to a few billion determinants. The factorial growth of the number of determinants with the size of the basis set, however, makes the full CI method infeasible for all but the very smallest systems. Full CI calculations are thus not a routine computational procedure for including electron correlation, but they are useful references for developing more approximate methods, as full CI gives the best result that can be obtained in a given basis.

4.2.3 Truncated CI Methods

In order to develop a computationally tractable model, the number of excited determinants in the CI expansion (4.2) must be reduced. Truncating the excitation level at 1 (CI with Singles, CIS) does not give any improvement over the HF result as all matrix elements between the HF wave function and singly excited determinants are zero. CIS is equal to HF for the ground-state energy, although higher roots from the secular equations may be used as approximations to excited states. It has already been mentioned that only doubly excited determinants have matrix elements with the HF wave function different from zero, thus the lowest CI level which gives an improvement over the HF result is that which includes only doubly excited states, yielding the CID (CI with Doubles) model. Compared to the number of doubly excited determinants there are relatively few singly excited determinants (see for example Table 4.1), including these gives the CISD method. Computationally this is only a marginal increase in effort over CID. Although the singly excited determinants have zero matrix elements with the HF reference, they enter the wave function indirectly as they have non-zero matrix elements with the doubly excited determinants. In the large basis set limit the CISD method scales as M^6. The next level in improvement is inclusion of the triply excited determinants, giving the CISDT method, which is an M^8 method. Taking into account also quadruply excited determinants yields the CISDTQ method which is an M^{10} method. As shown below, the CISDTQ model in general gives results close to the full CI limit, but even truncating the excitation level at 4 produces so many configurations that it can only be applied to small molecules and small basis sets. The only CI method which is generally applicable for a large variety of systems is CISD. For computationally feasible systems (i.e. medium size molecules and basis sets), it typically recovers 80–90% of the available correlation energy. The percentage is highest for small molecules. As the molecule gets larger the CISD method recovers less and less of the correlation energy, as discussed in Section 4.5.

Since only doubly excited determinants have non-zero matrix elements with the HF state, these are the most important. This may be illustrated by considering a full CI

Table 4.2 Weights of excited configurations for the Neon atom

Excitation level	Weight
0	0.9644945073
1	0.0009804929
2	0.0336865893
3	0.0003662339
4	0.0004517826
5	0.0000185090
6	0.0000017447
7	0.0000001393
8	0.0000000011

calculation for the Ne atom in a [5s,4p,3d] basis, where the 1s electrons are omitted from the correlation treatment.[3] The contribution to the full CI wave function from each level of excitation is given in Table 4.2.

The weight is the sum of a_i^2 coefficients at the given excitation level, eq. (4.2). The CI method determines the coefficients from the variational principle, thus Table 4.2 shows that the doubly excited determinants are by far the most important in terms of energy. The singly excited determinants are the second most important, then follow the quadruples and triples. Excitations higher than 4 make only very small contributions, although there are actually many more of these highly excited determinants than the triples and quadruples, as illustrated in Table 4.1.

The relative importance of the different excitations may qualitatively be understood by noting that the doubles provide electron correlation for electron pairs. Quadruply excited determinants are important as they primarily correspond to products of double excitations. The singly excited determinants allow inclusion of multi-reference character in the wave function, i.e. they allow the orbitals to "relax". Although the HF orbitals are optimum for the single determinant wave function, that is no longer the case when many determinants are included. The triply excited determinants are doubly excited relative to the singles, and can then be viewed as providing correlation for the "multi-reference" part of the CI wave function.

While singly excited states make relatively small contributions to the correlation energy of a CI wave function, they are very important when calculating properties (Chapter 10). Molecular properties measure how the wave function changes when a perturbation, such as an external electric field, is added. The change in the wave function introduced by the perturbation makes the MOs no longer optimal in the variational sense. The first-order change in the MOs is described by the off-diagonal elements in the Fock matrix, these are essentially the gradient of the HF energy with respect to the MOs (eq. 3.67). In the absence of a perturbation, these are zero; the HF energy is stationary with respect to an orbital variation (eq. (3.38)). As shown in eqs. (4.7) and (4.8), the Fock matrix off-diagonal elements are CI matrix elements between the HF and singly excited states. For molecular properties, the singly excited states thus allow the CI wave function to "relax" the MOs, i.e. letting the wave function respond to the perturbation.

4.2.4 Direct CI methods

As illustrated above, even quite small systems at the CISD level results in millions of CSFs. The variational problem is to extract one or possibly a few of the lowest eigenvalues and -vectors of a matrix the size of millions squared. This cannot be done by standard diagonalization methods where all the eigenvalues are found. There are, however, iterative methods for extracting one, or a few, eigenvalues and -vectors of a large matrix. The CI problem eq. (4.6) may be written as

$$(\mathbf{H} - E\mathbf{I})\mathbf{a} = \mathbf{0} \qquad (4.14)$$

The $\mathbf{H}$ matrix contains the matrix element between the CSFs in the CI expansion, and the $\mathbf{a}$ vector the expansion coefficients. The idea in iterative methods is to generate a suitable guess for the coefficient vector and calculate $(\mathbf{H} - E\,\mathbf{I})\mathbf{a}$. This will in general not be zero, and the deviation may be used for adding a correction to $\mathbf{a}$, forming an iterative algorithm. If the interest is in the lowest eigenvalue, a suitable start eigenvector may be one which contains only the HF configuration, i.e. $\{1,0,0,0,\ldots\}$. Since the $\mathbf{H}$ matrix elements essentially are two-electron integrals in the MO basis (eq. (4.7)), the iterative procedure may be formulated as integral driven, i.e. a batch of integrals are read in (or generated otherwise) and used directly in the multiplication with the corresponding $\mathbf{a}$-coefficients. The CI matrix is therefore not needed explicitly, only the effect of its multiplication with a vector containing the variational parameters, and storage of the entire matrix is avoided. This is the basis for being able to handle CI problems of almost any size, and is known as *direct CI*. Note that it is not "direct" in the sense used to describe the direct SCF method, where all the AO integrals are calculated as needed. The direct CI approach just assumes that the CI matrix elements (e.g. two-electron integrals in the MO basis) are available as required, traditionally stored in a file on a disk. There are several variations on how the $\mathbf{a}$ vector is adjusted in each iteration, the most commonly used versions are based on the *Davidson algorithm*.[4]

4.3 Illustrating how CI Accounts for Electron Correlation, and the RHF Dissociation Problem

Consider the H_2 molecule in a minimum basis consisting of one s-function on each centre, χ_A and χ_B. A RHF calculation will produce two MOs, ϕ_1 and ϕ_2, being the sum and difference of the two AOs. The sum of the two AOs is a bonding MO, with increased probability of finding the electrons between the two nuclei, while the difference is an antibonding MO, with decreased probability of finding the electrons between the two nuclei.

The HF wave function will have two electrons in the lowest energy (bonding) MO.

$$
\begin{aligned}
\phi_1 &= N_1(\chi_A + \chi_B) \\
\phi_2 &= N_2(\chi_A - \chi_B) \\
\Phi_0 &= \begin{vmatrix} \phi_1(1)\bar{\phi}_1(1) \\ \phi_1(2)\bar{\phi}_1(2) \end{vmatrix}
\end{aligned}
\qquad (4.15)
$$

Here N_1 and N_2 are suitable normalization constants, and the bar above the MO indicates that the electron has a β spin function, no bar indicates an α spin function. In

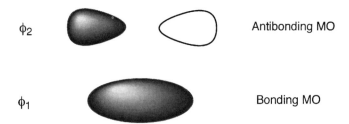

ϕ_2 Antibonding MO

ϕ_1 Bonding MO

Figure 4.5 Molecular orbitals for H_2

this basis there are one doubly (Φ_1) and four singly excited Slater determinants (Φ_{2-5}).

$$\Phi_1 = \begin{vmatrix} \phi_2(1) & \bar{\phi}_2(1) \\ \phi_2(2) & \bar{\phi}_2(2) \end{vmatrix}$$

$$\Phi_2 = \begin{vmatrix} \phi_1(1) & \bar{\phi}_2(1) \\ \phi_1(2) & \bar{\phi}_2(2) \end{vmatrix}$$

$$\Phi_3 = \begin{vmatrix} \bar{\phi}_1(1) & \phi_2(1) \\ \bar{\phi}_1(2) & \phi_2(2) \end{vmatrix} \tag{4.16}$$

$$\Phi_4 = \begin{vmatrix} \phi_1(1) & \phi_2(1) \\ \phi_1(2) & \phi_2(2) \end{vmatrix}$$

$$\Phi_5 = \begin{vmatrix} \bar{\phi}_1(1) & \bar{\phi}_2(1) \\ \bar{\phi}_1(2) & \bar{\phi}_2(2) \end{vmatrix}$$

Configurations Φ_4 and Φ_5 are clearly the $S_z = 1$ and $S_z = -1$ components of a triplet state. The plus combination of Φ_2 and Φ_3 is the $S_z = 0$ component of the triplet, and the minus combination is a singlet configuration, Figure 4.3. The H_2 molecule belongs to the $D_{\infty h}$ point group, and the two MOs transform as the $\sigma_g(\phi_1)$ and $\sigma_u(\phi_2)$ representations. The singly excited CSF ($\Phi_2 - \Phi_3$) has overall Σ_u symmetry, while the HF (Φ_0) and doubly excited determinants (Φ_1) have Σ_g. The full 6×6 CI problem therefore blocks into a 2×2 block of singlet Σ_g states, a 1×1 block of singlet Σ_u, and a 3×3 block of triplet Σ_u states, Figure 4.6. Owing to orthogonality of the spin functions the triplet block is already diagonal.

The full CI for the $^1\Sigma_g$ states involves only two configurations, the reference HF and the doubly excited determinant.

$$\Phi_0 = \phi_1(1)\bar{\phi}_1(2) - \bar{\phi}_1(1)\phi_1(2) = \phi_1\phi_1(\alpha\beta - \beta\alpha)$$
$$\Phi_1 = \phi_2(1)\bar{\phi}_2(2) - \bar{\phi}_2(1)\phi_2(2) = \phi_2\phi_2(\alpha\beta - \beta\alpha) \tag{4.17}$$

In eq. (4.17) the electron coordinate is given implicitly by the order in which the orbitals are written, i.e. $\phi_1\phi_1[\alpha\beta - \beta\alpha] = \phi_1(1)\phi_1(2)[\alpha(1)\beta(2) - \beta(1)\alpha(2)]$. Ignoring the spin functions (which may be integrated out since **H** is spin independent) and normalization, the determinants can be expanded in AOs.

$$\Phi_0 = (\chi_A(1) + \chi_B(1))(\chi_A(2) + \chi_B(2)) = \chi_A\chi_A + \chi_B\chi_B + \chi_A\chi_B + \chi_B\chi_A$$
$$\Phi_1 = (\chi_A(1) - \chi_B(1))(\chi_A(2) - \chi_B(2)) = \chi_A\chi_A + \chi_B\chi_B - \chi_A\chi_B - \chi_B\chi_A$$

$$\tag{4.18}$$

	$^1\Phi_0(\textstyle\sum_g)$	$^1\Phi_1(\textstyle\sum_g)$	$^1(\Phi_2-\Phi_3)(\textstyle\sum_u)$	$^3\Phi_4(\textstyle\sum_u)$	$^3(\Phi_2+\Phi_3)(\textstyle\sum_u)$	$^3\Phi_5(\textstyle\sum_u)$
$^1\Phi_0(\textstyle\sum_g)$			0	0	0	0
$^1\Phi_1(\textstyle\sum_g)$			0	0	0	0
$^1(\Phi_2-\Phi_3)(\textstyle\sum_u)$	0	0		0	0	0
$^3\Phi_4(\textstyle\sum_u)$	0	0	0		0	0
$^3(\Phi_2+\Phi_3)(\textstyle\sum_u)$	0	0	0	0		0
$^3\Phi_5(\textstyle\sum_u)$	0	0	0	0	0	

Figure 4.6 Structure of the full CI matrix for the H_2 system in a minimum basis

The first two terms on the right-hand side have both electrons on the same centre, they describe *ionic* contributions to the wave function, H^+H^-. The later two terms describe *covalent* contributions to the wave function, $H^{\cdot}H^{\cdot}$. The HF wave function thus contains equal amounts of ionic and covalent contributions. The full CI wave function may be written in terms of AOs as

$$\Psi_{CI} = a_0\Phi_0 + a_1\Phi_1 = a_0(\phi_1\phi_1) + a_1(\phi_2\phi_2)$$
$$\Psi_{CI} = (a_0 + a_1)(\chi_A\chi_A + \chi_B\chi_B) + (a_0 - a_1)(\chi_A\chi_B + \chi_B\chi_A) \tag{4.19}$$

The optimum values of the a_0 and a_1 coefficients are determined by the variational procedure. The HF wave function constrains both electrons to move in the same bonding orbital. By allowing the doubly excited state to enter the wave function, the electrons can better avoid each other, as the antibonding MO now is also available. The antibonding MO has a nodal plane (where $\phi_2 = 0$) perpendicular to the molecular axis (Figure 4.5), and the electrons are able to correlate their movements by being on opposite sides of this plane. This *left–right* correlation is a molecular equivalent of the atomic radial correlation discussed in Section 5.2.

Consider now the behaviour of the HF wave function Φ_0 (eq. (4.18)) as the distance between the two nuclei is increased toward infinity. Since the HF wave function is an equal mixture of ionic and covalent terms, the dissociation limit is 50% H^+H^- and 50% $H^{\cdot}H^{\cdot}$. In the gas phase all bonds dissociate homolytically, and the ionic contribution should be 0%. The HF dissociation energy is therefore much too high. This is a general problem of RHF type wave functions, the constraint of doubly occupied MOs is inconsistent with breaking bonds to produce radicals. In order for an RHF wave function to dissociate correctly, an even-electron molecule must break into two even-electron fragments, each being in the lowest electronic state. Furthermore, the orbital symmetries must match. There are only a few covalently bonded systems which obey these requirements (the simplest example is HHe^+). The wrong dissociation limit for RHF wave functions has several consequences.

(1) The energy for stretched bonds is too high. Most transition structures have partly formed/broken bonds, thus activation energies are too high at the RHF level.

(2) The excessively steep increase in energy as a function of the bond length causes the minimum on a potential energy curve to occur too "early" for covalently bonded systems, and equilibrium bond lengths are too short at the RHF level.
(3) The excessively steep increase in energy as a function of the bond length causes the curvature of the potential energy surface near the equilibrium to be too large, and vibrational frequencies, especially those describing bond stretching, are in general too high.
(4) The wave function contains too much "ionic" character, and RHF dipole moments (and also atomic charges) are in general too large.

It should be noted that dative bonds, like metal complexes and charge transfer species, in general have RHF wave functions which dissociate correctly, and the equilibrium bond lengths in these cases are normally too long.

The dissociation problem is solved in the case of a full CI wave function. As seen from eq. (4.19), the ionic term can be made to disappear by setting $a_1 = -a_0$. The full CI wave function generates the lowest possible energy (within the limitations of the chosen basis set) at all distances, with the optimum weights of the HF and doubly excited determinants determined by the variational principle. In the general case of a polyatomic molecule and a large basis set, correct dissociation of all bonds can be achieved if the CI wave function contains all determinants generated by a full CI in the valence orbital space. The latter corresponds to a full CI if a minimum basis is employed, but is much smaller than a full CI if an extended basis is used.

4.4   The UHF Dissociation, and the Spin Contamination Problem

The dissociation problem can also be "solved" by using a wave function of the UHF type. Here the α and β bonding MOs are allowed to "localize", thereby reducing the MO symmetries to $C_{\infty v}$.

$$\phi_1 = N(\chi_A + c\chi_B)\alpha$$
$$\bar{\phi}_1 = N(c\chi_A + \chi_B)\beta$$
$$\Phi_0^{UHF} = \begin{vmatrix} \phi_1(1) & \bar{\phi}_1^{\dagger}(1) \\ \phi_1(2) & \bar{\phi}_1(2) \end{vmatrix} \tag{4.20}$$

The optimum value of c is determined by the variational principle. If $c = 1$, the UHF wave function is identical to RHF. This will normally be the case near the equilibrium distance. As the bond is stretched, the UHF wave function allows each of the electrons to localize on a nucleus; c goes towards 0. The point where the RHF and UHF descriptions start to differ is often referred to as the RHF/UHF *instability* point. This is an example of symmetry breaking, as discussed in Section 3.8.3. The UHF wave function correctly dissociates into two hydrogen atoms, however, the symmetry breaking of the MOs has two other, closely connected, consequences: introduction of electron correlation and spin contamination. To illustrate these concepts, we need to look at the Φ_0 UHF determinant, and the six RHF determinants in eqs. (4.15) and (4.16) in more detail. We will again ignore all normalization constants.

The six RHF determinants can be expanded in terms of the AOs:

$$
\begin{aligned}
\Phi_0 &= [\chi_A\chi_A + \chi_B\chi_B + \chi_A\chi_B + \chi_B\chi_A](\alpha\beta - \beta\alpha) \\
\Phi_1 &= [\chi_A\chi_A + \chi_B\chi_B - \chi_A\chi_B - \chi_B\chi_A](\alpha\beta - \beta\alpha) \\
\Phi_2 &= [\chi_A\chi_A - \chi_B\chi_B](\alpha\beta - \beta\alpha) - [\chi_A\chi_B - \chi_B\chi_A](\alpha\beta + \beta\alpha) \\
\Phi_3 &= [\chi_A\chi_A - \chi_B\chi_B](\alpha\beta - \beta\alpha) + [\chi_A\chi_B - \chi_B\chi_A](\alpha\beta + \beta\alpha) \\
\Phi_4 &= [\chi_A\chi_B - \chi_B\chi_A](\alpha\alpha) \\
\Phi_5 &= [\chi_A\chi_B - \chi_B\chi_A](\beta\beta)
\end{aligned}
\tag{4.21}
$$

Subtracting and adding Φ_2 and Φ_3 produces a pure singlet ($^1\Phi_-$), and the $S_z = 0$ component of the triplet ($^3\phi_+$), wave functions.

$$
\begin{aligned}
{}^1\Phi_- &= \Phi_2 - \Phi_3 = [\chi_A\chi_A - \chi_B\chi_B](\alpha\beta - \beta\alpha) \\
{}^3\Phi_+ &= \Phi_2 + \Phi_3 = [\chi_A\chi_B - \chi_B\chi_A](\alpha\beta + \beta\alpha)
\end{aligned}
\tag{4.22}
$$

Performing the expansion of the Φ_0^{UHF} determinant (4.20) gives

$$
\begin{aligned}
\Phi_0^{\mathrm{UHF}} &= c[\chi_A\chi_A + \chi_B\chi_B](\alpha\beta - \beta\alpha) \\
&\quad + [\chi_A\chi_B\alpha\beta - c^2\chi_A\chi_B\beta\alpha] \\
&\quad + [c^2\chi_B\chi_A\alpha\beta - \chi_B\chi_A\beta\alpha]
\end{aligned}
\tag{4.23}
$$

Adding and subtracting factors of $\chi_A\chi_B\beta\alpha$ and $\chi_B\chi_A\alpha\beta$ allow this to be written as

$$
\begin{aligned}
\Phi_0^{\mathrm{UHF}} &= [c(\chi_A\chi_A + \chi_B\chi_B) + (\chi_A\chi_B + \chi_B\chi_A)](\alpha\beta - \beta\alpha) \\
&\quad + (1 - c^2)[\chi_A\chi_B\beta\alpha - \chi_B\chi_A\alpha\beta]
\end{aligned}
\tag{4.24}
$$

Since $0 \le c \le 1$ the first term shows that UHF orbitals reduce the ionic contribution relative to the covalent structures, compared to the RHF case, eq. (4.18). This is the same effect as for the CI procedure (eq. (4.19)), i.e. the first term shows that the UHF wave function partly includes electron correlation. The first term can be written as a linear combination of the Φ_0 and Φ_1 determinants, and describes a pure singlet state. The last part of the UHF determinant, however, has terms identical to two of those in the triplet $^3\Phi_+$ combination (4.22). If we had chosen the alternative set of UHF orbitals with the alpha spin being primarily on centre B in eq. (4.20), we would have obtained the other two terms in $^3\Phi_+$, i.e. the last term in (4.24) breaks the symmetry. The UHF determinant is therefore not a pure spin state, it contains both singlet and triplet spin states. This feature is known as *spin contamination*. For $c = 1$ the UHF wave function is identical to the RHF, and Φ_0^{UHF} is a pure singlet. For $c = 0$ the UHF wave function only contains the covalent terms, which is the correct dissociation behaviour, but also contains equal amounts of singlet and triplet character. When the bond distance is very large, the singlet and triplet states have identical energies, and the spin contamination has no consequence for the energy. In the intermediate region where the bond is not completely broken, however, spin contamination is important. Compared to full CI, the UHF energy is too high as the higher-lying triplet state is mixed into the wave function. The variational principle guarantees that the UHF energy is lower than or equal to the RHF energy (there are more variational parameters). The full CI energy is the lowest possible (for the given basis set) as it recovers 100% of the correlation energy. The UHF

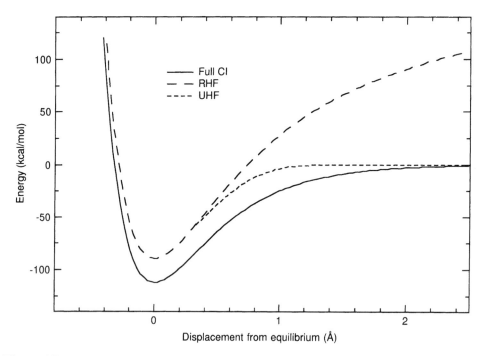

Figure 4.7   Bond dissociation curves for H_2

wave function thus lowers the energy by introducing some electron correlation, but at the same time raises the energy by including higher energy spin states. At the single determinant level (UHF) the variational principle guarantees that the first effect dominates. If the second effect were dominating, the UHF would collapse to the RHF solution. The instability point can thus be viewed as the geometry where the correlation effect becomes larger than the spin contamination. Pictorially the dissociation curves look as shown in Figure 4.7.

Another way of viewing spin contamination is to write the UHF wave function as a linear combination of pure R(O)HF determinants, e.g. for a singlet state.

$$^1\Phi^{UHF} = a_1 \, {}^1\Phi^{RHF} + a_3 \, {}^3\Phi^{ROHF} + a_5 \, {}^5\Phi^{ROHF} + \ldots \qquad (4.25)$$

Since the UHF wave function is multi-determinantal in terms of R(O)HF determinants, it follows that it to some extent includes electron correlation (relative to the RHF reference).

The amount of spin contamination is given by the expectation value of the $\mathbf{S}^2$ operator, $\langle \mathbf{S}^2 \rangle$. The theoretical value for a pure spin state is $S_z(S_z + 1)$, i.e. 0 for a singlet ($S_z = 0$), 0.75 for a doublet ($S_z = 1/2$), 2.00 for a triplet ($S_z = 1$) etc. A UHF "singlet" wave function will contain some amounts of triplet, quintet etc. states, increasing the $\langle \mathbf{S}^2 \rangle$ value from its theoretical value of zero for a pure spin state. Similarly, a UHF "doublet" wave function will contain some amounts of quartet, sextet etc. states. Usually the contribution from the next higher spin state from the desired is

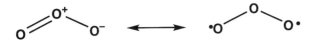

Figure 4.8   Resonance structures for ozone

the most important. The $\langle \mathbf{S}^2 \rangle$ value for a UHF wave function is operationally calculated from the spatial overlap between all pairs of α and β spinorbitals

$$\langle \mathbf{S}^2 \rangle = S_z(S_z + 1) + N_\beta - \sum_{ij}^{\text{MO}} \langle \phi_i^\alpha | \phi_j^\beta \rangle^2 \tag{4.26}$$

If the α and β orbitals are identical, there is no spin contamination, and the UHF wave function is identical to the RHF.

By including electron correlation in the wave function the UHF method introduces more biradical character into the wave function than RHF. The spin contamination part is also purely biradical in nature, i.e. a UHF treatment in general will overestimate the biradical character. Most singlet states are well described by a closed-shell wave function near the equilibrium geometry, and in those cases it is not possible to generate a UHF solution which has a lower energy than the RHF. There are systems, however, for which this does not hold. An example is the ozone molecule, where two types of resonance structure can be drawn, Figure 4.8.

The biradical resonance structure for ozone requires two singly occupied MOs, and it is clear that an RHF type wave function, which requires all orbitals to be doubly occupied, cannot describe this. A UHF type wave function, however, allows the α and β orbitals to be spatially different, and can to a certain extent incorporate both resonance structures. Systems with biradical character will often have a (singlet) UHF wave function different from an RHF.

As mentioned above, spin contamination in general increases as a bond is stretched. This has some important consequences for transition structures which contain elongated bonds. While most singlet systems have identical RHF and UHF descriptions near the equilibrium geometry, it will normally be possible to find a lower energy UHF solution in the TS region. However, since the spin contamination is not constant along the reaction coordinate, and since the UHF overestimates the biradical character, it is possible that the TS actually becomes a minimum on the UHF energy surface. In other words, the spin contamination may severely distort the shape of the potential energy surface. This may qualitatively be understood by considering the "singlet" UHF wave function as a linear combination of a singlet and a triplet state (eq. (4.25)), as shown in Figure 4.9.

The degree of mixing is determined by the energy difference between the pure singlet and triplet states (as shown for example by second-order perturbation theory, see Section 4.8), which in general decreases along the reaction coordinate. Even if the mixing is not large enough to actually transform a TS to a minimum, it is clear that the UHF energy surface will be much too flat in the TS region. Activation energies calculated at the UHF level will always be lower than the RHF value, but may be either higher or lower than the "correct" value, depending on the amount of spin contamination, since RHF normally overestimates activation energies.

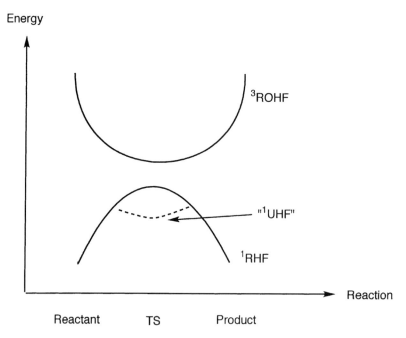

Figure 4.9 Mixing of pure singlet and triplet states may generate artificial minima on the UHF energy surface

    From the above it should be clear that UHF wave functions which are spin contaminated (more than a few percent deviation of $\langle \mathbf{S}^2 \rangle$ from the theoretical value of $S_z(S_z + 1)$) have disadvantages. For closed-shell systems an RHF procedure is therefore normally preferred. For open-shell systems, however, the UHF method has been heavily used. It is possible to use an ROHF type wave function for open-shell systems, but this leads to computational procedures which are somewhat more complicated than for the UHF case when electron correlation is introduced.
    The main problem with the UHF method is the spin contamination, and there have been several proposals on how to remove the unwanted states. There are three strategies which can be considered for removing the contamination: during the SCF procedure, after the SCF has converged, or after electron correlation has been added to the UHF solution. A popular method of removing unwanted states is to project them out with a suitable projection operator (in the picture of the wave function being described in the coordinate system consisting of determinants, the components of the wave function along the higher spin states are removed). As mentioned above, the next higher spin state is usually the most important, and in many cases it is a quite good approximation to only remove this state. After projection, the wave function is then renormalized. If only the first contaminant is removed, this may in extreme cases actually increase the $\langle \mathbf{S}^2 \rangle$, value. Performing the projection during the SCF procedure produces a wave function for which it is difficult to formulate a satisfactory theory for including electron correlation by means of perturbation or coupled cluster methods (Sections 4.8 and 4.9). Projections of the converged UHF wave function will lower the energy (although the PUHF energy is no longer variational), since the contributions of the higher-lying states are removed,

and only the correlation effect remains. However, the problems of artificial distortion of the energy surface is even more pronounced at the PUHF level, than with the UHF method itself. For example, it is often found that a false minimum is generated just after the RHF/UHF instability point on a bond dissociation curve. Furthermore, the derivatives of the PUHF energy are not continuous at the RHF/UHF instability point. Projection of the wave function after electron correlation has been added, however, turns out to be a viable pathway. This has mainly been used in connection with perturbation methods, to be described in Section 4.8.2.

4.5 Size Consistency and Size Extensivity

As mentioned above, full CI is impossible, except for very small systems. The only general applicable method is CISD. Consider now a series of CISD calculations in order to construct the interaction potential between two H_2 molecules as a function of the distance between them. Relative to the HF wave function, there will be determinants which correspond to single excitations on only one of the H_2 fragments (S-type determinants), single excitations on both (D-type determinants), and double excitations only on one of the H_2 fragments (also D-type determinants). This will be the case at all intermolecular distances, also when the separation is very large. In that case, however, the system is just two H_2 molecules, and we could consider calculating the energy instead as twice the energy of one H_2 molecule. A CISD calculation on one H_2 molecule would generate singly and doubly excited determinants, and multiplying this by two, would generate determinants which are quadruply excited for the combined H_4 system. A CISD calculation of two H_2 molecules separated by say 100 Å will not give the same energy as twice the results from a CISD calculation on one H_2 molecule (this will be lower). This problem is referred to a *Size Inconsistency*. A very similar, but not identical concept, is *Size Extensivity*. Size consistency is only defined if the two fragments are non-interacting (separated by say 100 Å) while size extensivity implies that the method scales properly with the number of particles, i.e. the fragments can be interacting (separated by say 5 Å). Full CI is size consistent (and extensive), but all forms of truncated CI are not. The lack of size extensivity is the reason why CISD recovers less and less electron correlation as the systems grow larger.

4.6 Multi-configuration Self-consistent Field

The *Multi-configuration Self-consistent Field* (MCSCF) method can be considered as a CI where not only the coefficients in front of the determinants are optimized by the variational principle, but also the MOs used for constructing the determinants are made optimum.[5] The MCSCF optimization is iterative just like the SCF procedure (if the "multi-configuration" is only one, it is simply HF). Since the number of MCSCF iterations required for achieving convergence tends to increase with the number of configurations included, the size of MCSCF wave function that can be treated is somewhat smaller than for CI methods.

When deriving the Hartree–Fock equations it was only required that the variation of the energy with respect to an orbital variation should be zero. This is equivalent to the first derivatives of the energy with respect to the MO expansion coefficients being equal to zero. The Hartree–Fock equations can be solved by an iterative SCF method, and there are many techniques for helping the iterative procedure to converge (Section 3.8). There is, however, no guarantee that the solution found by the SCF procedure is a <u>minimum</u> of the energy as a function of the MO coefficients. To determine that, the matrix of second derivatives of the energy with respect to the MO coefficients can be calculated and diagonalized. In order to be a minimum it should have all positive eigenvalues. In practice only the lowest eigenvalue is normally determined, if this is positive the solution is a minimum. This is rarely checked for SCF wave functions, in the large majority of cases the SCF procedure converges to a minimum without problems. MCSCF wave functions, on the other hand, are much harder to make converge, and much more prone to converge on solutions which are not minima. MCSCF wave function optimizations are therefore normally carried out by expanding the energy to second order in the variational parameters (orbital and configurational coefficients), analogously to the second-order SCF procedure described in Section 3.8.1, and using Newton–Raphson based methods described in Chapter 14 to force convergence to a minimum.

MCSCF methods are rarely used for calculating large fractions of the correlation energy. The orbital relaxation usually does not recover much electron correlation, it is more efficient to include additional determinants and keep the MOs fixed (CI) if the interest is just in obtaining a large fraction of the correlation energy. Single determinant HF wave functions normally give a qualitatively correct description of the electron structure; however, there are many examples where this is not the case. MCSCF methods can be considered as an extension of single determinant methods that gives a qualitatively correct description. Consider again the ozone molecule with the two resonance structures shown in Figure 4.8. Each type of resonance structure essentially translates into a different determinant. If more than one <u>non-equivalent</u> resonance structure is important, this means that the wave function cannot be described even qualitatively correctly at the RHF single determinant level (benzene, for example, has two <u>equivalent</u> cyclohexatriene resonance structures, and is adequately described by an RHF wave function). A UHF wave function allows some biradical character, with the disadvantages mentioned above. Alternatively, a second restricted type CSF (consisting of two determinants) with two singly occupied MOs may be included in the wave function. The simplest MCSCF for ozone will contain two configurations (often denoted TCSCF), with the optimum MOs and configurational weights determined by the variational principle. The CSFs entering an MCSCF expansion are pure spin states, and MCSCF wave functions therefore do not suffer from the problem of spin contamination.

Our definition of electron correlation uses the RHF energy as the reference. For ozone both the UHF and the TCSCF wave functions have lower energies, and include some electron correlation. This type of "electron correlation" is somewhat different from the picture presented at the start of this chapter. In a sense it is a consequence of our chosen zero point for the correlation energy, the RHF energy. The energy lowering introduced by adding enough flexibility in the wave function to be able to qualitatively describe the system is sometimes called the *static* electron correlation. This is essentially the effect of allowing orbitals to become (partly) singly occupied instead of forcing double occupation, i.e. describing near-degeneracy effects (two or more configurations having almost the same energy). The remaining energy lowering by correlating the motion of the electrons is called *dynamic* correlation. The problem is that there is no rigorous way of separating these effects. In the ozone example the energy lowering by going from

RHF to UHF, or to a TCSCF, is almost pure static correlation. Increasing the number of configurations in an MCSCF will recover more and more of the dynamical correlation, until at the full CI limit, the correlation treatment is exact. As mentioned above, MCSCF methods are mainly used for generating a qualitatively correct wave function, i.e. recovering the "static" part of the correlation.

The major problem with MCSCF methods is selecting the necessary configurations to include for the property of interest. One of the most popular approaches is the *Complete Active Space Self-consistent Field* (CASSCF) method (also called *Full Optimized Reaction Space*, FORS). Here the selection of configurations is done by partitioning the MOs into *active* and *inactive* spaces. The active MOs will typically be some of the highest occupied and some of the lowest unoccupied MOs from a RHF calculation. The inactive MOs either have 2 or 0 electrons, i.e. always doubly occupied or empty. Within the active MOs a full CI is performed and all the proper symmetry adapted configurations are included in the MCSCF optimization. Which MOs to include in the active space must be decided manually, by considering the problem at hand and the computational expense. If several points on the potential energy surface are desired, the MCSCF active space should include all those orbitals which change significantly, or for which the electron correlation is expected to change. A common notation is $[n, m]$-CASSCF, indicating that n electrons are distributed in all possible ways in m orbitals.

As for any full CI expansion, the CASSCF becomes unmanageably large even for quite small active spaces. A variation of the CASSCF procedure is the *Restricted Active Space Self-consistent Field* (RASSCF) method.[6] Here the active MOs are further divided into three sections, RAS1, RAS2 and RAS3, each having restrictions on the occupation numbers (excitations) allowed. The RAS1 space consists of MOs which are doubly occupied in the HF reference determinant, the RAS2 has both occupied and unoccupied orbitals, while the RAS3 space consists of MOs which are empty in the HF determinant. Configurations in the RAS2 space are generated by a full CI, analogous to

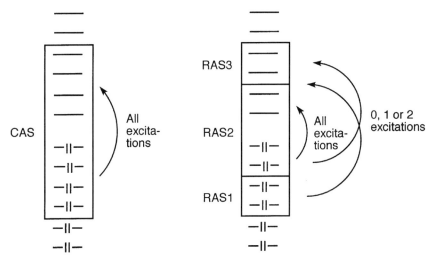

Figure 4.10   Illustration of the CAS and RAS orbital partitions

Table 4.3 Number of configurations generated in a $[n, n]$-CASSCF wave function

n	Number of CSFs
2	3
4	20
6	175
8	1 764
10	19 404
12	226 512
14	2 760 615

the CASSCF approach, and additional configurations are generated by allowing for example a maximum of two electrons to be excited <u>from</u> the RAS1 and a maximum of two electrons to be excited <u>to</u> the RAS3 space. In essence, a typical RASSCF procedure thus generates configurations by a combination of a full CI in a small number of MOs (RAS2) and a CISD in a somewhat larger MO space (RAS1 and RAS3).

The full CI expansion within the active space severely restricts the number of orbitals and electrons that can be treated by CASSCF methods. Table 4.3 shows how many singlet CSFs are generated for an $[n, n]$-CASSCF wave function (eq. (4.13)), without any reductions arising from symmetry.

The factorial increase in the number of CSFs effectively limits the active space for CASSCF wave functions to less than 10–12 electrons/orbitals. Selecting the "important" orbitals to correlate therefore becomes very important. The goal of MCSCF methods is usually not to recover a large fraction of the total correlation energy, but rather to recover all the <u>changes</u> that occur in the correlation energy for the given process. Selecting the active space for an MCSCF calculation requires some insight into the problem. There are a few rules of thumb that may be of help in selecting a proper set of orbitals for the active space.

(1) For each occupied orbital, there will typically be one corresponding virtual orbital. This leads naturally to $[n, m]$-CASSCF wave functions where n and m are identical or nearly so.

(2) The orbital energies from an RHF calculation may be used for selecting the important orbitals. The highest occupied and lowest unoccupied are usually the most important orbitals to include in the active space. This can be partly be justified by the formula for the second-order perturbation energy correction (Section 4.8.1), the smaller the orbital energy difference, the larger the contribution to the correlation energy. Using RHF orbital energies for selecting the active space may be problematic in two situations. The first is when extended basis sets are used, where there will be many virtual orbitals with low energies, and the exact order is more or less accidental. Furthermore, RHF virtual orbitals basicly describe electron attachment (via Koopmans' theorem, Section 3.4), and are therefore not particular well suited for describing electron correlation. An inspection of the form of the orbitals may reveal which to chose, they should be the ones which resemble the

occupied orbitals in terms of basis function contribution. The second problem is more fundamental. If the real wave function has significant multi-configurational character, then the RHF may be qualitatively wrong, and selecting the active orbitals based on a qualitatively wrong function may lead to erroneous results. The problem is that we wish to include the important orbitals for describing the multi-determinant nature, but these are not known until the final wave function is known.

(3) A way of overcoming this problem, is to use the concept of natural orbitals. The natural orbitals are those which diagonalize the density matrix, and the eigenvalues are the occupation numbers. Orbitals with occupation numbers significantly different from 0 or 2 (for a closed-shell system) are usually those which are the most important to include in the active space. An RHF wave function will have occupation numbers of exactly 0 or 2, and some electron correlation must be included to obtain orbitals with non-integer occupation numbers. This may for example be done by running a preliminary MP2 or CISD calculation prior to the MCSCF. Alternatively, a UHF (when different from the RHF) type wave function may also be used. The total UHF density, which is the sum of the α and β density matrices, will also provide fractional occupation numbers since UHF includes some electron correlation. The procedure may still fail. If the underlying RHF wave function is poor, the MP2 correction may also give poor results, and selecting the active MCSCF orbitals based on such MP2 occupation numbers may again lead to erroneous results. In practice, however, selecting active orbitals based on for example MP2 occupation numbers appears to be quite efficient, and better than using RHF orbital energies.

In a CASSCF type wave function the CI coefficients do not have the same significance as for a single reference CI based on HF orbitals. In a full CI (as in the active space of the CASSCF) the orbitals may be rotated among themselves without affecting the total wave function. A rotation of the orbitals, however, influences the magnitude of the coefficients in front of each CSF. While the HF coefficient in a single reference CISD gives some indication of the "multi-reference" nature of the wave function, this is not the case for a CASSCF wave function, where the corresponding CI coefficient is arbitrary.

It should be noted that CASSCF methods inherently tend to give an unbalanced description, since all the electron correlation recovered is in the active space, but none in the inactive space, or between the active and inactive electrons.[7] This is not a problem if all the valence electrons are included in the active space, but this is only possible for small systems. If only part of the valence electrons are included in the active space, the CASSCF methods tend to overestimate the importance of "biradical" structures. Consider for example acetylene where the hydrogens have been bent $60°$ away from linearity (this may be considered a model for *ortho*-benzyne). The in-plane "π-orbital" now acquires significant biradical character. The true structure may be described as a linear combination of the three configurations shown in Figure 4.11.

The structure on the left is biradical, while the two others are ionic, corresponding to both electrons being at the same carbon. The simplest CASSCF wave function which qualitatively can describe this system has two electrons in two orbitals, giving the three configurations shown above. The dynamical correlation between the two active electrons will tend to keep them as far apart as possible, i.e. favouring the biradical structure. Now

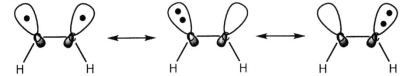

Figure 4.11   Important configurations for a bend acetylene model

Table 4.4   Natural orbital occupation numbers for the distorted acetylene model in Figure 4.11. Only the occupation numbers for the six "central" orbitals are shown

	n_5	n_6	n_7	n_8	n_9	n_{10}
RHF	2.00	2.00	2.00	0.00	0.00	0.00
UHF	2.00	1.72	1.30	0.70	0.28	0.01
[2,2]-CASSCF	2.00	2.00	1.62	0.38	0.00	0.00
[4,4]-CASSCF	2.00	1.85	1.67	0.33	0.14	0.00
[10,10]-CASSCF	1.97	1.87	1.71	0.30	0.13	0.02

consider a full valence CASSCF wave function with 10 electrons in 10 orbitals. This will analogously tend to separate the two electrons in each bond with one being at each end. The correlation of the electrons in the C–H bonds, for example, will place more electron density on the carbon atoms. This in turn will favour the ionic structures in Figure 4.11 and disfavour the biradical, i.e. the dynamical correlation of the other electrons may take advantage of the empty orbital in the ionic structures but not in the biradical. These general considerations may be quantified by considering the natural orbital occupancies for increasingly larger CASSCF wave functions, as shown in Table 4.4 for the 6-31G(d,p) basis. The [4,4]-CASSCF also includes the two out-of-plane π-orbitals in the active space, while the [10,10]-CASSCF generates a full-valence CI wave function. The unbalanced description for the [2,2]-CASSCF is reminiscent of the spin contamination problem for UHF wave functions, although the effect is much less pronounced. Nevertheless, the overestimation may be severe enough to alter the qualitative shape of energy surfaces, for example turning transition structures into minima, as illustrated in Figure 4.9. MCSCF methods are therefore not "black box" methods like for example HF and MP (Section 4.8.1); selecting a proper number of configurations, and the correct orbitals, to give a balanced description of the problem at hand requires some experimentation and insight.

4.7   Multi-reference Configuration Interaction

The CI methods described so far consider only CSFs generated by exciting electrons from a single determinant. This corresponds to having a HF type wave function as the reference. However, a MCSCF wave function may also be chosen as the reference. In that case a CISD involves excitations of one or two electrons out of <u>all</u> the determinants which enter the MCSCF, defining the *Multi-reference Configuration Interaction* (MRCI) method. Compared to the single reference CISD, the number of configurations is increased by a factor roughly equal to the number of configurations included in the MCSCF. Large-scale MRCI wave functions (many configurations in the MCSCF) can generate very accurate wave functions, but are also computationally very intensive. Since MRCI methods truncate the CI expansion, they are not size extensive.

Even truncating the (MR)CI expansion at the singles and doubles level frequently generates more configurations than can be handled readily. A further truncation is sometimes performed by selecting only those configurations which have an "interaction" with the reference configuration(s) above a selected threshold, where the "interaction" is evaluated by second-order perturbation theory (Section 4.8). Such *state selected* CI (or MCSCF) methods all involve a preset cutoff below which configurations are neglected. This may cause problems for comparing energies of different geometries, since the potential energy surface may become discontinuous, i.e. at some point the importance of a given configuration may drop below the threshold, and the contribution suddenly disappear.

4.8   Many-body Perturbation Theory

The idea in perturbation methods is that the problem at hand only differs slightly from a problem which has already been solved (exactly or approximately). The solution to the given problem should therefore in some sense be close to the solution of the already known system. This is described mathematically by defining a Hamilton operator which consists of two part, a reference ($\mathbf{H}_0$) and a perturbation ($\mathbf{H}'$). The premise of perturbation methods is that the $\mathbf{H}'$ operator in some sense is "small" compared to $\mathbf{H}_0$. In quantum mechanics, perturbational methods can be used for adding corrections to solutions which employ an independent particle approximation, and the theoretical framework is then called *Many-Body Perturbation Theory* (MBPT).

Let us assume that the Schrödinger equation for the reference Hamilton operator is solved.

$$\mathbf{H} = \mathbf{H}_0 + \lambda \mathbf{H}'$$
$$\mathbf{H}_0 \Phi_i = E_i \Phi_i, \; i = 0, \, 1, \, 2, \ldots, \infty \tag{4.27}$$

The solutions for the unperturbed Hamilton operator from a complete set (since $\mathbf{H}_0$ is hermitian) which can be chosen to be orthonormal, and λ is a (variable) parameter determining the strength of the perturbation. At present we will only consider cases where the perturbation is time-independent, and the reference wave function is non-degenerate. To keep the notation simple, we will furthermore only consider the lowest energy state. The perturbed Schrödinger equation is

$$\mathbf{H}\Psi = W\Psi \tag{4.28}$$

If $\lambda = 0$, then $\mathbf{H} = \mathbf{H}_0$, $\Psi = \Phi_0$ and $W = E_0$. As the perturbation is increased from zero to a finite value, the new energy and wave function must also change continuously, and they can be written as a Taylor expansion in powers of the perturbation parameter λ.

$$W = \lambda^0 W_0 + \lambda^1 W_1 + \lambda^2 W_2 + \lambda^3 W_3 + \ldots$$
$$\Psi = \lambda^0 \Psi_0 + \lambda^1 \Psi_1 + \lambda^2 \Psi_2 + \lambda^3 \Psi_3 + \ldots \tag{4.29}$$

For $\lambda = 0$, it is seen that $\Psi_0 = \Phi_0$ and $W_0 = E_0$, these are the *unperturbed*, or *zero-*

order wave function and energy. The Ψ_1, Ψ_2 ... and W_1, W_2 ... are the *first-*, *second-* etc. *order* <u>corrections</u>. The λ parameter will eventually be set equal to 1, and the nth order energy or wave function become a sum of all terms up to order n. It is convenient to chose the perturbed wave function to be *intermediately normalized*, i.e. the overlap with the unperturbed wave function should be 1. This has the consequence that all correction terms are orthogonal to the reference wave function.

$$\langle \Psi | \Phi_0 \rangle = 1$$
$$\langle \Psi_0 + \lambda \Psi_1 + \lambda^2 \Psi_2 + \ldots | \Phi_0 \rangle = 1$$
$$\langle \Psi_0 | \Phi_0 \rangle + \lambda \langle \Psi_1 | \Phi_0 \rangle + \lambda^2 \langle \Psi_2 | \Phi_0 \rangle + \ldots = 1 \tag{4.30}$$
$$\langle \Psi_{i \neq 0} | \Phi_0 \rangle = 0$$

Once all the correction terms have been calculated, it is trivial to normalize the total wave function.

With the expansions (4.29) the Schrödinger equation (4.28) becomes

$$(\mathbf{H}_0 + \lambda \mathbf{H}')(\lambda^0 \Psi_0 + \lambda^1 \Psi_1 + \lambda^2 \Psi_2 + \ldots) =$$
$$(\lambda^0 W_0 + \lambda^1 W_1 + \lambda^2 W_2 + \ldots)(\lambda^0 \Psi_0 + \lambda^1 \Psi_1 + \lambda^2 \Psi_2 + \ldots) \tag{4.31}$$

Since this holds for any value of λ, we can collect terms with the same power of λ to give

$$\lambda^0 : \mathbf{H}_0 \Psi_0 = W_0 \Psi_0$$
$$\lambda^1 : \mathbf{H}_0 \Psi_1 + \mathbf{H}' \Psi_0 = W_0 \Psi_1 + W_1 \Psi_0$$
$$\lambda^2 : \mathbf{H}_0 \Psi_2 + \mathbf{H}' \Psi_1 = W_0 \Psi_2 + W_1 \Psi_1 + W_2 \Psi_0 \tag{4.32}$$
$$\lambda^n : \mathbf{H}_0 \Psi_n + \mathbf{H}' \Psi_{n-1} = \sum_{i=0}^{n} W_i \Psi_{n-i}$$

These are zero-, first-, second-, nth-order perturbation equations. The zero-order equation is just the Schödinger equation for the unperturbed problem. The first-order equation contains two unknowns, the first-order correction to the energy, W_1, and the first-order correction to the wave function, Ψ_1. The nth-order energy correction can be calculated by multiplying from the left by Φ_0 and integrating, and using the "*turnover rule*" $\langle \Phi_0 | \mathbf{H}_0 | \Psi_i \rangle = \langle \Psi_i | \mathbf{H}_0 | \Phi_0 \rangle^*$.

$$\langle \Phi_0 | \mathbf{H}_0 | \Psi_n \rangle + \langle \Phi_0 | \mathbf{H}' | \Psi_{n-1} \rangle = \sum_{i=0}^{n-1} W_i \langle \Phi_0 | \Psi_{n-i} \rangle + W_n \langle \Phi_0 | \Psi_0 \rangle$$
$$E_0 \langle \Phi_0 | \Psi_n \rangle + \langle \Phi_0 | \mathbf{H}' | \Psi_{n-1} \rangle = W_n \langle \Phi_0 | \Psi_0 \rangle \tag{4.33}$$
$$W_n = \langle \Phi_0 | \mathbf{H}' | \Psi_{n-1} \rangle$$

From this it would appear that the $(n-1)$th-order wave function is required for calculating the nth-order energy. However, by using the turnover rule and the nth and lower-order perturbation equations (4.32), it can be shown that knowledge of the nth-order wave function actually allows a calculation of the $(2n+1)$th-order energy.

$$W_{2n+1} = \langle \Psi_n | \mathbf{H}' | \Psi_n \rangle - \sum_{k,l=1}^{n} W_{2n+1-k-1} \langle \Psi_k | \Psi_l \rangle \tag{4.34}$$

Up to this point we are still dealing with undetermined quantities, energy and wave function corrections at each order. The first-order equation is one equation with two unknowns. Since the solutions to the unperturbed Schrödinger equation generates a complete set of functions, the unknown first-order correction to the wave function can be expanded in these functions. This is known as *Rayleigh–Schrödinger* perturbation theory, and the λ^1 equation in (4.32) becomes

$$\Psi_i = \sum_i c_i \Phi_i$$

$$(\mathbf{H}_0 - W_0)(\sum_i c_i \Phi_i) + (\mathbf{H}' - W_1)\Phi_0 = 0 \tag{4.35}$$

Multiplying from the left by Φ_0^* and integrating yields

$$\sum_i c_i \langle \Phi_0 | \mathbf{H}_0 | \Phi_i \rangle - W_0 \sum_i c_i \langle \Phi_0 | \Phi_i \rangle + \langle \Phi_0 | \mathbf{H}' | \Phi_0 \rangle - W_1 \langle \Phi_0 | \Phi_0 \rangle = 0$$

$$\sum_i c_i E_i \langle \Phi_0 | \Phi_i \rangle - c_0 E_0 + \langle \Phi_0 | \mathbf{H}' | \Phi_0 \rangle - W_1 = 0 \tag{4.36}$$

$$c_0 E_0 - c_0 E_0 + \langle \Phi_0 | \mathbf{H}' | \Phi_0 \rangle - W_1 = 0$$

$$W_1 = \langle \Phi_0 | \mathbf{H}' | \Phi_0 \rangle$$

since the Φ_is are orthonormal (this also follows directly from eq. (4.34)). The latter equation shows that the first-order correction to the energy is an average of the perturbation operator over the unperturbed wave function.

The first-order correction to the wave function can be obtained by multiplying (4.32) from the left by a function other than $\Phi_0(\Phi_j)$ and integrating to give

$$\sum_i c_i \langle \Phi_j | \mathbf{H}_0 | \Phi_i \rangle - W_0 \sum_i c_i \langle \Phi_j | \Phi_i \rangle + \langle \Phi_j | \mathbf{H}' | \Phi_0 \rangle - W_1 \langle \Phi_j | \Phi_0 \rangle = 0$$

$$\sum_i c_i E_i \langle \Phi_j | \Phi_i \rangle - c_j E_0 + \langle \Phi_j | \mathbf{H}' | \Phi_0 \rangle = 0$$

$$c_j E_j - c_j E_0 + \langle \Phi_j | \mathbf{H}' | \Phi_0 \rangle = 0 \tag{4.37}$$

$$c_j = \frac{\langle \Phi_j | \mathbf{H}' | \Phi_0 \rangle}{E_0 - E_j}$$

The expansion coefficients determine the first-order correction to the perturbed wave function (eq. (4.35)), and they can be calculated for the known unperturbed wave functions and energies. The coefficient in front of Φ_0 for Ψ_1 cannot be determined from the above formula, but the assumption of intermediate normalization (eq. (4.30)) makes $c_0 = 0$.

Starting from the second-order perturbation equation (4.32), analogous formulas can be generated for the second-order corrections. Using intermediate normalization

$(c_0 = d_0 = 0)$, the second-order energy correction is

$$\Psi_2 = \sum_i d_i \Phi_i$$

$$(\mathbf{H}_0 - W_0)\left(\sum_i d_i \Phi_i\right) + (\mathbf{H}' - W_1)\left(\sum_i c_i \Phi_i\right) - W_2 \Phi_0 = 0$$

$$\sum_i d_i \langle \Phi_0 | \mathbf{H}_0 | \Phi_i \rangle - W_0 \sum_i d_i \langle \Phi_0 | \Phi_i \rangle$$

$$+ \sum_i c_i \langle \Phi_0 | \mathbf{H}' | \Phi_i \rangle - W_1 \sum_i c_i \langle \Phi_0 | \Phi_i \rangle - W_2 \langle \Phi_0 | \Phi_0 \rangle = 0$$

$$\sum_i d_i E_i \langle \Phi_0 | \Phi_i \rangle - d_0 E_0 + \sum_i c_i \langle \Phi_0 | \mathbf{H}' | \Phi_i \rangle$$

(4.38)

$$- c_0 W_1 - W_2 = 0$$

$$d_0 E_0 - d_0 E_0 + \sum_i c_i \langle \Phi_0 | \mathbf{H}' | \Phi_i \rangle - W_2 = 0$$

$$W_2 = \sum_i c_i \langle \Phi_0 | \mathbf{H}' | \Phi_i \rangle = \sum_{i \neq 0} \frac{\langle \Phi_0 | \mathbf{H}' | \Phi_i \rangle \langle \Phi_i | \mathbf{H}' | \Phi_0 \rangle}{E_0 - E_i}$$

The last equation shows that the second-order energy correction may be written in terms of the first order wave function (c_i) and matrix elements over unperturbed states. The second-order wave function correction is

$$\sum_i d_i \langle \Phi_j | \mathbf{H}_0 | \Phi_i \rangle - W_0 \sum_i d_i \langle \Phi_j | \Phi_i \rangle$$

$$+ \sum_i c_i \langle \Phi_j | \mathbf{H}' | \Phi_i \rangle - W_1 \sum_i c_i \langle \Phi_j | \Phi_i \rangle - W_2 \langle \Phi_j | \Phi_0 \rangle = 0$$

$$\sum_i d_i E_i \langle \Phi_j | \Phi_i \rangle - d_j E_0 + \sum_i c_i \langle \Phi_j | \mathbf{H}' | \Phi_i \rangle - c_j W_1 = 0$$

(4.39)

$$d_j E_j - d_j E_0 + \sum_i c_i \langle \Phi_j | \mathbf{H}' | \Phi_i \rangle - c_j \langle \Phi_0 | \mathbf{H}' | \Phi_0 \rangle = 0$$

$$d_j = \sum_{i \neq 0} \frac{\langle \Phi_j | \mathbf{H}' | \Phi_i \rangle \langle \Phi_i | \mathbf{H}' | \Phi_0 \rangle}{(E_0 - E_j)(E_0 - E_i)} - \frac{\langle \Phi_j | \mathbf{H}' | \Phi_0 \rangle \langle \Phi_0 | \mathbf{H}' | \Phi_0 \rangle}{(E_0 - E_j)^2}$$

The formulas for higher-order corrections become increasingly complex. The main point, however, is that all corrections can be expressed in terms of matrix elements of the perturbation operator over the unperturbed wave functions, and the unperturbed energies.

4.8.1 Møller–Plesset Perturbation Theory

So far the theory has been completely general. In order to apply perturbation theory to the calculation of correlation energy, the unperturbed Hamilton operator must be selected. The most common choice is to take this as a sum over Fock operators, leading to *Møller–Plesset* (MP) perturbation theory.[8] The sum of Fock operators counts the (average) electron–electron repulsion twice (eq. (3.43)), and the perturbation becomes the exact $\mathbf{V}_{ee}$ operator minus twice the $\langle\mathbf{V}_{ee}\rangle$ operator. The operator associated with this difference is often referred to as the *fluctuation potential*. This choice is not really consistent with the basic assumption that the perturbation should be small compared to $\mathbf{H}_0$. However, it does fulfill the other requirement that solutions to the unperturbed Schrödinger equation should be known. Furthermore, this is the only choice which leads to a size extensive method, which is a desirable feature.

$$\mathbf{H}_0 = \sum_{i=1}^{N} \mathbf{F}_i = \sum_{i=1}^{N} \left(\mathbf{h}_i + \sum_{j=1}^{N} (\mathbf{J}_{ij} - \mathbf{K}_{ij}) \right) = \sum_{i=1}^{N} \mathbf{h}_i + 2\langle\mathbf{V}_{ee}\rangle$$

$$= \sum_{i=1}^{N} \mathbf{h}_i + \sum_{i=1}^{N}\sum_{j=1}^{N} \langle\mathbf{g}_{ij}\rangle$$

$$\mathbf{H}' = \mathbf{H} - \mathbf{H}_0 = \mathbf{V}_{ee} - \sum_{i=1}^{N}\sum_{j=1}^{N} (\mathbf{J}_{ij} - \mathbf{K}_{ij}) = \mathbf{V}_{ee} - 2\langle\mathbf{V}_{ee}\rangle$$

$$= \sum_{i=1}^{N}\sum_{j>i}^{N} \mathbf{g}_{ij} - \sum_{i=1}^{N}\sum_{j=1}^{N} \langle\mathbf{g}_{ij}\rangle$$

(4.40)

The zero-order wave function is the HF determinant, and the zero-order energy is just a sum of MO energies. The first-order energy correction is the average of the perturbation operator over the zero-order wave function (eq. (4.36)).

$$W_1 = \langle\Phi_0|\mathbf{H}'|\Phi_0\rangle = \langle\mathbf{V}_{ee}\rangle - 2\langle\mathbf{V}_{ee}\rangle = -\langle\mathbf{V}_{ee}\rangle \qquad (4.41)$$

This yields a correction for the overcounting of the electron–electron repulsion at zero-order. Comparing eq. (4.40) with the expression for the total energy in eq. (3.32), it is seen that the first-order energy (sum of W_0 and W_1) is exactly the HF energy. Using the notation $E(\mathrm{MP}n)$ to indicate the correction at order n, and MPn to indicate the total energy up to order n, we have

$$\mathrm{MP0} = E(\mathrm{MP0}) = \sum_{i=1}^{N} \varepsilon_i$$

$$\mathrm{MP1} = \mathrm{MP0} + E(\mathrm{MP1}) = E(\mathrm{HF})$$

(4.42)

Electron correlation energy thus starts at order 2 with this choice of $\mathbf{H}_0$.

In developing perturbation theory it was assumed that the solutions to the unperturbed problem formed a complete set. This is general means that there must be an infinite number of functions, which is impossible in actual calculations. The lowest energy solution to the unperturbed problem is the HF wave function, additional higher energy solutions are excited Slater determinants, analogously to the CI metnod. When a finite basis set is employed it is only possible to generate a finite number of excited determinants. The expansion of the many-electron wave function is therefore truncated.

Let us look at the expression for the second-order energy correction, eq. (4.38). This involves matrix elements of the perturbation operator between the HF reference and all possible excited states. Since the perturbation is a two-electron operator, all matrix elements involving triple, quadruple etc. excitations are zero. When canonical HF

orbitals are used, matrix elements with singly excited states are also zero, which follows
from.

$$\langle\Phi_0|\mathbf{H}'|\Phi_i^a\rangle = \langle\Phi_0|\mathbf{H} - \sum_{j=1}^{N}\mathbf{F}_j|\Phi_i^a\rangle$$

$$= \langle\Phi_0|\mathbf{H}|\Phi_i^a\rangle - \sum_{j=1}^{N}\langle\Phi_0|\mathbf{F}_j|\Phi_i^a\rangle \qquad (4.43)$$

$$= \langle\Phi_0|\mathbf{H}|\Phi_i^a\rangle - \varepsilon_a\langle\Phi_0|\Phi_i^a\rangle$$

The first bracket is zero due to Brillouins theorem (Section 4.2.1), and the second set of
brackets is zero due to the orbitals being eigenfunctions of the Fock operators and being
orthogonal to each other. The second-order correction to the energy, which is the first
contribution to the correlation energy, therefore only involves a sum over doubly excited
determinants. These can be generated by promoting two electrons from occupied
orbitals i and j to virtual orbitals a and b. The summation must be restricted so that each
excited state is only counted once.

$$W_2 = \sum_{i<j}^{\text{occ}}\sum_{a<b}^{\text{vir}}\frac{\langle\Phi_0|\mathbf{H}'|\Phi_{ij}^{ab}\rangle\langle\Phi_{ij}^{ab}|\mathbf{H}'|\Phi_0\rangle}{E_0 - E_{ij}^{ab}} \qquad (4.44)$$

The matrix elements between the HF and a doubly excited state are given by two-
electron integrals over MOs (eq. (4.7)). The difference in total energy between
two Slater determinants becomes a difference in MO energies (essentially Koopmans'
theorem), and the explicit formula for the second-order Møller–Plesset correction
is

$$E(\text{MP2}) = \sum_{i<j}^{\text{occ}}\sum_{a<b}^{\text{vir}}\frac{[\langle\phi_i\phi_j|\phi_a\phi_b\rangle - \langle\phi_i\phi_j|\phi_b\phi_a\rangle]^2}{\varepsilon_i + \varepsilon_j - \varepsilon_a - \varepsilon_b} \qquad (4.45)$$

Once the two-electron integrals over MOs are available, the second-order energy
correction can be calculated as a sum over such integrals. There are of the order of M^4
integrals, thus the calculation of the energy (only) increases as M^4 with the system size.
However, the transformation of the integrals from the AO to the MO basis grows as M^5
(Section 4.2.1). MP2 is an M^5 method, but fairly inexpensive as not all two-electron
integrals over MOs are required. Only those corresponding to the combination of two
occupied and two virtual MOs are needed. In practical calculations this means that the
MP2 energy for systems with ~ 100–150 basis functions can be calculated at a cost
similar to or less than what is required for calculating the HF energy. MP2 typically
accounts for ~ 80–90% of the correlation energy, and it is the most economical method
for including electron correlation.

The formula for the first-order correction to the wave function (eq. (4.37)) similarly
only contains contributions from doubly excited determinants. Since knowledge of the
first-order wave function allows calculation of the energy up to third order ($2n + 1 = 3$,
eq. (4.34)), it is immediately clear that the third-order energy also only contains
contributions from doubly excited determinants. Qualitatively speaking, the MP2
contribution describes the correlation between pairs of electrons while MP3 describes
the interaction between pairs. The formula for calculating this contribution is somewhat

more complex than for second order, and involves a computational effort which formally increases as M^6. The third-order energy typically accounts for $\sim 90-95\%$ of the correlation energy.

The formula for the second-order correction to the wave function (eq. (4.39)) contains products of the type $\langle\Phi_j|\mathbf{H}'|\Phi_i\rangle\langle\Phi_i|\mathbf{H}'|\Phi_0\rangle$. The Φ_0 is the HF determinant and the <u>last</u> bracket can only be non-zero if Φ_i is a doubly excited determinant. This means that the <u>first</u> bracket only can be non-zero if Φ_j is either a singly, doubly, triply or quadruply excited determinant ($\mathbf{H}'$ is a two-electron operator). The second-order wave function allows calculation of the fourth- and fifth-order energies, these terms therefore have contributions from determinants which are singly, doubly, triply or quadruply excited. The computational cost of the fourth-order energy without the contribution from the triply excited determinants, MP4(SDQ), increases as M^6, while the triples contribution increases as M^7. MP4 is still a computationally feasible model for many molecular systems, requiring a time similar to CISD. In typical calculations the T contribution to MP4 will take roughly the same amount of time as the SDQ contributions, but the triples are often the most important at fourth order. The full fourth-order energy typically accounts for $\sim 95-98\%$ of the correlation energy.

The fifth-order correction to the energy also involve S, D, T and Q contributions, and the sixth-order term introduces quintuple and sextuple excitations. The working formulas for the MP5 and MP6 contributions are now so complex that actual calculations are only possible for small systems. The computational effort for MP5 increases as M^8 and for MP6 as M^9. There is very little experience with the performance of MP*n* beyond MP4.

As shown in Table 4.2, the most important contribution to the energy in a CI procedure comes from doubly excited determinants. This is also shown by the perturbation expansion, the second- and third-order energy corrections only involve doubles. At fourth order the singles, triples and quadruples enter the expansion for the first time. This is again consistent with Table 4.2, which shows that these types of excitation are of similar importance.

CI methods determine the energy by a variational procedure, and the energy is consequently an upper bound to the exact energy. There is no such guarantee for perturbation methods; it is possible that the energy will be lower than the exact energy. This is rarely a problem. Limitations in the basis set often mean that the error in total energy is several a.u. (thousands of kcal/mol) anyway. Furthermore, the interest is normally not in total energies, but in energy differences. Having a variational upper bound for two energies does not give any bound for the difference between these two numbers. The main interest is therefore in the error remaining relatively constant for different systems. The lack of size extensivity of CI methods is disadvantageous in this respect. The MP perturbation method <u>is</u> size extensive, but other forms of MBPT are not. It is now generally recognized that size extensivity is an important property, and the MP form of MBPT is used almost exclusively. Combined with the low cost relative to CI methods this often makes MP calculations a good method for including electron correlation.

The main limitation of perturbation methods is the assumption that the zero-order wave function is a reasonable approximation to the real wave function, i.e. the perturbation operator is sufficiently "small". The poorer the HF wave function describes the system, the larger are the correction terms, and more terms must be included to achieve a given level of accuracy. If the reference state is a poor description of the system, the convergence may be so slow or erratic that perturbation methods cannot be used. Actually it is difficult to prove that the perturbation expansion is convergent, although many systems show a behaviour which suggests that it is the case. This may to some extent be deceptive, as it has been demonstrated that the convergence properties depend on the size of the basis set,[9] and the majority of studies have employed small or medium sized basis sets. A convergent series in a DZP type basis for example may become divergent or oscillating in a larger basis, especially if diffuse functions are present.

In the ideal case the HF, MP2, MP3 and MP4 results show a monotonic convergence towards a limiting value, with the corrections being of the same sign and numerically smaller as the order of perturbation increases. Unfortunately, this is not the typical behaviour.[10] Even in systems where the reference is well described by a single determinant, oscillations in a given property as a function of perturbation order are often observed. This is not completely understood, but may at least partly be due to the fact that the choice of the unperturbed Hamilton operator does not make the perturbation particularly small. An analysis by Cremer and He[11] indicates that a smooth convergence (of the total energy) is only expected for systems containing well-separated electron pairs, and that oscillations occur when this is not the case. These encompass systems containing lone pairs and/or multiple bonds, covering the large majority of molecules. Another study for the neon atom[12] indicates that convergence of the MP series requires that the energy spectrum is well-separated (i.e. the HF reference is significantly below the first excited state), not only for the unperturbed ($\mathbf{H} = \mathbf{H}_0$) and fully perturbed ($\mathbf{H} = \mathbf{H}_0 + \mathbf{H}'$) systems, but also for the "negative" perturbed case ($\mathbf{H} = \mathbf{H}_0 - \mathbf{H}'$). This condition was found not to hold when a basis set containing diffuse functions was used, and consequently the MP series was divergent. At low order of perturbation theory the divergence may not be obvious, but instead shows up as an oscillation in the total energy.

In practice only low orders of perturbation theory can be carried out, and it is often observed that the HF and MP2 results differ considerably, the MP3 result moves back towards the HF and the MP4 moves away again. For "well-behaved" systems the correct answer is normally somewhere between the MP3 and MP4 results. MP2 typically overshoots the correlation effect, but often gives a better answer than MP3, at least if medium sized basis sets are used. Just as the first term involving doubles (MP2) tends to overestimate the correlation effect, it is often observed that MP4 overestimates the effect of the singles and triples contributions, since they enter the series for the first time at fourth order.

When the reference wave function contains substantial multi-reference character, a perturbation expansion based on a single determinant will display poor convergence. If the reference wave function suffers from symmetry breaking (Section 3.8.3), the MP method is almost guaranteed to give absurd results. The ability to calculate terms up to MP4, however, allows an internal evaluation of the performance. If the MP2, MP3 and MP4 results appear to converge, it is possible at least to give a rough estimate of the "MP∞" result (equivalent to full CI). If the MP2, MP3 and MP4 results on the other hand oscillate wildly, then one knows that none of the results can be trusted.

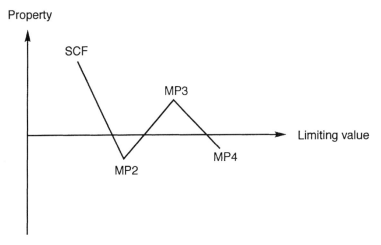

Figure 4.12   Typical oscillating behaviour of results obtained with the MP method

4.8.2   Unrestricted and Projected Møller–Plesset Methods

When the reference is an RHF type wave function the dissociation limit will normally be incorrect. As a bond is stretched, the RHF gives an increasingly poorer description of the wave function, and consequently the perturbation series will eventually break down. The use of a UHF wave function allows a correct dissociation limit in terms of energy, but at the cost of introducing spin contamination (Section 4.4). It is straightforward to derive an MP method based on a UHF reference wave function (UMP), in this case the unperturbed Hamilton operator is a sum of the α and β Fock operators. Addition of electron correlation decreases the spin contamination of the wave function (in the full CI limit the spin contamination is zero) but the improvement is usually small at low orders (2–4) of perturbation theory. As illustrated in Section 4.4, the UHF energy is lower than the RHF owing to the inclusion of some electron correlation (mainly static), but also contains some amount of higher energy spin states. Since MP methods recover a large part of the electron correlation (both static and dynamical), the net effect at the UMP level is an increase in energy due to spin contamination. In the dissociation limit, this has no consequence, as the different spin states have equal energies. In the intermediate region, where the bond is not completely broken, it is usually observed that the RMPn energy is <u>lower</u> than the UMPn energy, although the RHF energy is <u>higher</u> than the UHF (see also Section 11.5.2). The spin contamination in UHF wave functions causes a UMPn expansion to converge much more slowly than an RMPn.[13] For open-shell systems, where RHF cannot be used, this would suggest that the reference wave function should be of the ROHF type, instead of the UHF. Formulation of ROHF based perturbation methods, however, is somewhat more difficult than for the UHF case. The reason is that for an ROHF wave function is it not possible to chose a set of MOs that makes the matrix of Lagrange multipliers diagonal (eqs. (3.40) and (3.41)). There is therefore not a unique set of canonical MOs to be used in the perturbation expansion, which again has the consequence that several choices of the unperturbed Hamilton operator are possible[14]. Different ROMP methods therefore give different energies, and there is no firm theoretical ground for choosing one over the other. In practice, however, different choices of the unperturbed Hamilton operator lead to similar results, and perturbation calculations based on ROHF type wave functions are now becoming more and more common.

While projection methods for removing spin contamination are not recommended at the HF level, they work quite well at the UMP level. Formulas have been derived for removing all contaminants at the UMP2 level, and also the first few states at the UMP3 and UMP4 levels.[15] The associated acronyms are PUMP and PMP, denoting slightly different methods, although they in practice give similar results. For singlet wave functions with bond lengths only slightly longer than the RHF/UHF instability point, such PUMP methods tend to give results very similar to those based on an RHF wave function. At longer bond lengths the RMP perturbation series eventually breaks down, while the PUMP series approaches the correct dissociation limit. It would therefore appear the PUMP methods should always be preferred. There are, however, also some computational factors to consider. First, UMP methods are by nature a factor of ~ 2 more expensive since there are twice as many MO coefficients. Second, the projection itself also uses CPU time. This is especially true if many of the higher spin states need to be removed, or for projection at the MP4 level. Third, it is difficult to formulate derivatives of projected wave functions, which at present limits PUMP methods to the calculation of energies. A rule of thumb says that for uncomplicated systems the RMP4 treatment gives acceptable accuracy (relative errors of the order of a few kcal/mol) up to bond lengths ~ 1.5 times the equilibrium length. Longer bonds are better treated by PUMP methods, see also Section 11.5.2. Most transition structures have bond lengths shorter than ~ 1.5 times the equilibrium length, and RMP4 often gives quite accurate activation energies.

Just as single reference CI can be extended to MRCI, it is also possible to use perturbation methods with a multi-determinant reference wave function. Formulating MR-MBPT methods, however, is not straightforward. The main problem here is similar to that of ROMP methods, the choice of the unperturbed Hamilton operator. Several different choices are possible, which will give different answers when the theory is carried out only to low order. Nevertheless, there are now several different implementations of MP2 type expansions based on a CASSCF reference, denoted CASMP2 or CASPT2.[16] Experience of their performance is still somewhat limited.

4.9 Coupled Cluster Methods

Perturbation methods add all types of corrections (S, D, T, Q etc.) to the reference wave function to a given order (2, 3, 4 etc.). The idea in *Coupled Cluster* (CC) methods is to include all corrections of a given type to infinite order.[17] The (intermediate normalized) coupled cluster wave function is written as

$$\Psi_{cc} = e^{\mathbf{T}} \Phi_0$$

$$e^{\mathbf{T}} = 1 + \mathbf{T} + \frac{1}{2}\mathbf{T}^2 + \frac{1}{6}\mathbf{T}^3 + \ldots = \sum_{k=0}^{\infty} \frac{1}{k!}\mathbf{T}^k \qquad (4.46)$$

where the cluster operator $\mathbf{T}$ is given by

$$\mathbf{T} = \mathbf{T}_1 + \mathbf{T}_2 + \mathbf{T}_3 + \ldots + \mathbf{T}_N \qquad (4.47)$$

The $\mathbf{T}_i$ operator acting on a HF reference wave function generates all ith excited Slater determinants.

$$\mathbf{T}_1 \Phi_0 = \sum_i^{occ} \sum_a^{vir} t_i^a \Phi_i^a$$

$$\mathbf{T}_2 \Phi_0 = \sum_{i<j}^{occ} \sum_{a<b}^{vir} t_{ij}^{ab} \Phi_{ij}^{ab}$$

(4.48)

It is customary to use the term *amplitudes* for the expansion coefficients t, which are equivalent to the a_i coefficients in eq. (4.1).

From eqs. (4.46) and (4.47) the exponential operator may be written as

$$e^{\mathbf{T}} = \mathbf{1} + \mathbf{T}_1 + \left(\mathbf{T}_2 + \frac{1}{2}\mathbf{T}_1^2\right) + \left(\mathbf{T}_3 + \mathbf{T}_2\mathbf{T}_1 + \frac{1}{6}\mathbf{T}_1^3\right)$$

$$+ \left(\mathbf{T}_4 + \mathbf{T}_3\mathbf{T}_1 + \frac{1}{2}\mathbf{T}_2^2 + \frac{1}{2}\mathbf{T}_2\mathbf{T}_1^2 + \frac{1}{24}\mathbf{T}_1^4\right) + \cdots$$

(4.49)

The first term generates the reference HF and the second all singly excited states. The first parenthesis generates all doubly excited states, which may be considered as *connected* ($\mathbf{T}_2$) or *disconnected* ($\mathbf{T}_1^2$). The second parenthesis generates all triply excited states, which again may be either "true" ($\mathbf{T}_3$) or "product" triples ($\mathbf{T}_2\mathbf{T}_1$, $\mathbf{T}_1^3$). The quadruply excited states can similarly be viewed as composed of five terms, a true quadruple and four product terms. Physically a connected type such as $\mathbf{T}_4$ corresponds to four electrons interacting simultaneously, while a disconnected term such as $\mathbf{T}_2^2$ corresponds to two non-interacting pairs of interacting electrons.

With the coupled cluster wave function (4.46) the Schrödinger equation becomes

$$\mathbf{H}e^{\mathbf{T}}\Phi_0 = Ee^{\mathbf{T}}\Phi_0$$

(4.50)

Multiplying from the left by Φ_0^* and integrating gives

$$\langle\Phi_0|\mathbf{H}e^{\mathbf{T}}|\Phi_0\rangle = E_{cc}\langle\Phi_0|e^{\mathbf{T}}\Phi_0\rangle$$

$$\langle\Phi_0|\mathbf{H}e^{\mathbf{T}}|\Phi_0\rangle = E_{cc}\langle\Phi_0|(\mathbf{1} + \mathbf{T}_1 + \mathbf{T}_2 + \cdots)\Phi_0\rangle$$

$$E_{cc} = \langle\Phi_0|\mathbf{H}e^{\mathbf{T}}|\Phi_0\rangle$$

(4.51)

Expanding out the exponential in eq. (4.46) and using the fact that the Hamilton operator contains only one- and two-electron operators (eq. (3.24)) we get

$$E_{cc} = \langle\Phi_0|\mathbf{H}\left(\mathbf{1} + \mathbf{T}_1 + \mathbf{T}_2 + \tfrac{1}{2}\mathbf{T}_1^2\right)\Phi_0\rangle$$

$$E_{cc} = \langle\Phi_0|\mathbf{H}|\Phi_0\rangle + \langle\Phi_0|\mathbf{H}|\mathbf{T}_1\Phi_0\rangle + \langle\Phi_0|\mathbf{H}|\mathbf{T}_2\Phi_0\rangle + \tfrac{1}{2}\langle\Phi_0|\mathbf{H}|\mathbf{T}_1^2\Phi_0\rangle$$

$$E_{cc} = E_0 + \sum_i^{occ} \sum_a^{vir} t_i^a \langle\Phi_0|\mathbf{H}|\Phi_i^a\rangle + \sum_{i<j}^{occ} \sum_{a<b}^{vir} (t_{ij}^{ab} + t_i^a t_j^b - t_i^b t_j^a)\langle\Phi_0|\mathbf{H}|\Phi_{ij}^{ab}\rangle$$

(4.52)

When using HF orbitals for constructing the Slater determinants, the first matrix elements are zero (Brillouins theorem) and the second matrix elements are just two-electron integrals over MOs (eq. (4.7)).

$$E_{cc} = E_0 + \sum_{i<j}^{occ} \sum_{a<b}^{vir} (t_{ij}^{ab} + t_i^a t_j^b - t_i^b t_j^a)(\langle \phi_i \phi_j | \phi_a \phi_b \rangle - \langle \phi_i \phi_j | \phi_b \phi_a \rangle) \qquad (4.53)$$

The coupled cluster correlation energy is therefore determined completely by the singles and doubles amplitudes and the two-electron MO integrals.

Only equation for the amplitudes is obtained by multiplying the Schrödinger equation (4.50) from the left by a singly excited determinant $(\Phi_m^e)^*$ and integrating.

$$\langle \Phi_m^e | \mathbf{H} e^{\mathbf{T}} | \Phi_0 \rangle = E_{cc} \langle \Phi_m^e | e^{\mathbf{T}} \Phi_0 \rangle$$
$$\langle \Phi_m^e | \mathbf{H} | (1 + \mathbf{T}_1 + \mathbf{T}_2 + \tfrac{1}{2} \mathbf{T}_1^2 + \mathbf{T}_3 + \mathbf{T}_1 \mathbf{T}_2 + \tfrac{1}{6} \mathbf{T}_1^3) \Phi_0 \rangle = E_{cc} \langle \Phi_m^e | \mathbf{T}_1 \Phi_0 \rangle$$
$$\langle \Phi_m^e | \mathbf{H} | \Phi_0 \rangle + \langle \Phi_m^e | \mathbf{H} | \mathbf{T}_1 \Phi_0 \rangle + \langle \Phi_m^e | \mathbf{H} | \mathbf{T}_2 \Phi_0 \rangle + \tfrac{1}{2} \langle \Phi_m^e | \mathbf{H} | \mathbf{T}_1^2 \Phi_0 \rangle$$
$$+ \langle \Phi_m^e | \mathbf{H} | \mathbf{T}_3 \Phi_0 \rangle + \langle \Phi_m^e | \mathbf{H} | \mathbf{T}_1 \mathbf{T}_2 \Phi_0 \rangle + \tfrac{1}{6} \langle \Phi_m^e | \mathbf{H} | \mathbf{T}_1^3 \Phi_0 \rangle = E_{cc} \langle \Phi_m^e | \mathbf{T}_1 \Phi_0 \rangle$$

$$(4.54)$$

Only the indicated terms survive in the expansion when the orthogonality of the Slater determinants and the nature of the Hamilton operator is considered. The first term in the last equation is again zero due to Brillouins theorem, and the remaining forms a coupled set of equations of all singles, doubles and triples amplitudes (substituting in the expression for the energy, eq. (4.52)). Other equations connecting amplitudes may be obtained by multiplying from the left by a double, triple etc. excited determinant and integrating.

4.9.1 Truncated Coupled Cluster Methods

So far everything is exact. If all cluster operators up to $\mathbf{T}_N$ are inlcudes in $\mathbf{T}$, all possible excited determinants are generated and the coupled cluster wave function is equivalent to full CI. This is as already stated impossible for all but the smallest systems. The cluster operator must therefore be truncated at some excitation level. When the $\mathbf{T}$ operator is truncated, some of the terms in the amplitude equations will become zero, and the amplitudes derived from these approximate equations will no longer be exact. The energy calculated from these approximate singles and doubles amplitudes (eq. (4.53)) will therefore also be approximate. How severe the approximation is depends on how many terms are included in $\mathbf{T}$. Including only the $\mathbf{T}_1$ operator does not give any improvement over HF, as matrix elements between the HF and singly excited states are zero. The lowest level of approximation is therefore $\mathbf{T} = \mathbf{T}_2$, referred to as *Coupled Cluster Doubles* (CCD) Compared to the number of doubles, there are relatively few singly excited states. Using $\mathbf{T} = \mathbf{T}_1 + \mathbf{T}_2$ gives the CCSD model which is only slightly more demanding than CCD, and yields a more complete model. Both CCD and CCSD involve a computational effort which scales as M^6 in the limit of a large basis set. The next higher level has $\mathbf{T} = \mathbf{T}_1 + \mathbf{T}_2 + \mathbf{T}_3$, giving the CCSDT model[18] This involves a computational effort which scales as M^8, and is more demanding than CISDT. It (and higher-order methods like CCSDTQ) can consequently only be used for small systems, and CCSD is the only generally applicable coupled cluster method.

Let us look in a bit more detail at the CCSD method. In this case we have (eqs. (4.46) and (4.47)).

$$
e^{\mathbf{T}_1+\mathbf{T}_2} = 1 + \mathbf{T}_1 + \left(\mathbf{T}_2 + \tfrac{1}{2}\mathbf{T}_1^2\right) + \left(\mathbf{T}_2\mathbf{T}_1 + \tfrac{1}{6}\mathbf{T}_1^3\right) + \left(\tfrac{1}{2}\mathbf{T}_2^2 + \tfrac{1}{2}\mathbf{T}_2\mathbf{T}_1^2 + \tfrac{1}{24}\mathbf{T}_1^4\right) + \dots
$$

$$(4.55)$$

The CCSD energy is given by the general CC equation (4.53), and amplitude equations are derived by multiplying (4.50) with a singly excited determinant and integrating (analogously to eq. (4.54)).

$$
\langle\Phi_m^e|\mathbf{H}|\left(1 + \mathbf{T}_1 + \left(\mathbf{T}_2 + \tfrac{1}{2}\mathbf{T}_1^2\right) + \left(\mathbf{T}_2\mathbf{T}_1 + \tfrac{1}{6}\mathbf{T}_1^3\right)\right)\Phi_0\rangle = E_{\mathrm{CCSD}}\langle\Phi_m^e|\mathbf{T}_1\Phi_0\rangle
$$

$$
\langle\Phi_m^e|\mathbf{H}|\Phi_0\rangle + \langle\Phi_m^e|\mathbf{H}|\mathbf{T}_1\Phi_0\rangle + \langle\Phi_m^e|\mathbf{H}|\left(\mathbf{T}_2 + \tfrac{1}{2}\mathbf{T}_1^2\right)\Phi_0\rangle
$$

$$
+ \langle\Phi_m^e|\mathbf{H}|\left(\mathbf{T}_2\mathbf{T}_1 + \tfrac{1}{6}\mathbf{T}_1^3\right)\Phi_0\rangle = E_{\mathrm{CCSD}}\sum_{ia} t_i^a \langle\Phi_m^e|\Phi_i^a\rangle
$$

$$
\langle\Phi_m^e|\mathbf{H}|\Phi_0\rangle + \sum_{ia} t_i^a \langle\Phi_m^e|\mathbf{H}|\Phi_i^a\rangle + \sum_{ijab}\left(t_{ij}^{ab} + t_i^a t_j^b - t_j^b t_j^a\right)\langle\Phi_m^e|\mathbf{H}|\Phi_{ij}^{ab}\rangle
$$

$$
+ \sum_{ijkabc}\left(t_{ij}^{ab}t_k^c + \dots + t_i^a t_j^b t_k^c + \dots\right)\langle\Phi_m^e|\mathbf{H}|\Phi_{ijk}^{abc}\rangle = E_{\mathrm{CCSD}}t_m^e
$$

$$(4.56)$$

The notation $(t_i^a t_j^b t_k^c + \dots)$ indicates that several other terms involving permutations of the indices are omitted. Multiplying eq. (4.50) with a doubly excited determinant gives

$$
\langle\Phi_{mn}^{ef}|\mathbf{H}|\left(1 + \mathbf{T}_1 + \left(\mathbf{T}_2 + \tfrac{1}{2}\mathbf{T}_1^2\right) + \left(\mathbf{T}_2\mathbf{T}_1 + \tfrac{1}{6}\mathbf{T}_1^3\right)\right.
$$

$$
\left. + \left(\tfrac{1}{2}\mathbf{T}_2^2 + \tfrac{1}{2}\mathbf{T}_2\mathbf{T}_1^2 + \tfrac{1}{24}\mathbf{T}_1^4\right)\right)\Phi_0\rangle = E_{\mathrm{CCSD}}\langle\Phi_{mn}^{ef}|\left(\mathbf{T}_2 + \tfrac{1}{2}\mathbf{T}_1^2\right)\Phi_0\rangle
$$

$$
\langle\Phi_{mn}^{ef}|\mathbf{H}|\Phi_0\rangle + \langle\Phi_{mn}^{ef}|\mathbf{H}|\mathbf{T}_1\Phi_0\rangle
$$

$$
+ \langle\Phi_{mn}^{ef}|\mathbf{H}|\left(\mathbf{T}_2 + \tfrac{1}{2}\mathbf{T}_1^2\right)\Phi_0\rangle + \langle\Phi_{mn}^{ef}|\mathbf{H}|\left(\mathbf{T}_2\mathbf{T}_1 + \tfrac{1}{6}\mathbf{T}_1^3\right)\Phi_0\rangle
$$

$$
+ \langle\Phi_{mn}^{ef}|\mathbf{H}|\left(\tfrac{1}{2}\mathbf{T}_2^2 + \tfrac{1}{2}\mathbf{T}_2\mathbf{T}_1^2 + \tfrac{1}{24}\mathbf{T}_1^4\right)\Phi_0\rangle = E_{\mathrm{CCSD}}\sum_{ijab}\left(t_{ij}^{ab} + t_i^a t_j^b - t_i^b t_j^a\right)\langle\Phi_{mn}^{ef}|\Phi_{ij}^{ab}\rangle
$$

$$
\langle\Phi_{mn}^{ef}|\mathbf{H}|\Phi_0\rangle + \sum_{ia} t_i^a \langle\Phi_{mn}^{ef}|\mathbf{H}|\Phi_i^a\rangle + \sum_{ijab}\left(t_{ij}^{ab} + t_i^a t_j^b - t_i^b t_j^a\right)\langle\Phi_{mn}^{ef}|\mathbf{H}|\Phi_{ij}^{ab}\rangle
$$

$$
+ \sum_{ijkabc}\left(t_{ij}^{ab}t_k^c + \dots + t_i^a t_j^b t_k^c + \dots\right)\langle\Phi_{mn}^{ef}|\mathbf{H}|\Phi_{ijk}^{abc}\rangle
$$

$$
+ \sum_{ijklabcd}\left(t_{ij}^{ab}t_{kl}^{cd} + \dots + t_{ij}^{ab}t_k^c t_l^d + \dots + t_i^a t_j^b t_k^c t_l^d + \dots\right)\langle\Phi_{mn}^{ef}|\mathbf{H}|\Phi_{ijkl}^{abcd}\rangle
$$

$$
= E_{\mathrm{CCSD}}\left(t_{mn}^{ef} + t_m^e t_n^f - t_m^f t_n^e\right)
$$

$$(4.57)$$

The equations (4.56) and (4.57) involve matrix elements between singles and triples and between doubles and quadruples. However, since the Hamilton operator only contains one- and two-electron operators, these are actually identical to matrix elements between the reference and a doubly excited state. Consider for example $\langle\Phi_m^e|\mathbf{H}|\Phi_{ijk}^{abc}\rangle$. Unless m equals either i, j or k, and e equals either a, b, or c, there will be one overlap integral between different MOs which makes the matrix element zero. If for example $m = k$ and $e = c$, then the MO integral over these indices factor out as 1, and the rest is equal to a matrix element $\langle\Phi_0|\mathbf{H}|\Phi_{ij}^{ab}\rangle$. Similarly, the matrix element $\langle\Phi_{mn}^{ef}|\mathbf{H}|\Phi_{ijkl}^{abcd}\rangle$ between a doubly and a quadruply excited determinant is only non-zero if mn matches up with two of the $ijkl$ indices, and ef matches up with $abcd$. Again such non-zero matrix elements are equal to matrix elements between the reference and a doubly excited determinant, eq. (4.7).

All the matrix elements can be evaluated in terms of MO integrals, and when the expression for the energy (4.53) is substituted into (4.56) and (4.57), they form coupled non-linear equations for the singles and doubles amplitudes. The equations contain terms up to quartic in the amplitudes, e.g. $(t_i^a)^4$ (since $\mathbf{H}$ contains one- and two-electron operators), and must be solved by iterative techniques. Once the amplitudes are known, the energy and wave function can be calculated. The important aspect in coupled cluster methods is that excitations of higher order than the truncation of the $\mathbf{T}$ operator enter the amplitude equation. Quadruply excited states, for example, are generated by the $\mathbf{T}_2^2$ operator in CCSD, and they enter the amplitude equations with a weight given as a product of doubles amplitudes. Quadruply excited states influence the doubles amplitudes, and thereby also the CCSD energy. It is the inclusion of these products of excitations that makes coupled cluster theory size extensive.

In the above the coupled cluster equations have been derived by multiplying the Schrödinger equation with $\langle\Phi_0|$, $\langle\Phi_m^e|$ and $\langle\Phi_{mn}^{ef}|$. An alternative way of deriving the coupled cluster equations is to multiply with $\langle\Phi_0|e^{-\mathbf{T}}$, $\langle\Phi_m^e|e^{-\mathbf{T}}$ and $\langle\Phi_{mn}^{ef}|e^{-\mathbf{T}}$. Just as $e^{\mathbf{T}}$ is an excitation operator working on the function to the right, $e^{-\mathbf{T}}$ is a de-excitation operator working on the function to the left. Thus $\langle\Phi_0|e^{-\mathbf{T}}$ tries to generate de-excitations from the reference, which is impossible, i.e. $\langle\Phi_0|e^{-\mathbf{T}} = \langle\Phi_0|$. However $\langle\Phi_m^e|e^{-\mathbf{T}}$ generates in addition to the singly excited state also the reference wave function. Similarly $\langle\Phi_{mn}^{ef}|e^{-\mathbf{T}}$ generates additionally the reference and singly excited states. The main advantage is that the coupled cluster equations are obtained directly without having to substitute in the expression for the energy in the amplitude equations.

4.10 Connections between Coupled Cluster, Configuration Interaction and Perturbation Theory

The general cluster operator is given by

$$e^{\mathbf{T}} = 1 + \mathbf{T}_1 + \left(\mathbf{T}_2 + \tfrac{1}{2}\mathbf{T}_1^2\right) + \left(\mathbf{T}_3 + \mathbf{T}_2\mathbf{T}_1 + \tfrac{1}{6}\mathbf{T}_1^3\right)$$
$$+ \left(\mathbf{T}_4 + \mathbf{T}_3\mathbf{T}_1 + \tfrac{1}{2}\mathbf{T}_2^2 + + \tfrac{1}{2}\mathbf{T}_2\mathbf{T}_1^2 + \tfrac{1}{24}\mathbf{T}_1^4\right) + \cdots \tag{4.58}$$

where terms have been collected according to the excitation they generate. Each of the operators in a given parenthesis generates all the excited determinants of given type. Both $\mathbf{T}_2$ and $\mathbf{T}_1^2$ generate all doubly excited determinants, and the terms in (4.58) generate all determinants which are inlcuded in a CISDTQ calculation. This cluster expansion can be viewed as a method of dividing up the contributions from each excitation type. The total contribution from double excitations is the sum of two terms, one which is the square of the singles contributions and the remaining is (by definition) the connected doubles. Similarly the total contribution from triple excitations is a sum of three terms, the cube of the singles contributions, the product of the singles and doubles contributions, and the remaining is the connected triples.

When canonical HF orbitals are used the T_1 effect is small, although not zero since singles enter indirectly via the doubly excited states (note that if non-canonical orbitals are used, the T_1 term can be large). From CI we know that the effect of doubles is the most important, Section 4.2.3. In coupled cluster theory the doubles contribution is divided into T_1^2 and T_2. If T_1 is small, then T_1^2 must also be small, and the most important term is T_2. For the triples excitations, T_1^3 must be negligible, and $T_1 T_2$ is small owing to T_1. The most important contribution is therefore from connected triples T_3. For the quadruple excitations, all the terms involving T_1 must again be small, and since T_2 is large, we expect the disconnected quadruples T_2^2 to be the dominant term. This again suggests that the connected quadruples term T_4 is small. Higher-order excitations will always contain terms appearing as powers and/or products of T_2 and T_3, which normally will be dominanting. Higher-order connected terms, T_n with $n > 4$, are therefore expected to have small effects. This is consistent with the physical picture that connected T_n operators correspond to n electrons interacting simultaneously. As n becomes large, this is increasingly improbable. It should be noted however, that the higher-order cluster operators (T_4, T_5, ...) become more and more important as the number of electrons increases.

The principal deficiency of CISD is the lack of the T_2^2 term, which is the main reason for CISD not being size extensive. Furthermore, this term becomes more and more important as the number of electrons increases, and CISD therefore recovers a smaller and smaller percentage of the correlation energy as the system increases. There are various approximate corrections for this lack of size extensivity which can be added to standard CISD. The most widely known of these is the *Davidson correction*, sometimes denoted CISD + Q(Davidson), where the quadruples contribution is approximated as

$$\Delta E_Q = (1 - a_0^2) \Delta E_{CISD} \qquad (4.59)$$

with a_0 being the coefficient in front of the HF reference. If the renormalization of the wave function is also taken into account, the $(1 - a_0^2)$ quantity is divided by a_0^2, and the corresponding correction is called the *renormalized Davidson correction*. The effect of higher-order excitations is thus estimated from the correlation energy obtained at the CISD level times a factor which measures how important the single determinant reference is at the CISD level. The Davidson correction does not yield zero for two-electron systems, where CISD is equivalent to full CI, and it is likely that it overestimates the higher-order corrections for systems with few electrons. More complicated correction schemes have also been proposed[19] but are rarely used.

Coupled cluster is closely connected with Møller–Plesset perturbation theory, as mentioned at the start of this section. The infinite Taylor expansion of the exponential operator (eq. (4.46)) ensures that the contributions from a given excitation level are included to infinite order. Perturbation theory indicates that doubles are the most important, they are the only contributors to MP2 and MP3. At fourth order, there are contributions from singles, doubles, triples and quadruples. The MP4 quadruples contribution is actually the disconnected $\mathbf{T}_2^2$ term in the coupled cluster language, and the triples contribution corresponds to $\mathbf{T}_3$. This is consistent with the above analysis, the most important is $\mathbf{T}_2$ (and products thereof) followed by $\mathbf{T}_3$. The CCD energy is equivalent to MP ∞ (D) where all disconnected contributions of products of doubles are included. If the perturbation series is reasonably converged at fourth order, we expect that CCD $\sim$ MP4(DQ), and CCSD $\sim$ MP4(SDQ). The MP2, MP3 and MP4(SDQ) results may be obtained in the first iteration for the CCSD amplitudes, allowing a direct test of the convergence of the MP series. This also points out the principal limitation of the CCSD method, the neglect of the connected triples. Including them in the $\mathbf{T}$ operator leads to the CCSDT method which, as mentioned above, is too demanding computationally for all but the smallest systems. Alternatively the triples contribution may be evaluated by perturbation theory and added to the CCSD results. Several such hybrid methods exist;[20] two of the most common are known by the acronyms CCSD + T(CCSD) and CCSD(T). In both cases the triples contribution is calculated from the formula given by MP4, but using the CCSD amplitudes instead of the perturbation coefficients for the wave function corrections (eqs. (4.37) and (4.39)). For the CCSD(T) method an additional term arising from fifth-order perturbation theory, describing the coupling between singles and triples, is also included. This is computationally inexpensive to calculate, and the CCSD(T) method is preferred over CCSD + CCSD(T). Higher-order hybrid methods such as CCSD(TQ), where the connected quadruples contribution is estimated by fifth-order perturbation theory, are also possible, but they are again so demanding that they can only be used for small systems.[21]

As mentioned, the singles make a fairly small contribution to the correlation energy when canonical HF orbitals are used. *Brueckner* theory[22] is a variation of coupled cluster where the orbitals used for constructing the Slater determinants are optimized so that the contribution from singles is exactly zero, i.e. $t_i^a = 0$. The lowest level of Brueckner theory includes only doubles, giving the acronym BD. Although BD in theory should be slightly better than CCSD, since it includes orbital relaxation, it gives in practice essentially identical results (differences between BD and CCSD are of fifth order or higher in term of perturbation theory). This is presumably rooted in the fact that the singles in CCSD introduce orbital relaxation.[23] The computational cost is also very similar for CCSD and BD.[24] Similarly BD(T) is essentially equivalent to CCSD(T),[25] and BD(TQ) to CCSD(TQ).

Since the singly excited determinants effectively relax the orbitals in a CCSD calculation, non-canonical HF orbitals can also be used in coupled cluster methods. This allows for example the use of open-shell singlet states (which require two Slater determinants) as reference for a coupled cluster calculation.[26]

Another commonly used method is *Quadratic* CISD (QCISD). It was originally derived from CISD by including enough higher-order terms to make it size extensive.[27] It has later been shown that the resulting equations are identical to CCSD where some of the terms have been omitted.[28] The omitted terms are computationally inexpensive, and there appears to be no reason for using the less complete QCISD over CCSD (or QCISD(T) in place of CCSD(T)), although in practice they normally give very similar results.[29] There are a few other methods which may be considered either as CISD with addition of extra terms to make them approximately size extensive, or as approximate versions of CCSD. Some of the methods falling into this category are *Averaged*

Coupled–Pair Functional (ACPF) and *Coupled Electron Pair Approximation* (CEPA). The simplest form of CEPA, CEPA-0, is also known as *Linear Coupled Cluster Doubles* (LCCD).

Recently two new intermediate coupled cluster methods for calculating molecular properties have been defined, known as CC2 and CC3.[30] The single excitations allow the MOs to relax from their HF form, but do not give any direct contribution to the energy due to the Brillouin theorem. For studying properties which measure the response of the energy to a perturbation, the HF orbitals are no longer optimum, and the singles are at least as important as the doubles. The CC2 method is derived from CCSD by including only the doubles contribution arising from the lowest (non-zero) order in perturbation theory, where the perturbation is defined as in MP theory (i.e. as the true electron–electron potential minus twice the average repulsion). The amplitude equations corresponding to multiplication of a doubly excited determinant in the CCSD equations (eq. (4.57)) thereby reduce to an MP2-like expression, and the t_2 amplitudes may be expressed directly in terms of the t_1 amplitudes and MO integrals. The iterative procedure therefore only involves the t_1 amplitudes. CC2 may loosely be defined as MP2 with the added feature of orbital relaxation arising from the singles. Similarly, CC3 is an approximation to the full CCSDT model, where the triples contribution is approximated by the expression arising from the lowest non-vanishing order in perturbation theory. The triples amplitudes can then be expressed directly in terms of the singles and doubles amplitudes, and MO integrals. Both in terms of computational cost and accuracy, the following progression is expected, although the CC2 and CC3 models are so new that there are few data for comparison.

$$HF \ll CC2 < CCSD < CC3 < CCSDT$$

Analogously to MP methods, coupled cluster theory may also be based on a UHF reference wave function. The resulting UCC methods again suffer from spin contamination of the underlying UHF, but the infinite nature of coupled cluster methods is substantially better at reducing spin contamination relative to UMP.[31] Projection methods analogous to those of the PUMP case have been considered but are not commonly used. ROHF based coupled cluster methods have also been proposed, but appear to give results very similar to UCC, especially at the CCSD(T) level.[32]

Standard coupled cluster theory is based on a single determinant reference wave function. If suffers from the same problem as MP: it works best if the zeroth-order wave function is sufficiently "good". Due to the summation of contributions to infinite order, however, coupled cluster is somewhat more tolerant of a poor reference wave function than MP methods. Since the singly excited determinants allow the MOs to relax in order to describe the multi-reference character in the wave function, the magnitude of the singles amplitude is an indicator of how good the HF single determinant is as a reference. The T_1-*diagnostic* defined as the norm of the singles amplitude vector divided by the square root of the number of electrons has been suggested as an internal evaluation of the quality of a CCSD wave function.[33]

$$T_1 = \frac{|\mathbf{t}_1|}{\sqrt{N}} \tag{4.60}$$

Specifically, if $T_1 < 0.02$, the CCSD(T) method is expected to give results close the full CI limit for the given basis set. If T_1 is larger than 0.02, it indicates that the reference wave function has significant multi-determinant character, and multi-reference coupled cluster should preferentially be employed. Such methods are being developed[34] but have not yet seen any extensive use.

4.11 Methods Involving Interelectronic Distances

The necessity for going beyond the HF approximation is the fact that electrons are further apart than described by the product of their orbital densities, i.e. their motions are correlated. This arises from the electron–electron repulsion operator, which is a sum of terms of the type

$$\frac{1}{|\mathbf{r}_1 - \mathbf{r}_2|} = \frac{1}{r_{12}} \tag{4.61}$$

Without these terms the Schrödinger equation can be solved exactly, with the solution being a Slater determinant composed of orbitals.

The electron–electron repulsion operator has a singularity for $r_{12} = 0$ which results in the exact wave function having a *cusp* (discontinuous derivative).[35]

$$\left(\frac{\partial \Psi}{\partial r_{12}}\right)_{r_{12}=0} = \tfrac{1}{2}\Psi(r_{12}=0) \tag{4.62}$$

In other words, the exact wave function behaves asymptotically as a constant $+ 1/2r_{12}$ when r_{12} is small. It would therefore seem natural that the interelectronic distance would be a necessary variable for describing electron correlation. For two-electron systems, extremely accurate wave functions may be generated by taking a trial wave function consisting of an orbital product times an expansion in electron coordinates such as

$$\Psi(\mathbf{r}_1, \mathbf{r}_2) = e^{-\alpha_1 r_1} e^{-\alpha_2 r_2} \sum_{klm} C_{klm}(\mathbf{r}_1 + \mathbf{r}_2)^k (\mathbf{r}_1 - \mathbf{r}_2)^l r_{12}^m \tag{4.63}$$

and variationally optimize the α_i and C_{klm} parameters. Such expansions are known as *Hylleraas* type wave functions. For the hydrogen molecule it is possible to converge the total energy to $\sim 10^{-9}$ a.u., which is more accurate than can be determined experimentally. In fact, the prediction that the experimental dissociation energy for H_2 was wrong, based on calculations, was one of the first hallmarks of quantum chemistry.[36] Such wave functions unfortunately become impractical for more than 3–4 electrons.

All electron correlation methods based on expanding the N-electron wave function in terms of Slater determinants built from orbitals (one-electron functions) suffer from an agonizingly slow convergence. Literally millions or billions of determinants are required for obtaining results which in an absolute sense are close to the exact results. This is due to the fact that products of one-electron functions are poor at describing the cusp behaviour of the wave function when two electrons are close together. At the second-order perturbation level (i.e. MP2) it may be shown that the error in the correlation energy behaves asymptotically as $(L + 1/2)^{-4}$, where l is the highest angular momentum in the basis set. For a general wave function the convergence is $(L + 1/2)^{-4} + (L + 1/2)^{-5} + (L + 1/2)^{-6} + \dots$. This means that the total energy will converge as $(L + 1)^{-3} + (L + 1)^{-4} + (L + 1)^{-5} + \dots$, if the basis set is saturated up to angular momentum L.[37] For sufficiently large values of L the convergence is thus $\sim (L + 1)^{-3}$, which is quite slow.

In order to achieve a high accuracy, it would seem desirable to explicitly include terms in the wave functions which are linear in the interelectronic distance. This is the idea in the R12 methods developed by Kutzelnigg and co-workers.[38] The first order correction to the HF wave function only involves doubly excited determinants (eqs. (4.35) and (4.37)). In R12 methods additional terms are included which essentially are the HF determinant multiplied with r_{ij} factors.

$$\Psi_{R12} = \Phi_{HF} + \sum_{ijab} a_{ijab} \Phi_{ij}^{ab} + \sum_{ij} b_{ij} r_{ij} \Phi_{HF} \qquad (4.64)$$

The exact definition is slightly more complicated, since the wave function has to be properly antisymmetrized and projected onto the actual basis, but for illustration the above form is sufficient. Such R12 wave functions may then be used in connection with the CI, MBPT or CC methods described above. Consider for example a CI calculation with an R12 type wave function. The energy is given as

$$E = \langle \Psi_{R12} | \mathbf{H} | \Psi_{R12} \rangle \qquad (4.65)$$

and the a_{ijab} and b_{ij} parameters in (4.64) are optimized variationally. The overwhelming problem is that matrix elements from (4.65) now involve integrals depending on three and four electron coordinates. Consider for example the following terms arising from the r_{ij} operator written out in terms of the one- and two-electron operators (eq. (3.24)).

$$\langle \Phi_{HF} | \mathbf{H} | r_{ij} \Phi_{HF} \rangle = \langle \Phi_{HF} | \mathbf{h} | r_{ij} \Phi_{HF} \rangle + \langle \Phi_{HF} | \mathbf{g} | r_{ij} \Phi_{HF} \rangle$$
$$\langle r_{ij} \Phi_{HF} | \mathbf{H} | r_{ij} \Phi_{HF} \rangle = \langle r_{ij} \Phi_{HF} | \mathbf{h} | r_{ij} \Phi_{HF} \rangle + \langle r_{ij} \Phi_{HF} | \mathbf{g} | r_{ij} \Phi_{HF} \rangle \qquad (4.66)$$

the **g**-operator leads to integrals over molecular orbitals of the type

$$\left\langle \phi_i(1)\phi_j(2)\phi_k(3) \left| \frac{r_{12}}{r_{13}} \right| \phi_i(1)\phi_j(2)\phi_k(3) \right\rangle$$

$$\left\langle \phi_i(1)\phi_j(2)\phi_k(3) \left| \frac{r_{12}r_{23}}{r_{13}} \right| \phi_i(1)\phi_j(2)\phi_k(3) \right\rangle \qquad (4.67)$$

$$\left\langle \phi_i(1)\phi_j(2)\phi_k(3)\phi_l(4) \left| \frac{r_{12}r_{13}}{r_{23}} \right| \phi_i(1)\phi_j(2)\phi_k(3)\phi_l(4) \right\rangle$$

Not only are such integrals difficult to calculate, but when the MOs are expanded in a basis set consisting of M AOs, there will be on the order of M^6 three-electron integrals and on the order of M^8 four-electron integrals. Such methods are therefore inherently more expensive than for example the full CCSDT model.

The trick for turning the R12 method into a viable tool is to avoid calculating the three- and four-electron integrals, without jeopardizing the accuracy. In a complete basis, a three-electron integral may be written in terms of products of two-electron integrals by inserting a "resolution of the identity" between the two operators.

$$1 = \sum_p^\infty |\phi_p\rangle\langle\phi_p| = \sum_{pqr}^\infty |\phi_p\phi_q\phi_r\rangle\langle\phi_p\phi_q\phi_r|$$

$$\left\langle \phi_i\phi_j\phi_k \left| \frac{r_{12}}{r_{13}} \right| \phi_i\phi_j\phi_k \right\rangle = \sum_{pqr}^\infty \langle\phi_i\phi_j\phi_k|r_{12}|\phi_p\phi_q\phi_r\rangle\left\langle\phi_p\phi_q\phi_r\left|\frac{1}{r_{13}}\right|\phi_i\phi_j\phi_k\right\rangle$$

$$= \sum_{pqr}^\infty (\delta_{kr}\langle\phi_i\phi_j|r_{12}|\phi_p\phi_q\rangle)\left(\delta_{qj}\left\langle\phi_p\phi_r\left|\frac{1}{r_{13}}\right|\phi_i\phi_k\right\rangle\right)$$ (4.68)

$$= \sum_p^\infty \langle\phi_i\phi_j|r_{12}|\phi_p\phi_j\rangle\left\langle\phi_p\phi_k\left|\frac{1}{r_{13}}\right|\phi_i\phi_k\right\rangle$$

The first reduction occurs since the r_{12} and r_{13}^{-1} operators only involve two-electron coordinates, the second reduction is due to the two delta functions. Three- and four-electron integrals can therefore be written as a sum over products of integrals involving only two electron coordinates. In a finite basis set, the resolution is not exact, and the identities in eq. (4.68) become approximations. The beauty of the R12 methods is that this error can be controlled, and is not significantly larger than the inherent basis set error, once the basis set reaches a certain size. The significance of R12 methods is that the energy error in terms of angular momentum of the basis set now behaves approximately as $(L+1)^{-7}$, a significant improvement over standard methods. The drawback is that R12 methods cannot be used with small basis sets, only in well-polarized basis sets is the resolution obeyed with a reasonable accuracy. In practice this means that a fairly dense basis set including up to at least f-functions must be used. However, if accurate results are desired, a large polarized basis set is required anyway, and R12 methods converge much faster in terms of basis set extension than traditional methods.[39] It should be noted that in the limit of a complete basis set the MP2-R12 (for example) will give the same result as a traditional MP2 calculation, i.e. the R12 approach speeds up the basis set convergence, but does not change the fundamental characteristics of the MP2 method. The improved convergence, however, does not come for free: there is a significantly larger number (and different types) of two-electron integrals which must be calculated and handled.

4.12 Direct Methods

Conventional HF methods rely on storing the two-electron integrals over atomic orbitals on disk, and reading them in each SCF iteration, while direct methods generate the integrals as they are needed (Section 3.8.5). This is an easy change in algorithm since the HF energy is expressed directly in terms of AO integrals. Methods involving electron correlation, however, require matrix elements between Slater determinants, which can be expressed in terms of integrals over MOs (eq. (4.7)). Conventional methods for the integral transformation (Section 4.2.1) read the AOs, perform the multiplications with the MO coefficients (eq. (4.11)), and write the MO integrals to disk. These can then be read in and used in the correlation treatment. Although the number of MO integrals typically is somewhat smaller than the number of AO integrals (for example MO integrals involving four virtual orbitals may not be needed), the disk space requirements are still significant if more than a few hundred basis functions are used. To eliminate the disk space requirements, and remove the relatively inefficient data transfer step for reading/writing to disk, it is desirable also to have direct algorithms for electron correlation method. Direct in this context means that the integrals are calculated as needed and then discarded. The need for integrals over MOs instead of AOs, however, makes the development of direct methods in electron correlation somewhat more complicated than at the HF level.

Consider for example the MP2 energy expression.[40]

$$\text{MP2} = \sum_{i<j}^{\text{occ}} \sum_{a<b}^{\text{vir}} \frac{[\langle \phi_i \phi_j | \phi_a \phi_b \rangle - \langle \phi_i \phi_j | \phi_b \phi_a \rangle]^2}{\varepsilon_i + \varepsilon_j - \varepsilon_a - \varepsilon_b} \tag{4.69}$$

with the MO integrals given as

$$\langle \phi_i \phi_j | \phi_a \phi_b \rangle = \sum_\alpha \sum_\beta \sum_\gamma \sum_\delta c_{\alpha i} c_{\beta j} c_{\gamma a} c_{\delta b} \langle \chi_\alpha \chi_\beta | \chi_\gamma \chi_\delta \rangle \tag{4.70}$$

Since each MO integral in principle contains contributions from <u>all</u> the AO integrals, a straightforward calculation of an MO integral each time it is needed will involve a generation of all the AO integrals. In other words, it would be necessary to recalculate the AO integrals $\sim O^2 V^2$ times (O and V being the number of occupied and virtual orbitals), compared to the 15–20 times in an SCF calculation. The MP2 method would therefore change from being an M^5 to an M^8 method, which is clearly an unacceptably large penalty for a direct method.

The M^8 dependence is a consequence of performing the four index transformation with all four indices at once. As shown in Section 4.2.1, it is advantageous to perform the transformation one index at a time.

$$\begin{aligned}
\langle \phi_i \phi_j | \phi_a \phi_b \rangle &= \sum_\delta c_{\delta b} \langle \phi_i \phi_j | \phi_a \chi_\delta \rangle \\
\langle \phi_i \phi_j | \phi_a \chi_\delta \rangle &= \sum_\gamma c_{\gamma a} \langle \phi_i \phi_j | \chi_\gamma \chi_\delta \rangle \\
\langle \phi_i \phi_j | \chi_\gamma \chi_\delta \rangle &= \sum_\beta c_{\beta j} \langle \phi_i \chi_\beta | \chi_\gamma \chi_\delta \rangle \\
\langle \phi_i \chi_\beta | \chi_\gamma \chi_\delta \rangle &= \sum_\alpha c_{\alpha i} \langle \chi_\alpha \chi_\beta | \chi_\gamma \chi_\delta \rangle
\end{aligned} \tag{4.71}$$

By choosing the right order of the transformation the scaling can be reduced considerably. In eq. (4.71) the indices corresponding to the occupied orbitals may be transformed before the virtuals. There are of the order of M^4 AO integrals, $\langle \chi_\alpha \chi_\beta | \chi_\gamma \chi_\delta \rangle$, but only OM^3 quarter transformed integrals, $\langle \phi_i \chi_\beta | \chi_\gamma \delta_\sigma \rangle$. Instead of storing and reading the AO integrals from the SCF step, they can be recalculated in the transformation step, reducing the storage from M^4 to OM^3. The subsequent quarter transformations require less storage, i.e. the next transformation with an occupied index reduces the number of integrals to $O^2 M^2$, the third to $O^2 VM$, and the last to $O^2 V^2$. Since the MP2 energy can be written as a sum of contributions from each occupied orbital, we can furthermore treat one occupied orbital at a time, i.e. first sum all contributions of $\langle \phi_1 \chi_\beta | \chi_\gamma \chi_\delta \rangle$, then $\langle \phi_2 \chi_\beta | \chi_\gamma \chi_\delta \rangle$ etc. This reduces the necessary storage to only order M^3. It may be further reduced to OVM by proper scheduling of the evaluation order of the remaining three indices. The OVM number of integrals is much less than the original M^4, and will in many cases fit into the memory. The net result is that disk storage is effectively eliminated, or at least greatly reduced. If only one occupied orbital is treated at a time, O integral evaluations are required, however, the more memory that is available, the more occupied orbitals can be treated in a single sweep, decreasing the number of integral evaluations.

The above is an example of how direct algorithms may be formulated for methods involving electron correlation. It illustrates that it is not as straightforward to apply direct methods at the correlated level as at the SCF level. However, the steady increase in CPU performance, and especially the evolution of multiprocessor machines, favours direct (and semi-direct where some intermediate results are stored on disk) algorithms. Recently direct methods have also been implemented at the coupled cluster level.[41]

4.13 Localized Orbital Methods

Ab initio calculations involving electron correlation essentially always build on a set of canonical HF orbitals. As illustrated in this chapter, this leads to a computational effort which increases as a rather high power of the system size, i.e. M^5–M^8. Considering that the fundamental physical force is only between pairs of particles, this scaling is "unphysical". One of the reason for the high scaling is that fact the canonical orbitals in general are delocalized over the whole molecule, i.e. essentially all orbitals make a (small) contribution to the wave function for a specific part of the molecule. This suggests that localized orbitals may be a better starting point, since a single or only a few orbitals then would contribute the large majority at a given point, and the remaining contributions could simply be neglected. Such local MP2 and local CC methods have started to appear, but are not yet commonly used.[42] These methods are somewhat more complicated to formulate as the Fock matrix is only diagonal in the canonical orbitals. Nevertheless, methods based on localized orbitals hold the promise of a near-linear scaling with problem size in the large-scale limit. It is at present not clear exactly how large the systems need to be to reach the "large-scale" limit.

4.14 Summary of Electron Correlation Methods

The only generally applicable methods are: CISD, MP2, MP3, MP4, CCSD and CCSD(T). CISD is variational, but not size extensive, while MP and CC methods are non-variational but size extensive. CISD and MP are in principle non-iterative methods, although the matrix diagonalization involved in CISD usually is so large that it has to be done iteratively. Solution of the coupled cluster equations must be done by an iterative technique since the parameters enter in a non-linear fashion. In terms of the most expensive step in each of the methods they may be classified according to how they formally scale in the large system limit, shown in Table 4.5.

We have so far been careful to used the wording "formal scaling". As already discussed, HF is formally an M^4 method but in practice the scaling may be reduced all the way down to M^1. Similarly, MP2 is formally an M^5 method. However, an MP2 calculation consists of three main parts: the HF calculation, the AO to MO integral

Table 4.5 Limiting scaling in terms of basis set size M for different methods

Scaling	Cl methods	MP methods	CC methods
M^5		MP2	CC2 (iterative)
M^6	CISD	MP3, MP4(SDQ)	CCSD (iterative)
M^7		MP4	CCSD(T), CC3 (iterative)
M^8	CISDT	MP5	CCSDT (iterative)
M^9		MP6	
M^{10}	CISDTQ	MP7	CCSDTQ (iterative)

transformation, and the MP2 energy calculation. Only the second part has a formal scaling of M^5, the other are (formal) M^4 steps. In the large system limit the transformation required for the MP2 procedure <u>will</u> become the most expensive step, however, in practice, where calculations may be restricted to a few hundred basis functions, it is often observed that the MP2 step takes less time than the HF. The formal scaling only indicates what the rate limiting step will be in the large system limit. Whether this limit is actually reached in practical calculations is something different.

The lower value of M^5 scaling for methods involving electron correlation arises from the transformation of the two-electron integrals from the AO to MO basis, but if the transformation is carried out with one of the indices belonging to an occupied MO first, the scaling is actually the number of occupied orbitals (O) times M^4. If we consider making the system larger by doubling the fundamental unit (for example calculations on a series of increasingly larger water clusters), keeping the basis set per atom constant, O scales linearly with M, and we arrive at the M^5 scaling. This assumption (increasing system size) is the basis for Table 4.5. More often, however, a series of calculations are performed on the same system with increasingly larger basis sets. In this case the number of electrons (occupied orbitals) are constant and the scaling is M^4. Many of the commonly employed methods for electron correlation (including for example MP2, MP3, MP4, CISD, CCSD and CCSD(T)) scale in fact as M^4 when the number of occupied orbitals is constant.

In terms of accuracy with a medium sized basis set the following order is often observed.

$$\text{HF} \ll \text{MP2} < \text{CISD} < \text{MP4(SDQ)} \sim \text{CCSD} < \text{MP4} < \text{CCSD(T)} \qquad (4.72)$$

All of these are single determinant based methods. Multi-reference methods cannot easily be classified as the quality of the results depends heavily on the size of the reference. A two-configurational references is only a slight improvement over HF, but including all configurations generates a full CI. The ordering above is only valid when the HF reference is a "good" zero-order description of the system. The more multi-reference character in the wave function, the better the "infinite"-order coupled cluster performs relative to perturbation methods.

MP3 has not been included in the above comparison. As already mentioned, MP3 results are often inferior to those at MP2. In fact MP2 often gives surprisingly good results, especially if large basis sets are used.[43] Furthermore, it should be kept in mind that the MP perturbation series in many cases may actually be divergent, although corrections carried out to low order (i.e. 2–4) rarely display excessive oscillations.

HF results are by modern standards more and more approaching model calculations, like semi-empirical methods such as AM1 and PM3. Minimal basis HF calculations often give results which are worse than AM1 or PM3, but at a computational cost of maybe 100 times as much. Medium and large basis set HF calculations usually do not give absolute results which are particular close to experimental values, but since the errors to a certain degree are systematic (like all vibrational frequencies being overestimated by $\sim 10\%$), they can be used with more or less "empirical" corrections to treat systems for which correlated calculations are not possible. The distinct advantage of *ab initio* methods is their ability to treat all systems at an equal level of accuracy, independently of whether experimental data exist or not. A detailed assessment of the level of accuracy that can be expected at a given level of theory is difficult to establish as it is heavily dependent on the quality of the basis set. Given a sufficiently large basis set, however, the CCSD(T) method is able to meet the goal of an accuracy of $\sim 1\,\text{kcal/mol}$ for most systems. Even with less complete methods (like MP4) and medium size basis sets such as DZP or TZ2P, it is often possible to get accuracies of the order of a few kcal/mol.

The use of CI methods has been declining in recent years, to the profit of MP and especially CC methods. It is now recognized that size extensivity is important for obtaining accurate results. Excited states, however, are somewhat difficult to treat by perturbation or coupled cluster methods, and CI or MCSCF based methods have been the preferred methods here. More recently propagator or equation of motion (Section 10.9) methods have been developed for coupled cluster wave functions, which allows calculation of exited state properties.

Finally, a few words on the size of system that can be treated. The limiting parameters will again be taken as the number of basis functions, although as noted above, a more detailed breakdown in terms of occupied and virtual MOs can be done. Note also that a given limit in terms of basis functions may translate either into a large molecular system with a small basis per atom, or a small molecular system with a very large basis set on each atom. The ordering in eq. (4.72) suggests three levels of electron correlation: none (HF), MP2 or extended (MP4 or CCSD(T)). HF methods are in general possible with up to ~ 5000 basis functions, MP2 is fairly routine with up to ~ 800 basis functions, while the advanced correlation methods are limited to $\sim 300\text{-}400$ basis functions. With a DZP basis these values translate into roughly 200, 30 and 10 CH_2 fragments. The limits hold just for calculating the energy at a single geometry. If more advanced features are desired, such as optimizing the geometry or calculating frequencies, the limits drop to roughly half of the above.

With the continuing advances in computer hardware and more efficient algorithms, these limits are gradually being shifted upwards. Owing to the rather steep scaling with system size, however, they will (barring a fundamental breakthrough) give a rough idea of the size of systems which can be handled also in the future. Currently the speed of computer hardware improves by a factor of 2 in a timespan of about 18 months. In other words, a factor of 10 in terms of performance for the same price is gained roughly every five years. Due to a scaling between 4 and 7 for the different methods, however, an increase by a factor of 10 in raw speed only translates into an increase in system size of 1.7 (M^4 scaling) or 1.4 (M^7 scaling). Linear scaling methods in Hartree–Fock methods of course will benefit fully from increased computational speed.

4.15   Excited States

The development of HF and correlated methods in the previous chapters has focused on the electronic ground state. In some cases it is also of interest to consider electronically excited states. It is useful to distinguish between two cases, depending on whether the excited state has the same or a different symmetry than the lower state(s). The different symmetry case is easy to handle, as the lowest energy state of a given symmetry may be handled completely analogously to the ground state. A HF wave function may be obtained by a proper specification of the occupied orbitals, and the resulting wave function improved by adding electron correlation by for example CI MP or CC methods. The only possible caveat may be that the state is an open shell, which often requires a (small) MCSCF wave function for an adequate zero-order description.

Excited states having lower energy solutions of the same symmetry are somewhat more difficult to treat. It will in general be difficult to generate a HF type wave function for such states, as the variational optimization will collapse to the lowest energy solution of the given symmetry. The lack of a proper HF solution means that perturbation and coupled cluster methods are not well suited to calculating excited states, although excited state properties (for example excitation energies) may be calculated directly using propagator methods (Section 10.9). Propagator methods can be based for example on a coupled cluster wave function. It is, however, relatively easy to generate higher energy states by CI methods; this simply corresponds to using the $(n + 1)$th eigenvalue from the diagonalization of the CI matrix as a description of the nth excited state (the second root is the first excited state etc.). Such a CI procedure will normally employ a set of HF orbitals from a calculation on the lowest energy state, and the CI procedure is therefore biased against the excited states.

The simplest description of an excited state is the orbital picture where one electron has been moved from an occupied to an unoccupied orbital, i.e. an S-type determinant as illustrated in Figure 4.1. The lowest level of theory for a qualitative description of excited states is therefore a CI including only the singly excited determinants, denoted CIS. CIS gives wave functions of roughly HF quality for excited states, since no orbital optimization is involved. For valence excited states, for example those arising from excitations between π-orbitals in an unsaturated system, this may be a reasonable description. There are, however, normally also quite low-lying states which essentially correspond to a double excitation, and those require at least inclusion of the doubles as well, i.e. CISD.

A more balanced description requires MCSCF based methods where the orbitals are optimized for each particular state, or optimized for a suitable average of the desired states (*state averaged* MCSCF). It should be noted that such excited state MCSCF solutions correspond to saddle points in the parameter space for the wave function, and second-order optimization techniques are therefore almost mandatory. In order to obtain accurate excitation energies it is normally necessarily to also include dynamical correlation, for example by using the CASPT2 method.

References

1. A. Szabo and N. S. Ostlund, *Modern Quantum Chemistry*, McGraw-Hill, 1982; R. McWeeny, *Methods of Molecular Quantum Mechanics*, Academic Press, 1992; W. J. Hehre, L. Radom, J. A. Pople and P. v. R. Schleyer, *Ab Initio Molecular Orbital Theory*, Wiley, 1986; J. Simons, *J. Phys. Chem.*, **95** (1991), 1017; R. J. Bartlett, J. F. Stanton, *Rev. Comput. Chem.*, **5** (1994), 65.

2. J. Olsen, P. Jørgensen, H. Koch, A. Balkova and R. J. Bartlett, *J. Chem. Phys.*, **104** (1996), 8007.

3. J. Olsen, P. Jørgensen and J. Simons, *Chem. Phys. Lett.*, **169** (1990), 463.

4. E. Davidson, *J. Comput. Phys.*, **17** (1975), 87.

5. B. O. Roos, in *Lecture Notes in Quantum Chemistry*, Ed. B. O. Roos, Springer-Verleg, 1992.

6. J. Olsen, B. O. Roos, P. Jørgensen and H. J. Aa. Jensen, *J. Chem. Phys.*, **89** (1998), 2185.

7. W. T. Borden and E. R. Davidson, *Acc. Chem. Res.*, **29** (1996), 67.

8. C. Møller and M. S. Plesset, *Phys. Rev.*, **46** (1934), 618.

9. J. Olsen, O. Christiansen, H. Koch and P. Jørgensen, *J. Chem. Phys.*, **105** (1996), 5082.

10. T. H. Dunning, Jr. and K. A. Peterson, *J. Chem. Phys.*, **108** (1998), 4761.

11. D. Cremer and Z. He, *J. Phys. Chem.*, **100** (1996), 6173.

12. O. Christiansen, J. Olsen, P. Jørgensen, H. Koch and P.-Å. Malmqvist, *Chem. Phys. Lett.*, **261** (1996), 369.

13. N. C. Handy, P. J. Knowles and K. Somasundram, *Theo. Chem. Acta*, **68** (1985), 87.

14. P. M. Kozlowski and E. R. Davidson, *J. Chem. Phys.*, **100** (1994), 3672.

15. H. B. Schlegel, *J. Phys. Chem.*, **92** (1988), 3075; P. J. Knowles and N. C. Handy, *J. Phys. Chem.*, **92** (1988), 3097.

16. B. O. Roos, K. Andersson, M. P. Fülscher, P.-Å. Malmqvist, L. Serrano-Andres, K. Pierloot and M. Merchan, *Adv. Chem. Phys.*, **93** (1996), 216.

17. R. J. Bartlett, *J. Phys. Chem.*, **93** (1989), 1697.

18. J. D. Watts and R. J. Bartlett, *Int. J. Quantum Chem.*, **S27** (1993), 51.

19. J. M. L. Martin, J. P. Francois and R. Gijbels, *Chem. Phys. Lett.*, **172** (1990), 346.

20. G. E. Scuseria and T. J. Lee, *J. Chem. Phys.*, **93** (1990), 5851.

21. K. Raghavachari, J. A. Pople, E. S. Replogle and M. Head-Gordon, *J. Phys. Chem.*, **94** (1990), 5579; R. J. Bartlett, J. D. Watts, S. A. Kucharski and J. Noga, *Chem. Phys. Lett.*, **165** (1990), 513.

22. K. A. Brueckner, *Phys. Rev.*, **96** (1954), 508; J. F. Stanton, J. Gauss and R. J. Bartlett, *J. Chem. Phys.*, **97** (1992), 5554.

23. E. A. Salter, H. Sekino and R. J. Bartlett, *J. Chem. Phys.*, **87** (1987), 502.

24. C. Hampel, K. A. Peterson and H.-J. Werner, *Chem. Phys. Lett.*, **190**, (1992), 1.

25. T. J. Lee, R. Kobayashi, N. C. Handy and R. D. Amos, *J. Chem. Phys.*, **96** (1992), 8931.

26. A. Balkova and R. J. Bartlett, *Chem. Phys. Lett.*, **193** (1992), 364.

27. J. A. Pople, M. Head-Gordon and K. Raghavachari, *J. Chem. Phys.*, **87** (1987), 5968.

28. G. E. Scuseria and H. F. Schaefer III, *J. Chem. Phys.*, **90** (1989), 3700.

29. T. J. Lee, A. P. Rendall and P. R. Taylor, *J. Phys. Chem.*, **94** (1990), 5463; for an exception see M. Böhme and G. Frenkine *Chem. Phys. Lett.*, **224** (1994), 195.

30. O. Christiansen, H. Koch and P. Jørgensen, *Chem. Phys. Lett.*, **243** (1995), 409; O. Christiansen, H. Koch and P. Jørgensen, *J. Chem. Phys.*, **103** (1995), 7429.

31. J. F. Stanton, *J. Chem. Phys.*, **101** (1994), 371.

32. J. P. Watts, J. Gauss and R. J. Bartlett, *J. Chem. Phys.*, **98** (1993), 8718.

33. T. J. Lee and P. R. Taylor, *Int. J. Quant. Chem.*, **S23** (1989), 199.

34. P. G. Szalay and R. J. Bartlett, *J. Chem. Phys.*, **101** (1994), 4936.

35. T. Kato, *Commun. Pure Appl. Math.*, **10** (1957), 151.

36. W. Kolos and L. Wolniewics, *J. Chem. Phys.*, **49** (1968), 404.
37. W. Kutzelnigg and J. D. Morgan III, *J. Chem. Phys.*, **96** (1992), 4484.
38. W. Kutzelnigg and W. Klopper, *J. Chem. Phys.*, **94** (1991), 1985.
39. W. Klopper, *J. Chem. Phys.*, **102** (1995), 6168.
40. M. Head-Gordon, J. A. Pople and M. J. Frisch, *Chem. Phys. Lett.*, **153** (1988), 503.
41. H. Koch, P. Jørgensen and T. Helgaker, *J. Chem. Phys.*, **104** (1996), 9528.
42. A. K. Wilson and J. Almlöf, *Theo. Chim. Acta*, **95** (1997), 49.
43. T. Helgaker, J. Gauss, P. Jørgensen and J. Olsen, *J. Chem. Phys.*, **106** (1997), 6430.

5 Basis Sets

Ab initio methods try to derive information by solving the Schrödinger equation without fitting parameters to experimental data. Actually, *ab initio* methods also make use of experimental data, but in a somewhat more subtle fashion. Many different approximate methods exist for solving the Schrödinger equation, and the one to use for a specific problem is usually chosen by comparing the performance against known experimental data. Experimental data thus guides the selection of the computational model, rather than directly entering the computational procedure.

One of the approximations inherent in essentially all *ab initio* methods is the introduction of a basis set. Expanding an unknown function, such as a molecular orbital, in a set of known functions is not an approximation, if the basis is complete. However, a complete basis means that an infinite number of functions must be used, which is impossible in actual calculations. An unknown MO can be thought of as a function in the infinite coordinate system spanned by the complete basis set. When a finite basis is used, only the components of the MO along those coordinate axes corresponding to the selected basis can be represented. The smaller the basis, the poorer the representation. The type of basis functions used also influence the accuracy. The better a single basis function is able to reproduce the unknown function, the fewer are basis functions necessary for achieving a given level of accuracy. Knowing that the computational effort of *ab initio* methods scales formally as at least M^4, it is of course of prime importance to make the basis set as small as possible without compromising the accuracy.[1]

5.1 Slater and Gaussian Type Orbitals

There are two types of basis functions (also called *Atomic Orbitals*, AO, although in general they are not solutions to an atomic Schrödinger equation) commonly used in electronic structure calculations: *Slater Type Orbitals* (STO) and *Gaussian Type Orbitals* (GTO). Slater type orbitals[2] have the functional form

$$\chi_{\zeta,n,l,m}(r,\theta,\varphi) = NY_{l,m}(\theta,\varphi)r^{n-1}e^{-\zeta r} \qquad (5.1)$$

N is a normalization constant and $Y_{l,m}$ are the usual spherical harmonic functions. The exponential dependence on the distance between the nucleus and the electron mirrors the exact orbitals for the hydrogen atom. However, STOs do not have any radial nodes,

nodes in the radial part are introduced by making linear combinations of STOs. The exponential dependence ensures a fairly rapid convergence with increasing number of functions, however, as noted in section 3.5, the calculation of three- and four-centre two-electron integrals cannot be performed analytically. STOs are primarily used for atomic and diatomic systems where high accuracy is required, and in semi-empirical methods where all three- and four-centre integrals are neglected.

Gaussian type orbitals[3] can be written in terms of polar or cartesian coordinates

$$
\chi_{\zeta,n,l,m}(r,\theta,\varphi) = N Y_{l,m}(\theta,\varphi) r^{(2n-2-l)} e^{-\zeta r^2}
$$
$$
\chi_{\zeta,l_x,l_y,l_z}(x,y,z) = N x^{l_x} y^{l_y} z^{l_z} e^{-\zeta r^2}
$$

(5.2)

where the sum of l_x, l_y and l_z determines the type of orbital (for example $l_x + l_y + l_z = 1$ is a p-orbital). Although a GTO appears similar in the two sets of coordinates, there is a subtle difference. A d-type GTO written in terms of the spherical functions has five components $(Y_{2,2}, Y_{2,1}, Y_{2,0}, Y_{2,-1}, Y_{2,-2})$, but there appear to be six components in the Cartesian coordinates $(x^2, y^2, z^2, xy, xz, yz)$. The latter six functions, however, may be transformed to the five spherical d-functions and one additional s-function $(x^2 + y^2 + z^2)$. Similarly, there are 10 Cartesian "f-functions" which may be transformed into seven spherical f-functions and one set of spherical p-functions. Modern programs for evaluating two-electron integrals are geared to Cartesian coordinates, and they generate pure spherical d-functions by transforming the six Cartesian components to the five spherical functions. When only one d-function is present per atom the saving by removing the extra s-function is small, but if many d-functions and/or higher angular moment functions (f-, g-, h- etc. functions) are present, the saving can be substantial. Furthermore, the use of only the spherical components reduces the problems of linear dependence for large basis sets, as discussed below.

The r^2 dependence in the exponential makes the GTO inferior to the STOs in two aspects. At the nucleus the GTO has zero slope, in contrast to the STO which has a "cusp" (discontinous derivative), and GTOs have problems representing the proper behaviour near the nucleus. The other problem is that the GTO falls off too rapidly far from the nucleus compared with an STO, and the "tail" of the wave function is consequently represented poorly. Both STOs and GTOs can be chosen to form a complete basis, but the above considerations indicate that more GTOs are necessary for achieving a certain accuracy compared with STOs. A rough guideline says that three times as many GTOs as STOs are required for reaching a given level of accuracy. The increase in number of basis functions, however, is more than compensated for by the ease by which the required integrals can be calculated. In terms of computational efficiency, GTOs are therefore preferred, and used almost universally as basis functions in electronic structure calculations. Furthermore, essentially all applications take the GTOs to be centred at the nuclei. For certain types of calculation the centre of a basis function may be taken not to coincide with a nucleus, for example being placed at the centre of a bond.

5.2 Classification of Basis Sets

Having decided on the type of function (STO/GTO) and the location (nuclei), the most important factor is the number of functions to be used. The smallest number of functions possible is a *minimum basis set*. Only enough functions are employed to contain all the electrons of the neutral atom(s). For hydrogen (and helium) this means a single s-function. For the first row in the periodic table it means two s-functions (1s and 2s) and one set of p-functions $(2p_x, 2p_y$ and $2p_z)$. Lithium and beryllium formally only require two s-functions, but a set of p-functions is usually also added. For the second row elements, three s-functions (1s, 2s and 3s) and two sets of p-functions (2p and 3p) are used.

The next improvement in the basis sets is a doubling of all basis functions, producing a *Double Zeta* (DZ) type basis. The term zeta stems from the fact that the exponent of STO basis functions is often denoted by the greek letter ζ. A DZ basis thus employs two s-functions for hydrogen (1s and 1s'), four s-functions (1s, 1s', 2s and 2s') and two p-functions (2p and 2p') for first row elements, and six s-functions and four p-functions for second row elements. The importance of a DZ over a minimum basis can be illustrated by considering the bonding in HCN (Figure 5.1). The C–H bond will primarily consist of the hydrogen s-orbital and the p_z-orbital on C. The π-bond between C and N will consist of the p_x (and p_y) orbitals of C and N. The π-bond will have a more diffuse electron distribution than the C–H σ-bond. The optimum exponent for the carbon p-orbital will thus be smaller for the x-direction than for the z-direction. If only a single set of p-orbitals is available (minimum basis) a compromise will be necessary. A DZ basis, however, has two sets of p-orbitals with different exponents. The tighter function (larger exponent) can enter the C–H σ-bond with a large coefficient, while the more diffuse function (small exponent) can be used primarily for describing the C–N π-bond. Doubling the number of basis functions thus allows for a much better description of the fact that the electron distribution is different in different directions.

The chemical bonding occurs between valence orbitals. Doubling the 1s-functions in for example carbon allows for a better description of the 1s-electrons. However, the 1s-orbital is essentially independent of the chemical environment, being very close to the atomic case. A variation of the DZ type basis only doubles the number of valence orbitals, producing a *split valence basis*. In actual calculations a doubling of the core orbitals would rarely be considered, and the term DZ basis is also used for split valence basis sets (or sometimes VDZ, for valence double zeta).

The next step up in basis set size is a *Triple Zeta* (TZ). Such a basis contains three times as many functions as the minimum basis, i.e. six s-functions and three p-functions for the first row elements. Some of the core orbitals may again be saved by only splitting the valence, producing a *triple split valence* basis set. Again the term TZ is used to cover both cases. The names *Quadruple Zeta* (QZ) and *Quintuple Zeta* (5Z, not QZ) for the next levels of basis sets are also used, but large sets are often given explicitly in terms of the number of basis functions of each type.

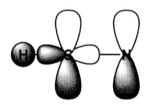

Figure 5.1   A double zeta basis allows for different bonding in different directions

So far only the number of s- and p-functions for each atom (first or second row in the periodic table) has been discussed. In most cases higher angular momentum functions are also important, these are denoted *polarization functions*. Consider again the bonding in HCN in Figure 5.1. The C–H bond is primarily described by the hydrogen s-orbital(s) and the carbon s- and p_z-orbitals. It is clear that the electron distribution <u>along</u> the bond will be different than that <u>perpendicular</u> to the bond. If only s-functions are present on the hydrogen, this cannot be described. However, if a set of p-orbitals is added to the hydrogen, the p_z component can be used for improving the description of the H–C bond. The p-orbital introduces a polarization of the s-orbital(s). Similarly, d-orbitals can be used for polarizing p-orbitals, f-orbitals for polarizing d-orbitals etc. Once a p-orbital has been added to a hydrogen s-orbital, it may be argued that the p-orbital now should be polarized by adding a d-orbital, which should be polarized by an f-orbital, etc. For single determinant wave functions, where electron correlation is not considered, the first set of polarization functions (i.e. p-functions for hydrogen and d-functions for heavy atoms) is by far the most important, and will in general describe all the important charge polarization effects.

If methods including electron correlation are used, higher angular momentum functions are essential. Electron correlation describes the energy lowering by the electrons "avoiding" each other, beyond the average effect taken into account by HF methods. Two types of correlation can be identified, an "in–out" and an "angular" correlation. The "in–out" or *radial correlation* refers to the situation where one electron is close to, and the other far from, the nucleus. To describe this, the basis set needs functions of the same type, but with different exponents. The *angular correlation* refers to the situation where two electrons are on opposite sides of the nucleus. To describe this, the basis set needs functions of same magnitude exponents, but different angular momenta. For example, to describe the angular correlation of an s-function, p-functions (and d-, f-, g-functions etc.) are needed. The angular correlation is of similar importance as the radial correlation, and higher angular momentum functions are consequently essential for correlated calculations. Although these properly should be labelled correlation functions, they also serve as polarization functions for HF wave functions, and it is common to denote them as polarization functions. Normally only the correlation of the valence electrons is considered, and the exponent of the polarization functions should be of the same magnitude as the valence s- and p-functions. In contrast to HF methods, the higher angular momentum functions (beyond the first set of polarization functions) are quite important. Or alternatively formulated, the convergence in terms of angular momentum is slower for correlated wave functions than at the HF level. For a basis set which is complete up to angular momentum L, numerical analysis suggests that the asymptotic convergence at the HF level is exponential (i.e. $\sim\exp(-L)$), while it is $\sim (L+1)^{-3}$ at correlated levels.[4]

Polarization functions are added to the chosen sp-basis. Adding a single set of polarization functions (p-functions on hydrogens and d-functions on heavy atoms) to the DZ basis forms a *Double Zeta plus Polarization* (DZP) type basis. There is a variation where polarization functions are only added to non-hydrogen atoms. This does not mean that polarization functions are not important on hydrogens. However, hydrogens often have a "passive" role, sitting at the end of bond which does not take an active part in the property of interest. The errors introduced by not including hydrogen polarization functions are often rather constant and, as the interest is usually in energy differences, they tend to cancel out. As hydrogens often account for a large number of atoms in the system, a saving of three basis functions for each hydrogen is significant. If hydrogens play an important role in the property of interest, it is of course not a good idea to neglect polarization functions on hydrogens.

Similarly to the sp-basis sets, multiple sets of polarization functions with different exponents may be added. If two sets of polarization functions are added to a TZ sp-basis, a *Triple Zeta plus Double Polarization* (TZ2P) type basis is obtained. For larger basis sets with many polarization functions the explicit composition in terms of number and types of functions is usually given. At the HF level there is usually little gained by expanding the basis set beyond TZ2P, and even a DZP type basis set usually gives "good" results (compared to the HF limit). Correlated methods, however, require more, and higher, angular momentum, polarization functions to achieve the same level of convergence.

Before moving on we need to introduce the concept of *basis set balance*. In principle many sets of polarization functions may be added to a small sp-basis. This is not a good idea. If an insufficient number of sp-functions has been chosen for describing the fundamental electron distribution, the optimization procedure used in obtaining the wave function (and possibly also the geometry) may try to compensate for inadequacies in the sp-basis by using higher angular momentum functions, producing artefacts. A rule of thumb says that the number of functions of a given type should at most be one less than the type with one lower angular momentum. A 3s2p1d basis is balanced, but a 3s2p2d2f1g basis is too heavily polarized. It may not be necessary to polarize the basis all the way up, thus a 5s4p3d2f1g basis is balanced, but if it is known (for example by comparison with experimental data) that f- and g-functions are unimportant, they may be left out. Furthermore, it may be that 2 d-functions are sufficient for the given purpose, although a 5s4p1d basis would be considered underpolarized.

Another aspect of basis set balance is the occasional use of *mixed* basis sets, for example a DZP quality on the atoms in the "interesting" part of the molecule and a minimum basis for the "spectator" atoms. Another example would be addition of polarization functions for only a few hydrogens which are located "near" the reactive part of the system. For a large molecule this may lead to a substantial saving in the number of basis functions. It should be noted that this may bias the results and can create artefacts. For example, a calculation on the H_2 molecule with a minimum basis at one end and a DZ basis at the other end will predict that H_2 has a dipole moment, since the variational principle will preferentially place the electrons near the centre with the most basis functions. The majority of calculations are therefore performed with basis sets of the same quality (minimum, DZP, TZ2P, ...) on all atoms, possibly cutting polarization and/or diffuse (small exponent) functions on hydrogens. Even so, it may be argued that small basis sets inherently tend to be unbalanced. Consider for example the LiF molecule in a minimum or DZ type basis. This will have a very ionic structure, Li^+F^-, with nearly all the valence electrons being located at the fluorine. In terms of number of basis functions per electron, the Li basis is thus of a much higher quality than the F basis, and thereby unbalanced. Of course this effect diminishes as the size of the atomic basis set increases.

Except for very small systems it is impractical to saturate the basis set so that the absolute error in the energy is reduced below chemical accuracy, for example 1 kcal/mol. The important point in choosing a balanced basis set is to keep the error as constant as possible. The use of mixed basis sets should therefore only be done after careful consideration. Furthermore, the use of small basis sets for systems containing elements with substantially different numbers of valence electrons (like LiF) may produce artefacts.

Having decided on the number of basis functions (from a consideration of the property of interest and the computational cost), the question becomes: how are the values for the exponents in the basis functions chosen? The values for s- and p-functions are typically determined by performing variational HF calculations for atoms, using the exponents as variational parameters. The exponent values which give the lowest energy are the "best", at least for the atom. In some cases the optimum exponents are chosen on the basis of minimizing the energy of a wave function which includes electron correlation. The HF procedure cannot be used for determining exponents of polarization functions for atoms. By definition these functions are unoccupied in atoms, and therefore make no contribution to the energy. Suitable polarization exponents may be chosen by performing variational calculations on molecular systems (where the HF energy does depend on polarization functions) or on atoms with correlated wave functions. Since the main function of higher angular momentum functions is to recover electron correlation, the latter approach is usually preferred. Often only the optimum exponent is determined for a single polarization function, and multiple polarization functions are generated by splitting the exponents symmetrically around the optimum value for a single function. The splitting factor is typically taken in the range 2–4. For example if a single d-function for carbon has an exponent value of 0.8, two polarization functions may be assigned with exponents of 0.4 and 1.6 (splitting factor of 4).

5.3 Even- and Well-tempered Basis Sets

The optimization of basis function exponents is an example of a highly non-linear optimization (Chapter 14). When the basis set becomes large, the optimization problem is no longer easy. The basis functions start to become linearly dependent (the basis approaches completeness) and the energy becomes a very flat function of the exponents. Furthermore, the multiple local minima problem is encountered. An analysis of basis sets which have been optimized by variational methods reveals that the ratio between two successive exponents is approximately constant. Taking this ratio to be constant reduces the optimization problem to only two parameters for each type of basis function, independently of the size of the basis. Such basis sets were labelled *even-tempered* basis sets, with the *i*th exponent given as $\zeta_i = \alpha \beta^i$ where α and β are fixed constants for a given type of function and nuclear charge. It was later discovered that the optimum α and β constants to a good approximation can be written as functions of the size of the basis set, M.[5]

$$\zeta_i = \alpha \beta^i, \quad i = 1, 2, \dots, M$$
$$\ln(\ln \beta) = b \ln M + b' \tag{5.3}$$
$$\ln \alpha = a \ln(\beta - 1) + a'$$

The constants a, a', b and b' depend only on the atom type and the type of function. Even-tempered basis sets have the advantage that it is easy to generate a sequence of basis sets which are guaranteed to converge towards a complete basis. This is useful if the attempt is to extrapolate a given property to the basis set limit. The disadvantage is that the convergence is somewhat slow, and an explicitly optimized basis set of a given size will usually give a better answer than an even-tempered basis of the same size.

Even-tempered basis sets have the same ratio between exponents over the whole range. From chemical considerations it is usually preferable to cover the valence region better than the core region. This may be achieved by *well-tempered* basis sets.[6] The idea is similar to the even-tempered basis sets, the exponents are generated by a suitable formula containing only a few parameters to be optimized. The exponents in a well-tempered basis of size M are generated as:

$$\zeta_i = \alpha\beta^{i-1}\left(1 + \gamma\left(\frac{i}{M}\right)^\delta\right), \quad i = 1, 2, \dots, M \tag{5.4}$$

The parameters α, β, γ and δ are optimized for each atom. The exponents are the same for all types of angular momentum functions, and s-, p- and d-functions (and higher angular momentum) consequently have the same radial part.

Optimization of basis sets is not something the average user need to worry about. Optimized basis sets of many different sizes and qualities are available either in the forms of tables, or built into the computer programs. The user "merely" has to select a suitable basis set. However, if the interest is in specialized properties the basis set may need to be tailored to meet the specific needs. For example if the property of interest is an accurate value for the electron density at the nucleus (for example for determining the Fermi contact contribution to spin–spin coupling, see section 10.7.2) then basis functions with very large exponents are required. Alternatively, for calculating hyperpolarizabilites very diffuse functions are required. In such cases the basis function optimization is in terms of the property of interest, and not in terms of energy. Basis functions are added until the change upon addition of one extra function is less than a given threshold.

5.4   Contracted Basis Sets

One disadvantage of all energy optimized basis sets is the fact that they primarily depend on the wave function in the region of the inner shell electrons. The 1s-electrons account for a large part of the total energy, and minimizing the energy will tend to make the basis set optimum for the core electrons, and less than optimum for the valence electrons. However, chemistry is mainly dependent on the valence electrons. Furthermore, many properties (for example polarizability) depend mainly on the wave function "tail" (far from the nucleus), which energetically is unimportant. An energy optimized basis set which gives a good description of the outer part of the wave function therefore needs to be very large, with the majority of the functions being used to describe the 1s-electrons with an accuracy comparable to that for the outer electrons in an energetic sense. This is not the most efficient way of designing basis sets for describing the outer part of the wave function. Instead energy optimized basis sets are usually augmented explicitly with *diffuse* functions (basis functions with small exponents). Diffuse functions are needed whenever loosely bound electrons are present (for example in anions or excited states) or when the property of interest is dependent on the wave function tail (for example polarizability).

The fact that many basis functions go into describing the energetically important, but chemically unimportant, core electrons is the foundation for *contracted* basis sets. Consider for example a basis set consisting of 10 s-functions (and some p-functions) for carbon. Having optimized these 10 exponents by variational calculations on the carbon atom, maybe six of the 10 functions are found primarily to be used for describing the 1s-orbital, and two of the four remaining describe the "inner" part of the 2s-orbital. The important chemical region is the outer valence. Out of the 10 functions, only two are actually used for describing the chemically interesting phenomena. Considering that the computational cost increases as the fourth power (or higher) of the number of basis functions, this is very inefficient. As the core orbitals change very little depending on the chemical bonding situation, the MO expansion coefficients in front of these inner basis functions also change very little. The majority of the computational effort is therefore spent describing the chemically uninteresting part of the wave function, which furthermore is almost constant.

Consider now making the variational coefficients in front of the inner basis functions constant, i.e. they are no longer parameters to be determined by the variational principle. The 1s-orbital is thus described by a <u>fixed</u> linear combination of say six basis functions. Similarly the remaining four basis functions may be contracted into only two functions, for example by fixing the coefficient in front of the inner three functions. In doing this the number of basis functions to be handled by the variational procedure has been reduced from 10 to three.

Combining the full set of basis functions, known as the *primitive* GTOs (PGTOs), into a smaller set of functions by forming fixed linear combinations is known as basis set *contraction*, and the resulting functions are called *contracted* GTOs (CGTOs).

$$\chi(\text{CGTO}) = \sum_{i}^{k} a_i \chi_i (\text{PGTO}) \qquad (5.3)$$

The previously introduced acronyms DZP, TZ2P etc., refer to the number of contracted basis functions. Contraction is especially useful for orbitals describing the inner (core) electrons, since they require a relatively large number of functions for representing the wave function cusp near the nucleus, and furthermore are largely independent of the environment. Contracting a basis set will always increase the energy, since it is a restriction of the number of variational parameters, and makes the basis set less flexible, but will also reduce the computational cost significantly. The decision is thus how much loss in accuracy is acceptable compared to the gain in computational efficiency.

The *degree of contraction* is the number of PGTOs entering the CGTO, typically varying between 1 and 10. The specification of a basis set in terms of primitive and contracted functions is given by the notation (10s4p1d/4s1p) → [3s2p1d/2s1p]. The basis in parentheses is the number of primitives with heavy atoms (first row elements) before the slash and hydrogen after. The basis in the square brackets is the number of contracted functions. Note that this does not tell how the contraction is done, it only indicates the size of the final basis (and thereby the size of the variational problem in HF calculations).

There are two different ways of contracting a set of primitive GTOs to a set of contracted GTOs: *segmented* and *general* contraction. Segmented contraction is the oldest method, and the one used in the above example. A given set of PGTOs is partitioned into smaller sets of functions which are made into CGTO by determining suitable coefficients. A 10s basis may be contracted to 3s by taking the inner six functions as one CGTO, the next three as the second CGTO and the one remaining PGTO as the third "contracted" GTO.

$$\chi_1 \left(\mathrm{CGTO}\right) = \sum_{i=1}^{6} a_i \chi_i \left(\mathrm{PGTO}\right)$$

$$\chi_2(\mathrm{CGTO}) = \sum_{i=7}^{9} a_i \chi_i \left(\mathrm{PGTO}\right) \tag{5.4}$$

$$\chi_3(\mathrm{CGTO}) = \chi_{10} \left(\mathrm{PGTO}\right)$$

In a segmented contraction each primitive as a rule is only used in <u>one</u> contracted function. In some cases it may be necessary to duplicate one or two PGTOs in two adjacent CGTOs. The contraction coefficients can be determined by a variational optimization, for example from an atomic HF calculation.

In a general contraction <u>all</u> primitives (on a given atom) and of a given angular momentum enter all the contracted functions having that angular momentum, but with different contraction coefficients.

$$\chi_1(\mathrm{CGTO}) = \sum_{i=1}^{10} a_i \chi_i \left(\mathrm{PGTO}\right)$$

$$\chi_2(\mathrm{CGTO}) = \sum_{i=1}^{10} b_i \chi_i \left(\mathrm{PGTO}\right) \tag{5.5}$$

$$\chi_3(\mathrm{CGTO}) = \sum_{i=1}^{10} c_i \chi_i \left(\mathrm{PGTO}\right)$$

One popular way of obtaining general contraction coefficients is from Atomic Natural Orbitals (ANO), to be discussed in section 5.4.4. The difference between segmented and general contraction may be illustrated as shown in Figure 5.2.

There are many different contracted basis sets available in the literature or built into programs, and the average user usually only needs to select a suitable quality basis for the calculation. Below is a short description of some basis sets which often are used in routine calculations.

5.4.1 Pople Style Basis Sets

STO-nG basis sets   Slater Type Orbital consisting of n PGTOs.[7] This is a minimum type basis where the exponents of the PGTO are determined by <u>fitting</u> to the STO, rather than optimizing them by a variational procedure. Although basis sets with $n = 2-6$ have been derived, it has been found that using more than three PGTOs to represent the STO gives little improvement, and the STO-3G basis is a widely used minimum basis. This type of basis set has been determined for many elements of the periodic table. The designation of the carbon/hydrogen STO-3G basis is (6s3p/3s) $\rightarrow$ [2s1p/1s].

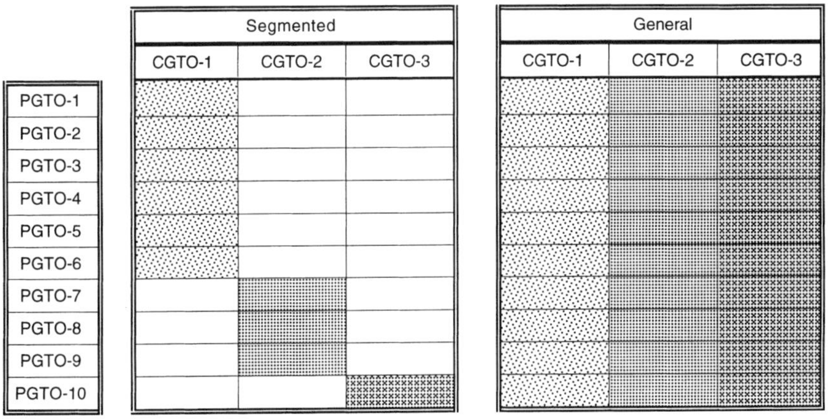

Figure 5.2 Segmented and general contraction

k-nlmG basis sets These basis sets have been designed by Pople and co-workers, and are of the split valence type, with the *k* in front of the dash indicating how many PGTOs are used for representing the core orbitals. The *nlm* after the dash indicate both how many functions the valence orbitals are split into, and how many PGTOs are used for their representation. Two values (e.g. *nl*) indicate a split valence, while three values (e.g. *nlm*) indicate a triple split valence. The values underlined before the G (for Gaussian) indicate the s- and p-functions in the basis; the polarization functions are placed underlined after the G. This type of basis sets has the further restriction that the same exponent is used for both the s- and p-functions in the valence. This increases the computational efficiency, but of course decreases the flexibility of the basis set. The exponents in the PGTO have been optimized by variational procedures.

3-21G This is a split valence basis, where the core orbitals are a contraction of three PGTOs, the inner part of the valence orbitals is a contraction of two PGTOs and the outer part of the valence is represented by one PGTO.[8] The designation of the carbon/ hydrogen 3-21G basis is (6s3p/3s) → [3s2p/2s]. Note that the 3-21G basis contains the same number of primitive GTOs as the STO-3G, however, it is much more flexible as there are twice as many valence functions which can combine freely to make MOs.

6-31G This is also a split valence basis, where the core orbitals are a contraction of six PGTOs, the inner part of the valence orbitals is a contraction of three PGTOs and the outer part of the valence represented by one PGTO.[9] The designation of the carbon/ hydrogen 6-31G basis is (10s4p/4s) → [3s2p/2s]. In terms of contracted basis functions it contains the same number as 3-21G, but the representation of each functions is better since more PGTOs are used.

6-311G This is a triple split valence basis, where the core orbitals are a contraction of six PGTOs and the valence split into three functions, represented by three, one, and one PGTOs, respectively.[10]

To each of these basis sets can be added diffuse[11] and/or polarization functions. Diffuse functions are normally s- and p-functions and consequently go before the G. They are denoted by + or ++, with the first + indicating one set of diffuse s- and p-functions on heavy atoms, and the second + indicating that a diffuse s-function is also added to hydrogens. The arguments for adding only diffuse functions on non-hydrogen atoms is the same as that for adding only polarization functions on non-hydrogens (Section 5.2). Polarization functions are indicated after the G, with a separate designation for heavy atoms and hydrogens. The 6–31+G(d) is a split valence basis with one set of diffuse sp-functions on heavy atoms only and a single d-type polarization function on heavy atoms. A 6-311++G(2df,2pd) is similarly a triple split valence with additional diffuse sp-functions, and two d- and one f-functions on heavy atoms and diffuse s- and two p- and one d-functions on hydrogens. The largest standard Pople style basis set is 6-311++G(3df, 3pd). These types of basis sets have been derived for hydrogen and the first row elements, and some of the basis sets have also been derived for second and higher row elements.

If only one set of polarization functions is used, an alternative notation in terms of * is also widely used. The 6-31G* basis is identical to 6-31G(d), and 6-31G** is identical to 6-31G(d,p). A special note should be made for the 3-21G* basis. The 3-21G basis is basicly too small to support polarization functions (it becomes unbalanced). However, the 3-21G basis by itself performs poorly for hypervalent molecules, such as sulfoxides and sulfones. This can be substantially improved by adding a set of d-functions. The 3-21G* basis has only d-functions on second row elements (it is sometimes denoted 3-21G(*) to indicate this), and should not be considered a polarized basis. Rather, the addition of a set of d-functions should be considered an *ad hoc* repair of a known flaw.

5.4.2 Dunning–Huzinaga Basis Sets

Huzinaga determined uncontracted energy optimized basis sets up to (10s6p) for first row elements.[13] This was later extended to (14s9p) by van Duijneveldt[14] and up to (18s, 13p) by Partridge.[15] Dunning used the Huzinaga primitive GTOs to derive various contraction schemes (DH type basis sets).[16] A DZ type basis can be made by a contraction of the (9s5p/4s) PGTO to [4s2p/2s]. The contraction scheme is 6,1,1,1 for s-functions, 4,1 for the p-functions, and 3,1 for hydrogen. A widely used split valence type basis is a contraction of the same primitive set to [3s2p/2s] where the s-contraction is 7,2,1 (note that one primitive enters twice). A widely used TZ type basis (actually only a triple split valence) is a contraction of the (10s6p/5s) to [5s3p/3s], with the contraction scheme 5,3,1,1,1 for s-functions, 4,1,1 for p-functions, and 3,1,1 for hydrogens. Again a duplication of one of the s-primitives has been allowed.

McLean and Chandler developed a similar set of contracted basis sets from Huzinaga's primitive optimized set for second row elements.[17] A DZ type basis is derived by contracting (12s8p) → [5s3p], and a TZ type is derived by contracting (13s9p) → [6s4p]. The latter contraction is 6,3,1,1,1,1 for the s-functions and 4,2,1,1,1 for the p-functions, and is often used in connection with the Pople 6-31G when second row elements are present.

The Dunning–Huzinaga type basis sets do not have the restriction of the Pople style basis sets of equal exponents for the s- and p-functions, and they are therefore somewhat more flexible, but computationally also more expensive. The major determining factor, however, is the number of basis functions and less the exact description of each function. Normally there is little difference in the performance of different DZ or different TZ type basis sets.

The primary reason for the popularity of the Pople and DH style basis sets is the extensive calibration available. There have been so many calculations reported with these basis sets that it is possible to get a fairly good idea of the level of accuracy that can be attained with a given basis. This is of course a self-sustaining procedure, the more calculations that are reported with a given basis, the more popular it becomes, since the calibration set becomes larger and larger.

5.4.3 MINI, MIDI and MAXI basis sets

Tatewaki and Huzinaga optimized minimum basis sets for a large part of the periodic system.[18] The MINI-i ($i = 1-4$) basis are all minimum basis sets with three PGTOs in the 2s CGTO and varying numbers of PGTOs in the 1s and 2p CGTOs. In terms of PGTOs the MINI-1 is (3s,3s,3p), MINI-2 is (3s,3s,4p), MINI-3 is (4s,3s,3p) and MINI-4 is (4s,3s,4p). These MINI basis sets in general perform better than STO-3G, but it should be kept in mind that they still are minimum basis sets. The MIDI-i basis sets are identical to MINI-i, except that the outer valence function is decontracted. The MAXI-i basis sets all employ four PGTOs for the 2s CGTO and between five and seven PGTOs for the 1s and 2p CGTOs. The valence orbitals are split into three or four functions, and MAXI-1 is (9s5p) $\rightarrow$ [4s3p] (contraction 5,2,1,1 and 3,1,1), MAXI-3 (10s6p) $\rightarrow$ [5s4p] (contraction 6,2,1,1,1 and 3,1,1,1) and MAXI-5 is (11s7p) $\rightarrow$ [5s4p] (contraction 7,2,1,1,1 and 4,1,1,1).

5.4.4 Atomic Natural Orbitals Basis Sets

All of the above basis sets are of the segmented contraction type. Modern contracted basis sets aimed at producing very accurate wave functions often employ a general contraction scheme. The ANO and cc basis sets below are of the general contraction type.

The idea in the *Atomic Natural Orbitals* (ANO) type basis sets is to contract a large PGTO set to a fairly small number of CGTOs by using *natural orbitals* from a correlated calculation on the free atom, typically at the CISD level.[19] The natural orbitals are those which diagonalize the density matrix, and the eigenvalues are called *orbital occupation numbers* (see Section 9.5). The orbital occupation number is the number of electrons in the orbital. For an RHF wave function, ANOs would be identical to the canonical orbitals with occupation numbers of exactly 0 or 2. When a correlated wave function is used, however, the occupation number may have any value between 0 and 2. The ANO contraction selects the important combinations of the PGTOs from the magnitude of the occupation numbers. A large primitive basis, typically generated as an even-tempered sequence, may generate several different contracted basis sets by gradually lowering the selection threshold for the occupation number. The nice feature of the ANO contraction is that it more of less "automatically" generates balanced basis sets, e.g. for neon the ANO procedure generates the following basis set: [2s1p], [3s2p1d], [4s3p2d1f] and [5s4p3d2f1g]. Furthermore, in such a sequence the smaller ANO basis sets are true subsets of the larger, since the same primitive set of functions is used.

5.4.5 Correlation Consistent Basis Sets

The primary disadvantage of ANO basis sets is that a very large number of PGTOs is necessary for converging towards the basis set limit. Dunning has proposed a somewhat smaller set of primitives which yield results comparable to those for the ANO basis sets.[20] The *correlation consistent* (*cc*, convention is to use small letters for the acronym, to distinguish it from Coupled Cluster, CC) basis sets are geared toward recovering the correlation energy of the valence electrons. The name correlation consistent refers to the fact that the basis sets are designed so that functions which contribute similar amounts of correlation energy are included at the same stage, independently of the function type. For example, the first d-function provides a large energy lowering, but the contribution from a second d-function is similar to that from the first f-function. The energy lowering from a third d-function is similar to that from the second f-function and the first g-function. Addition of polarization functions should therefore be done in the order: 1d, 2d1f and 3d2f1g. An additional feature of the cc basis sets is that the energy error from the sp-basis should be comparable to (or at least not exceed by) the correlation error arising from the incomplete polarization space, and the sp-basis therefore also increases as the polarization space is extended. Several different sizes of cc basis sets are available in terms of final number of contracted functions. These are known by their acronyms: *cc-pVDZ, cc-pVTZ, cc-pVQZ, cc-pV5Z* and *cc-pV6Z* (*correlation consistent polarized Valence Double/Triple/Quadruple/Quintuple/Sextuple Zeta*). The composition in terms of primitive and contracted functions is shown in Table 5.1. A step up in terms of quality increases each type of basis function by one, and adds a new type of higher-order polarization function.

The energy optimized cc basis sets can be augmented by additional diffuse functions, adding the prefix *aug-* to the acronym.[21] The augmentation consists of adding one extra function with a smaller exponent for each angular momentum, i.e. the aug-cc-pVDZ has additionally 1s-, 1p- and 1d-functions, the cc-pVTZ has 1s1p1d1f extra for non-hydrogens and so on. The cc basis sets may also be augmented by addtional *tight* functions (large exponents) if the interest is in recovering core–core and core–valence electron correlation, producing the acronyms cc-pCVXZ (X = D, T, Q, 5). The cc-pCVDZ has additionally 1s and 1p tight functions, the cc-pCVTV has 2s2p1d tight functions, the cc-pCVQZ has 3s3p2d1f and the cc-pCV5Z has 4s4p3d2f1g for non-hydrogens.[22]

The main advantage of the ANO and cc basis sets is the ability to generate a sequence of basis sets which converges toward the basis set limit. For example, from a series of

Table 5.1 Correlation consistent basis sets

Basis	Primitive functions	Contracted functions
cc-pVDZ	9s,4p,1d/4s,1p	3s,2p,1d/2s,1p
cc-pVTZ	10s,5p,2d,1f/5s,2p,1d	4s,3p,2d,1f/3s,2p,1d
cc-pVQZ	12s,6p,3d,2f,1g/6s,3p,2d,1f	5s,4p,3d,2f,1g/4s,3p,2d,1f
cc-pV5Z	14s,9p,4d,3f,2g,1h/8s,4p,3d,2f,1g	6s,5p,4d,3f,2g,1h/5s,4p,3d,2f,1g
cc-pV6Z	16s,10p,5d,4f,3g,2h,1i/ 10s,5p,4d,3f,2g,1h	7s,6p,5d,4f,3g,2h,1i/ 6s,5p,4d,3f,2g,1h

\

calculations with the 3-21G, 6-31G(d,p), 6-311G(2d,2p) and 6-311++G(3df,3pd) basis sets it may not be obvious whether the property of interest is "converged" with respect to further increases in the basis, and it is difficult to estimate what the basis set limit would be. This is partly because different PGTOs are used in each of these segmented basis sets, and the lack of higher angular momentum functions. From the same (large) set of primitive GTOs, however, increasingly large ANO basis sets may be generated by a general contraction scheme which allows an estimate of the basis set limiting value. Similarly the cc basis sets consistently reduce errors (both HF and correlation) for each step up in quality. In test cases it has been found that the cc-pVDZ basis can provide $\sim 65\%$ of the total (valence) correlation energy, the cc-pVTV $\sim 85\%$, cc-pVQZ $\sim 93\%$, cc-pV5Z $\sim 96\%$ and cc-pV6Z $\sim 98\%$, with similar reductions of the HF error. Given the systematic nature of the cc basis sets, several different schemes have been proposed for extrapolation to the infinite basis set limit.[23] At the HF level the convergence is expected to be exponential, and indeed a function of the form $A + Be^{-CL}(L = 2, 3, 4, 5, 6)$ usually yields an excellent fit. Such fitting functions have also been used at correlated levels, although theoretical analysis suggests that the correlation contribution (i.e. not the correlated result) should converge with an inverse power dependence, with the leading term being $(L + 1)^{-3}$. Several alternative forms such as $A + Be^{-(L-1)} + Ce^{-(L-1)(L-1)}$, $A + B(L + 1/2)^{-4} + C(L + 1/2)^{-6}$ and $A + B(L + 1/2)^{-C}$ have also been used. The theoretical assumption underlying an inverse power dependence is that the basis set is saturated in the radial part (e.g. the cc-pVTZ basis is complete in the s-, p-, d- and f-function spaces). This is not the case for the correlation consistent basis sets, even for the cc-pV6Z basis the errors due to insuficient numbers of s- to i-functions is comparable to that from neglect of functions with angular moment higher than i-functions.

The main difficulty in using the cc basis sets is that each step up in quality almost doubles the number of basis functions, or equivalently, the number of functions increase as the third power of the highest angular momentum function (i.e. $M \propto L^3$). Fitting functions of the above type contain three parameters, thus at least three calculations with increasingly larger basis sets are required. The simplest is cc-pVDZ, cc-pVTZ and cc-pVQZ, but the cc-pVDZ basis is too small to give a good extrapolated value, and a better sequence is cc-pVTZ, cc-pVQZ and cc-pV5Z. The requirement of performing calculations with at least the cc-pVQZ basis places severe constraints on the size of the systems that can be treated.

Perhaps the most interesting aspect of the analysis which led to the development of the correlation consistent basis sets is the fact that high angular momentum functions are necessary for achieving high accuracy. While d-polarization functions are sufficient for a DZ type basis, a TZ type should also include f-functions. Similarly, it is questionable to use a QZ type basis for the sp-functions without also including 3d-, 2f- and 1g-functions in order to systematically reduce the errors. It can therefore be argued that an extension of for example the 6-31G(d,p) to 6-311G(d,p) is inconsistent as a second set of d-orbitals (and a second set of p-orbitals for hydrogen) and a set of f-functions (d-functions for hydrogen) will give contributions similar to those of the extra set of sp-functions. Similarly, a extension of the 6-311G(2df,2pd) basis to 6-311G(3df,3pd) may be considered inconsistent, as the third d-function is expected to be as important as a fourth valence set of sp-functions, a second set of f-functions and the first set of g-functions, all of which are neglected.

In the search for a basis set converged value, other approximations should be keep in mind. Basis sets with many high angular momentum functions are normally designed for recovering a large fraction of the correlation energy. In the majority of cases, only the electron correlation of the valence electrons is considered (frozen core approximation), since the core orbitals usually are insensitive to the molecular environment. As the valence space approaches completeness in terms of basis functions, the error from the frozen core approximation will at some point become comparable to the remaining valence error. From studies of small molecules, where good experimental data are available, it is suggested that the effect of core-electron correlation for unproblematic systems is comparable to the change observed upon enlarging the cc-pV5Z basis, i.e. of a magnitude similar to that for the introduction of h-functions.[24] Improvements beyond the cc-pV6Z basis set have been argued to produce changes of similar magnitude as those expected from relativistic corrections (for first row elements), and further increases to cc-pV7Z and cc-pV8Z basis sets would be comparable to corrections due to breakdown of the Born–Oppenheimer approximation. Within the non-relativistic realm it would therefore appear that basis sets larger than cc-pV6Z would be of little use, except for extrapolating to the non-relativistic, clamped nuclei limit for testing purposes.

There is a practical aspect of using large basis sets, especially those including diffuse functions, which requires special attention, namely the problem of *linear dependence* of the basis functions. Linear dependence means that one (or more) of the basis functions can be written as a linear combination of the other, i.e. the basis set is essentially overcomplete. A diffuse function has a small exponent and consequently extends far away from the nucleus on which it is located. A equally diffuse function located on a nearby atom will therefore span almost the same space. A measure of the degree of linear dependence in a basis set can be obtained from the eigenvalues of the overlap matrix $\mathbf{S}$ (eq. (3.50)). A truely linearly dependent basis will have at least one eigenvalue of exactly zero, and the smallest eigenvalue of the $\mathbf{S}$ matrix is therefore an indication of how close the actual basis set is to linear dependence. As described in Section 13.2, solution of the SCF equations requires orthogonalization of the basis by means of the $\mathbf{S}^{-1/2}$ matrix (or a related matrix which makes the basis orthogonal). If one of the $\mathbf{S}$ matrix eigenvalues is close to zero, this means that the $\mathbf{S}^{-1/2}$ matrix is essentially singular, which in turn will cause numerical problems if trying to carry out an actual calculation. In practice there is therefore an uper limit on how close to completeness a basis set can be chosen to be, this limit is determined by the finite precision by which the calculations are carried out. If the selected basis set turns out to be too close to linear dependence to be handled, the linear combinations of basis functions with low eigenvalues in the $\mathbf{S}$ matrix may be discarded.

5.5 Extrapolation Procedures

In principle the large majority of systems can be calculated with a high accuracy by using a highly correlated method such as CCSD(T) and performing a series of calculations with systematically larger basis sets (like the correlation consistent sets) in order to extrapolate to the basis set limit. In practice even a single water molecule is demanding to treat in this fashion (Chapter 11). Various approximate procedures have therefore been developed for estimating the "infinite correlation, infinite basis" limit as efficiently as possible. These models rely on the fact that different properties converge with different rates as the level of sophistication increases, and that effects from extending the basis set to a certain degree are additive. There are four main steps in these procedures:

(1) Selection of a geometry.
(2) Selecting a basis set for calculating the HF energy.
(3) Estimating the electron correlation energy.
(4) Estimating the energy from translation, rotation and vibrations.

Given a certain target accuracy, the error from each of these four steps should be reduced below the desired tolerance. The error at a given level may defined as the change which would occur if the calculation were taken to the "infinite correlation, infinite basis" limit. A typical target accuracy is ~ 1 kcal/mol, so-called chemical accuracy.

Geometries converge relatively fast, already at the HF level with a DZP type basis the "geometry error" is often ~ 1 kcal/mol or less, and a MP2/DZP optimzed geometry is normally sufficient for most applications. The translational and rotational contributions are trivial to calculate, they depend only on the molecular mass and the geometry (Sections 12.2.1 and 2.2.2), and are very small in absolute values. The error from these can be neglected. The vibrational effect is mainly the zero-point energy, and it requires calculation of the frequencies. An accurate prediction of frequencies is fairly difficult, however, since the absolute value of the zero-point energy is small, a large relative error is tolerable. Furthermore, the errors in calculated frequencies are to a certain extent systematic, and can therefore be improved by a uniform scaling.[25]

The HF error depends only on the size of the basis set. The energy, however, behaves asymptotically as $\sim \exp(-L)$, L being the highest angular momentum in the basis set, i.e. already, with a basis set of TZ(2df) (4s3p2d1f) quality the results are quite stable. Combined with the fact that an HF calculation is the least expensive *ab initio* method, this means that the HF error is not the limiting factor.

The main problem is estimating the correlation effect. All electron correlation methods have a rather steep increase in computational cost as the size of the basis is enlarged, and the convergence in terms of the highest angular momentum in the basis is quite slow ($\sim L^{-3}$). A large contribution to the correlation energy is from pairs of electrons in the same spatial MO; this effect is reasonably well described at the MP2 level, but requires a large basis set in order to recover a large fraction of the absolute value. The remaining correlation energy is much harder to calculate; coupled cluster is the preferred method here, but since the absolute value is substantially smaller than the MP2 correlation energy, a smaller basis can be employed. This means that the relative error is quite large, but the absolute error is of the same magnitude as the correlation error from the MP2 calucalation with the large basis.

In the *Gaussian-1* (G1), *Gaussian-2* (G2) and *Complete Basis Set* (CBS) models, calculations from different levels of theory are combined with the goal of producing energy differences accurate to about 1 kcal/mol, as compared to experimental results. They have been calibrated on a reference set of 125 atomic and molecular properties (atomization energies, ionization potentials, electron and proton affinities), often referred to as the G2 or G2-1 data set.[26] A somewhat larger set of data, called G2-2, has been used more recently.[27] The ability to accurately calculate atomization energies (corresponding to dissociating molecules into isolated atoms) enables the prediction of absolute values of heat of formation, since the atomic values are known experimentally.

The main difference between the G1/G2 and CBS methods is the way in which they try to extrapolate the correlation energy, as described below. Both the G1/G2 and CBS methods come in different flavours, depending on the exact combinations of methods used for obtaining the above four contributions.

As an example, the G2(MP2) method[28] involves the following steps:

(1) The geometry is optimized at the HF/6-31G(d) level, and the vibrational frequencies are calculated. To correct for the known deficiencies at the HF level, these are scaled by 0.893 to produce zero-point energies.
(2) The geometry is reoptimized at the MP2/6-31G(d) level, which is used as the reference geometry.
(3) A MP2/6−311+G(3df,2p) calculation is carried out, which automatically yields the corresponding HF energy.
(4) The energy is calculated at the QCISD(T)/6-311G(d,p) level. This automatically generates the MP2 value as an intermediate result, and the difference between the QCISD(T) and MP2 energies is taken as an estimate of the higher-order correlation energy. The G2 method (not G2(MP2)) performs additional MP4 calculations with larger basis sets to get a better estimate of the higher-order correlation energy.
(5) To correct for electron correlation beyond QCISD(T) and basis set limitations, an empirical correction is added to the total energy.

$$\Delta E(\text{empirical}) = -0.00481\,N_\alpha - 0.00019\,N_\beta \qquad (5.6)$$

(it is assumed that the number of α-electrons is larger than or equal to the number of β-electrons). The numerical constants are determined by fitting to the reference data. It should be noted that this correction makes the G2 methods non-size-extensive.

The net effect of steps (3)–(5) is that a single calculation at the QCISD(T)/6−311+G(3df,2p) level is replaced by a series of calculations at lower levels, which in combination yields a comparable accuracy with significantly less computer time.[29]

The main difference between the G2 models is the way in which the electron correlation beyond MP2 is estimated. The G2 method itself performs a series of MP4 and QCISD(T) calculations, G2(MP2) only does a single QCISD(T) calculation with the 6-311 G(d,p) basis, while G2(MP2, SVP) (SVP stands for Split Valence Polarization) reduces the basis set to only 6-31 G(d).[30] An even more pruned version, G2(MP2,SV), uses the unpolarized 6-31 G basis for the QCISD(T) part, which increases the Mean Absolute Deviation (MAD) to 2.1 kcal/mol. That it is possible to achieve such good performance with this small a basis set for QCISD(T) partly reflects the importance of the large basis MP2 calculation and partly the absorption of errors in the empirical correction.

A comparison between G1, G2, G2(MP2) and G2(MP2,SVP) is shown in Table 5.2; for the reference G2 data set the mean absolute deviations in kcal/mol vary from 1.1 to 1.6 kcal/mol. There are other variations of the G2 methods in use, for example involving DFT methods for geometry optimization and frequency calculation[31] or CCSD(T) instead of QCISD(T),[32] with slightly varying performance and computational cost. The errors with the G2 method are comparable to those obtained directly from calculations at the CCSD(T)/cc-pVTZ level, at a significantly lower computational cost.[33]

Table 5.2 Computational levels in the G1/G2 models

Method	G1	G2	G2(MP2)	G2(MP2,SVP)
Geometry	MP2/6-31G(d)	P2/6-31G(d)	MP2/6-31G(d)	MP2/6-31G(d)
HF and MP2	6-311G(2df,p)	6-311+G(3df,2p)	6-311+G(3df,2p)	6-311+G(3df,2p)
Higher-order	MP4(SDTQ)/	MP4(SDTQ)/		
correlation	6-311G(d,p)	6-311G(d,p)		
	MP4(SDTQ)/	MP4(SDTQ)/		
	6-311+G(d,p)	6-311+G(d,p)		
	MP4(SDTQ)/	MP4(SDTQ)/		
	6-311G(2df,p)	6-311G(2df,p)		
	QCISD(T)/	QCISD(T)/	QCISD(T)/	QCISD(T)/
	6-311G(d,p)	6-311G(d,p)	6-311G(d,p)	6-31G(d)
Thermo	HF/6-31G(d)	HF/6-31G(d)	HF/6-31G(d)	HF/6-31G(d)
[scale factor]	[0.893]	[0.893]	[0.893]	[0.893]
Empirical factors				
for electron	yes	yes	yes	yes
correlation				
MAD error	1.5	1.1	1.5	1.6

Geometry: level at which the structure is optimized; higher-order correlation: method(s) for estimating higher-order correlation effects; thermo: level at which the thermodynamical corrections are calculated [vibrational scale factor]; MAD: Mean Absolute Deviation for reference data set in kcal/mol.

The main difference between the G1/G2 and CBS models is the extrapolation of the correlation energy. The G1/G2 methods assume basis set additivity, and add an empirical correction to recover some of the remaining correlation energy. The CBS procedures, on the other hand, attempt to perform an explicit extrapolation of the calculated values. The main part of the correlation energy is due to electron pairs, i.e. that described by doubly excited configurations. In terms of perturbation theory, this may again be divided into contributions from different orders, the most important coming from second order (MP2). By using pair natural orbitals (being eigenvectors of the density matrix, see Section 9.5) as the expansion parameter, and assuming that enough pairs have been included to reach the asymptotic limit, it may be shown that the MP2 energy calculated by a limited natural orbitals expansion (of size N_{ij}) behaves as $1/N_{ij}$, and can therefore be extrapolated to the complete basis set limit.

There are several different CBS methods, each having its own set of prescriptions and resulting computational cost and accuracy, they are known under the acronyms: CBS-4, CBS-q, CBS-Q and CBS-APNO. As an explicit example, we will take the CBS-Q model,[34] which is computationally similar to the G2(MP2) method.

(1) The geometry is optimized at the HF/6-31 G(d†) level (d† denotes that the exponents for the d-functions are taken from the 6-311G(d) basis), and the vibrational frequencies are calculated. To correct for the known deficiencies at the HF level, these are scaled by 0.918 to produce zero-point energies.
(2) The geometry is reoptimized at the MP2/6-31G(d†) level, which is used as the reference geometry.
(3) A MP2/6-311+G(2df,2p) calculation is carried out, which automatically yields the corresponding HF energy. The MP2 result is extrapolated to the basis set limit by the pair natural orbital method.

(4) The energy is calculated at the MP4(SDQ)/6-31G(d,p) and QCISD(T)/6-31+G(d†) levels to estimate the effect from higher-order electron correlation.

(5) Corrections due to remaining correlation effects are estimated by an empirical expression.

$$\Delta E(\text{empirical}) = -0.00533 \sum_i \left(\sum_\mu C_{\mu ii} \right)^2 |S|^2_{ij} \tag{5.7}$$

where the sum over $C_{\mu ii}$ is the trace of the first-order wave function coefficients for the natural orbital pair ii, $|S|_{ij}$ is the spatial overlap between the absolute values of MOs i and j, and the factor 0.00533 is determined by fitting to the reference data. This empirical correction is size extensive.

(6) For open-shell species the UHF method is used, which in some cases suffers from spin contamination. To correct for this an empirical correction based on the deviation of $\langle S^2 \rangle$ from the theroretical value is added.

$$\Delta E(\text{empirical}) = -0.0092\left[\langle S^2 \rangle - S_z(S_z - 1)\right] \tag{5.8}$$

where the factor of -0.0092 is derived by fitting. The use of the smaller basis for the QCISD(T) calculation means that the CBS-Q model is computationally faster than the G2(MP2), but nevertheless gives slightly lower errors.

A comparison between the four CBS models is shown in Table 5.3.

It should be noted that the G2-1 data set, with two exceptions (SO_2 and CO_2), only includes data for molecules containing one or two heavy (non-hydrogen) atoms. It is likely that the typical error for a given model to a certain extent depends on the size of the system, i.e. the G2 method is presumably not able to predict the heat of formation of

Table 5.3 Computational levels in the CBS models

Method	CBS-4	CBS-q	CBS-Q	CBS-APNO
Geometry	HF/3-21G(*)	HF/3-21G(*)	MP2/6-31G(d†)	QCISD/ 6-311G(d,p)
HF	6-311++G(2df,p)	6-311++G(2df,p)	6-311++G (2df,2p)	[6s6p3d2f /4s2p1d]
MP2	6-31+G(d†)	6-31+G(d†)	6-311++G (2df,2p)	[6s6p3d2f /4s2p1d]
Higher-order correlation	MP4(SDQ)/ 6-31G	MP4(SDQ)/ 6-31G(d†) QCISD(T)/ 6-31G	MP4(SDQ)/ 6-31+G(d,p) QCISD(T)/ 6-31+G(d†)	QCISD(T)/ 6-311+G(2df,p)
Thermo [scale factor]	HF/3-21G (0.917)	HF/3-21G (0.917)	HF/6-31G(d†) [0.918]	HF/6-311G(d,p) [0.925]
Empirical factors for electron correlation	yes	yes	yes	yes
Empirical factors for spin contamination	yes	no	yes	no
MAD error	2.1	1.6	1.0	0.5

say C_{60} (if it was computationally feasible) with an accuracy of ~ 1 kcal/mol. Furthermore, the properties included (atomization energies, ionization potentials, electron and proton affinities) all correspond to energy differences between well-separated system: atomization energies are energy differences between a molecule and isolated atoms, the other three properties correspond to removal or addition of a single electron or proton. As illustrated in Chapter 11, such energy differences are easier to calculate than those between systems containing half broken/formed bonds. As with any scheme which has been parameterized on experimental data, it is questionable to assume that the typical accuracy for a selected set of properties will be true in general. A good performance for the G2 data set does not necessarily indicate that the same level of accuracy can be obtained over a wide variety of geometries, for example those including transition structures. A modified version of the G2 method, denoted G2Q, involving geometry optimization and frequency calculation at the QCISD/6-311 G(d,p) level has been advocated by Durant and Rohlfing for use with transition structures.[35]

There are a few other correction procedures that may be considered as extrapolation schemes. The *Scaled External Correlation* (SEC) and *Scaled All Correlation* (SAC) methods scale the correlation energy by a factor such that calculated dissociation energy agrees with the experimental value.[36]

$$E_{\text{SEC/SAC}} = E_{\text{ref}} + \frac{E_{\text{corr}} - E_{\text{ref}}}{F}$$
$$F = \frac{D_{\text{e}}(\text{corr}) - D_{\text{e}}(\text{ref})}{D_{\text{e}}(\text{exp}) - D_{\text{e}}(\text{ref})} \tag{5.9}$$

The acronym SEC refers to the case where the reference wave function is of the MCSCF type and the correlation energy is calculated by an MR-CISD procedure. When the reference is a single determinant (HF) the SAC nomenclature is used. In the latter case the correlation energy may be calculated for example by MP2, MP4 or CCSD, producing acronyms like MP2-SAC, MP4-SAC and CCSD-SAC. In the SEC/SAC procedure the scale factor F is assumed constant over the whole surface. If more than one dissociation channel is important, a suitable average F may be used.

The *Parameterized Configuration Interaction (PCI-X)* method[37] simply takes the correlation energy and scales it by a constant factor X (typical value ~ 1.2), i.e. it is assumed that the given combination of method and basis set recovers a constant fraction of the correlation energy.

The introduction of various empirical corrections, such as scale factors for frequencies and energy corrections based on the number of electrons and degree of spin contamination, blurs the distinction between whether they should be considered *ab initio*, or as belonging to the semi-empirical class of methods, such as AM1 and PM3. Nevertheless, the accuracy that these methods are capable of delivering makes it possible to calculate absolute stabilities (heat of formation) for small and medium sized systems which rival (or surpass) experimental data, often at a substantial lower cost than for actually performing the experiments.

5.6 Isogyric and Isodesmic Reactions

The most difficult part in calculating absolute stabilities (heat of formation) is the correlation energy. For calculating energies relative to isolated atoms, the goal of the

$$CH_4 \longrightarrow CH_3 + H$$

Figure 5.3 Dissociation of CH_4

$$CH_4 + H \longrightarrow CH_3 + H_2$$

Figure 5.4 An example of an isogyric reaction

G2/CBS models, essentially all the correlation energy of the bond being broken must be recovered. This in turn necessitates large basis sets and sophisticated correlation methods. This is also the reason why *ab initio* energies are not converted into heats of formation, as is normally done for semi-empirical methods (eq. (3.89)), since the resulting values are poor unless a very high level of theory is employed.

In many cases, however, it is possible to choose reference systems that are less demanding than the isolated atoms. Consider for example calculating the C–H dissociation energy of CH_4. In a direct calculation this is given as the difference in total energy between CH_4 and $CH_3 + H$, Figure 5.3. In order to calculate an accurate value for this energy difference, essentially all the electron correlation (and HF) energy for the C–H bond must be recovered.

Consider now the reaction in Figure 5.4. The difference between the reactions in Figures 5.3 and 5.4 is that the latter has the same number of electron pairs on both sides; such reactions are called *isogyric*. The task of calculating all the correlation energy of a C–H bond is replaced by calculating the difference in correlation energy between a C–H and H–H bond. The latter will benefit from cancellation of errors, and therefore stabilize much earlier in terms of theoretical level. Isogyric reactions can thus be used for obtaining relative values; in the above example the CH_4 dissociation energy is given relative to that of H_2. By using the experimental value for H_2, the CH_4 dissociation energy may be calculated quite accurately even at relatively low levels of theory.

The concept may be taken one step further. It is often possible to set up reactions where not only the number of electron pairs is constant, but also the formal types of bonds is the same on both sides. Consider for example calculating the stability of propene by the reaction in Figure 5.5. In this case the number of C=C, C–C and C–H bonds is the same on both sides, and the "reaction" energy is therefore relatively easy to calculate since the electron correlation to a large extent is the same on both sides. Such reactions, which conserve both the number and types of bonds, are called *isodesmic* reactions. Combining the calculated energy difference for the left- and right-hand sides with experimental values for $H_2C=CH_2, H_3C–CH_3$ and CH_4, the (absolute) stability of propene can be obtained reasonably accurately at a quite low levels of theory. This does, however, require that the experimental values for the chosen reference compounds are available. Furthermore, there are several possible ways of constructing isodesmic or

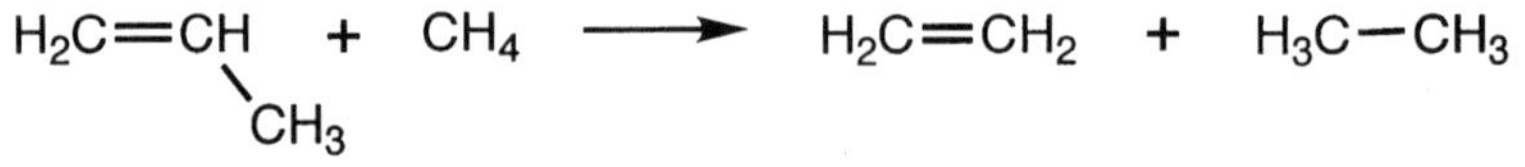

$$H_2C{=}CH{-}CH_3 + CH_4 \longrightarrow H_2C{=}CH_2 + H_3C{-}CH_3$$

Figure 5.5 An example of an isodesmic reaction

isogyric reactions (e.g. replacing H with Cl in Figure 5.4), i.e. such methods are not unique.

5.7 Effective Core Potential Basis Sets

For systems involving elements from the third row or higher in the periodic table there is a large number of core electrons which in general are unimportant in a chemical sense. However, it is necessary to use a large number of basis functions to expand the corresponding orbitals, otherwise the valence orbitals will not be properly described (due to a poor description of the electron–electron repulsion). In the lower half of the periodic table relativistic effects furthermore complicate matters, see Chapter 8. These two problems may be "solved" simultaneously by introducing an *Effective Core Potential* (ECP) (also called *Pseudopotential*) to represent all the core electrons.[38] This is in the spirit of semi-empirical methods; the core electrons are modelled by a suitable function, and only the valence electrons are treated explicity. In many cases this gives quite good results at a fraction of the cost of a calculation involving all electrons. Part of the relativistic effects may also be taken care of (especially the scalar effects), without having to perform the full relativistic calculation. ECPs have also been designed for second row elements, although the saving is only marginal relative to all-electron calculations.

There are four major steps in designing ECP type basis sets. First a good quality all-electron wave function is generated for the atom. This will typically be a numerical Hartree–Fock, or a relativistic Dirac–Hartree–Fock, calculation. The valence orbitals are then replaced by a set of nodeless pseudo-orbitals. The regular valence orbitals will have a series of radial nodes in order to make them orthogonal to the core orbitals, and the pseudo-orbitals are designed so that they behave correctly in the outer part, but do not have a nodal structure in the core region. The core electrons are then replaced by a potential so that solution of the Schrödinger (or Dirac) equation produces valence orbitals matching the pseudo-orbitals. Since relativistic effects are mainly important for the core electrons, this potential effectively includes relativity. The potential will be different for each angular momentum, and will normally be obtained in a tabulated form. In the final step, this numerical potential is fitted to a suitable set of analytical functions, normally a set of Gaussian functions.

$$U_{\mathrm{ECP}}(r) = \sum_i a_i r^{n_i} \mathrm{e}^{-\alpha_i r^2} \tag{5.10}$$

The parameters a_i, n_i and α_i depend on the angular momentum (s-, p-, d- etc.) and are determined by least squares fit. Typically between two and seven Gaussian functions are used in the fit; many Gaussians improve the fit (and consequently the resulting orbitals) at the price of increased computational time.

For transition metals it is clear that the outer $(n + 1)$s-, $(n + 1)$p- and (n)d-orbitals constitute the valence space. While such "full-core" potentials give reasonable geometries, it has been found that the energetics are not always satisfactory. Better results can be obtained by also including the orbitals in the next lower shell in the valency space, albeit at an increase in the computational cost.

The gain by using ECPs is largest for atoms in the lower part of the periodic table, especially those where relativistic effects are important. Since fully relativistic results are scarce, the performance of ECPs is somewhat difficult to evaluate by comparison with other calculations,[39] but they often reproduce known experimental results, thereby justifying the approach.

5.8 Basis Set Superposition Errors

By far the most common type of basis set is centered on the nuclei. As a complete basis set cannot be used in practice, the M^4 (or worse) increase in computational effort limits practical calculations to ~hundreds or at best a few thousand basis functions. For most systems this means that absolute errors in the energy from basis set incompleteness are quite large, maybe several a.u. (thousands of kcal/mol). However, usually the interest is in relative energies, and the primary goal is therefore to make the error as constant as possible. This is one of the reasons why it is important to choose a "balanced" basis set. The first, perhaps obvious condition, is that the same basis must be used when comparing energies, i.e. comparing energies of two isomers where the 6-31 G basis has been used for one of them, and the DH basis for the other, is meaningless, although both basis sets are of double zeta quality.

Fixing the position of the basis functions to the nuclei allows for a compact basis set, otherwise sets of basis functions positioned at many points in the geometrical space would be needed. When comparing energies at different geometries, however, the nuclear fixed basis set introduces an error. The quality of the basis set is not the same at all geometries, owing to the fact that the electron density around one nucleus may be described by functions centred at another nucleus. This is especially troublesome when calculating small effects, such as energies of van der Waals complexes and hydrogen bonds. Consider for example the hydrogen bond between two water molecules. The simplest approach consists in calculating the energy of the dimer and substracting twice the energy of an isolated molecule (assuming a size extensive method). The electron distribution within each water molecule in the dimer is very close to that of the monomer. In the dimer, however, basis functions from one molecule can help compensate for the basis set incompleteness on the other molecule, and vice versa. The dimer will therefore be artificially lowered in energy, and the strength of the hydrogen bond overestimated. This effect is known as *Basis Set Superposition Error* (BSSE). In the limit of a complete basis set, the BSSE will be zero; adding additional basis functions does not give any improvement. The conceptually simplest approach for eliminating BSSE is therefore to add more and more basis functions, until the interaction energy no longer changes. Unfortunately, this requires very large basis sets. Since non-bonded interactions are weak, the desired accuracy is often ~ 0.1 kcal/mol. Using the cc basis sets, the water dimer interaction energy stabilizes at this level with the aug-cc-pVTZ basis (184 basis functions for H_2O) at the HF level, but requires (at least) the aug-cc-pV5Z basis (574 basis functions) at the MP2 level.[40] As inclusion of electron correlation is mandatory for calculating the dispersion interaction between molecules, even the water dimer potential is computationally challenging.

An approximate way of assessing BSSE is the *Counterpoise* (CP) correction.[41] In this method the BSSE is estimated as the difference between monomer energies with the regular basis and the energies calculated with the full set of basis functions for the whole complex. Consider two molecules A and B, each having regular nuclear centred basis sets denoted by subscripts a and b, and the complex AB having the combined basis set ab. The geometries of the two isolated molecules and of the complex are first optimized or otherwise assigned. The geometries of the A and B molecules in the complex will usually be slightly different from those of the isolated species, and the complex geometry will be denoted by a $*$. The dimer energy minus the monomer energies is the directly calculated complexation energy.

$$\Delta E_{\text{complexation}} = E(\text{AB})^*_{ab} - E(\text{A})_a - E(\text{B})_b \qquad (5.11)$$

To estimate how much of this complexation energy is due to BSSE, four additional energy calculations are needed. Using basis set a for A, and basis set b for B, the energies of each of the two fragments are calculated <u>with the geometry they have in the complex</u>. Two additional energy calculations of the fragments at the complex geometry are then carried out with the full ab basis set. For example, the energy of A is calculated in the presence of both the normal a basis functions <u>and</u> with the b basis functions of fragment B located at the corresponding nuclear positions, but <u>without</u> the B nuclei present. Such basis functions located at fixed points in space are often referred to as *ghost orbitals*. The fragment energy for A will be lowered owing to these ghost functions, since the basis becomes more complete. The CP correction is defined as

$$\Delta E_{\text{CP}} = E(\text{A})^*_{ab} + E(\text{B})^*_{ab} - E(\text{A})^*_a - E(\text{B})^*_b \qquad (5.12)$$

The counterpoise corrected complexation energy is given as $\Delta E_{\text{complexation}} - \Delta E_{\text{CP}}$. For regular basis sets this typically stabilizes at the basis set limiting value much earlier than the uncorrected value, but this is not necessarily the case if diffuse functions are included in the basis set.

There are variations of this method. For example may it be argued that the full set of ghost orbitals should not be used, since some of the functions in the complex are used for describing the electrons of the other component, and only the virtual orbitals are available for "artificial" stabilization. However, it appears that the method of full counterpoise corection (using all basis functions as ghost orbitals) gives the best results. Note that ΔE_{CP} is an approximate correction, it gives an <u>estimate</u> of the BSSE effect, but it does not provide either an upper or lower limit.

It is usually observed that the CP correction for methods including electron correlation is larger and more sensitive to the size of the basis set, than that at the HF level. This is in line with the fact that the HF wave function converges much faster with respect to the size of the basis set than correlated wave functions.

There have been attempts to develop methods where the BSSE is excluded explicitly in the computational expressions, an example of this is the *Chemical Hamiltonian Approach* (CHA),[42] but such methods are not yet commonly used.

The BSSE is always present, also in calculating energies of "normal" species, for example the differential stability of ethanol and dimethyl ether, or the conformational difference between staggered and eclipsed ethane. Indeed, part of what is often referred to as the "basis set effect" (the change in a relative energy when the basis is enlarged) should more correctly be considered as an intramolecular BSSE.[43] For intermolecular (non-bonded) interactions the CP correction is well defined, although it may not be as accurate as desired. For intramolecular cases, however, it is difficult to define a unique procedure for estimating the BSSE, and it is almost always ignored.

5.9 Pseudospectral Methods

The goal of *pseudospectral* methods is to reduce the formal M^4 dependence of the Coulomb and Exchange operators in the basis set representation (two-electron integrals, eq. (3.51)) to M^3.[44] This can be accomplished by switching between a grid representation in the physical space (the three-dimensional cartesian space) and the spectral representation in the function space (the basis set).

Consider the following Coulomb contribution to the $F_{\alpha\beta}$ element of the Fock matrix (eq. (3.51)); similar considerations hold for the exchange contribution.

$$\langle \chi_\alpha | \mathbf{F} | \chi_\beta \rangle \leftarrow \sum_j^N \langle \chi_\alpha | \mathbf{J}_j | \chi_\beta \rangle = \sum_j^N \langle \chi_\alpha \phi_j | \mathbf{g} | \chi_\beta \phi_j \rangle \tag{5.13}$$

Written out in terms of the actual integrals, this is

$$\langle \chi_\alpha | \mathbf{J}_j | \chi_\beta \rangle = \int\int \chi_\alpha(\mathbf{r}_1) \mathbf{J}_j(\mathbf{r}_2) \chi_\beta(\mathbf{r}_1) \, d\mathbf{r}_1 \, d\mathbf{r}_2$$

$$\langle \chi_\alpha \phi_j | \mathbf{g} | \chi_\beta \phi_j \rangle = \int\int \chi_\alpha(\mathbf{r}_1) \phi_j(\mathbf{r}_2) \frac{1}{|\mathbf{r}_1 - \mathbf{r}_2|} \chi_\beta(\mathbf{r}_1) \phi_j(\mathbf{r}_2) \, d\mathbf{r}_1 \, d\mathbf{r}_2 \tag{5.14}$$

For a specific point in space for coordinate 1, $\mathbf{r}_g$, the integration over coordinate 2 may be carried out

$$\langle \chi_\alpha(\mathbf{r}_g) \phi_j | \mathbf{g} | \chi_\beta(\mathbf{r}_g) \phi_j \rangle = \int \chi_\alpha(\mathbf{r}_g) \chi_\beta(\mathbf{r}_g) \, d\mathbf{r}_1 \left(\int \frac{\phi_j^2(\mathbf{r}_2)}{|\mathbf{r}_2 - \mathbf{r}_g|} \, d\mathbf{r}_2 \right) \tag{5.15}$$

and the integral written in terms of atomic quantities

$$\int \frac{\phi_j^2(\mathbf{r}_2)}{|\mathbf{r}_2 - \mathbf{r}_g|} \, d\mathbf{r}_2 = \sum_{\gamma\delta}^M c_{\gamma j} c_{\delta j} \int \chi_\gamma(\mathbf{r}_2) \frac{1}{|\mathbf{r}_2 - \mathbf{r}_g|} \chi_\delta(\mathbf{r}_2) \, d\mathbf{r}_2 = \sum_{\gamma\delta}^M c_{\gamma j} c_{\delta j} A_{\gamma\delta}(\mathbf{r}_g) \tag{5.16}$$

The $A_{\gamma\delta}$ integral is just a three-centre one-electron integral, which can be evaluated analytically. The integration over coordinate 1 may then be approximated as a sum over a finite set of grid points in the physical space.

$$\langle \chi_\alpha \phi_j | \mathbf{g} | \chi_\beta \phi_j \rangle \simeq \sum_{g=1}^G \chi_\alpha(\mathbf{r}_g) \chi_\beta(\mathbf{r}_g) \sum_{\gamma=1}^M \sum_{\delta=1}^M c_{\gamma j} c_{\delta j} A_{\gamma\delta}(\mathbf{r}_g) \tag{5.17}$$

As the number of grid points increases, this approximation becomes better. The reduction in the formal scaling from M^4 to $\sim M^3$ comes from the fact that the summations involve GM^2 operations, G being the number of grid points, which typically will be linearly dependent on the number of basis functions M, i.e. $GM^2 \sim M^3$.

Unfortunately the above formula does not work well unless a very large number of grid points is used. This is due to an effect known as *aliasing*; the physical space Coulomb operator $\mathbf{J}(\mathbf{r}_g)$ acting on the basis function χ_β produces a result which has components outside the basis set. In practice the $\mathbf{J}(\mathbf{r}_g)\chi_\beta$ product is therefore fitted to a larger *de-aliasing* basis set, typically constructed from the original basis set by adding functions with exponents intermediate between those already present, and polarization functions with one higher angular momentum than already present.

$$\mathbf{J}_j(\mathbf{r}_g)\chi_\beta(\mathbf{r}_g) \simeq \sum_\sigma^{M^*} W^*_{\sigma\beta}\chi^*_\sigma(\mathbf{r}_g) \qquad (5.18)$$

The full set of de-aliasing basis functions is denoted χ^*_σ, and contains M^* functions. The weights $W^*_{\sigma\beta}$ are assigned based on a least squares fitting procedure. A similar scheme may be constructed for the exchange operator $\mathbf{K}$. By careful control of the grid size, the de-aliasing basis, and by analytical evaluation of the one-centre (and sometimes also the two-centre) Coulomb and exchange contributions, which are computationally insignificant compared to the three- and four-centre integrals, pseudospectral methods may provide energies at the same accuracy as fully analytical methods. For small to medium size systems the gain at the HF level may be a factor of 2–5, but the M^3 scaling should eventually make such methods significantly faster than M^4 all-integral methods. The use of techniques such as integral screening and linear scaling methods (section 3.8.6), however, reduces the formal scaling substantially below M^4, pontentially down to M^1, in which case pseudospectral methods are no longer competitive.

References

1. D. Feller and E. R. Davidson, *Rev. Comput. Chem.*, **1** (1990), 1; T. Helgaker and P. R. Taylor, *Modern Electronic Structure Theory*, Part II, ed. D. Yarkony, World Scientific 1995, p. 727–856.
2. J. C. Slater, *Phys. Rev.*, **36** (1930), 57.
3. S. F. Boys, *Proc. R. Soc. (London)* A, **200** (1950), 542.
4. W. Klopper and W. Kutzelnigg, *J. Mol. Struct.*, **135** (1986), 339; W. Kutzelnigg and J. D. Morgan III, *J. Chem. Phys.*, **96** (1992), 4484.
5. M. W. Schmidt and K. Ruedenberg, *J. Chem. Phys.*, **71** (1979), 3951.
6. S. Huzinaga, M. Klobukowski and H. Tatewski, *Can. J. Chem.*, **63** (1985), 1812.
7. W. J. Hehre, R. F. Stewart and J. A. Pople, *J. Chem. Phys.*, **51** (1969), 2657.
8. J. S. Binkley and J. A. Pople, *J. Am. Chem. Soc.*, **102** (1980), 939.
9. W. J. Hehre, R. Ditchfield and J. A. Pople, *J. Chem. Phys.*, **56** (1972), 2257.
10. R. Krishnan, J. S. Binkley, R. Seeger and J. A. Pople, *J. Chem. Phys.*, **72** (1980), 650.
11. M. J. Frisch, J. A. Pople and J. S. Binkley, *J. Chem. Phys.*, **80** (1984), 3265.
12. M. M. Francl, W. J. Pietro, W. J. Hehre, J. S. Binkley, M. S. Gordon, D. J. DeFrees and J. A. Pople, *J. Chem. Phys.*, **77** (1982), 3654.
13. S. Huzinaga, *J. Chem. Phys.*, **42** (1965), 1293.
14. F. B. van Duijneveldt, *IBM Tech. Res. Rep. RJ945* (1971).
15. H. Partridge, *J. Chem. Phys.*, **90** (1989), 1043.
16. T. H. Dunning, *J. Chem. Phys.*, **55** (1971), 716.
17. A. D. McLean and G. S. Chandler, *J. Chem. Phys.*, **72** (1980), 5639.
18. H. Tatewaki and S. Huzinaga, *J. Comp. Chem.*, **3** (1980), 205.
19. J. Almlöf and P. R. Taylor, *J. Chem. Phys.*, **92** (1990), 551; J. Almlöf and P. R. Taylor, *Adv. Quantum Chem.*, **22** (1991), 301.
20. T. H. Dunning Jr, *J. Chem. Phys.*, **90** (1989), 1007; A. K. Wilson, T. van Mourik and T. H. Dunning Jr, *J. Mol. Struct.*, **388** (1996), 339.
21. R. A. Kendall, T. H. Dunning Jr and R. J. Harrison, *J. Chem. Phys.*, **96** (1992), 6796.
22. D. E. Woon and T. H. Dunning, Jr, *J. Chem. Phys.*, **103** (1995), 4572.

23. A. K. Wilson and T. H. Dunning, Jr, *J. Chem. Phys.*, **106** (1997), 8718; D. Feller and K. A. Peterson, *J. Chem. Phys.*, **108** (1998), 154; A. Halkier, T. Helgaker, P. Jørgensen, W. Klopper, H. Koch, J. Olsen and A. K. Wilson, *Chem. Phys. Lett*, **286** (1998), 243.

24. K. A. Petersoo, K. A. Wilson, D. E. Woon and T. H. Dunning, Jr, *Theor. Chem. Acc.*, **97** (1997), 251.

25. A. P. Scott and L. Radom, *J. Phys. Chem.*, **100** (1996), 16502.

26. L. A. Curtiss, K. Raghavachari, G. W. Tucks and J. A. Pople, *J. Chem. Phys.*, **94** (1991), 7221.

27. L. A. Curtiss, K. Raghavachari, P. C. Redfern and J. A. Pople, *J. Chem. Phys.*, **106** (1997), 1063.

28. J. A. Curtiss, K. Raghavachari and J. A. Pople, *J. Chem. Phys.*, **98** (1993), 1293.

29. L. A. Curtiss, J. E. Carpenter, K. Raghavachari and J. A. Pople, *J. Chem. Phys.*, **96**, (1992), 9030.

30. L. A. Curtiss, P. C. Redfern, B. J. Smith and L. Radom, *J. Chem. Phys.*, **104** (1996), 5148.

31. C. W. Bauschlicher and H. Partridge, *J. Chem. Phys.*, **103** (1995), 1788; A. M. Mebel, K. Morokuma and M. C. Lin, *J. Chem. Phys.*, **103** (1995), 7414.

32. L. A. Curtiss, K. Raghavachari and J. A. Pople, *J. Chem. Phys.*, **103** (1995), 4192.

33. J. M. L. Martin, *J. Chem. Phys.*, **100** (1994), 8186.

34. J. W. Ochtershi, G. A. Petersson and J. A. Montgomer Jr., *J. Chem. Phys.*, **104** (1996), 2598.

35. J. L. Durant Jr and G. M. Rohlfing, *J. Chem. Phys.*, **98** (1993), 8031.

36. J. Rossi and D. G. Truhlar, *Chem. Phys. Lett.*, **234** (1995), 64.

37. P. E. M. Siegbahn, M. Svensson and P. J. E. Boussard, *J. Chem. Phys.*, **102** (1995), 5377.

38. G. Frenking, I. Antes, M. Böhme, S. Dapprioh, A. W. Ehlers, V. Jonas, A. Nauhaus, M. Otto, R. Stegmann, A. Veldkamp and S. F. Vyboishchikov, *Rev. Comput. Chem.*, **8** (1996), 63; T. R. Cundari, M. T. Benson, M. L. Lutz and S. O. Sommerer, *Rev. Comput. Chem.*, **8** (1996), 145.

39. See however K. Dyall, *J. Chem. Phys.*, **96** (1991), 1210.

40. A. Halkier, H. Koch, P. Jørgensen, O. Christiansen, I. M. B. Nielsen and T. Helgaker, *Theor. Chim. Acta.*, **97** (1997), 150.

41. F. B. van Duijneveldt, J. G. C. M. van Duijneveldt-van de Rijdt and J. H. van Lenthe, *Chem. Rev.*, **94** (1994), 1873.

42. I. Mayer and A. Vibok, *Mol. Phys.*, **92** (1997), 503. See also E. Gianinetti, M. Raimondi and E. Tornaghi, *Int. J, Quant. Chem.*, **60** (1996), 157.

43. F. Jensen, *Chem. Phys. Lett.*, **261** (1996), 633.

44. B. H. Greedy, T. V. Russo, D. T. Mainz, R. A. Friesner, J.-M. Langlois, W. A. Goddard III, R. E. Donnally and M. N. Ringalda, *J. Chem. Phys.*, **101** (1994), 4028.

6 Density Functional Theory

The basis for *Density Functional Theory* (DFT) is the proof by Hohenberg and Kohn[1] that the ground-state electronic energy is determined completely by the electron density ρ, see Appendix B for details. In other words, there exists a one-to-one correspondence between the electron density of a system and the energy. The significance of this is perhaps best illustrated by comparing to the wave function approach. A wave function for an N-electron system contains $3N$ coordinates, three for each electron (four if spin is included). The electron density is the square of the wave function, integrated over $N-1$ electron coordinates, this only depends on three coordinates, independently of the number of electrons. While the complexity of a wave function increases with the number of electrons, the electron density has the same number of variables, independently of the system size. The "only" problem is that although it has been proven that each different density yields a different ground-state energy, the functional connecting these two quantities is not known. The goal of DFT methods is to design functionals connecting the electron density with the energy.[2]

A note on semantics: a function is a prescription for producing a number from a set of variables (coordinates). A functional is similarly a prescription for producing a number from a function, which in turn depends on variables. A wave function and the electron density are thus functions, while an energy depending on a wave function or an electron density is a functional. We will denote a function depending on a set of variables with parentheses, $f(x)$, while a functional depending on a function is denoted with brackets, $F[f]$.

Comparing this with the wave mechanics approach, it seems clear that the energy functional may be divided into three parts, kinetic energy, $T[\rho]$, attraction between the nuclei and electrons, $E_{ne}[\rho]$, and electron-electron repulsion, $E_{ee}[\rho]$ (the nuclear–nuclear repulsion is a constant in the Born–Oppenheimer approximation). Furthermore, with reference to Hartree–Fock theory (eq. (3.33)), the $E_{ee}[\rho]$ term may be divided into a Coulomb and an Exchange part, $J[\rho]$ and $K[\rho]$, implicitly including correlation energy in all terms. The $E_{ne}[\rho]$ and $J[\rho]$ functionals are given by their classical expressions, where the factor of 1/2 in $J[\rho]$ allows the integration to run over all space for both variables.

$$E_{ne}[\rho] = \sum_a \int \frac{Z_a \rho(\mathbf{r})}{|\mathbf{R}_a - \mathbf{r}|} d\mathbf{r}$$

$$J[\rho] = \frac{1}{2} \int \int \frac{\rho(\mathbf{r})\rho(\mathbf{r}')}{|\mathbf{r} - \mathbf{r}'|} d\mathbf{r}\, d\mathbf{r}' \tag{6.1}$$

Early attempts at deducing functionals for the kinetic and exchange energies considered a non-interacting uniform electron gas. For such a system it may be shown that $T[\rho]$ and $K[\rho]$ are given as

$$T_{TF}[\rho] = C_F \int \rho^{5/3}(\mathbf{r}) d\mathbf{r}$$

$$K_D[\rho] = -C_x \int \rho^{4/3}(\mathbf{r}) d\mathbf{r}$$

$$C_F = \frac{3}{10}(3\pi^2)^{2/3} \tag{6.2}$$

$$C_x = \frac{3}{4}\left(\frac{3}{\pi}\right)^{1/3}$$

The energy functional $E_{TF}[\rho] = T_{TF}[\rho] + E_{ne}[\rho] + J[\rho]$ is known as *Thomas–Fermi* (TF) theory, including the $K_D[\rho]$ exchange part (first derived by Block[3] but commonly associated with the name of Dirac[4] (constitutes the *Thomas–Fermi–Dirac* (TFD) model.

The assumption of a non-interacting uniform electron gas does not hold very well for atomic and molecular systems. Total energies are in error by 15–50%, but more serious is it that neither TF nor TFD theories predict bonding, molecules simply do not exist. The T and K functionals may be improved by addition of terms which depend not only on the density itself, but also on its derivative(s). This is equivalent of considering a non-uniform electron gas, and performing a Taylor-like expansion of the density. Addition of such gradient correction terms improves the results, for example bonding is now allowed, but in general it has been found that this is not a viable approach for constructing DFT models capable of yielding results comparable to those obtained by wave mechanics methods.

The foundation for the use of DFT methods in computational chemistry was the introduction of orbitals by Kohn and Sham.[5] The main problem in Thomas–Fermi models is that the kinetic energy is represented poorly. The basic idea in the Kohn and Sham (KS) formalism is splitting the kinetic energy functional into two parts, one of which can be calculated exactly, and a small correction term.

Assume for the moment a Hamilton operator of the following form with $0 \le \lambda \le 1$.

$$\mathbf{H}_\lambda = \mathbf{T} + \mathbf{V}_{ext}(\lambda) + \lambda\mathbf{V}_{ee} \tag{6.3}$$

The $\mathbf{V}_{ext}$ operator is equal to $\mathbf{V}_{ne}$ for $\lambda = 1$, for intermediate λ values, however, it is assumed that the external potential $\mathbf{V}_{ext}(\lambda)$ is adjusted so that the same density is obtained for both $\lambda = 1$ (the real system) and $\lambda = 0$ (a hypothetical system with non-interacting electrons). For the $\lambda = 0$ case the exact solution to the Schrödinger equation is given as a Slater determinant composed of (molecular) orbitals, ϕ_i, for which the exact kinetic energy functional is given as

$$T_S = \sum_{i=1}^{N} \langle \phi_i | -\tfrac{1}{2} \mathbf{V}^2 | \phi_i \rangle \qquad (6.4)$$

The subscript S denotes that it is the kinetic energy calculated from a Slater determinant. The $\lambda = 1$ case corresponds to interacting electrons, and eq. (6.4) is therefore only an approximation to the real kinetic energy, but a substantial improvement over the TF formula (eq. (6.2)).

Another way of justifying the use of eq. (6.4) for calculating the kinetic energy is by reference to natural orbitals (eigenvectors of the density matrix, Section 9.5). The exact kinetic energy can be calculated from the natural orbitals (NO) arising from the exact density matrix.

$$T[\rho_{\text{exact}}] = \sum_{i=1}^{\infty} n_i \langle \phi_i^{\text{NO}} | -\tfrac{1}{2} \mathbf{V}^2 | \phi_i^{\text{NO}} \rangle$$

$$\rho_{\text{exact}} = \sum_{i=1}^{\infty} n_i |\phi_i^{\text{NO}}|^2 \qquad (6.5)$$

$$N = \sum_{i=1}^{\infty} n_i$$

The orbital occupation numbers n_i (eigenvalues of the density matrix) will be between 0 and 1, corresponding to the number of electrons in the orbital. Note that the representation of the exact density normally will require an infinite number of natural orbitals. The first N occupation numbers (N being the total number of electrons in the system) will normally be close to 1, and the remaining close to 0.

Since the exact density matrix is not known, the (approximate) density is written in terms of a set of auxiliary one-electron functions, orbitals, as

$$\rho(\mathbf{r}) = \sum_{i=1}^{N} |\phi_i(\mathbf{r})|^2 \qquad (6.6)$$

This corresponds to eq. (6.5) with occupation numbers of exactly 1 or 0. The "missing" kinetic energy from eq. (6.4) is thus due to the occupation numbers deviating from being exactly 1 or 0.

The key to Kohn–Sham theory is thus the calculation of the kinetic energy under the assumption of non-interacting electrons (in the same sense as HF orbitals in wave mechanics describe non-interacting electrons) from eq. (6.4). In reality the electrons are interacting, and eq. (6.4) does not provide the total kinetic energy. However, just as HF theory provides $\sim 99\%$ of the correct answer, the difference between the exact kinetic energy and that calculated by assuming non-interacting orbitals is small. The remaining kinetic energy is absorbed into an exchange–correlation term, and a general DFT energy expression can be written as

$$E_{\text{DFT}}[\rho] = T_S[\rho] + E_{\text{ne}}[\rho] + J[\rho] + E_{\text{xc}}[\rho] \qquad (6.7)$$

By equating E_{DFT} to the exact energy, this expression may be taken as the definition of

E_{xc}, it is the part which remains after subtraction of the non-interacting kinetic energy, and the E_{ne} and J potential energy terms.

$$E_{\mathrm{xc}}[\rho] = (T[\rho] - T_S[\rho]) + (E_{\mathrm{ee}}[\rho] - J[\rho]) \qquad (6.8)$$

The first parenthesis in eq. (6.8) may be considered the kinetic correlation energy, while the second contains both exchange and potential correlation energy.

The exchange energy is by far the largest contributor to E_{xc}. For the neon atom, for example, the exchange energy is -12.11 a.u., while the correlation energy is -0.39 a.u. (as calculated by wave mechanics methods). Since the exchange energy dominates E_{xc}, one may reasonably ask why we do not calculate this term "exactly" from orbitals (analogously to the kinetic energy), by the formula known from wave mechanics (eq. (3.33)), and only calculate the computationally difficult part, the correlation energy, by DFT. Although this has been tried, it gives poor results. The basic problem is that the DFT definitions of exchange and correlation energies are not completely equivalent to their wave mechanics counterparts.[6] The correlation energy in wave mechanics is defined as the difference between the exact energy and the corresponding Hartree–Fock value, and the exchange energy is the total electron–electron repulsion minus the Coulomb energy. These energies have both a short- and long-range part (in terms of the distance between two electrons). The long-range correlation is essentially the "static" correlation energy (i.e. the "multi-reference" part, see Section 4.6) while the short-range part is the "dynamical" correlation. The long-range part of the exchange energy in wave mechanics effectively cancels the long-range part of the correlation energy. The definitions of exchange and correlation in DFT, however, are local (short range), they only depend on the density at a given point and in the immediate vicinity (via derivatives of the density). The cancellation at long range is (or should be) implicitly built into the exchange–correlation functional. Calculating the exchange energy by wave mechanics and the correlation by DFT thus destroys the cancellation.

The strength of DFT is that only the total density needs to be considered. In order to calculate the kinetic energy with sufficient accuracy, however, orbitals have to be reintroduced. Nevertheless, as discussed in Section 6.5, DFT has a computational cost which is similar to HF theory, with the possibility of providing more accurate (exact, in principle) results.

The major problem in DFT is deriving suitable formulas for the exchange–correlation term. Assume for the moment that such a functional is available, the problem is then similar to that encountered in wave mechanics HF theory: determine a set of orthogonal orbitals which minimize the energy (the requirement of orthogonal orbitals basicly enforces the Pauli principle). Since the $J[\rho]$ (and $E_{\mathrm{xc}}[\rho]$) functional depends on the total density, a determination of the orbitals involves an iterative sequence. The orbital orthogonality constraint may be enforced by the Lagrange method (section 14.6), again in analogy with wave mechanics HF methods (eq. (3.34)).

$$L[\rho] = E_{\mathrm{DFT}}[\rho] - \sum_{ij}^{N} \lambda_{ij}[\langle \phi_i | \phi_j \rangle - \delta_{ij}] \qquad (6.9)$$

Requiring the variation of L to vanish provides a set of equations involving an effective one-electron operator ($\mathbf{h}_{\mathrm{KS}}$), similar to the Fock operator in wave mechanics (eq. (3.36)).

$$\mathbf{h}_{KS}\phi_i = \sum_j^N \lambda_{ij}\phi_j$$

$$\mathbf{h}_{KS} = -\frac{1}{2}\nabla^2 + \mathbf{V}_{\text{eff}} \qquad (6.10)$$

$$\mathbf{V}_{\text{eff}}(\mathbf{r}) = \mathbf{V}_{\text{ne}}(\mathbf{r}) + \int \frac{\rho(\mathbf{r}')}{|\mathbf{r}-\mathbf{r}'|}d\mathbf{r}' + \mathbf{V}_{\text{xc}}(\mathbf{r})$$

We may again chose a unitary transformation which makes the matrix of the Lagrange multiplier diagonal, producing a set of canonical *Kohn–Sham* (KS) orbitals. The resulting pseudo-eigenvalue equations are known as the Kohn–Sham equations.

$$\mathbf{h}_{KS}\,\phi_i = \varepsilon_i\phi_i \qquad (6.11)$$

The Lagrange multipliers can again be associated with molecular orbital energies, and the highest occupied orbital energy is the ionization potential (Koopmans' theorem), but only if the <u>exact</u> exchange–correlation functional is employed. Since this is not the case in actual calculations, the orbital energies in practice do not carry quite the same significance as in HF theory.[7] The unknown KS orbitals may be determined by numerical methods, or expanded in a set of basis functions, analogously to the HF method (Section 3.5).

Although it is clear that there are many similarities between wave mechanics HF theory and DFT, there is an important difference. If the <u>exact</u> $E_{\text{xc}}[\rho]$ was known, DFT would provide the <u>exact</u> total energy, <u>including</u> electron correlation. DFT methods therefore have the potential of including the computationally difficult part in wave mechanics, the correlation energy, at a computational effort similar to that for determining the uncorrelated HF energy. Although this is certainly the case for approximations to $E_{\text{xc}}[\rho]$ (as illustrated below), this is not necessarily true for the <u>exact</u> $E_{\text{xc}}[\rho]$. It may well be that the exact $E_{\text{xc}}[\rho]$ functional is so complicated that the computational effort for solving the KS equations will be similar to that required for solving the Schrödinger equation (exactly) with a wave mechanics approach. Indeed, unless one believes that the Schrödinger equation contains superfluous information, this is likely to be the case. Since exact solutions are generally not available in either approach, the important question is instead: what is the computational cost for generating a solution of a certain accuracy. In this respect DFT methods have very favourable characteristics.

It is possible to prove that the exchange–correlation potential is a unique functional, valid for all systems, but an explicit functional form of this potential has been elusive. The difference between DFT methods is the choice of the functional form of the exchange–correlation energy. There is little guidance from theory how such functionals should be chosen, and consequently many different potentials have been proposed. Functional forms are often designed to have a certain limiting behaviour (for example including the uniform electron gas limit), and fitting parameters to known accurate data. Which functional is the better will have to be settled by comparing the performance with experiments or high-level wave mechanics calculations.

One of the more recent approaches for designing E_{xc} functionals is based on "inverting" eq. (6.7). An accurate electron density may be calculated by advanced wave mechanics methods (e.g. coupled cluster), and a set of KS orbitals which yield this density can be determined by a "constrained search" method involving a minimization of the kinetic energy. The idea was originally introduced by Levy and Perdew[8] and has since been elaborated on by Parr and others.[9] Recent implementations and applications have been reported by Baerends *et al.* and Handy *et al.*[10] This enables a calculation of the exchange–correlation energy (by subtracting T, V_{ne} and J from the total DFT energy), thereby giving the direct dependence of $\mathbf{V}_{xc}$ on the density ρ. Since the DFT area is fairly new in computational chemistry, there are at present no clear "standard" methods, like MP2 and CISD in traditional *ab initio* theory, and the calibration of different methods is much less developed. Some examples of functionals which currently are used quite extensively are given below, but since DFT is an active area of research, new and improved potentials are likely to emerge.

It is customary to separate E_{xc} into two parts, a pure exchange E_x and a correlation part E_c, although it is not clear that this is a valid assumption (cf. the above discussion of the definition of exchange and correlation). Each of these energies is often written in terms of the energy per particle (energy density), ε_x and ε_c.

$$E_{xc}[\rho] = E_x[\rho] + E_c[\rho] = \int \rho(\mathbf{r})\varepsilon_x[\rho(\mathbf{r})]d\mathbf{r} + \int \rho(\mathbf{r})\varepsilon_c[\rho(\mathbf{r})]d\mathbf{r} \qquad (6.12)$$

The corresponding potential required in eq. (6.10) is given as the derivative of the energy with respect to the density.

$$\mathbf{V}_{xc}(\mathbf{r}) = \frac{\partial E_{xc}[\rho]}{\partial \rho(\mathbf{r})} = \varepsilon_{xc}[\rho(\mathbf{r})] + \rho(\mathbf{r})\frac{\partial \varepsilon_{xc}(\mathbf{r})}{\partial \rho} \qquad (6.13)$$

As mentioned in the start of Chapter 4, the correlation between electrons of parallel spin is different from that between electrons of opposite spin. The exchange energy is "by definition" given as a sum of contributions from the α and β spin densities, as exchange energy only involves electrons of the same spin. The kinetic energy, the nuclear-electron attraction and Coulomb terms are trivially separable.

$$\begin{aligned} E_x[\rho] &= E_x^\alpha[\rho_\alpha] + E_x^\beta[\rho_\beta] \\ E_c[\rho] &= E_c^{\alpha\alpha}[\rho_\alpha] + E_c^{\beta\beta}[\rho_\beta] + E_c^{\alpha\beta}[\rho_\alpha, \rho_\beta] \end{aligned} \qquad (6.14)$$

The total density is the sum of the α and β contributions, $\rho = \rho_\alpha + \rho_\beta$, and for a closed-shell singlet these are identical ($\rho_\alpha = \rho_\beta$). Functionals for the exchange and correlation energies may be formulated in terms of separate spin-densities; however, they are often given instead as functions of the spin polarization ζ (normalized difference between ρ_α and ρ_β), and the radius of the effective volume containing one electron, r_S.

$$\zeta = \frac{\rho^\alpha - \rho^\beta}{\rho^\alpha + \rho^\beta}, \quad \frac{4}{3}\pi r_S^3 = \rho^{-1} \qquad (6.15)$$

In the formulas below it is implicitly assumed that the exchange energy is a sum over both the α and β densities.

6.1 Local Density Methods

In the *Local Density Approximation* (LDA) it is assumed that the density locally can be treated as a uniform electron gas, or equivalently that the density is a slowly varying function. The exchange energy for a uniform electron gas is given by the Dirac formula (eq. (6.2)).

$$E_x^{LDA}[\rho] = -C_x \int \rho^{4/3}(\mathbf{r})d\mathbf{r}$$
$$\varepsilon_x^{LDA}[\rho] = -C_x \rho^{1/3} \tag{6.16}$$

In the more general case, where the α and β densities are not equal, LDA (where the sum of the α and β densities is raised to the 4/3 power) has been virtually abandoned and replaced by the *Local Spin Density Approximation* (LSDA) (which is given as the sum of the individual densities raised to the 4/3 power, eq. (6.17)).

$$E_x^{LSDA}[\rho] = -2^{1/3}C_x \int [\rho_\alpha^{4/3} + \rho_\beta^{4/3}]d\mathbf{r}$$
$$\varepsilon_x^{LSDA}[\rho] = -2^{1/3}C_x[\rho_\alpha^{1/3} + \rho_\beta^{1/3}] \tag{6.17}$$

LSDA may also be written in terms of the total density and the spin polarization.

$$\varepsilon_x^{LSDA}[\rho] = -\tfrac{1}{2}C_x\rho^{1/3}[(1+\zeta)^{4/3} + (1-\zeta)^{4/3}] \tag{6.18}$$

For closed-shell systems LSDA is equal to LDA, and since this is the most common case, LDA is often used interchangeably with LSDA, although this is not true in the general case (eqs. (6.16) and (6.17)). The X_α method proposed by Slater in 1951[11] can be considered as an LDA method where the correlation energy is neglected and the exchange term is given as

$$\varepsilon_{X_\alpha}[\rho] = -\tfrac{3}{2}\alpha C_x\rho^{1/3} \tag{6.19}$$

With $\alpha = 2/3$ this is identical to the Dirac expression. The original X_α method used $\alpha = 1$, but a value of 3/4 has been shown to give better agreement for atomic and molecular systems. The name Slater is often used as a synonym for the L(S)DA exchange energy involving the electron density raised to the 4/3 power (1/3 power for the energy density).

   The correlation energy of a uniform electron gas has been determined by Monte Carlo methods for a number of different densities. In order to use these results in DFT calculations, it is desirable to have a suitable analytic interpolation formula. This has been constructed by Vosko, Wilk and Nusair (VWN)[12] and is in general considered to be a very accurate fit. It interpolates between the unpolarized ($\zeta = 0$) and spin polarized ($\zeta = 1$) limits by the following functional.

$$\varepsilon_c^{VWN}(r_S, \zeta) = \varepsilon_c(r_S, 0) + \varepsilon_a(r_S)\left[\frac{f(\zeta)}{f''(0)}\right][1 - \zeta^4] + [\varepsilon_c(r_S, 1) - \varepsilon_c(r_S, 0)]f(\zeta)\zeta^4$$
$$f(\zeta) = \frac{(1+\zeta)^{4/3} + (1-\zeta)^{4/3} - 2}{2(2^{1/3} - 1)} \tag{6.20}$$

The $\varepsilon_c(r_S, \zeta)$ and $\varepsilon_a(r_S)$ functionals are parameterized as in eq. (6.21).

$$\varepsilon_{c/a}(x) = A \left\{ \begin{array}{l} \ln \dfrac{x^2}{X(x)} + \dfrac{2\ell}{Q} \tan^{-1}\left(\dfrac{Q}{2x+\ell}\right) - \\ \dfrac{\ell x_0}{X(x_0)}\left[\ln \dfrac{(x-x_0)^2}{X(x)} + \dfrac{2(\ell+2x_0)}{Q}\tan^{-1}\left(\dfrac{Q}{2x+\ell}\right)\right] \end{array} \right\} \tag{6.21}$$

$$x = \sqrt{r_S}$$
$$X(x) = x^2 + \ell x + c$$
$$Q = \sqrt{4c - \ell^2}$$

The parameters A, x_0, ℓ and c are fitting constants, different for $\varepsilon_c(r_S, 0)$, $\varepsilon_c(r_S, 1)$, and $\varepsilon_a(r_S)$.

A modified form for $\varepsilon_{c/a}(r_S)$ has been given by Perdew and Wang,[13] and is used in connection with the PW91 functional described in Section 6.2.

$$\varepsilon_{c/a}^{PW91}(x) = -2a\rho(1 + \alpha x^2)\ln\left(1 + \dfrac{1}{2a(\beta_1 x + \beta_2 x^2 + \beta_3 x^3 + \beta_4 x^3)}\right) \tag{6.22}$$

Here a, α, β_1, β_2, β_3 and β_4 are suitable constants.

The LSDA approximation in general underestimates the exchange energy by $\sim 10\%$, thereby creating errors which are larger than the whole correlation energy. Electron correlation is furthermore overestimated, often by a factor close to 2, and bond strengths are as a consequence overestimated. Despite the simplicity of the fundamental assumptions, LSDA methods are often found to provide results with an accuracy similar to that obtained by wave mechanics HF methods.

6.2 Gradient Corrected Methods

Improvements over the LSDA approach have to consider a non-uniform electron gas. A step in this direction is to make the exchange and correlation energies dependent not only on the electron density, but also on derivatives of the density. Such methods are known as *Gradient Corrected* or *Generalized Gradient Approximation* (GGA) methods (a straightforward Taylor expansion does not lead to an improvement over LSDA, it actually makes things worse, thus the name generalized gradient approximation). GGA methods are also sometimes referred to as *non-local* methods, although this is somewhat misleading since the functionals depend only on the density (and derivatives) at a given point, not on a space volume as for example the Hartree–Fock exchange energy.

Perdew and Wang (PW86)[14] proposed modifying the LSDA exchange expression to that shown in eq. (6.23), where x is a dimensionless gradient variable, and a, b and c being suitable constants (summation over equivalent expressions for the α and β densities is implicitly assumed).

$$\varepsilon_x^{PW86} = \varepsilon_x^{LDA}(1 + ax^2 + bx^4 + cx^6)^{1/15}$$
$$x = \dfrac{|\nabla\rho|}{\rho^{4/3}} \tag{6.23}$$

Becke[15] proposed a widely used correction (B or B88) to the LSDA exchange energy, which has the correct $-r^{-1}$ asymptotic behaviour for the energy density (but not for the exchange potential).[16]

$$\varepsilon_x^{B88} = \varepsilon_x^{LDA} + \Delta\varepsilon_x^{B88}$$

$$\Delta_x^{B88} = -\beta\rho^{1/3}\frac{x^2}{1 + 6\beta x \sinh^{-1}x} \tag{6.24}$$

The β parameter is determined by fitting to known atomic data and x is defined in eq. (6.23).

Another functional form (not a correction) proposed by Becke and Roussel (BR)[17] has the form

$$\varepsilon_x^{BR} = -\frac{2 - 2e^{-ab} - abe^{-ab}}{4b}$$

$$a^3 e^{-ab} = 8\pi\rho$$

$$a(ab - 2) = b\frac{\nabla^2\rho - 2D}{\rho} \tag{6.25}$$

$$D = \sum_i^N |\nabla\phi_i|^2 - \frac{(\nabla\rho)^2}{4\rho}$$

This functional contains derivatives of the orbitals, not just the gradient of the total density, and is computationally slightly more expensive. Despite the apparent difference in functional form, exchange expressions (6.24) and (6.25) have been found to provide results of similar quality.

Perdew and Wang have proposed an exchange functional similar to B88 to be used in connection with the PW91 correlation functional given below (eq. (6.30)).

$$\varepsilon_x^{PW91} = \varepsilon_x^{LDA}\left(\frac{1 + xa_1\sinh^{-1}(xa_2) + (a_3 + a_4e^{-bx^2})x^2}{1 + xa_1\sinh^{-1}(xa_2) + a_5x^2}\right) \tag{6.26}$$

where a_{1-5} and b again are suitable constants and x is defined in eq. (6.23).

There have been various gradient corrected functional forms proposed for the correlation energy. One popular functional (not a correction) is due to Lee, Yang and Parr (LYP)[18] and has the form

$$\varepsilon_c^{LYP} = -a\frac{\gamma}{(1 + d\rho^{-1/3})} - ab\frac{\gamma e^{-c\rho^{-1/3}}}{9(1 + d\rho^{-1/3})\rho^{8/3}}$$

$$\times\left[\begin{array}{l}18(2^{2/3})C_F(\rho_\alpha^{8/3} + \rho_\beta^{8/3}) - 18\rho t_W \\ +\rho_\alpha(2t_W^\alpha + \nabla^2\rho_\alpha) + \rho_\beta(2t_W^\beta + \nabla^2\rho_\beta)\end{array}\right]$$

$$\gamma = 2\left[1 - \frac{\rho_\alpha^2 + \rho_\beta^2}{\rho^2}\right] \tag{6.27}$$

$$t_W^\sigma = \frac{1}{8}\left(\frac{|\nabla\rho_\sigma|^2}{\rho_\sigma} - \nabla^2\rho_\sigma\right)$$

where the a, b, c and d parameters are determined by fitting to data for the helium atom. The t_W functional is known as the local Weizsacker kinetic energy density. Note that the γ-factor becomes zero when all the spins are aligned ($\rho = \rho_\alpha$, $\rho_\beta = 0$), i.e. the LYP functional does not predict any parallel spin correlation in such a case (e.g. the LYP correlation energy in triplet He is zero). The appearance of the second derivative of the density can be removed by partial integration[19] to give eq. (6.28).

$$\varepsilon_c^{LYP} = -4a\frac{\rho_\alpha\rho_\beta}{\rho^2(1+d\rho^{-1/3})}$$
$$-ab\omega\left\{\frac{\rho_\alpha\rho_\beta}{18}\left[\begin{array}{c}144(2^{2/3})C_F(\rho_\alpha^{8/3}+\rho_\beta^{8/3})+(47-7\delta)|\nabla\rho|^2\\-(45-\delta)(|\nabla\rho_\alpha|^2+|\nabla\rho_\beta|^2)+2\rho^{-1}(11-\delta)(\rho_\alpha|\nabla\rho_\alpha|^2+\rho_\beta|\nabla\rho_\beta|^2)\end{array}\right]\right.$$
$$\left.+\tfrac{2}{3}\rho^2(|\nabla\rho_\alpha|^2+|\nabla\rho_\beta|^2-|\nabla\rho|^2)-(\rho_\alpha^2|\nabla\rho_\beta|^2+\rho_\beta^2|\nabla\rho_\alpha|^2)\right\}$$

$$\omega = \frac{e^{-c\rho^{-1/3}}}{(1+d\rho^{-1/3})\rho^{14/3}}$$

$$\delta = c\rho^{1/3}+\frac{d\rho^{-1/3}}{(1+d\rho^{-1/3})} \tag{6.28}$$

Perdew proposed a gradient correction to the LSDA result. It appeared in 1986[20] and is known by the acronym P86.

$$\varepsilon_c^{P86} = \varepsilon_c^{LDA} + \Delta\varepsilon_c^{P86}$$
$$\Delta\varepsilon_c^{P86} = \frac{e^\Phi C(\rho)|\nabla\rho|^2}{f(\zeta)\rho^{7/3}}$$
$$f(\zeta) = 2^{1/3}\sqrt{\left(\frac{1+\zeta}{2}\right)^{5/3}+\left(\frac{1-\zeta}{2}\right)^{5/3}} \tag{6.29}$$
$$\Phi = a\frac{C(\infty)|\nabla\rho|}{C(\rho)\rho^{7/6}}$$
$$C(\rho) = \mathscr{C}_1 + \frac{\mathscr{C}_2+\mathscr{C}_3r_S+\mathscr{C}_4r_S^2}{1+\mathscr{C}_5r_S+\mathscr{C}_6r_S^2+\mathscr{C}_7r_S^3}$$

where a and $\mathscr{C}_{1-7}$ are numerical constants.

This functional was later modified to the following form (also a correction to the LSDA energy) by Perdew and Wang in 1991 (PW91 or P91)[21]

$$\Delta\varepsilon_c^{PW91}[\rho] = \rho(H_0(t,r_S,\zeta)+H_1(t,r_S,\zeta))$$
$$H_0(t,r_S,\zeta) = b^{-1}f(\zeta)^3\ln\left[1+a\frac{t^2+At^4}{1+At^2+A^2t^4}\right]$$
$$H_1(t,r_S,\zeta) = \left(\frac{16}{\pi}\right)(3\pi^2)^{1/3}[C(\rho)-c]f(\zeta)^3t^2e^{-dx^2/f(\zeta)^2}$$
$$f(\zeta) = \tfrac{1}{2}((1+\zeta)^{2/3}+(1-\zeta)^{2/3})$$
$$t = \left(\frac{192}{\pi^2}\right)^{1/6}\frac{|\nabla\rho|}{2f(\zeta)\rho^{7/6}}$$
$$A = a[e^{-b\varepsilon_c(r_S,\zeta)/f(\zeta)^3}-1]^{-1}$$

where $\varepsilon(r_S, \zeta)$ is the PW92 parameterization of the LSDA correlation energy functional (eq. (6.22)), x and $C(\rho)$ are as defined in eqs. (6.23) and (6.29), and a, ℓ, c and d are suitable constants.

It should be noted that several of the proposed functionals violate fundamental restrictions, such as predicting correlation energies for one-electron systems (for example P86 and PW91) or failing to have the exchange energy cancel the Coulomb self-repulsion (Section 3.3, eq. (3.32)). One of the more recent functionals which does not have these problems is due to Becke (B95),[22] which has the form

$$
\begin{aligned}
\varepsilon_c^{B95} &= \varepsilon_c^{\alpha\beta} + \varepsilon_c^{\alpha\alpha} + \varepsilon_c^{\beta\beta} \\
\varepsilon_c^{\alpha\beta} &= [1 + a(x_\alpha^2 + x_\beta^2)]^{-1}\varepsilon_c^{PW91,\alpha\beta} \\
\varepsilon_c^{\sigma\sigma} &= [1 + \ell x_\sigma^2]^{-2}\frac{D_\sigma}{D_\sigma^{LDA}}\varepsilon_c^{PW91,\sigma\sigma}
\end{aligned}
\tag{6.31}
$$

$$
D_\sigma^{LDA} = 2^{5/3}C_F\rho_\sigma^{5/3}
$$

Here σ runs over α and β spins, x_σ and D_σ have been defined in eqs. (6.23) and (6.25), a and ℓ are fitting parameters, and ε_c^{PW91} is the Perdew–Wang parameterization of the LSDA correlation functional (eq. (6.22)).

6.3 Hybrid Methods

From the Hamiltonian in eq. (6.3) and the definition of the exchange–correlation energy in eq. (6.8), an exact connection can be made between the exchange–correlation energy and the corresponding potential connecting the non-interacting reference and the actual system, see appendix B for details. The resulting equation is called the *Adiabatic Connection Formula* (ACF)[23] and involves an integration over the parameter λ which "turns on" the electron–electron interaction.

$$
E_{xc} = \int_0^1 \langle \Psi_\lambda | \mathbf{V}_{xc}(\lambda) | \Psi_\lambda \rangle d\lambda
\tag{6.32}
$$

In the crudest approximation (taking $\mathbf{V}_{xc}$ to be linear in λ) the integral is given as the average of the values at the two end-points.

$$
E_{xc} \simeq \tfrac{1}{2}\langle \Psi_0 | \mathbf{V}_{xc}(0) | \Psi_0 \rangle + \tfrac{1}{2}\langle \Psi_1 | \mathbf{V}_{xc}(1) | \Psi_1 \rangle
\tag{6.33}
$$

In the $\lambda = 0$ limit, the electrons are non-interacting and there is consequently no correlation energy, only exchange energy. Furthermore, since the <u>exact</u> wave function in this case is a single Slater determinant composed of KS orbitals, the exchange energy is <u>exactly</u> that given by Hartree–Fock theory (eq. (3.33)). If the KS orbitals are <u>identical</u> to the HF orbitals, the "exact" exchange is precisely the exchange energy calculated by HF wave mechanics methods. The last term in eq. (6.33) is still unknown. Approximating it by the LSDA result defines the *Half-and-Half* (H+H) method.[24]

$$
E_{xc}^{H+H} = \tfrac{1}{2}E_x^{exact} + \tfrac{1}{2}(E_x^{LSDA} + E_c^{LSDA})
\tag{6.34}
$$

Since the GGA methods give a substantial improvement over LDA, a generalized version of the Half-and-Half method may be defined by writing the exchange energy as a suitable combination of LSDA, exact exchange and a gradient correction term. The correlation energy may similarly be taken as the LSDA formula plus a gradient correction term.

$$E_{xc}^{B3} = (1 - a)E_x^{LSDA} + aE_x^{exact} + b\Delta E_x^{B88} + E_c^{LSDA} + c\Delta E_c^{GGA} \qquad (6.35)$$

Models which include exact exchange are often called *hybrid* methods, the names *Adiabatic Connection Model* (ACM) and *Becke 3 parameter functional* (B3) are examples of such hybrid models defined by eq. (6.35).[25] The a, b and c parameters are determined by fitting to experimental data and depend on the form chosen for E_c^{GGA}, typical values are $a \sim 0.2$, $b \sim 0.7$ and $c \sim 0.8$. Owing to the substantially better performance of such parameterized functionals the Half-and-Half model is rarely used anymore. The B3 procedure has been generalized to include more fiiting parameters, however, the improvement is rather small.[26]

6.4 Performance

The specification of a DFT method requires selection of a suitable form for the exchange and correlation energies. Although one may in principle select an LSDA form for one of them and a gradient form for the other, this is not really consistent. Within the LSDA approximation, the exchange is given explicitly by the Dirac–Slater expression (6.17), and the only difference is the interpolation function used for reproducing the (very good) Monte Carlo results for the correlation energy. Since the VWN formula (6.20)/(6.21) generally is considered a good interpolating function, the term LSDA has become almost synonymous with the acronym SVWN. Gradient corrected methods have typically used the B88 exchange (6.24), or the B3/ACM hybrid (6.35), and either the LYP, P86 or PW91 correlation (6.27–6.30). Associated acronyms are BLYP, BP86, BPW91, B3LYP, B3P86 and B3PW91.

Gradient corrected methods usually perform much better than LSDA. For the G2-1 data set (see Section 5.5), omitting electron affinities, the mean absolute deviations shown in Table 6.1 are obtained. The improvement achieved by adding gradient terms is impressive, and hybrid methods (like B3PW91) perform almost as well as the elaborate G2 model for these test cases. For a somewhat larger set of reference data, called the G2-2 set, the data shown in Table 6.2 are obtained.[27]

In general it is found that GGA methods often give geometries and vibrational frequencies for stable molecules of the same or better quality than MP2, at a computational cost similar to HF. For systems containing multi-reference character, where MP2 usually fails badly, DFT methods are often found to generate results of a

Table 6.1 Comparison of the performance of DFT methods by mean absolute deviations (kcal/mol)

Method	G2	LSDA	B88	BPW91	B3PW91
Atomization Energies	1.2	35.7	3.9	5.7	2.4
Ionization Potentials	1.4	6.3	11.2	4.1	3.8
Proton Affinities	1.0	5.6	2.4	1.5	1.2

Table 6.2 Comparison of the performance of DFT methods (kcal/mol)

Method	Mean absolute deviation	Maximum absolute deviation
G2	1.6	8.2
G2(MP2)	2.0	10.1
G2(MP2, SVP)	1.9	12.5
SVWN	90.9	228.7
BLYP	7.1	28.4
BPW91	7.9	32.2
B3LYP	3.1	20.1
B3PW91	3.5	21.8

quality comparable to that obtained with coupled cluster methods[28] see also Section 11.7.3.

Another significant advantage is that DFT methods based on unrestricted determinants (analogously to UHF, Section 3.7) for open-shell systems are not very prone to "spin contamination", i.e. $\langle \mathbf{S}^2 \rangle$, is normally close to $S_z(S_z + 1)$ (see also Sections 4.4 and 11.5.3). This is basicly a consequence of electron correlation being included in the single determinantal wave function (by means of E_{xc}). Actually it has been argued that "spin contamination" is not well defined in DFT methods[29] and that $\langle \mathbf{S}^2 \rangle$, should <u>not</u> be equal to $S_z(S_z + 1)$. The argument is basicly that real systems display "spin polarization", e.g. there are points in space where ρ_β is larger than ρ_α (assuming that the number of α-electron is larger than the number of β- electrons). This effect cannot be achieved by a restricted open-shell type determinant (analogous to ROHF), only by an unrestricted treatment which allows the α- and β- orbitals to be different. Another consequence of the presence of E_{xc} is that restricted type determinants are much more stable towards symmetry breaking to an unrestricted determinant (Section 3.8.3) than Hartree–Fock wave functions. For ozone (Section 4.4), for example, it is not possible to find a lower energy solution corresponding to UHF for "pure" DFT methods (LSDA, BLYP, BPW91), although those including exact exchange (B3LYP, B3PW91) display a triplet instability.

Weak interactions due to dispersion (van der Waals type interactions) are poorly described by current functionals.[30] Owing to the general overestimation of bond strengths, LDA does predict an attraction between for example rare gas atoms, although not very accurately, while essentially all gradient corrected methods predict a purely repulsive interaction (at least when corrected for basis set superposition error). Hydrogen bonding, however, is mainly electrostatic, which is reasonably well accounted for by DFT methods. There are indications that relative energies are predicted less accurately by DFT methods, and that transition structures are sometimes poorly described, but as already mentioned, the number of systems for which DFT methods have been calibrated is still fairly small.

Finally, DFT methods are at present not well suited for excited states of the same symmetry as the ground state. The absence of a wave function makes it difficult to ensure orthogonality between the ground and excited states.

6.5 Computational Considerations

The KS orbitals can be determined by a numerical procedure, analogous to numerical HF methods. In practice such procedures are limited to small systems, and essentially all calculations employ an expansion of the KS orbitals in an atomic basis set.

$$\phi_i = \sum_{\alpha}^{M} c_{\alpha i} \chi_{\alpha} \tag{6.36}$$

The basis functions are normally the same as used in wave mechanics for expanding the HF orbitals, see Chapter 5 for details. Although there is no guarantee that the exponents and contraction coefficients determined by the variational procedure for wave functions are also optimum for DFT orbitals, the difference is presumably small since the electron densities derived by both methods are very similar.[31]

The variational procedure again leads to a matrix equation in the atomic orbital basis which can be written in the form (compare eq. (3.50)).

$$\mathbf{h}_{KS}\mathbf{C} = \mathbf{SC}\varepsilon$$
$$h_{\alpha\beta} = \langle \chi_{\alpha}|\mathbf{h}_{KS}|\chi_{\beta}\rangle$$
$$S_{\alpha\beta} = \langle \chi_{\alpha}|\chi_{\beta}\rangle \tag{6.37}$$
$$\mathbf{h}_{KS} = -\tfrac{1}{2}\nabla^2 + \mathbf{V}_{ne} + \int \frac{\rho(\mathbf{r}')}{|\mathbf{r} - \mathbf{r}'|}d\mathbf{r}' + \mathbf{V}_{xc}$$

The $\mathbf{h}_{KS}$ matrix is analogous to the Fock matrix in wave mechanics, and the one-electron and Coulomb parts are identical to the corresponding Fock matrix elements. The exchange–correlation part, however, is given in terms of the electron density, and possibly also involves derivatives of the density (or orbitals, as in the BR functional, eq. (6.25)).

$$\int \chi_{\alpha}(\mathbf{r})\mathbf{V}_{xc}[\rho(\mathbf{r}), \nabla\rho(\mathbf{r})]\chi_{\beta}(\mathbf{r})d\mathbf{r} \tag{6.38}$$

Since the $\mathbf{V}_{xc}$ functional depends on the integration variables implicitly via the electron density, these integrals cannot be evaluated analytically, but must be generated by a numerical integration.

$$\int \chi_{\alpha}(\mathbf{r})\mathbf{V}_{xc}[\rho(\mathbf{r}), \nabla\rho(\mathbf{r})]\chi_{\beta}(\mathbf{r})d\mathbf{r} \simeq \sum_{k=1}^{G} \mathbf{V}_{xc}[\rho(\mathbf{r}_k), \nabla\rho(\mathbf{r}_k)]\chi_{\alpha}(\mathbf{r}_k)\chi_{\beta}(\mathbf{r}_k)\nabla\mathbf{v}_k \tag{6.39}$$

As the number of grid points G goes to infinity, the approximation becomes exact. In practice the number of points is selected based on the desired accuracy in the final results, i.e. if the energy is only required with an accuracy of 10^{-3}, the number of integration points can be smaller than if the energy is required with an accuracy of 10^{-5}.[32] There are also some technical skills involved in selecting the optimum distribution of a given number of points to yield the best accuracy, i.e. the points should be dense where the function $\mathbf{V}_{xc}$ varies most. The grid is usually selected as being spherical around each nucleus, making it dense in the radial direction near the nucleus, and dense in the angular part in the valence space. For typical applications 1 000–10 000 points are used for each atom.[33] It should be noted that only the larger of such grids approach saturation, i.e. in general the energy will depend on the number (and location) of grid points. In order to compare energies for different systems, the same grid must therefore be used. The grid plays the same role for E_{xc} as the basis set for the other terms. Just as it is improper to compare energies calculated with different basis sets, it is not justified to compare DFT energies calculated with different grid sizes. Furthermore, an incomplete grid will lead to "grid superposition errors" analogous to basis set superposition errors (Section 5.8).

With an expansion of the orbitals in basis functions, the number of integrals necessary for solving the KS equations increases as M^4, owing to the Coulomb integrals in the J functional (and possibly also "exact" exchange in the hybrid methods). The number of grid points for the numerical E_{xc} integration (eq. (6.39)) increases linearly with system size, and the computational effort for the exchange–correlation term increases as GM^2, i.e. a cubic dependence on system size. When the Coulomb (and possibly "exact" exchange) term is evaluated directly from integrals over basis functions, DFT methods scale formally as M^4. However, as discussed in Section 3.8.6, the Coulomb (and exchange) part can be calculated with an effort which scales only as M^1 for large systems with for example fast multipole methods. The numerical integration required for the exchange and correlation parts may also be reduced to a computation cost which scales linearly with system size,[34] i.e. with modern techniques DFT is a true linear scaling method. This opens up the possibilities of performing accurate calculations on systems containing thousands of atoms, which is likely to have impacts on many areas outside traditional computational chemistry.

Nevertheless, the formal M^4 scaling has spawned approaches which reduce the dependence to M^3. This may be achieved by <u>fitting</u> the electron density to a linear combination of functions, and using the fitted density in evaluating the J integrals in the Coulomb term.

$$\rho \simeq \sum_{\alpha}^{M'} a'_{\alpha}\chi'_{\alpha} \tag{6.40}$$

The density fitting functions may or may not be the same as those used in expanding the orbitals. The fitting constants a'_{α} are chosen so that the Coulomb energy arising from the <u>difference</u> between the exact and fitted densities is minimized, subject to the constraint of charge conservation.[35] The J integrals then become

$$\int \chi_{\alpha}(1)\chi_{\beta}(1)\frac{1}{|\mathbf{r}_1 - \mathbf{r}_2|}\chi'_{\gamma}(2)d\mathbf{r}_1 d\mathbf{r}_2 \tag{6.41}$$

which only involves three basis functions, thereby reducing the computational effort to M^3.

In some cases the exchange–correlation potential $\mathbf{V}_{xc}$ is also fitted to a set of functions, similarly to the fitting of the density.

$$\mathbf{V}_{xc} \simeq \sum_{\alpha}^{M''} b''_{\alpha}\chi''_{\alpha} \tag{6.42}$$

Again the set of fitting functions may or may not be the same as the orbital and/or the density basis functions. Once the potential has been fitted, the exchange–correlation energy may be evaluated from integrals involving three functions, analogously to eq.

(6.41). This fitting does not reduce the formal scaling, since the exchange–correlation term already is of order M^3.

Many of these fitting schemes were derived before linear scaling techniques (Section 3.8.6) were fully developed, and it is not clear whether they have any advantages. For calculation of energy derivatives, they actually seem counterproductive, since the fitting procedures seriously complicate the computational expressions.[36]

The use of grid based techniques for the numerical integration of the exchange–correlation contribution has some disadvantages when derivatives of the energy are desired. For this reason there is interest in developing grid-free DFT methods where the exchange–correlation potential is expressed completely in terms of analytical integrals.[37]

In practice a DFT calculation involves an effort similar to that required for an HF calculation. Furthermore, DFT methods are one-dimensional just as HF methods are: increasing the size of the basis set allows a better and better description of the KS orbitals. Since the DFT energy depends directly on the electron density, it is expected that it has basis set requirements similar to those for HF methods, i.e. close to converged with a TZ(2df) type basis.

Should DFT methods be considered *ab initio* or semi-empirical? If *ab initio* is taken to mean the absence of fitting parameters, LSDA methods are *ab initio*, but gradient corrected methods may or may not be. The LSDA exchange energy contains no parameters, and the correlation functional is known accurately as a tabulated function of the density. The VWN correlation functional (eqs. (6.20) and (6.21)) is merely a suitable interpolation formula necessarily for practical calculations; the constants do not represent fitting parameters chosen to improve the performance for atomic and molecular systems. Some gradient corrected methods (e.g. the B88 exchange and the LYP correlation), however, contain parameters which are fitted to give the best agreement with experimental atomic data, but the number of parameters is significantly smaller than for semi-empirical methods. The semi-empirical PM3 method (Section 3.10.5), for example, has 18 parameters for each atom, while the B88 exchange functional only has one fitting constant, valid for the whole periodic table. Other functionals are derived entirely from theory, and can consequently be considered "pure" *ab initio*.

If *ab initio* is taken to mean that the method is based on theory which in principle is able to produce the exact results, DFT methods are *ab initio*. The only caveat is that current methods cannot yield the exact results, even in the limit of a complete basis set, since the functional form of the exact exchange–correlation energy is not known. Wave mechanics employ the exact Hamilton operator and make approximations for the wave function, while DFT makes approximations in the Hamilton operator, and it is easier to improve on the wave function description than to add correction to the operator. It is perhaps a little disturbing that seemingly very different functionals give results of similar quality.[38] Although gradient corrected DFT methods have been shown to give impressive results, even for theoretically difficult problems, the lack of a systematic way of extending a series of calculations to approach the exact result is a major drawback of DFT. The results converge towards a certain value as the basis set is increased, but theory does not allow an evaluation of the errors inherent in this limit (like the systematic overestimation of vibrational frequencies with wave mechanics HF methods). Furthermore, although a progression of methods such as LSDA, BPW91 and B3PW91 provides successively lower errors for a suitable set of reference data (such as used for calibrating the Gaussian-2 model), there is no guarantee that the same progression will provide better and better results for a specific property of a given system. Indeed, LSDA methods may in some cases provide better results, even in the limit of a large basis set, than either of the more "complete" gradient corrected models. The quality of a given result can therefore only be determined by comparing the performance for similar systems where experimental or high quality wave mechanics results are available. In this respect DFT resembles semi-empirical methods. Nevertheless, DFT methods, especially those involving gradient corrections and hybrid methods, are significantly more accurate (and the errors are much more uniform) than those of for example the MNDO family, and DFT is consequently a valuable tool for systems where a (very) high accuracy is not needed.

References

1. P. Hohenberg and W. Kohn, *Phys. Rev.,* **136** (1964), B864.
2. R. G. Parr and W. Yang *Density Functional Theory*, Oxford University Press, 1989; L. J. Bartolotti and K. Flurchick, *Rev. Comput. Chem.*, **7** (1996), 187; A. St-Amant, *Rev. Comput. Chem.*, **7** (1996), 217; T. Ziegler, *Chem. Rev.*, **91** (1991), 651; E. J. Baerends and O. V. Gritsenko, *J. Phys. Chem.*, **101** (1997), 5383.
3. F. Block, *Z. Physik*, **57** (1929), 545.
4. P. A. M. Dirac, *Proc. Cambridge Phil. Soc.*, **26** (1930), 376.
5. W. Kohn, L. J. Sham, *Phys. Rev.*, **140** (1965), A1133.
6. O. V. Gritsenko, P. R. T. Schippen and E. J. Baerends, *J. Chem. Phys.*, **107** (1997), 5007.
7. P. Politzer and F. Abu-Awwad, *Theor. Chem. Acta,* **99** (1998), 83.
8. M. Levy and J. P. Perdew, in *Density Functional Methods in Physics*, ed. R. M. Dreizler and J. da Providencia, Plenum, 1985.
9. Q. Zhao, R. C. Morrison and R. G. Parr, *Phys. Rev. A.*, **50** (1994), 2138.
10. O. V. Gritsenko, R. van Leeuwen and E. J. Baerends, *J. Chem. Phys.*, **104** (1996), 8535; D. J. Tozer, V. E. Ingamells and N. C. Handy, *J. Chem. Phys.*, **105** (1996), 9200.
11. J. C. Slater, *Phys. Rev.*, **81** (1951), 385.
12. S. J. Vosko, L. Wilk and M. Nusair, *Can. J. Phys.*, **58** (1980), 1200.
13. J. P. Perdew and Y. Wang, *Phys. Rev. B*, **45** (1992), 13244.
14. J. D. Perdew and Y. Wang, *Phys. Rev.*, B, **33** (1986), 8800.
15. A. D. Becke, *Phys. Rev.*, B, **38** (1988), 3098.
16. G. Ortiz and P. Ballone, *Phys. Rev.*, B, **43** (1991), 6376.
17. A. D. Becke and M. R. Roussel, *Phys. Rev. A*, **39** (1989), 3761.
18. C. Lee, W. Yang and R. G. Parr, *Phys. Rev. B*, **37** (1988), 785.
19. B. Miehlich, A. Savin, H. Stoll and H. Preuss, *Chem. Phys. Lett.*, **157** (1989), 200.
20. J. P. Perdew, *Phys. Rev. B.*, **33** (1986), 8822; **34** (1986), 7406.
21. J. P. Perdew, J. A. Chevary, S. H. Vosko, K. A. Jackson, M. R. Pederson, D. J. Singh and C. Fiolhais, *Phys. Rev. B.*, **46** (1992), 6671.
22. A. D. Becke, *J. Chem. Phys.*, **104** (1996), 1040.
23. J. Harris, *Phys. Rev. A*, **29** (1984), 1648.
24. A. D. Becke, *J. Chem. Phys.*, **98** (1992), 1372.
25. A. D. Becke, *J. Chem. Phys.*, **98** (1993), 5648.
26. A. D. Becke, *J. Chem. Phys.*, **107** (1997), 8554.
27. L. A. Curtiss, K. Raghavachari, P. C. Redfern and J. A. Pople, *J. Chem. Phys.*, **106** (1997), 1063.

28. N. Oliphant and R. J. Bartlett, *J. Chem. Phys.*, **100** (1994), 6550.
29. J. A. Pople, P. M. W. Gill and N. C. Handy, *Int. J. Quantum. Chem.*, **56** (1995), 303.
30. E. J. Meijer and M. Sprik, *J. Chem. Phys.*, **105** (1996), 8684.
31. N. Godbout, D. R. Salahub, J. Andzelm and E. Wimmer, *Can. J. Chem.*, **70** (1992), 560.
32. J. M. Perez-Jorda, A. D. Becke and E. San-Fabian, *J. Chem. Phys.*, **100** (1994), 6520.
33. P. M. W. Gill and B. G. Johnson, *Chem. Phys. Lett.*, **209** (1993), 506.
34. R. E. Stratmann, G. E. Scuseria and M. J. Frisch, *Chem. Phys. Lett.*, **257** (1996), 213.
35. B. I. Dunlap, J. W. D. Connolly and J. R. Sabin, *J. Chem. Phys.*, **71** (1979), 3396.
36. A. Kormornicki and G. Fitzgerald, *J. Chem. Phys.*, **98** (1993), 1398.
37. K. S. Werpetinski and M. Cook, *J. Chem. Phys.*, **106** (1997), 7124; Y. C. Zheng and J. E. Almlöf, *J. Mol. Struct.*, **388** (1996), 277.
38. R. Neumann, R. H. Nobes and N. C. Handy, *Mol. Phys.*, **87** (1996), 1.

7 Valence Bond Methods

Essentially all practical calculations for generating solutions to the electronic Schrödinger equation have been performed using molecular orbital methods. The zero-order wave function is constructed as a single Slater determinant and the MOs are expanded in a set of atomic orbitals, the basis set. In a subsequent step the wave function may be improved by adding electron correlation with either CI, MP or CC methods. There are two characteristics of such approaches: (1) the one-electron functions, the MOs, are delocalized over the whole molecule, and (2) an accurate treatment of the electron correlation requires many (millions or billions) "excited" Slater determinants. The delocalized nature of the MOs is partly a consequence of choosing the Lagrange multiplier matrix to be diagonal (canonical orbitals, eq. (3.41)); they may in a subsequent step be mixed to form localized orbitals (see Section 9.4) without affecting the total wave function. Such a localization, however, is not unique. Furthermore, delocalized MOs are at variance with the basic concept in chemistry, that molecules are composed of structural units (functional groups) which to a very good approximation are constant from molecule to molecule. The MOs for propane and butane, for example, are quite different, although "common" knowledge is that they contain CH_3 and CH_2 units which in terms of structure and reactivity are very similar for the two molecules. A description of the electronic wave function as having electrons in orbitals formed as linear combinations of all (in principle) atomic orbitals is also at variance with the chemical language of molecules being composed of atoms held together by bonds, where the bonds are formed by pairing unpaired electrons contained in atomic orbitals. Finally, when electron correlation is important (as is usually the case), the need to include many Slater determinants obscures the picture of electrons residing in orbitals.

There is an equivalent way of generating solutions to the electronic Schrödinger equation which conceptually is much closer to the experimentalists language, known as *Valence Bond* (VB) theory. We will start by illustrating the concepts for the H_2 molecule, and note how it differ from MO methods.

7.1 Classical Valence Bond

A single determinant MO wave function for the H_2 molecule within a minimum basis consisting of a single s-function on each nucleus is given as (see also

Section 4.3)

$$\Phi_0 = \begin{vmatrix} \phi_1(1) \, \bar{\phi}_1(1) \\ \phi_1(2) \, \bar{\phi}_1(2) \end{vmatrix}$$

$$\phi_1 = (\chi_A + \chi_B)\alpha$$

$$\bar{\phi}_1 = (\chi_A + \chi_B)\beta \tag{7.1}$$

where we have ignored the normalization constants. The Slater determinant may be expanded in AOs.

$$\Phi_0 = \phi_1 \bar{\phi}_1 - \bar{\phi}_1 \phi_1 = (\phi_1 \phi_1)[\alpha\beta - \beta\alpha]$$

$$\Phi_0 = (\chi_A + \chi_B)(\chi_A + \chi_B)[\alpha\beta - \beta\alpha] \tag{7.2}$$

$$\Phi_0 = (\chi_A \chi_A + \chi_B \chi_B + \chi_A \chi_B + \chi_B \chi_A)[\alpha\beta - \beta\alpha]$$

which shows that the HF wave function consists of equal amounts of ionic and covalent terms. In the dissociation limit only the covalent terms are correct, but the single determinant description does not allow the ratio of covalent to ionic terms to vary. In order to provide a correct description, a second determinant is necessary.

$$\Phi_1 = \begin{vmatrix} \phi_2(1) \, \bar{\phi}_2(1) \\ \phi_2(2) \, \bar{\phi}_2(2) \end{vmatrix}$$

$$\phi_2 = (\chi_A - \chi_B)\alpha$$

$$\bar{\phi}_2 = (\chi_A - \chi_B)\beta \tag{7.3}$$

$$\Phi_1 = (\chi_A \chi_A + \chi_B \chi_B - \chi_A \chi_B - \chi_B \chi_A)[\alpha\beta - \beta\alpha]$$

By including the doubly excited determinant Φ_1, built from the antibonding MO, the amount of covalent and ionic terms may be varied, and be determined completely by the variational principle (eq. (4.19)).

$$\Psi_{CI} = a_0 \Phi_0 + a_1 \Phi_1 = ((a_0 - a_1)(\chi_A \chi_B + \chi_B \chi_A)$$

$$+ (a_0 + a_1)(\chi_A \chi_A + \chi_B \chi_B))[\alpha\beta - \beta\alpha] \tag{7.4}$$

The 2-configurational CI wave function (7.4) allows a qualitatively correct description of the H_2 molecule at all distances, in the dissociation limit the weights of the two configurations become equal.

The classical VB wave function, on the other hand, is build from the atomic fragments by coupling the unpaired electrons to form a bond. In the H_2 case, the two electrons are coupled into a singlet pair, properly antisymmetrized. The simplest VB description, known as a *Heitler–London* (HL) function, includes only the two covalent terms in the HF wave function.

$$\Phi_{HL}(\text{cov}) = (\chi_A \chi_B + \chi_B \chi_A)[\alpha\beta - \beta\alpha] \tag{7.5}$$

Just as the single determinant MO wave function may be improved by including excited determinants, the simple VB–HL function may also be improved by adding terms which correspond to higher energy configurations for the fragments, in this case ionic structures.

$$\Phi_{HL}(\text{ion}) = (\chi_A \chi_A + \chi_B \chi_B)[\alpha\beta - \beta\alpha] \tag{7.6}$$

$$\Phi_{HL} = a_0 \Phi_{HL}(\text{cov}) + a_1 \Phi_{HL}(\text{ion}) \tag{7.7}$$

The final description, either in terms of a CI wave function written as a linear combination of two determinants built from delocalized MOs (eq. (7.4)), or as a VB wave function written in terms of two VB-HL structures composed of AOs (eq. (7.7)), is identical.

For the H_2 system, the amount of ionic HL structures determined by the variational principle is 44%, close to the MO-HF value of 50%. The need for including large amounts of ionic structures in the VB formalism is due to the fact that pure atomic orbitals are used.

Consider now a covalent VB function build from "atomic" orbitals which are allowed to distort from the pure atomic shape.

$$\Phi_{CF} = (\phi_A \phi_B + \phi_B \phi_A)[\alpha\beta - \beta\alpha]$$
$$\phi_A = \chi_A + c\chi_B \qquad\qquad (7.8)$$
$$\phi_B = \chi_B + c\chi_A$$

Such a VB function is known as a *Coulson–Fischer* (CF) type. The c constant is fairly small (for H_2 $c \sim 0.04$), but by allowing the VB orbitals to adopt the optimum shape, the need for ionic VB structures is strongly reduced. Note that the two VB orbitals in eq. (7.8) are not orthogonal, the overlap is given by

$$\langle\phi_A|\phi_B\rangle = (1 + c^2)\langle\chi_A|\chi_B\rangle + 2c(\langle\chi_A|\chi_A\rangle + \langle\chi_B|\chi_B\rangle)$$
$$\langle\phi_A|\phi_B\rangle = (1 + c^2)S_{AB} + 4c \qquad\qquad (7.9)$$

Compared to the overlap of the undistorted atomic orbitals used in the HL wave function, which is just S_{AB}, it is seen that the overlap is increased (c is positive), i.e. the orbitals distort so that they overlap better in order to make a bond. Although the distortion is fairly small (a few %) this effectively eliminates the need for including ionic VB terms. When c is variationally optimized, the MO-CI, VB-HL and VB-CF wave functions (eqs. (7.4), (7.7) and (7.8)) are all completely equivalent. The MO approach incorporates the flexibility in terms of an "excited" determinant, the VB-HL in terms of "ionic" structures, and the VB-CF in terms of "distorted" atomic orbitals.

In the MO-CI language, the correct dissociation of a single bond requires addition of a second doubly excited determinant to the wave function. The VB-CF wave function, on the other hand, dissociates smoothly to the correct limit, the VB orbitals simply reverting to their pure atomic shapes, and the overlap disappearing.

7.2   Spin Coupled Valence Bond

The generalization of a Coulson–Fischer type wave function to the molecular case with an arbitrary size basis set is known as *Spin Coupled Valence Bond* (SCVB) theory.[1]

Again it is instructive to compare this with the traditional MO approach, taking the CH_4 molecule as an example. The MO single determinant description (RHF, which is identical to UHF near the equilibrium geometry) of the valence orbitals is in terms of four delocalized orbitals, each occupied by two electrons with opposite spins. The C–H bonding is described by four different, orthogonal molecular orbitals, each expanded in a set of AOs.

$$\Phi_{\text{valence-MO}}^{\text{CH}_4} = \mathbf{A}[\phi_1\bar{\phi}_1\phi_2\bar{\phi}_2\phi_3\bar{\phi}_3\phi_4\bar{\phi}_4]$$

$$\phi_i = \sum_{\alpha=1}^{M} c_{\alpha i}\chi_\alpha \qquad (7.10)$$

Here $\mathbf{A}$ is the usual antisymmetrizer (eq. (3.21)) and an bar above an MO indicates that the electron has a β spin function, no bar indicates an α spin function.

The SCVB description, on the other hand, considers the four bonds in CH_4 as arising from coupling of a single electron at each of the four hydrogen atoms with a single unpaired electron at the carbon atom. Since the ground state of the carbon atom is a triplet, corresponding to the electron configuration $1s^2 2s^2 2p^2$, the first step is formation of four equivalent "hybrid" orbitals by mixing three parts p-function with one part s-function, generating four equivalent "sp^3-hybrid" orbitals. Each of these singly occupied hybrid orbitals can then couple with a hydrogen atom to form four equivalent C–H bonds. The electron spins are coupled so that the total spin is a singlet; this can be done in several different fashions. The coupling of four electrons to a total singlet state, for example, can be done either by coupling two electrons in a pair to a singlet, and then couple two singlet pairs, or by first coupling two electrons in a pair to a triplet, and subsequently couple two triplet pairs to an overall singlet, Figure 7.1.

The $\Theta_{S,i}^N$ symbol is used to designate the ith combination of spin functions coupling N electrons to give an overall spin of S, and there are f_S^N number of ways of doing this. The

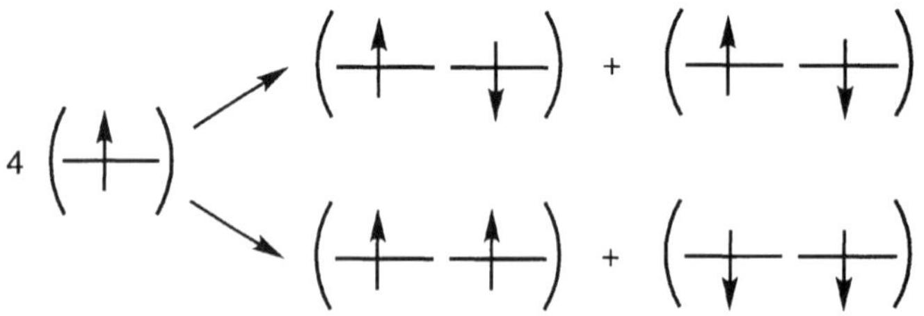

Figure 7.1 Two possible schemes for coupling four electrons to an overall singlet

Table 7.1 Number of possible spin coupling schemes for achieving an overall singlet state

N	f_0^N
2	1
4	2
6	5
8	14
10	42
12	132
14	429

value of f_S^N is given by

$$f_S^N = \frac{(2S+1)N!}{(\frac{1}{2}N+S+1)!(\frac{1}{2}N-S)!}$$

(7.11)

For a singlet wave function ($S = 0$), the number of coupling schemes for N electrons is given in Table 7.1.

For the eight valence electrons in CH_4 there are 14 possible spin couplings resulting in an overall singlet state. The full SCVB function may be written (again neglecting normalization) as

$$\Phi_{valence-SCVB}^{CH_4} = \sum_{i=1}^{14} a_i \mathbf{A}\{[\phi_1\phi_2\phi_3\phi_4\phi_5\phi_6\phi_7\phi_8]\Theta_{0,i}^N\}$$

$$\phi_i = \sum_{\alpha=1}^{M} c_{\alpha i} \chi_\alpha$$

(7.12)

There are now eight different spatial orbitals, ϕ_i, four of which are essentially carbon sp^3-hybrid orbitals, the other four being close to atomic hydrogen s-orbitals. The expansion of each of the VB orbitals in terms of <u>all</u> the basis functions located on <u>all</u> the nuclei allows the orbitals to distort from the pure atomic shape. The SCVB wave function is variationally optimized, both with respect to the VB orbital coefficient, $c_{\alpha i}$, and the spin coupling coefficients, a_i. The result is that a complete set of optimum "distorted" atomic orbitals is determined together with the weight of the different spin couplings. Each spin coupling term (in the so-called *Rumer* basis) is closely related to the concept of a resonance structure used in organic chemistry textbooks. An SCVB calculation of CH_4 gives as a result that one of the spin coupling schemes completely dominates the wave function, namely that corresponding to the electron pair in each of the C–H bonds being singlet coupled. This is the quantum mechanical analogue of the graphical representation of CH_4 shown in Figure 7.2. Each of the lines represents a singlet coupled electron pair between two orbitals which strongly overlap to form a bond, and the drawing in Figure 7.2 is the only important "resonance" form.

Consider now the π-system in benzene. The MO approach will generate linear combinations of the atomic p-orbitals, producing six π-orbitals delocalized over the whole molecule with four different orbital energies (two sets of degenerate orbitals), Figure 7.3. The stability of benzene can be attributed to the large gap between the HOMO and LUMO orbitals.

An SCVB calculation considering only the coupling of the six π-electrons gives a somewhat different picture. The VB π-orbitals are strongly localized on each carbon,

Figure 7.2   A representation of the dominating spin coupling in CH_4

Figure 7.3   Molecular orbital energies in benzene

Figure 7.4   Representations of important spin coupling schemes in benzene

resembling p-orbitals which are slightly distorted in the direction of the nearest neighbour atoms. It is now found that five spin coupling combinations are important, these are shown in Figure 7.4, where a bold line indicates two electrons coupled into a singlet pair.

Each of the two first VB structures contributes $\sim 40\%$ to the wave function, and each of the remaining three contributes $\sim 6\%$.[2] The stability of benzene in the SCVB picture is due to <u>resonance</u> between these VB structures. It is furthermore straightforward to calculate the resonance energy by comparing the full SCVB energy with that calculated from a VB wave function omitting certain spin coupling functions.

The MO wave function for CH_4 may be improved by adding configurations corresponding to excited determinants, i.e. replacing occupied MOs with virtual MOs. Allowing all excitations in the minimal basis valence space and performing the full optimization corresponds to a [8,8]-CASSCF wave function (Section 4.6). Similarly, the SCVB wave function in eq. (7.10) may be improved by adding ionic VB structures like CH_3^-/H^+ and CH_3^+/H^-; this corresponds to exciting an electron from one of the singly occupied VB orbitals into another VB orbital, thereby making it doubly occupied. The importance of these excited/ionic terms can again be determined by the variational principle. If all such ionic terms are included, the fully optimized SCVB+CI wave function is for all practical purposes identical to that obtained by the MO-CASSCF approach (the only difference is a possible slight difference in the description of the carbon 1s core orbital). Both types of wave function provide essentially the same total energy, and thus include the same amount of electron correlation. The MO-CASSCF wave function attributes the electron correlation to interaction of 1764 configurations, the HF reference and 1763 excited configurations, with each of the 1763 configurations providing only a small amount of the correlation energy. The SCVB wave function (which includes only one resonance structure), however, contains 90+% of the correlation energy, and only a few % is attributed to "excited" structures. The ability of

SCVB wave functions to include electron correlation is due to the fact that the VB orbitals are strongly localized, and since they are occupied by only one electron, they have the built-in feature of electrons avoiding each other. In a sense, an SCVB wave function is the best wave function that can be constructed in terms of products of spatial orbitals. By allowing the orbitals to become non-orthogonal, the large majority (80–90%) of what is called electron correlation in an MO approach can be included in a single determinant wave function composed of spatial orbitals, multiplied by proper spin coupling functions.

There are a number of technical complications associated with optimizing the SCVB wave function due to the non-orthogonal orbitals. The MO-CI or MO-CASSCF approaches simplify considerably owing to the orthogonality of the MOs, and thereby also of the Slater determinants. Computationally the optimization of an SCVB wave function, where N electrons are coupled in all possible ways, is similar to that required for constructing an $[N,N]$-CASSCF wave function. This effectively limits the size of SCVB wave functions to the coupling of 12–16 electrons. The actual optimization of the wave function is usually done by a second-order expansion of the energy in terms of orbital and spin coupling coefficients, and by employing a Newton–Raphson type scheme, analogously to MCSCF methods (Section 4.6). The non-orthogonal orbitals have the disadvantage that it is difficult to add dynamical correlation on top of a SCVB wave function by perturbation or coupled cluster theory, although (non-orthogonal) CI methods are straightforward. SCVB+CI approaches may also be used to describe excited states, analogously to MO-CI methods.

It should be emphasized again that the results obtained from an $[N, N]$-CASSCF and a corresponding N-electron SCVB wave function (or SCVB+CI and MRCI) are virtually identical. The difference is the way the results can be analysed. Molecules in the SCVB picture are composed of atoms held together by bonds, where bonds are formed by (singlet) coupling of the electron spins between (two) overlapping orbitals. These orbitals are strongly localized, usually on a single atom, and are basically atomic orbitals slightly distorted by the presence of the other atoms in the molecule. The VB description of a bond as the result of two overlapping orbitals is in contrast with the MO approach where a bond between two atoms arises as a sum of (small) contributions from many delocalized molecular orbitals. Furthermore, the weight of the different ways spin coupling can take place in a SCVB wave function has a direct analogy with chemical concepts such as "resonance" structures.

The SCVB method is a valuable tool for providing insight into the problem. This is to a certain extent also possible from an MO type wave function by localizing the orbitals or from an analysis of natural orbitals, see Sections 9.4 and 9.5 for details. However, there is no unique method for producing localized orbitals, and different methods may give different orbitals. Natural orbitals are analogously to canonical orbitals usually delocalized over the whole molecule. The SCVB orbitals, in contrast, are uniquely determined by the variational procedure, and there is no freedom to transform them further by making linear combinations without destroying the variational property.

The primary feature of SCVB is the use of non-orthogonal orbitals, which allows a much more compact representation of the wave function. An MO-CI wave function of a certain quality may involve many thousand Slater determinants, while a similar quality VB wave function may be written as only a handful of "resonating" VB structures.

Furthermore, the VB orbitals, and spin couplings, of a C–H bond in say propane and butane are very similar, in contrast to the vastly different MO descriptions of the two systems. The VB picture is thus much closer to the traditional descriptive language, with molecules composed of functional groups. The widespread availability of programs for performing CASSCF calculations, and the fact that CASSCF calculations are computationally more efficient owing to the orthogonality of the MOs, have prompted developments of schemes for transforming CASSCF wave functions to VB structures, denoted CASVB.[3] A corresponding procedure using orthogonal orbitals (which introduce large weights of ionic structures) has also been reported.[4]

7.3   Generalized Valence Bond

The SCVB wave function allows all possible spin couplings to take place, and has no restrictions on the form of the orbitals. The *Generalized Valence Bond* (GVB) method can be considered as a reduced version of the full problem where only certain subsets of spin couplings are allowed.[5] For a typical case of a singlet system, the GVB method has two (non-orthogonal) orbitals assigned to each bond, and each pair of electrons in a bond is required to couple to a singlet pair. The coupling of such singlet pairs will then give an overall singlet spin state. This is known as *Perfect Pairing* (PP), and is one of all the possible spin coupling schemes. Such two-electron two-orbital pairs are also known under the name of *geminal pairs*. Just as an orbital is a wave function for one electron, a geminal is a wave function for two electrons. In order to reduce the computational problem, *Strong Orthogonality* (SO) is normally imposed on the GVB wave function. This means that orbitals belonging to different pairs are required to be orthogonal. While the PP coupling typically is the largest contribution to the full SCVB wave function, the SO constraint is often a quite poor approximation, and may lead to artefacts. For diazomethane, for example, the SCVB wave function is dominated (91%) by the PP coupling, leading to the conclusion that the molecule has essentially normal C=N and N=N π-bonds, perpendicular to the plane defined by the CH_2 moiety.[6] Taking into account also the in-plane bonding, this suggest that diazomethane is best described with a triple bond between the two nitrogens, thereby making the central nitrogen "hypervalent", as illustrated in Figure 7.5.

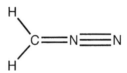

Figure 7.5   A representation of the SCVB wave function for diazomethane

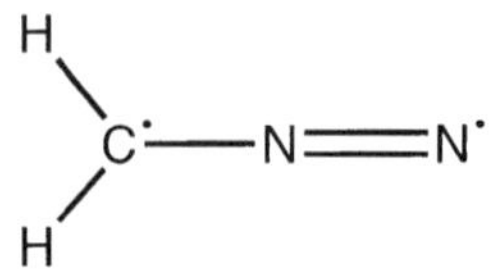

Figure 7.6   A representation of the GVB wave function for diazomethane

There are strong overlaps between the VB orbitals, the <u>smallest</u> overlap (between the carbon and terminal nitrogen) is ~ 0.4, and that between the two orbitals on the central nitrogen is ~ 0.9. The GVB-SOPP approach, however, forces these geminal pairs to be orthogonal, leading to the conclusion that the electronic structure of diazomethane has a very strong diradical nature, as illustrated in Figure 7.6.

References

1 D. L. Cooper, J. Gerratt and M. Raimondi, *Chem. Rev.*, **91** (1991), 929; J. Gerratt, D. L. Cooper, P. B. Karadakov and M. Raimondi, *Chem. Soc. Rev.*, **26** (1997), 87.
2 D. L. Cooper, T. Thorsteinsson and J. Gerratt, *Int. J. Quant. Chem.*, **65** (1997), 439.
3 D. L. Cooper, T. Thorsteinsson and J. Gerratt, *Int. J. Quantum. Chem.*, **65** (1997), 439.
4 K. Hirao, H. Nakano, K. Nakayama and M. Dupuis, *J. Chem. Phys.*, **105** (1996), 9227.
5 W. A. Goddard III and L. B. Harding, *Ann. Rev. Phys. Chem.*, **29** (1978), 363.
6 D. L. Cooper, J. Gerratt, M. Raimondi and S. C. Wright, *Chem. Phys. Lett.*, **138** (1987), 296.

8 Relativistic Methods

The central theme in relativity is that the speed of light, c, is constant in all inertia frames (coordinate systems which move with respect to each other). Augmented with the requirement that physical laws should be identical in such frames, this has as a consequence that time and space coordinates become "equivalent". A relativistic description of a particle thus require four coordinates, three space and one time coordinate.[1] The latter is usually multiplied by c to have units identical to the space variables. A change between different coordinate systems is described by a *Lorentz* transformation, which may mix space and time coordinates. The postulate that physical laws should be identical in all coordinate systems is equivalent to the requirement that equations describing the physics must be invariant (unchanged) to a Lorentz transformation. Considering the time-dependent Schrödinger equation (8.1), it is clear that it is not Lorentz invariant since the derivative with respect to space coordinates is of second order, but the time derivative is only of first order. The fundamental structure of the Schrödinger equation is therefore not relativistically correct.

$$\left[-\frac{1}{2m}\left(\frac{\partial^2}{\partial x^2} + \frac{\partial^2}{\partial y^2} + \frac{\partial^2}{\partial z^2} \right) + \mathbf{V} \right] \Psi = i \frac{\partial \Psi}{\partial t} \tag{8.1}$$

(For use below, we have elected here to explicitly write the electron mass as m, although it is equal to one in atomic units).

One of the consequences of the constant speed of light is that the mass of a particle which moves at a substantial fraction of c increases over the rest mass.

$$m = m_0 \left[\sqrt{1 - \frac{v^2}{c^2}} \right]^{-1} \tag{8.2}$$

The energy of a 1s-electron in a hydrogen-like system (one nucleus and one electron) is $-Z^2/2$, and classically this is equal to minus the kinetic energy, $1/2\,mv^2$, due to the virial theorem ($E = -T = 1/2\,V$). In atomic units the classical velocity of a 1s-electron is thus Z ($m = 1$). The speed of light in these units is 137.036, and it is clear that relativistic effects cannot be neglected for the core electrons in heavy nuclei. For nuclei with large Z, the 1s-electrons are relativistic and thus heavier, which has the effect that the 1s-orbital shrinks in size, by the same factor by which the mass increases (eq. (8.2)).

In order to maintain orthogonality, the higher s-orbitals also contract. This provides a more effective screening of the nuclear change for the higher angular momentum orbitals, which consequently increase in size. For p-orbitals the spin–orbit interaction, which mixes s- and p-orbitals, counteracts the inflation. The net effect is that p-orbitals are relatively unaffected, while d- and f-orbitals become larger. Relativistic effects for geometries and energetics are normally negligible for the first three rows in the periodic table (up to Kr, $Z = 36$, corresponding to a "mass correction" of 1.04), the fourth row represents an intermediate case, while relativistic corrections are necessary for the fifth and sixth rows, and lanthanide/actenide metals. For effects involving electron spin (e.g. spin–orbit coupling), which are purely relativistic in origin, there is no non-relativistic counterpart, and the "relativistic correction" is of course everything.

Although an in-depth treatment is outside the scope of this book, it may be instructive to point out some of the features and problems in a relativistic quantum description of atoms and molecules.

For a free electron Dirac proposed that the (time-dependent) Schrödinger equation should be replaced by

$$[c\boldsymbol{\alpha} \cdot \mathbf{p} + \boldsymbol{\beta}mc^2]\Psi = i\frac{\partial \Psi}{\partial t} \tag{8.3}$$

where $\boldsymbol{\alpha}$ and $\boldsymbol{\beta}$ are 4×4 matrices, $\boldsymbol{\alpha}$ being written in terms of the three Pauli 2×2 spin matrices $\boldsymbol{\sigma}$, and $\boldsymbol{\beta}$ in term of a 2×2 unit matrix $\mathbf{I}$.

$$\boldsymbol{\alpha}_{x,y,z} = \begin{pmatrix} 0 & \boldsymbol{\sigma}_{x,y,z} \\ \boldsymbol{\sigma}_{x,y,z} & 0 \end{pmatrix}, \quad \boldsymbol{\beta} = \begin{pmatrix} \mathbf{I} & 0 \\ 0 & -\mathbf{I} \end{pmatrix}$$

$$\boldsymbol{\sigma}_x = \begin{pmatrix} 0 & 1 \\ 1 & 0 \end{pmatrix}, \quad \boldsymbol{\sigma}_y = \begin{pmatrix} 0 & -i \\ i & 0 \end{pmatrix}, \quad \boldsymbol{\sigma}_z = \begin{pmatrix} 1 & 0 \\ 0 & -1 \end{pmatrix}, \quad \mathbf{I} = \begin{pmatrix} 1 & 0 \\ 0 & 1 \end{pmatrix} \tag{8.4}$$

Except for a factor of 1/2, the $\boldsymbol{\sigma}_{x,y,z}$ matrices can be viewed as representations of the $\mathbf{s}_x$, $\mathbf{s}_y$ and $\mathbf{s}_z$ spin operators, respectively, when the α and β spin functions are taken as (1,0) and (0,1) vectors.

$$\mathbf{s}_z = \frac{1}{2}\boldsymbol{\sigma}_z$$

$$\mathbf{s}_z\begin{pmatrix} 1 \\ 0 \end{pmatrix} = \frac{1}{2}\begin{pmatrix} 1 \\ 0 \end{pmatrix} \tag{8.5}$$

$$\mathbf{s}_z\begin{pmatrix} 0 \\ 1 \end{pmatrix} = -\frac{1}{2}\begin{pmatrix} 0 \\ 1 \end{pmatrix}$$

The α function is an eigenfunction of the $\mathbf{s}_z$ operator with an eigenvalue of $1/2$, and the β function similarly has an eigenvalue of $-1/2$.

The Dirac equation is of the same order in all variables (space and time), since the momentum operator $\mathbf{p}$ $(= -i\nabla)$ involves a first-order differentiation with respect to the space variables. It should be noted that the free electron rest energy in eq. (8.3) is mc^2, equal to 0.511 MeV, while this situation is defined as zero in the non-relativistic case. The zero point of the energy scale is therefore shifted by 0.511 MeV, a large amount compared with the binding energy of 13.6 eV for a hydrogen atom. The two energy scales may be aligned by subtracting the electron rest energy, which corresponds to replacing the $\boldsymbol{\beta}$ matrix in eq. (8.3) by $\boldsymbol{\beta'}$.

$$\boldsymbol{\beta'} = \begin{pmatrix} 0 & 0 \\ 0 & -2\mathbf{I} \end{pmatrix} \tag{8.6}$$

The Dirac equation corresponds to satisfying the requirements of special relativity in connection with the quantum behaviour of the electron. Special relativity considers only systems which move with a constant velocity with respect to each other, which hardly can be considered a good approximation for the movement of an electron around a nucleus. A relativistic treatment of accelerated systems is described by general relativity, which is a gravitational theory. For atomic systems, however, the gravitational interaction between electrons and nuclei (or between electrons) is insignificant compared to the electrostatic interaction. Furthermore, a consistent theory describing the quantum aspects of gravitation has not yet been developed.

The Dirac equation is four-dimensional, and the relativistic wave function consequently contains four components. Two of the degrees of freedom are accounted for by assigning an intrinsic magnetic moment (spin), while the other two are interpreted as two different particles, electron and positron. The positronic solutions show up as "negative" energy states, having energies $\sim -2mc^2$, as illustrated in Figure 8.1. Note that the spacing between bound states has been exaggerated, as the binding energy is of the order of eV while $2mc^2$ is of the order of MeV.

It is conventional to write the relativistic wave function as

$$\Psi = \begin{pmatrix} \Psi_{L\alpha} \\ \Psi_{L\beta} \\ \Psi_{S\alpha} \\ \Psi_{S\beta} \end{pmatrix} \tag{8.7}$$

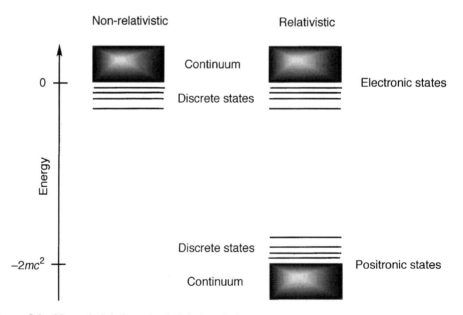

Figure 8.1   Non-relativistic and relativistic solutions

where Ψ_L and Ψ_S are the *large* and *small* components of the wave function, and α and β indicate the usual spin functions. Note that the spatial parts of $\Psi_{L\alpha}$ and $\Psi_{L\beta}$, and $\Psi_{S\alpha}/\Psi_{S\beta}$, are not necessarily identical. For electrons (positrons), the large (small) component reduces to the solutions of the Schrödinger equation when $c \to \infty$ (the non-relativistic limit), and the small (large) component disappears. The small component of the electronic wave function corresponds to coupling with the positronic states.

8.1 Connection Between the Dirac and Schrödinger Equations

In the presence of electric and magnetic fields the Dirac equation is modified to

$$[c\boldsymbol{\alpha} \cdot \boldsymbol{\pi} + \boldsymbol{\beta}'mc^2 + \mathbf{V}]\Psi = i\frac{\partial\Psi}{\partial t} \tag{8.8}$$

where $\mathbf{V}$ is the electrostatic potential and the generalized momentum operator $\boldsymbol{\pi}$ for an electron includes the vector potential $\mathbf{A}$ associated with the magnetic field $\mathbf{B}$ ($\nabla\times$ is the curl operator).

$$\boldsymbol{\pi} = \mathbf{p} + \mathbf{A}$$
$$\mathbf{B} = \nabla \times \mathbf{A} \tag{8.9}$$

In the time-independent case the Dirac equation may be written as

$$[c\boldsymbol{\alpha} \cdot \boldsymbol{\pi} + (\boldsymbol{\beta}'mc^2 + \mathbf{V})]\Psi = E\Psi \tag{8.10}$$

Since $\boldsymbol{\alpha}$ and $\boldsymbol{\beta}'$ are block matrices in terms of $\boldsymbol{\sigma}$ and $\mathbf{I}$, eq. (8.10) can be factored out in two equations.

$$c(\boldsymbol{\alpha} \cdot \boldsymbol{\pi})\Psi_S + \mathbf{V}\Psi_L = E\Psi_L$$
$$c(\boldsymbol{\sigma} \cdot \boldsymbol{\pi})\Psi_L + (-2mc^2 + \mathbf{V})\Psi_S = E\Psi_S \tag{8.11}$$

Here Ψ_L and Ψ_S are (large and small) two-component wave functions which include the α and β spin functions. The latter equation can be solved for Ψ_S.

$$\Psi_S = (E + 2mc^2 - \mathbf{V})^{-1} c(\boldsymbol{\sigma} \cdot \boldsymbol{\pi})\Psi_L \tag{8.12}$$

The inverse quantity can be factorized as

$$(E + 2mc^2 - \mathbf{V})^{-1} = (2mc^2)^{-1}\left(1 + \frac{E - \mathbf{V}}{2mc^2}\right)^{-1} = (2mc^2)^{-1}\mathbf{K} \tag{8.13}$$

Eq. (8.12) may then be written as

$$\Psi_S = \mathbf{K}\frac{\boldsymbol{\sigma} \cdot \boldsymbol{\pi}}{2mc}\Psi_L \tag{8.14}$$

while the top equation in (8.11) becomes

$$\left[\frac{1}{2m}(\boldsymbol{\sigma} \cdot \boldsymbol{\pi})\mathbf{K}(\boldsymbol{\sigma} \cdot \boldsymbol{\pi}) + (-E + \mathbf{V})\right]\Psi_L = 0 \tag{8.15}$$

In the non-relativistic limit ($c \to \infty$) the $\mathbf{K}$ factor is 1, and the first term is $(\boldsymbol{\sigma} \cdot \boldsymbol{\pi})(\boldsymbol{\sigma} \cdot \boldsymbol{\pi})$.

Using the vector identity $(\boldsymbol{\sigma} \cdot \boldsymbol{\pi})(\boldsymbol{\sigma} \cdot \boldsymbol{\pi}) = \boldsymbol{\pi} \cdot \boldsymbol{\pi} + i\boldsymbol{\sigma}(\boldsymbol{\pi} \times \boldsymbol{\pi})$, this may be written as

$$\left[\frac{1}{2m}(\boldsymbol{\pi} \cdot \boldsymbol{\pi} + i\boldsymbol{\sigma} \cdot (\boldsymbol{\pi} \times \boldsymbol{\pi})) + \mathbf{V}\right]\Psi_{\mathrm{L}} = E\Psi_{\mathrm{L}} \tag{8.16}$$

The term involving $\boldsymbol{\pi} \cdot \boldsymbol{\pi} = \boldsymbol{\pi}^2$ is the usual kinetic energy operator. The vector product $(\boldsymbol{\pi} \times \boldsymbol{\pi})$ gives

$$\begin{aligned} \boldsymbol{\pi} \times \boldsymbol{\pi} &= (\mathbf{p} + \mathbf{A}) \times \mathbf{p} + \mathbf{A}) \\ &= \mathbf{p} \times \mathbf{p} + \mathbf{p} \times \mathbf{A} + \mathbf{A} \times \mathbf{p} + \mathbf{A} \times \mathbf{A} \end{aligned} \tag{8.17}$$

The first and last terms are zero (since $\mathbf{a} \times \mathbf{a} = 0$). With $\mathbf{p} = -i\boldsymbol{\nabla}$ the other two yield

$$\begin{aligned} (\mathbf{p} \times \mathbf{A} + \mathbf{A} \times \mathbf{p})\Psi &= -i\boldsymbol{\nabla} \times (\mathbf{A}\Psi) - i\mathbf{A} \times (\boldsymbol{\nabla}\Psi) \\ &= -i(\boldsymbol{\nabla} \times \mathbf{A})\Psi - i(\boldsymbol{\nabla}\Psi) \times \mathbf{A} - i\mathbf{A} \times (\boldsymbol{\nabla}\Psi) \end{aligned} \tag{8.18}$$

The two last terms cancel (since $\mathbf{a} \times \mathbf{b} = -\mathbf{b} \times \mathbf{a}$), and the curl of the vector potential is the magnetic field, eq. (8.9). The final result is

$$\left[\frac{\boldsymbol{\pi}^2}{2m} + \mathbf{V} + \frac{\boldsymbol{\sigma} \cdot \mathbf{B}}{2m}\right]\Psi_{\mathrm{L}} = E\Psi_{\mathrm{L}} \tag{8.19}$$

Except for the $\boldsymbol{\sigma} \cdot \mathbf{B}$ term, called the *Zeeman* interaction, this is exactly the (non-relafivistic) Schrödinger equation. The extra term represents the interaction of an (external) magnetic field with an intrinsic magnetic moment associated with the electron, i.e. the electron spin. As noted in eq. (8.5), $\boldsymbol{\sigma}$ represents the spin operator (except for a factor of 1/2), and the $\boldsymbol{\sigma} \cdot \mathbf{B}/2m$ interaction can (in atomic units) also be written as $\mathbf{s} \cdot \mathbf{B}$, with $\mathbf{s}$ being the spin operator. In a more refined treatment, by including quantum field corrections, it turns out that the electron magnetic moment is not exactly equal to the spin. It is conventional to write the interaction as $g_{\mathrm{e}} \mu_{\mathrm{B}} \mathbf{s} \cdot \mathbf{B}$ where the Bohr magneton $\mu_{\mathrm{B}} (= e\hbar/2m)$ has a value of 1/2 in atomic units and the electronic g-factor g_{e} is approximately equal to 2.0023. Although electronic spin is often said to arise from relativistic effects, the above shows that spin arises naturally in the non-relativistic limit of the Dirac equation. It may also be argued that electron spin is actually present in the non-relativistic case, as the kinetic energy operator $\mathbf{p}^2/2m$ is mathematically equivalent to $(\boldsymbol{\sigma} \cdot \mathbf{p})^2/2m$. If the kinetic energy is written as $(\boldsymbol{\sigma} \cdot \mathbf{p})^2/2m$ in the Schrödinger Hamiltonian, then electron spin is present in the non-relativistic case, although this would only have consequences in the presence of a magnetic field.

In the non-relativistic limit the small component of the wave function (eq. (8.14)) is

$$\Psi_{\mathrm{S}} = \frac{\boldsymbol{\sigma} \cdot \boldsymbol{\pi}}{2mc}\Psi_{\mathrm{L}} \tag{8.20}$$

For a hydrogenic wave function ($\Psi_{\mathrm{L}} \sim \mathrm{e}^{-Zr}$), this gives in atomic units (setting $m = 1$).

$$\Psi_{\mathrm{S}} \sim \frac{Z}{2c}\Psi_{\mathrm{L}} \tag{8.21}$$

For a hydrogen atom the small component accounts for only $\sim 0.4\%$ of the total wave function and only $10^{-3}\%$ of the electron density, but for a uranium 1s-electron it is a third of the wave function and $\sim 10\%$ of the density.

We may obtain relativistic corrections by expanding the $\mathbf{K}$ factor in eq. (8.15).

$$\mathbf{K} = \left(1 + \frac{E - \mathbf{V}}{2mc^2}\right)^{-1} \approx 1 - \frac{E - \mathbf{V}}{2mc^2} + \dots \tag{8.22}$$

This is only valid when $E - \mathbf{V} \ll 2mc^2$, however, all atoms have a region close to the nucleus where this is not fulfilled (since $\mathbf{V} \to -\infty$ for r $\to$ 0). Inserting (8.22) in (8.15), and assuming a Coulomb potential $-Z/r$ (i.e. $\mathbf{V}$ is the attraction to a nucleus), gives after renormalization of the (large component) wave function and some rearrangement the following terms

$$\left[\frac{\boldsymbol{\pi}^2}{2m} + \mathbf{V} - \frac{\boldsymbol{\pi}^4}{8m^3c^2} + \frac{Z\mathbf{s} \cdot \mathbf{L}}{2m^2c^2r^3} + \frac{Z\pi\delta(\mathbf{r})}{2m^2c^2}\right]\Psi_L = E\Psi_L \tag{8.23}$$

Eq. (8.23) is called the *Pauli equation*. The first two terms are the usual non-relativistic kinetic and potential energy operators, the $\boldsymbol{\pi}^4$ term is called the *mass–velocity* correction, and is due to the dependence of the electron mass on the velocity. The next is the *spin–orbit* term ($\mathbf{L}$ is the angular moment $\mathbf{r} \times \boldsymbol{\pi}$), which corresponds to an interaction of the intrinsic magnetic moment (spin) with the magnetic field generated by the movement of the electron. The last term involving the delta function is the *Darwin* correction, which corresponds to a correction that can be interpreted as the electron making a high-frequency oscillation around its mean position, sometimes referred to as *Zwitterbewegung*. The mass–velocity and Darwin corrections are often collectively called the *scalar* relativistic corrections. Since they have opposite signs they do to a certain extent cancel each other.

Owing to the divergence of the $\mathbf{K}$ expansion near the nuclei, the mass–velocity and Darwin corrections can only be used as first-order corrections. An alternative method is to partition eq. (8.13) as in eq (8.24), which avoids the divergence near the nucleus.

$$(E + 2mc^2 - \mathbf{V})^{-1} = (2mc^2 - \mathbf{V})^{-1}\left(1 + \frac{E}{2mc^2 - \mathbf{V}}\right)^{-1} = (2mc^2 - \mathbf{V})^{-1}\mathbf{K}' \tag{8.24}$$

In contrast to eq. (8.13), the factor $E/(2mc^2 - \mathbf{V})$ is always much smaller than 1. $\mathbf{K}'$ may now be expanded in powers of $E/(2mc^2 - \mathbf{V})$, analogously to eq. (8.22). Keeping only the zero-order term (i.e. setting $\mathbf{K}' = 1$) gives the *Zero-Order Regular Approximation* (ZORA) method.[2]

$$\left[\boldsymbol{\pi}\frac{c^2}{2m^2c^2 - \mathbf{V}}\boldsymbol{\pi} + \left(\frac{c^2}{2m^2c^2 - \mathbf{V}}\right)^2\frac{2Z\mathbf{s} \cdot \mathbf{L}}{r^3} + \mathbf{V}\right]\Psi_L = E\Psi_L \tag{8.25}$$

Note that in this case the spin–orbit coupling is included already in zero order. Including the first-order term from an expansion of $\mathbf{K}'$ defines the *First-Order Regular Approximation* (FORA) method.

The Dirac equation automatically includes effects due to electron spin, while this must be introduced in a more or less *ad hoc* fashion in the Schrödinger equation (the Pauli principle). Furthermore, once the spin–orbit interaction is included, the total electron spin is no longer a "good" quantum number, an orbital no longer contains an integer number of α and β spin functions. The proper quantum number is now the total angular momentum obtained by vector addition of the orbital and spin moments.

8.2 Many-particle systems

A fully relativistic treatment of more than one particle has not yet been developed. For many particle systems it is assumed that each electron can be described by a Dirac operator $(c\boldsymbol{\alpha} \cdot \boldsymbol{\pi} + \boldsymbol{\beta}' mc^2)$ and the many-electron operator is a sum of such terms, in analogy with the kinetic energy in non-relativistic theory. Furthermore, potential energy operators are added to form a total operator equivalent to the Hamilton operator in non-relativistic theory. Since this approach gives results which agree with experiments, the assumptions appear justified.

The Dirac operator incorporates relativistic effects for the kinetic energy. In order to describe atomic and molecular systems, the potential energy operator must also be modified. In non-relativistic theory the potential energy is given by the Coulomb operator.

$$\mathbf{V}(\mathbf{r}_{12}) = \frac{q_1 q_2}{r_{12}} \tag{8.26}$$

According to this equation, the interaction between two charged particles depends only on the distance between them, <u>but not on time</u>. This cannot be correct when relativity is considered, as it implies that the attraction/repulsion between two particles occurs <u>instantly</u> over the distance r_{12}. However, nothing can move faster than the speed of light. The interaction between distant particles must be "later" than between particles which are close, and the potential is "retarded" (delayed). The relativistic interaction requires a description, *Quantum ElectroDynamics* (QED), which involves exchange of photons between charged particles. The photons travel at the speed of light and carry the information equivalent to the classical Coulomb interaction. The relativistic potential energy operator becomes very complicated, and cannot be written in closed form. For actual calculations it may be expanded in a Taylor series in $1/c$, and for chemical purposes it is normally only necessary to include terms up to $1/c^2$. In this approximation the potential energy operator for the electron–electron repulsion becomes.

$$\mathbf{V}_{ee}(\mathbf{r}_{12}) = \frac{1}{r_{12}} - \frac{1}{r_{12}} \left[\boldsymbol{\alpha}_1 \cdot \boldsymbol{\alpha}_2 + \frac{(\boldsymbol{\alpha}_1 \times \mathbf{r}_{12})(\boldsymbol{\alpha}_2 \times \mathbf{r}_{12})}{2r_{12}^2} \right] \tag{8.27}$$

Note that the subscript on the $\boldsymbol{\alpha}$ matrices refers to the particle, and $\boldsymbol{\alpha}$ here includes all of the $\boldsymbol{\alpha}_x$, $\boldsymbol{\alpha}_y$ and $\boldsymbol{\alpha}_z$ components in eq. (8.4). The first correction term in the square brackets is called the *Gaunt* interaction, and the whole term in the square brackets is the *Breit* interaction. The Dirac matrices appear since they represent the velocity operators in a relativistic description. The Gaunt term is a magnetic interaction (spin) while the other term represents a retardation effect. Eq. (8.27) is more often written in the form

$$\mathbf{V}_{ee}^{\text{Coulomb–Breit}}(\mathbf{r}_{12}) = \frac{1}{r_{12}} - \frac{1}{2r_{12}} \left[\boldsymbol{\alpha}_1 \cdot \boldsymbol{\alpha}_2 + \frac{(\boldsymbol{\alpha}_1 \cdot \mathbf{r}_{12})(\boldsymbol{\alpha}_2 \cdot \mathbf{r}_{12})}{r_{12}^2} \right] \tag{8.28}$$

Relativistic corrections to the nuclear–electron attraction ($\mathbf{V}_{ne}$) are of order $1/c^3$ (owing to the much smaller velocity of the nuclei) and are normally neglected.

An expansion in powers of $1/c$ is a standard approach for deriving relativistic correction terms. Taking into account electron (**s**) and nuclear spins (**I**), and indicating explicitly an external electric potential by means of the field ($\mathbf{F} = -\nabla\phi$, or $-\nabla\phi - \partial\mathbf{A}/\partial t$ if time dependent), an expansion up to order $1/c^2$ of the Dirac Hamiltonian including the

Coulomb–Breit potential gives the following set of operators, where the QED correction to the electronic spin has been introduced by means of the $g_e\,\mu_B$ factor.

One-electron operators:

$$\mathbf{H}_e^{\text{Zeeman}} = g_e\,\mu_B \sum_{i=1}^{N} \left[\mathbf{s}_i \cdot \mathbf{B}_i - \frac{1}{2mc^2}(\mathbf{s}_i \cdot \mathbf{B}_i)\boldsymbol{\pi}^2 \right]$$

$$\mathbf{H}_e^{\text{mv}} = -\frac{1}{8m^3c^2} \sum_{i=1}^{N} \boldsymbol{\pi}^4 \qquad (8.29)$$

$$\mathbf{H}_e^{\text{so}} = -\frac{g_e\,\mu_B}{4mc^2} \sum_{i=1}^{N} (\mathbf{s}_i \cdot \boldsymbol{\pi}_i \times \mathbf{F}_i - \mathbf{s}_i \cdot \mathbf{F}_i \times \boldsymbol{\pi}_i)$$

where $\mathbf{F}_i$ and $\mathbf{B}_i$ indicate the (electric and magnetic) fields at the position of particle i. $\mathbf{H}_e^{\text{Zeeman}}$ has the $\mathbf{s} \cdot \mathbf{B}$ term from (8.19) and a relativistic correction, and $\mathbf{H}_e^{\text{mv}}$ is the mass–velocity correction, is also present in eq. (8.23). $\mathbf{H}_e^{\text{so}}$ is a spin–orbit type correction with respect to an external electric field. In these expressions it should be noted that the generalized momentum contains magnetic fields via the vector potential, $\boldsymbol{\pi} = \mathbf{p} + \mathbf{A}$. The kinetic energy operator, $\boldsymbol{\pi}^2/2m$, for example, gives raise to a $\mathbf{B} \cdot \mathbf{L}$ term, which is the orbital part of the Zeeman effect, as shown in Section 10.7.

Two electron operators:

$$\mathbf{H}_{ee}^{\text{so}} = -\frac{g_e\,\mu_B}{2m^2c^2} \sum_{i=1}^{N}\sum_{j\neq i}^{N} \frac{\mathbf{s}_i \cdot (\mathbf{r}_{ij} \times \boldsymbol{\pi}_i)}{r_{ij}^3}$$

$$\mathbf{H}_{ee}^{\text{SOO}} = \frac{g_e\,\mu_B}{m^2c^2} \sum_{i=1}^{N}\sum_{j\neq i}^{N} \frac{\mathbf{s}_i \cdot (\mathbf{r}_{ij} \times \boldsymbol{\pi}_j)}{r_{ij}^3}$$

$$\mathbf{H}_{ee}^{\text{SS}} = \frac{g_e^2\,\mu_B^2}{2c^2} \left\{ \sum_{i=1}^{N}\sum_{j\neq i}^{N} \left(\frac{\mathbf{s}_i \cdot \mathbf{s}_j}{r_{ij}^3} - 3\frac{(\mathbf{s}_i \cdot \mathbf{r}_{ij})(\mathbf{r}_{ij} \cdot \mathbf{s}_j)}{r_{ij}^5} \right) - \frac{8\pi}{3} \sum_{i=1}^{N}\sum_{j=1}^{N}(\mathbf{s}_i \cdot \mathbf{s}_j)\delta(\mathbf{r}_{ij}) \right\}$$

$$\mathbf{H}_{ee}^{\text{OO}} = -\frac{1}{4m^2c^2} \sum_{i=1}^{N}\sum_{j\neq 1}^{N} \left(\frac{\boldsymbol{\pi}_i \cdot \boldsymbol{\pi}_j}{r_{ij}} + \frac{(\boldsymbol{\pi}_i \cdot \mathbf{r}_{ij})(\mathbf{r}_{ij} \cdot \boldsymbol{\pi}_i)}{r_{ij}^3} \right)$$

$$\mathbf{H}_{ee}^{\text{Darwin}} = -\frac{\pi}{2m^2c^2} \sum_{i=1}^{N}\sum_{j=1}^{N} \delta(\mathbf{r}_{ij}) \qquad (8.30)$$

The sums run over all values of i and j (possibly excluding the $i = j$ term), and there is consequently a factor of 1/2 included to avoid overcounting. $\mathbf{H}_{ee}^{\text{SO}}$ is a spin–orbit operator, describing the interaction of the electronic spin with the magnetic field generated by its own movement. $\mathbf{H}_{ee}^{\text{SOO}}$ is the *spin–other-orbit* operator, describing the interaction of an electronic spin with the magnetic field generated by the movement of the other electrons. $\mathbf{H}_{ee}^{\text{SS}}$ and $\mathbf{H}_{ee}^{\text{OO}}$ are *spin–spin* and *orbit–orbit* terms, accounting for additional magnetic interactions. The $\mathbf{H}_{ee}^{\text{SS}}$ operator is for example responsible for making the three individual components of a triplet state non-degenerate, the splitting (*zero field splitting*) typically being of the order of a few cm^{-1}. The Darwin and spin–spin interactions have a term involving a delta function which arises from the divergence of the (electron–electron) potential energy operator, i.e. $\nabla \cdot (1/\mathbf{r}) = 4\pi\delta(\mathbf{r})$. Such terms are often called contact interactions, since they depend on the two particles being at the same position ($\mathbf{r} = 0$). In the spin–spin case it is normally called the *Fermi-Contact* (FC) term.

Corrections involving nuclei (with the nuclear spin $\mathbf{I}$ replacing the electron spin $\mathbf{s}$) are analogous to the above one- and two-particle terms in eqs. (8.29–8.30), with the exception of those involving the nuclear mass, which disappears in the Born–Oppenheimer approximation (which may be be considered as the $M_{\text{nucleus}} \rightarrow \infty$ limit).

Operators involving one nucleus:

$$\mathbf{H}_{\text{n}}^{\text{Zeeman}} = -\mu_{\text{N}} \sum_{A} g_A \mathbf{I}_A \cdot \mathbf{B}_A \qquad (8.31)$$

Here $\mu_{\text{N}}(= e\hbar/2m_p = 2.723 \times 10^{-4}$ in atomic units) is the nuclear magneton, analogous to μ_{B}, but with the proton mass instead of the electron mass. The g_A factor depends on the specific nucleus and is analogous to g_{e}, i.e. the nuclear magnetic moment is given as $g_A \mu_{\text{N}} \mathbf{I}_A$. The remaining terms corresponding to those in (8.29) involve the nuclear mass and disappear owing to the Born–Oppenheimer approximation.

Operators involving one nucleus and one electron:

$$\mathbf{H}_{\text{ne}}^{\text{SO}} = \frac{g_{\text{e}}\mu_{\text{B}}}{2mc^2} \sum_{i=1}^{N} \sum_{A} Z_A \frac{\mathbf{s}_i \cdot (\mathbf{r}_{iA} \times \boldsymbol{\pi}_i)}{r_{iA}^3}$$

$$\mathbf{H}_{\text{ne}}^{\text{PSO}} = \frac{\mu_{\text{N}}}{mc^2} \sum_{i=1}^{N} \sum_{A} g_A \frac{\mathbf{I}_A \cdot (\mathbf{r}_{iA} \times \boldsymbol{\pi}_i)}{r_{iA}^3}$$

$$\mathbf{H}_{\text{ne}}^{\text{SS}} = -\frac{g_{\text{e}}\mu_{\text{B}}\mu_{\text{N}}}{c^2} \sum_{i=1}^{N} \sum_{A} g_A \left(\frac{\mathbf{s}_i \cdot \mathbf{I}_A}{r_{iA}^3} - 3\frac{(\mathbf{s}_i \cdot \mathbf{r}_{iA})(\mathbf{r}_{iA} \cdot \mathbf{I}_A)}{r_{iA}^5} - \frac{8\pi}{3}(\mathbf{s}_i \cdot \mathbf{I}_A)\delta(\mathbf{r}_{iA}) \right)$$

$$\mathbf{H}_{\text{ne}}^{\text{Darwin}} = \frac{\pi}{2m^2c^2} \sum_{i=1}^{N} \sum_{A} Z_A \delta(\mathbf{r}_{iA}) \qquad (8.32)$$

The equivalent of the spin–other-orbit operator in eq. (8.30) splits into two contributions, one involving the interaction of the electron spin with the magnetic field generated by the movement of the nuclei, and one describing the interaction of the nuclear spin with the magnetic field generated by the movement of the electrons. Only the latter survives in the Born–Oppenheimer approximation, and is normally called the *Paramagnetic Spin–Orbit* (PSO) operator. The $\mathbf{H}_{\text{ne}}^{\text{SO}}$ operator is the one-electron part of the spin–orbit interaction, while the $\mathbf{H}_{\text{ee}}^{\text{SO}}$ and $\mathbf{H}_{\text{ee}}^{\text{SOO}}$ operators in eq. (8.30) define the two-electron part. The one-electron term dominates and the two-electron contributions are often neglected or accounted for approximately by introducing an effective nuclear charge in $\mathbf{H}_{\text{ne}}^{\text{SO}}$ (corresponding to a screening of the nucleus by the electrons). The effect of spin–orbit operators is to mix states having different total spin, as for example singlet and triplet states.

The spin–spin term is analogous to that in (8.30), while the whole term describing the orbit–orbit interaction disappears owing to the Born–Oppenheimer approximation. The spin–orbit and Darwin terms are the same as given in eq. (8.23), except for the quantum field correction factor of $g_e \mu_B$.

Operators involving two nuclei:

$$\mathbf{H}_{nn}^{ss} = \frac{\mu_N^2}{2c^2} \sum_A \sum_B g_A g_B \left(\frac{\mathbf{I}_A \cdot \mathbf{I}_B}{\mathbf{r}_{AB}^3} - 3 \frac{(\mathbf{I}_A \cdot \mathbf{r}_{AB})(\mathbf{r}_{AB} \cdot \mathbf{I}_B)}{\mathbf{r}_{AB}^5} \right) \qquad (8.33)$$

The only term surviving the Born–Oppenheimer approximation is the direct spin-spin coupling, as all the others involve nuclear masses. Furthermore, there is no Fermi-contact term since nuclei cannot occupy the same position. Note that the direct spin–spin coupling is independent of the electronic wave function, it depends only on the molecular geometry.

Terms up to order $1/c^2$ are normally sufficient for explaining experimental data. There is one exception, however, namely the interaction of the nuclear quadrupole moment with the electric field gradient, which is of order $1/c^3$. Although nuclei often are modelled as point charges in quantum chemistry, they do in fact have a finite size. The internal structure of the nucleus leads to a quadrupole moment for nuclei with spin larger than 1/2 (the dipole and octopole moments vanish by symmetry). As discussed in section 10.1.1, this leads to an interaction term which is the product of the quadrupole moment with the field gradient ($\mathbf{F}' = \nabla \mathbf{F}$) created by the electron distribution.

$$\mathbf{H}_Q = - \sum_A \mathbf{Q}_A \mathbf{F}' \qquad (8.34)$$

All of the terms in eqs. (8.29–8.34) may be used as perturbation operators in connection with non-relativistic theory, as discussed in more detail in Chapter 10. It should be noted, however, that some of the operators are inherently divergent, and should not be used beyond a first-order perturbation correction.

8.3 Four-component Calculations

Although relativistic effects can be included in the Schrödinger equation by addition of operators describing corrections to the non-relativistic wave function, it is perhaps more satisfying to include relativistic effects by solving the Dirac equation directly. The simplest approximate wave function is a single determinant constructed from four-component one-electron functions, called *spinors*, having large and small components multiplied with the two spin functions. The spinors are the relativistic equivalents of the spin orbitals in non-relativistic theory. With such a wave function the relativistic equation corresponding to the Hartree–Fock equation is the *Dirac–Fock* equation, which in its time-independent form (setting $\mathbf{A} = 0$ in eq. (8.8)) can be written as

$$(c\boldsymbol{\alpha} \cdot \mathbf{p}\Psi + \boldsymbol{\beta}'c^2 + \mathbf{V})\Psi = E\Psi \qquad (8.35)$$

where the electron rest mass has been set equal to one. The requirement that the wave function should be stationary with respect to a variation in the orbitals, results in an equation which is formally the same as in non-relativistic theory, $\mathbf{FC} = \mathbf{SC\varepsilon}$ (eq. (3.50)). However, the presence of solutions for the positronic states means that the desired solution is no longer the global minimum (Figure 8.1), and care must be taken that the procedure does not lead to variational collapse. An essential ingredient in this is the choice of the basis set. Since practical calculations necessarily must use basis sets which are far from complete, the large and small component basis sets must be proper balanced. The large component corresponds to the normal non-relativistic wave function, and has similar basis set requirements. The small component basis set is chosen to obey the *kinetic balance* condition, which follows from (8.20).

$$\chi_\mu^{\text{small}} = \frac{(\boldsymbol{\sigma} \cdot \mathbf{p})}{2c} \chi_\mu^{\text{large}} \tag{8.36}$$

The presence of the momentum operator means that the small component basis set must contain functions which are derivatives of the large basis set. The use of kinetic balance ensures that the relativistic solution smoothly reduces to the non-relativistic wave function as c is increased.

When the Dirac operator is invoked, the point charge model of the nucleus becomes problematic. For a non-relativistic hydrogen atom, the orbitals have a <u>cusp</u> (discontinuous derivative) at the nucleus, however, the relativistic solutions have a <u>singularity</u>. A singularity is much harder to represent in an approximate treatment (such as an expansion in a Gaussian basis) than a cusp. Consequently a (more realistic) finite size nucleus is often used in relativistic methods. A finite nucleus model removes the singularity of the orbitals, which now assume a Gaussian type behaviour within the nucleus. Experiments or theory, however, do not provide a good model for how the positive charge is distributed <u>within</u> the nucleus. The wave function and energy will of course depend on the exact form used for describing the nuclear charge distribution. A popular choice is either a uniformly charged sphere, where the radius is proportional to the nuclear mass to the 1/3 power, or a Gaussian charge distribution (which facilitates the calculation of the additional integrals) with the exponent depending on the nuclear mass. Note that this implies that the energy (and derived properties) depends on the specific isotope, not just the atomic charge, i.e. the results for say ^{37}Cl will be (slightly) different from those for ^{35}Cl. The difference between a finite and a point charge nuclear model is large in terms of total energy ($\sim$ 1 a.u.), however, the exact shape for the finite nucleus is not important. For valence properties any "reasonable" model gives essentially the same results.

The differences due to relativity can be described as:

(1) Differences in the dynamics due to the velocity-dependent mass of the electron. This alters the size of the orbitals.
(2) New (magnetic) interactions in the Hamilton operator due to electron spin. This destroys the picture of an orbital having a definite spin.
(3) Introduction of "negative" energy (positron) states. The coupling between the electronic and positronic states introduce a "small" component in the electronic wave function. The result is that the shape of the orbitals change, relativistic orbitals, for example, do not have nodes.
(4) Modification of the potential operator due to the finite speed of light. In the lowest order approximation this corresponds to addition of the Breit operator to the Coulomb interaction.

Results from fully relativistic calculations are scarce, and there is no clear consensus on which effects are the most important. The Breit (Gaunt) term is believed to be small, and many relativistic calculations neglect this term, or include it as a perturbational term evaluated from the converged wave function. For geometries, the relativistic contraction of the s-orbitals normally means that bond lengths become shorter.

Working with a full four-component wave function and the Dirac–Fock operator is significantly more complicated than solving the Roothaan–Hall equations. The spin dependence can no longer be separated out, and the basis set for the small component of the wave function must contain derivatives of the corresponding large component basis. This means that the basis set becomes roughly four times as large as in the non-relativistic case for a comparable accuracy. Furthermore, the presence of magnetic terms (spin) in the Hamilton operator means that the wave function in general contains both real and imaginary parts, yielding a factor of two in complexity. In practice a (single determinant) Dirac–Fock–Coulomb calculation is about a factor of 100 more expensive than the corresponding non-relativistic Hartree–Fock case, although implementation of suitable integral screening techniques is likely to reduce this factor.[3] Since heavy atom systems by definition contain many electrons, even small systems (in terms of the number of atoms) are demanding. A relativistic calculation for a single Radon atom with a DZP quality basis, for example, is computationally equivalent to a non-relativistic calculation of a $C_{13}H_{28}$ alkane, for a comparable quality in terms of basis set limitations. To further complicate matters, there are many more systems which cannot be adequately described by a single determinant wave function in a relativistic treatment owing to spin–orbit coupling, and therefore require MCSCF type wave functions.

Since working with the full four-component wave function is so demanding, different approximate methods have been developed where the small component of the wave function is "eliminated" to a certain order in $1/c$ or approximated (like the *Foldy–Wouthuysen*[4] or *Douglas–Kroll* transformations[5] thereby reducing the four-component wave function to only two components. A description of such methods is outside the scope of this book.

Table 8.1 illustrates the magnitude of relativistic effects for dihydrides of the sixth main group in the periodic table, where the relativistic calculations are of the Dirac–Fock–Coulomb type (i.e. a single determinant wave function and neglecting the Breit interaction).[6] The relativistic correction to the total energy is significant, even for first row species like H_2O the difference is 0.055 a.u. (35 kcal/mol). It increases rapidly down the periodic table, reaching $\sim 7\%$ of the total energy for H_2Po. Nevertheless, the equilibrium distances and angles change relatively little. Similarly, the atomization energy (for breaking both X–H bonds completely) is remarkably insensitive to the large changes in the total energies. This is of course due to a high degree of cancellation of error, the major relativistic correction is associated with the inner-shell electrons of the heavy atom, with the correction being almost constant for the atom and the molecule. For the lighter elements the effect on the atomization energies is almost solely due to the spin–orbit interaction in the triplet X atom (e.g. $H_2O \rightarrow {}^3O + 2\,{}^2H$) which is not present in the singlet H_2X molecule.

Similar results have been obtained for the fourth group tetrahydrides, CH_4, SiH_4, SbH_4, GeH_4 and PbH_4,[7] where the Gaunt term was shown to give corrections typically an order of magnitude less than the other relativistic changes. The general conclusion is that relativistic effects for geometries and energetics normally can be neglected for molecules containing only first and second row elements. This will also be true for third row elements, unless a high accuracy is required. Although the geometry and atomization energy changes for H_2S and H_2Se may be considered significant, it should

Table 8.1 Properties of the sixth group dihydrides. Total energies in a.u., ∇E_{atom} in kcal/mol

	Method	Total energy	R_{eq} (Å)	θ_{eq}	∇E_{atom}
H_2O	non-relativistic	-76.05440	0.93905	107.75	153.87
	relativistic	-76.10945	0.93902	107.68	153.49
H_2S	non-relativistic	-398.64141	1.34291	94.23	122.87
	relativistic	-399.74889	1.34276	94.14	121.80
H_2Se	non-relativistic	-2400.97745	1.4530	93.14	109.81
	relativistic	-2429.60586	1.4504	92.87	106.61
H_2Te	non-relativistic	-6612.79787	1.6557	92.57	93.81
	relativistic	-6794.86918	1.6485	91.99	84.81
H_2Po	non-relativistic.	-20676.70909	1.7539	92.21	83.71
	relativistic	-22232.53167	1.7433	90.59	53.40

be noted that errors due to incomplete basis sets and neglect of electron correlation are much larger than the relativistic corrections. The experimental geometries for H_2S and H_2Se, for example, are 1.3356 Å and $92.12°$, and 1.4600 Å and $90.57°$, respectively. While the relativistic contraction of the H–Se bond is 0.0026 Å, the basis set and electron correlation error is 0.0096 Å. Relativistic effects typically become comparable to those from electron correlation at atomic numbers ~ 40–50. For molecules involving atoms beyond the fourth row in the periodic table, relativistic effects cannot be neglected for quantitative work.

Relativistic methods can be extended to include electron correlation by methods analogous to those for the non-relativistic cases, e.g. CI, MCSCF, MP and CC. Such methods are currently at the development stage.[8]

References

1 R. E. Moss, *Advanced Molecular Quantum Mechanics*, Chapman and Hall, 1973; P. Pyykko, *Chem. Rev.*, **88** (1988), 563; J. Almlöf and O. Gropen, *Rev. Comput. Chem.*, **8** (1996), 203; K. Balasubramanian, *Relativistic Effects in Chemistry*, Wiley, 1997.
2 E. van Lenthe, E. J. Baerends and J. G. Snijder, *J. Chem. Phys.*, **99** (1993), 4597; J. G. Snijder and A. J. Sadlej, *Chem. Phys. Lett.*, **252** (1996), 51.
3 T. Saue, K. Faegri, T. Helgaker and O. Gropen, *Mol. Phys.*, **91** (1997), 937.
4 L. L. Foldy and S. A. Wouthuysen, *Phys. Rev.*, **78** (1950), 29.
5 M. Douglas and N. M. Kroll, *Ann. Phys. NY*, **82** (1974), 89.
6 L. Pisani and E. Clementi, *J. Chem. Phys.*, **101** (1994), 3079.
7 O. Visser, L. Visscher, P. J. C. Aerts and W. C. Nieuwpoort, *Theor. Chim. Acta*, **81** (1992), 405.
8 L. Visscher, T. J. Lee and K. G. Dyall, *J. Chem. Phys.*, **105** (1996), 8769.

9 Wave Function Analysis

The previous chapters have focused on different methods for obtaining more or less accurate solutions to the Schrödinger equation. The natural "by-product" of determining the electronic wave function is the energy; however, there are many other properties that may be derived. Although the quantum mechanical description of a molecule is in terms of positive nuclei surrounded by a cloud of negative electrons, chemistry is still formulated as "atoms" held together by "bonds". This raises questions such as: given a wave function how can we define an atom and its associated electron population, or how do we determine whether two atoms are bonded?

Atomic charge is an example of a property often used for discussing/rationalizing structural and reactivity differences.[1] There are three commonly used methods for assigning a charge to a given atom.

(1) Partitioning the wave function in terms of the basis functions.
(2) Fitting schemes.
(3) Partitioning the wave function based on the wave function itself.

9.1 Population Analysis Based on Basis Functions

The electron density (probability of finding an electron) at a certain position $\mathbf{r}$ from a single molecular orbital containing one electron is given as the square of the MO.

$$\rho_i(\mathbf{r}) = \phi_i^2(\mathbf{r}) \tag{9.1}$$

Assuming that the MO is expanded in a set of normalized, but non-orthogonal, basis functions, this can be written as (eq. (3.48)).

$$\phi_i = \sum_{\alpha}^{AO} c_{\alpha i} \chi_\alpha$$

$$\phi_i^2 = \sum_{\alpha\beta}^{AO} c_{\alpha i} c_{\beta i} \chi_\alpha \chi_\beta \tag{9.2}$$

Integrating and summing over all occupied MOs gives the total number of electrons, N.

$$\sum_i^{MO} \int \phi_i^2 d\mathbf{r} = \sum_i^{MO} \sum_{\alpha\beta}^{AO} c_{\alpha i} c_{\beta i} \int \chi_\alpha \chi_\beta \, d\mathbf{r} = \sum_i^{MO} \sum_{\alpha\beta}^{AO} c_{\alpha i} c_{\beta i} S_{\alpha\beta} = N \qquad (9.3)$$

We may generalize this by introducing an occupation number (number of electrons), n, for each MO. For a single determinant wave function this will either be 0, 1 or 2, while it may be a fractional number for a correlated wave function (Section 9.5).

$$\sum_i^{MO} n_i \int \phi_i^2 \, d\mathbf{r} = \sum_{\alpha\beta}^{AO} \left(\sum_i^{MO} n_i c_{\alpha i} c_{\beta i} \right) S_{\alpha\beta} = \sum_{\alpha\beta}^{AO} D_{\alpha\beta} S_{\alpha\beta} = N \qquad (9.4)$$

The sum of the product of MO coefficients and the occupation numbers is the density matrix defined in Section 3.5 (eq. (3.51)). The sum over the product of the density and overlap matrix elements is the number of electrons.

The *Mulliken Population Analysis*[2] uses the **DS** matrix for distributing the electrons into atomic contributions (note that it is the matrix of products of elements, not elements of the product matrix). A diagonal element $D_{\alpha,\alpha} S_{\alpha,\alpha}$ is the number of electrons in the α AO, and an off-diagonal element $D_{\alpha\beta} S_{\alpha\beta}$ is (half) the number of electrons shared by AOs α and β (there is an equivalent $D_{\beta\alpha} S_{\beta\alpha}$ element). The contributions from all AOs located on a given atom A may be summed up to give the number of electrons associated with atom A. This requires a decision on how a contribution involving basis functions on different atoms should be divided. The simplest, and the one used in the Mulliken scheme, is to partition the contribution equally between the two atoms. The Mulliken electron population is thereby defined as

$$\rho_A = \sum_{\alpha \varepsilon A}^{AO} \sum_\beta^{AO} D_{\alpha\beta} S_{\alpha\beta} \qquad (9.5)$$

The *gross* charge on atom A is the sum of the nuclear and electronic contributions.

$$Q_A = Z_A - \rho_A \qquad (9.6)$$

The Mulliken method is just one of a whole family of population analyses. Another commonly used is the *Löwdin* partitioning.[3] The **DS** matrix product (eq. (9.4)) may be rewritten as

$$\sum \mathbf{DS} = N$$
$$\sum \mathbf{S}^{1/2} \mathbf{D} \left(\mathbf{S}^{1/2} \mathbf{S}^{1/2} \right) = \mathbf{S}^{1/2} N$$
$$\sum \mathbf{S}^{1/2} \mathbf{DS}^{1/2} \left(\mathbf{S}^{1/2} \mathbf{S}^{-1/2} \right) = \mathbf{S}^{1/2} N \mathbf{S}^{-1/2} \qquad (9.7)$$
$$\sum \mathbf{S}^{1/2} \mathbf{DS}^{1/2} = N$$

The Löwdin method uses the $\mathbf{S}^{1/2} \mathbf{DS}^{1/2}$ matrix for analysis, and it is equivalent to a population analysis of the density matrix in the orthogonalized basis set (Section 13.2) formed by transforming the original set of functions by $\mathbf{S}^{-1/2}$.

$$\chi' = \mathbf{S}^{-1/2} \chi \qquad (9.8)$$

The Mulliken and Löwdin methods give different atomic charges, but mathematically there is nothing to indicate which of these partitionings gives the "best" result. There are some common problems with all population analyses based on partitioning the wave function in terms of basis functions.

(1) The diagonal elements may be larger than 2. This implies more than two electrons in an orbital, violating the Pauli principle.
(2) The off-diagonal elements may become negative. This implies a negative number of electrons between two basis functions, which clearly is physically impossible.
(3) There is no reason for dividing the off-diagonal contributions equally between the two orbitals. It may be argued that the most "electronegative" (which then needs to be defined) atom (orbital) should receive most of the shared electrons.
(4) Basis functions centered on atom A may have a small exponent, so that they effectively describe the wave function far from atom A. Nevertheless, the electron density is counted as only belonging to A.
(5) The dipole, quadrupole etc. moments are in general not conserved, i.e. a set of population atomic charges does not reproduce the original multipole moments.

The Mulliken scheme suffers from all of the above, while the Löwdin method solves problems (1), (2) and (3). In the orthogonalized basis all off-diagonal elements are 0, and the diagonal elements are restricted to values between 0 and 2.

Problem (4) is especially troublesome, as a few examples will demonstrate. Consider for example a water molecule. A reasonable description of the wave function can be obtained by an HF single determinant with a DZP basis set. An equally good wave function (in terms of energy) may be constructed by having a very large number of basis functions centred on oxygen, and none on the hydrogens. This latter will according to the above population analysis give a $+1$ charge on hydrogen, and a -2 charge on oxygen. Worse, another, equally good wave function, may be constructed by having a large number of basis functions only on the hydrogens. This will give charges of -5 for each of the hydrogens and $+10$ for the oxygen! Or the basis functions can be taken to be non-nuclear centred, in which case the electrons are not associated with any nuclei at all, i.e. atomic charges of $+1$ and $+10$! The fundamental problem is that basis functions often describe electron density near a nucleus other than the one they are centred on. An s-type Gaussian function on oxygen with an exponent of 0.15, for example, has a maximum in the radial distribution $(r^2\phi^2)$ that peaks at 0.97 Å, i.e. at the distance where the hydrogen nuclei are located. Atomic charges calculated from a Mulliken or Löwdin analysis will therefore not converge to a constant value as the size of the basis set is increased. Enlarging the basis set involves addition of more and more diffuse basis functions, often leading to unpredictable changes in the atomic charges. This is a case where a "better" theoretical procedure actually is counter-productive. Basis function derived population analyses are therefore most useful for comparing trends in electron distributions when small or medium sized basis sets (which only contain relatively tight functions) are used.

The density matrix can also be used to generate information about bond strengths. A quantitative measure is given by the *Bond Order* (BO). It was originally defined[4] from bond distances as

$$BO = e^{-(r - r_0)/a} \qquad (9.10)$$

Defining ethane, ethylene and acetylene to have bond orders of 1, 2 and 3, the constant a is found to have a value of approximately 0.3 Å. For bond orders less than 1 (i.e. breaking and forming single bonds) it appears that 0.6 Å is a more appropriate proportionality constant. A "Mulliken" style measure of the bond strength between atoms A and B can be defined from the density matrix as[5] (note that this involves the elements of the product of the **D** and **S** matrices).

$$BO_{AB} = \sum_{\alpha \epsilon A} \sum_{\beta \epsilon B} (\mathbf{DS})_{\alpha\beta} (\mathbf{DS})_{\beta\alpha} \tag{9.11}$$

This will again be basis set dependent, but not nearly as sensitive as atomic charges.

Population analysis with semi-empirical methods requires a special comment. These methods normally employ the ZDO approximation, i.e. the overlap **S** is a unit matrix. The population analysis can therefore be performed directly on the density matrix. In some cases, however, a Mulliken population analysis is performed with **DS**, which requires an explicit calculation of the **S** matrix.

9.2 Population Analysis Based on the Electrostatic Potential

One area where the concept of atomic charges is deeply rooted is force field methods (Chapter 2). A significant part of the non-bonded interaction between polar molecules is described in terms of electrostatic interactions between fragments having an internal asymmetry in the electron distribution. The fundamental interaction is between the *ElectroStatic Potential* (ESP) generated by one molecule (or fraction of) and the charged particles of another. The electrostatic potential at position **r** is given as a sum of contributions from the nuclei and the electronic wave function.

$$\mathbf{V}_{ESP}(\mathbf{r}) = \sum_{A}^{nuclei} \frac{Z_A}{|\mathbf{r} - \mathbf{R}_A|} - \int \frac{|\Psi(\mathbf{r}_i)|^2}{|\mathbf{r} - \mathbf{r}_i|} d\mathbf{r}_i \tag{9.12}$$

The first part of the potential is trivially calculated from the nuclear charges and their positions, but the electronic contribution requires a knowledge of the wave function. This is not available in force field methods, and the simplest way of modelling the electrostatic potential is to assign partial charges to each atom. Atomic charges may be treated as regular force field parameters, and assigned values based on fitting to experimental data, such as dipole, quadrupole, octopole etc. moments, but there are rarely enough data to allow a unique assignment.

A common way of deriving partial atomic charges in force fields is to chose a set of parameters which in a least squares sense generate the best fit to the actual electrostatic potential as calculated from an electronic wave function.[6] The electrostatic potential stretches far beyond the molecular dimension (the Coulomb interaction falls off as R^{-1} (charge) or as R^{-3} (dipole)), but the most important region is just beyond the van der Waals distance. The potential is sampled by placing a suitable grid of points around each nucleus with distances from just outside the van der Waals radius to about twice that distance. A typical sampling will have a few hundred points for each atom. The atomic charges are determined as those parameters which reproduce the electrostatic potential as closely as possible at these points, subject to the constraint that the sum is equal to the total molecular charge. In some cases the atomic charges may also be constrained to reproduce for example the dipole moment. The various schemes for deriving atomic charges differ in the number and location of points used in the fitting, and whether additional constraints beyond preservation of charge are added. In many cases the fitted set of charges is uniformly increased by 10–20% to model the fact that polarization in condensed phases will increase the effective dipole moment relative to the isolated molecule case (Section 2.2.6), or the charges are derived by fitting to an ESP which naturally overestimates the charge polarization (for example HF/6-31G(d), Section 4.3).

The electrostatic potential depends directly on the wave function and therefore converges as the size of the basis set and amount of electron correlation is increased. Since the potential depends directly on the electron density ($\rho = |\Psi|^2$), it is in general found to be fairy insensitive to the level of sophistication, i.e. already an HF calculation with a DZP type basis set gives quite good results. One might thus anticipate that atomic charges based on fitting to the electrostatic potential would lead to well defined values. This, however, is not the case. One of the problems is that a straight forward fitting tends to give conformationally dependent charges. The three hydrogens in a freely rotating methyl group, for example, may end up having significantly different charges, or two conformations may give two widely different sets of fitted parameters. This is a problem in connection with force field methods which rely on the fundamental assumption that parameters are transferable between similar fragments, and consequently atoms which are easily interchanged (e.g. by bond rotation) should have identical parameters. Conformationally dependent charges can be modelled in force field methods by introducing cross terms between the electrostatic and bonded (bond stretch, angle bend and torsional) terms, or by including a polarization term for the atoms. Such additions lead to significantly more complicated force fields, with subsequent loss of efficiency.

One way of eliminating the problem with conformationally dependent charges is to add additional constraints, for example forcing the three hydrogens in a methyl group to have identical charges[7] or averaging over different conformations.[8] The more fundamental problem (which probably is also part of the conformational problem) is that the fitting procedure in general is statistically underdetermined.[9] The difference between the true electrostatic potential and that generated by a set of atomic charges on say 80% of the atoms, is not significantly reduced by having fitting parameters on all atoms. The electrostatic potential experienced outside the molecule is mainly determined by the atoms near the surface, and consequently the charges on atoms buried within a molecule cannot be assigned with any great confidence. Even for a medium sized molecule it may only be statistically valid to assign charges to perhaps half the nuclei. A full set of atomic charges thus forms a redundant set, many different sets of charges may be chosen, all of which are capable of reproducing the true electrostatic potential to almost the same accuracy. It has furthermore been shown that increasing the number of sampling points, or the quality of the wave function, does not alleviate the problem. Although a very large number of sampling points (several thousands) may be chosen to be fitted by relatively few (perhaps 20–30) parameters, the fact that the sampling points are highly correlated makes the problem underdetermined.

Another problem of atomic charges determined by fitting is related to the absolute accuracy. Although inclusion of charges on all atoms does not significantly improve the results over those determined from a reduced set of parameters, the <u>absolute</u> deviation between the true and fitted electrostatic potentials can be quite large. Interaction

Figure 9.1   Generation of dipole and quadrupole moments by charges

energies as calculated by an atom centred charge model in a force field may be off by several kcal/mol per atom in certain regions of space just outside the molecular surface, an error of one or two orders of magnitude larger than the van der Waals interaction. In order to improve the description of the electrostatic interaction, additional non-nuclear centred charges may be added,[10] or dipole, quadrupole etc. moments may be added at nuclear or bond positions.[11] These descriptions are essentially equivalent since a dipole may be generated as two opposite charged monopoles, a quadrupole as four monopoles etc. as illustrated in Figure 9.1.

The *Distributed Multipole Analysis* (DMA) developed by Stone uses the fact that the electrostatic potential arising from the charge overlap between two basis functions can be written in terms of a multipole expansion around a point between the two nuclei.[12] These moments can be calculated <u>directly</u> from the density matrix and the basis functions, and are not a result of a fitting procedure. The multipole expansion is, furthermore, finite; the highest non-vanishing term is given as the sum of the angular momenta for the two basis functions, e.g. the product of two p-functions has only moments up to quadrupole. For Gaussian orbitals the expansion point is given as

$$\mathbf{R}_C = \frac{\alpha \mathbf{R}_A + \beta \mathbf{R}_B}{\alpha + \beta} \tag{9.13}$$

where $\mathbf{R}_A$ and $\mathbf{R}_B$ are the positions of the two nuclei, and α and β are the exponents of the basis functions (this follows since the product of two Gaussians is a single Gaussian located between the two originals, eq. (3.59)). If such distributed multipoles are assigned for each pair of basis functions, the electrostatic potential as seen from outside the charge distribution is reproduced exactly. This, however, would mean that $\sim M^2$ different sites (M being the size of the basis set) would be required. In practice only the nuclei and possibly the midpoints of all bonds are selected as multipole points, and all the pair expansion points are moved to the nearest multipole point. By moving the origin, the termination after a finite number of terms is destroyed, and an infinite sum over all moments must be used for an exact representation. Since most of the pair expansion points are rather close to either a nucleus or the centre of a bond, the higher-order moments are usually quite small. Furthermore, since the majority of the electron density can be represented with just s- and p-functions for elements belonging to the first or second row of the periodic table, it follows that a representation in terms of charges, dipoles and quadrupoles located on all nuclei and bond centres gives a quite accurate representation of the electrostatic potential. If only nuclear centered multipoles are used, an expansion up to quadrupoles will typically generate an electrostatic potential of the same quality as a model based on fitted atomic charges. Unfortunately additions of non-nuclear centred charges or multipoles significantly increase the computational time for force field calculations as they add to the non-bonded terms, the number of which grows as N^2. Since the electrostatic interaction is long ranged, it is often the most important for determining intermolecular interactions, but essentially all force fields to date use a simple atomic charge model.

9.3 Population Analysis Based on the Wave Function

The examples in Section 9.1 illustrate that it would be desirable to base a population analysis on properties of the wave function itself, and not on the basis set chosen for representing the function. Some textbooks state that it is impossible to define a unique atomic charge since there is no quantum mechanical operator associated with charge. This is not true; the electronic charge operator is simply the negative of the number operator (the charge from an electron is -1). The problem is in the definition of an "atom" within a molecule. If the total molecular volume somehow can be divided into subsections, each belonging to a certain nucleus, then the square of the wave function can be integrated to give the number of electrons present in each of these volumes. This again requires a <u>choice</u> to be made of whether a certain point in space belongs to one nucleus or to another.

Perhaps the most rigorous way of dividing a molecular volume into atomic subspaces is the *Atoms In Molecules* (AIM) method of Bader.[13] The electron density is the square of the wave function integrated over $N - 1$ coordinates (it does not matter which coordinates since all electrons are identical).

$$\rho(\mathbf{r}_1) = \int |\Psi(\mathbf{r}_2, \mathbf{r}_3, \ldots, \mathbf{r}_N)|^2 \, d\mathbf{r}_2 \, d\mathbf{r}_3 \ldots d\mathbf{r}_N \qquad (9.14)$$

The electron density is a function of three spatial coordinates, and it may be analysed in terms of its topology (maxima, minima and saddle points). In the large majority of cases it is found that the only maxima in the electron density occur at the nuclei (or very close to), which is reasonable since they are the only sources of positive charge. The nuclei act as *attractors* of the electron density. At each point in space the gradient of the electron density points in the direction of the strongest (local) attractor. This forms a rigorous way of dividing the physical space into atomic subspaces: starting from a given point in space a series of infinitesimal steps may be taken in the gradient direction until an attractor is encountered. The collection of all such points forms the atomic *basin* associated with the attractor (nucleus). If the negative of the electron density is considered, the attractors are local minima, and a basin is then defined as points which end up at the local minimum by a steepest descent minimization (see Section 14.1). In the outer direction (away from other nuclei) the gradient goes asymptotically to zero, and the atomic basin stretches into infinity in this direction.

The border between two three-dimensional atomic basins is a two-dimensional surface. Points on such dividing surfaces have the property that the gradient of the electron density is perpendicular to the normal vector of the surface, i.e. the radial part of the derivative of the electron density (the electronic "flux") is zero.

Once the molecular volume has been divided, the electron density may be integrated within each of the atomic basins to give a net atomic charge. As the dividing surface is

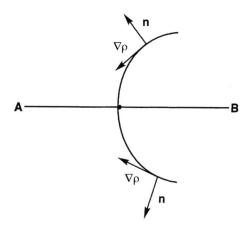

Figure 9.2   Dividing surface between to atomic basins

rigorously defined in terms of the wave function, the charges will converge to specific values as the quality of the wave function is increased. Furthermore, as only the electron density is involved, the results are fairly insensitive to the theoretical level used for generating the wave function. Analogously to the Mulliken and Löwdin procedures, the resulting charges will not reproduce the dipole, quadrupole etc. moments, nor will they yield a particularly good representation of the electrostatic potential. Despite the aesthetically attraction of deriving charges in terms of atomic basins, they are not well suited for modelling purposes.

There are two other problems with AIM atomic charges. One is the magnitude of the calculated atomic charges for polar bonds, the calculated internal redistribution is often significantly larger than "commonly accepted" values. This may qualitatively be understood from simple orbital overlap arguments.[14] Consider for example the CH_3F molecule, where the fluorine orbital involved in the bonding will be "tighter" (have a larger exponent) than the corresponding carbon orbital. The bonding molecular orbital can schematically be represented as shown in Figure 9.3. The dividing surface (vertical line) is at the minimum of the combined electron density, and Figure 9.3 shows that the carbon orbital contributes more electron density to the fluorine basin than the fluorine orbital contributes to the carbon basin. Consequently the charge separation will tend to be overestimated when atoms of unequal electronegativity are involved in a bond.

The other problem in the AIM approach is the presence of non-nuclear attractors in certain metallic systems, such as lithium and sodium clusters.[15] While these are of interest by themselves, they spoil the picture of electrons associated with nuclei forming atoms within molecules.

The division of the molecular volume into atomic basins follows from a deeper analysis based on the principle of stationary action. The shapes of the atomic basins, and the associated electron densities, in a functional group are very similar in different molecules. The <u>local</u> properties of the wave function are therefore transferable to a very good approximation, which rationalizes the basis for organic chemistry, that functional groups react similarly in different molecules. It may be shown that any observable

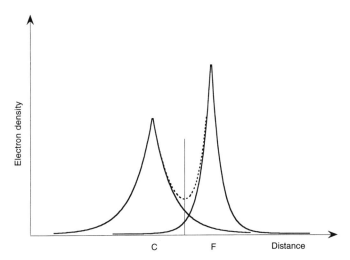

Figure 9.3 Illustration of a bonding orbital formed from two atomic orbitals

molecular property may be written as a sum of corresponding atomic contributions.

$$\langle \mathbf{A} \rangle = \sum_{i}^{\substack{\text{atomic} \\ \text{basins}}} \langle \mathbf{A} \rangle_{i} \tag{9.15}$$

The total energy, for example, may be written as a sum of atomic energies, and these atomic energies are again almost constant for the same structural units in different molecules. The atomic basins are probably the closest quantum mechanical analogy to the chemical concepts of atoms within a molecule.

The surface of an atomic basin has the radial derivative of the electron density equal to zero. In general there will also be points on such surfaces where the total derivative is zero, i.e. the tangential components are also zero, such a point is marked with a dot in Figure 9.2. The basin attractor is also such a stationary point on the electron density surface. The second derivative of the electron density, the Hessian, is a function of the three (cartesian) coordinates, i.e. it is a 3×3 matrix. At stationary points it may be diagonalized and the number of negative eigenvalues determined. The basin attractor is an overall maximum, it has three negative eigenvalues. Other stationary points are usually found between nuclei which are "bonded". Such points have a minimum in the electron density in the direction of the nuclei, and a maximum in the perpendicular directions, i.e. there is one positive and two negative eigenvalues in the Hessian. These are known as *bond critical points*. If the negative of the electron density is considered instead, the attractors are minima (all positive eigenvalues in the second derivative matrix) and the bond critical points are analogous to transition structures (one negative eigenvalue). Comparing with potential energy surfaces (see Section 12.1), the (negative) electron density surface may be analysed in terms of "reaction paths" connecting "transition structures" with minima. Such paths trace the maximum electron density connecting two nuclei, and may be taken as the molecular "bond". It should be noted that bond critical points are not necessarily located on a straight line connecting two nuclei, small strained rings like cyclopropane, for example, have bond paths which are significantly curved. Indeed the degree of bending tends to correlate with the strain energy.

The value of the electron density at the bond critical point correlates with the strength of the bond, the bond order. As mentioned above there are certain systems like metal clusters which have non-nuclear centred attractors. The corresponding bond critical points have electron densities at least an order of magnitude smaller than "normal" single bonds, and the value of the density at the local maximum is only slightly larger than at the bond critical point. The non-nuclear attractors are thus only weakly defined, and may be considered as a special kind of metal bonding, where a "sea" of electrons with weak local maxima surrounds the positive nuclei, which are strong local maxima. In certain cases bond critical points may also be found between atoms which are not bonded, but experience a strong steric repulsion. This corresponds to a situation where two atoms are forced to be closer than the sum of their van der Waals radii. Such systems usually have values of the electron density at the bond critical point which are at least an order of magnitude smaller than ordinary "bonded" atoms.

There are two other types of critical points, having either one or zero negative eigenvalues in the density Hessian. The former is usually found in the centre of a ring (e.g. benzene), and consequently denoted a *ring critical point*, the latter is typically found at the centre of a cage (e.g. cubane), and denoted a *cage critical point*. They corresponds to local minima in the electron density in two or three directions.

The derivative of the dipole moment with respect to the coordinates determines the intensity of IR absorptions (Section 10.1.5). A central quantity in this respect is the *Atomic Polar Tensor* (APT), which for a given atom is defined as

$$\mathbf{V}(\text{APT}) = \begin{pmatrix} \dfrac{\partial \mu_x}{\partial x} & \dfrac{\partial \mu_x}{\partial y} & \dfrac{\partial \mu_x}{\partial z} \\ \dfrac{\partial \mu_y}{\partial x} & \dfrac{\partial \mu_y}{\partial y} & \dfrac{\partial \mu_y}{\partial z} \\ \dfrac{\partial \mu_z}{\partial x} & \dfrac{\partial \mu_z}{\partial y} & \dfrac{\partial \mu_z}{\partial z} \end{pmatrix} \tag{9.16}$$

Such a matrix is not independent on the coordinate system, but the trace is. Cioslowski has proposed a definition of atomic charges as one-third of the trace over the APT, denoted *Generalized Atomic Polar Tensor* (GAPT) charges.[16] The charge on atom A is defined as

$$Q_A(\text{GAPT}) = \frac{1}{3}\left(\frac{\partial \mu_x}{\partial x_A} + \frac{\partial \mu_y}{\partial y_A} + \frac{\partial \mu_z}{\partial z_A}\right) \tag{9.17}$$

Since the dipole moment itself is the first derivative of the energy with respect to an external electric field (Section 10.1.1), a calculation of GAPT charges requires the second derivative of the energy. This is a computationally expensive method for generating atomic charges; however, if vibrational frequencies are calculated anyway, GAPT charges may be determined with very little additional effort. Dipole derivatives determine the intensity of IR absorptions (Section 10.1.5) and GAPT charges are therefore directly related to experimentally observable quantities. The fact that they are computationally expensive to generate, and that they are found to be sensitive to the amount of electron correlation in the wave function, has limited the general use of GAPT charges.

9.4 Localized Orbitals

A Hartree–Fock wave function can be written as a single Slater determinant, composed of a set of orthonormal MOs (eq. (3.20)).

$$\Phi = \frac{1}{\sqrt{N!}} \begin{vmatrix} \phi_1(1) & \phi_2(1) & \cdots & \phi_N(1) \\ \phi_1(2) & \phi_2(2) & \cdots & \phi_N(2) \\ \cdots\cdots & \cdots\cdots & \cdots & \cdots\cdots \\ \phi_1(N) & \phi_2(N) & \cdots & \phi_N(N) \end{vmatrix} \tag{9.18}$$

For computational purposes it is convenient to work with canonical MOs, i.e. those which make the matrix of Lagrange multipliers diagonal, and which are eigenfunctions of the Fock operator at convergence (eq. (3.41)). This corresponds to a specific choice of a unitary transformation of the occupied MOs. Once the SCF procedure has converged, however, we may chose other sets of orbitals by forming linear combinations of the canonical MOs. The total wave function, and thus all observable properties, are independent of such a rotation of the MOs.

$$\boldsymbol{\phi}' = \mathbf{U}\boldsymbol{\phi}$$
$$\phi_i' = \sum_{j=1}^{N} u_{ij}\phi_j \tag{9.19}$$

The traditional view of molecular bonds is that they are due to an increased probability of finding electrons between two nuclei, as compared to a sum of the contributions of the pure atomic orbitals. The canonical MOs are delocalized over the whole molecule and do not readily reflect this. There is, furthermore, little similarity between MOs for systems which by chemical measures should be similar, such as a series of alkanes. The canonical MOs therefore do not reflect the concept of functional groups.

One of the goals of *Localized Molecular Orbitals* (LMO) is to derive MOs which are approximately constant between structurally similar units in different molecules. A set of LMOs may be defined by optimizing the expectation value of an two-electron operator $\boldsymbol{\Omega}$.[17] The expectation value depends on the u_{ij} parameters in eq. (9.19), i.e. this is again a function optimization problem (Chapter 14). In practice, however, the localization is normally done by performing a series of 2×2 orbital rotations, as described in Chapter 13.

$$\langle\boldsymbol{\Omega}\rangle = \sum_{i=1}^{N} \langle \phi_i'\phi_i'|\boldsymbol{\Omega}|\phi_i'\phi_i'\rangle \tag{9.20}$$

Since all observable properties depend only on the total electron density, and not the individual MOs, there is no unique choice for $\boldsymbol{\Omega}$.

The *Boys* localization scheme[18] uses the square of the distance between two electrons as the operator, and minimizes the expectation value.

$$\langle\boldsymbol{\Omega}\rangle_{\text{Boys}} = \sum_{i=1}^{N} \left\langle \phi_i'\phi_i' \left| (\mathbf{r}_1 - \mathbf{r}_2)^2 \right| \phi_i'\phi_i' \right\rangle \tag{9.21}$$

This corresponds to determining a set of LMOs which minimize the spatial extent, i.e. they are as compact as possible.

The *Edmiston–Ruedenberg* localization scheme[19] uses the inverse of the distance between two electrons as the operator, and maximizes the expectation value.

$$\langle \mathbf{\Omega} \rangle_{\text{ER}} = \sum_{i=1}^{N} \left\langle \phi_i' \phi_i' \left| \frac{1}{|\mathbf{r}_1 - \mathbf{r}_2|} \right| \phi_i' \phi_i' \right\rangle \tag{9.22}$$

This corresponds to determining a set of LMOs which maximize the self-repulsion energy.

The *von Niessen* localization scheme[20] uses the delta function of the distance between two electrons as the operator, and maximizes the expectation value.

$$\langle \mathbf{\Omega} \rangle_{\text{VN}} = \sum_{i=1}^{N} \langle \phi_i' \phi_i' | \delta(\mathbf{r}_1 - \mathbf{r}_2) | \phi_i' \phi_i' \tag{9.23}$$

This corresponds to determining a set of LMOs which maximize the "self-charge".

The *Pipek–Mezey* localization scheme corresponds to maximizing the sum of the Mulliken atomic charges.[17] The contribution from the ith MO to atom A is given as (eq. (9.5))

$$\rho_i(\text{A}) = \sum_{\alpha \varepsilon \text{A}}^{\text{AO}} \sum_{\beta}^{\text{AO}} c_{\alpha i} c_{\beta i} S_{\alpha \beta} \tag{9.24}$$

The function to be maximized is

$$\langle \mathbf{\Omega} \rangle_{\text{PM}} = \sum_{\text{A}=1}^{\text{Atoms}} [\rho_i(\text{A})]^2 \tag{9.25}$$

There is little experience with the von Niessen method, but for most molecules the remaining three schemes tend to give very similar LMOs. The main exception is systems containing both σ- and π-bonds, such as ethylene. The Pipek–Mezey procedure preserves the σ/π-separation, while the Edmiston–Ruedenberg and Boys schemes produce bent "banana" bonds. Similarly, for planar molecules which contain lone pairs (like water), the Pipek–Mezey method produces one in-plane σ-type lone pair and one out-of-plane π-type lone pair, while the Edmiston–Ruedenberg and Boys schemes produce two equivalent "rabbit ear" lone pairs.

The operator used in the Boys localization scheme can be expanded as

$$(\mathbf{r}_1 - \mathbf{r}_2)^2 = \mathbf{r}_1^2 + \mathbf{r}_2^2 - 2\mathbf{r}_1 \cdot \mathbf{r}_2 \tag{9.26}$$

and it may be shown that minimization of $\langle \mathbf{\Omega} \rangle_{\text{Boys}}$ is equivalent to maximizing the following functional.

$$\langle \mathbf{\Omega}' \rangle_{\text{Boys}} = \sum_{i>j}^{N} \left[\langle \phi_i' | \mathbf{r} | \phi_i' \rangle - \langle \phi_j' | \mathbf{r} | \phi_j' \rangle \right]^2 \tag{9.27}$$

This corresponds to maximizing the distance between centroids of the orbitals. The dipole integrals in the molecular basis may be obtained from the corresponding AO integrals.

$$\langle \phi_i | \mathbf{r} | \phi_j \rangle = \sum_{\alpha} c_{\alpha i} \left(\sum_{\beta} c_{\beta j} \langle \chi_\alpha | \mathbf{r} | \chi_\beta \rangle \right) \tag{9.28}$$

This is a process which increases as the third power of the size of the basis set, M^3, and the optimization of the $\langle \Omega' \rangle_{\text{Boys}}$ function is therefore an M^3 method. The Edmiston–Ruedenberg localization, however, requires standard two-electron integrals over MOs, analogous to those used in electron correlation methods (eq. (4.11)), and it therefore involves a computational effort which increases as M^5. Since only integrals involving occupied MOs are needed, the transformation is not particularly time-consuming for reasonably sized systems,[21] but the Edmiston–Ruedenberg method will ultimately require a significant effort for large systems. The von Niessen method may also be shown to involve a computational effort which increases as M^5. The Pipek–Mezey charge localization only involves overlap integrals between basis functions, and consequently has an M^3 computational dependence.

Although the localization by energy criteria (Edmiston–Ruedenberg) may be considered more "fundamental" than one based on distance (Boys) or atomic charge (Pipek–Mezey), the difference in computational effort means that the Boys or Pipek–Mezey procedures are often used in practice, especially since there is normally little difference in the shape of the final LMOs.

Localized molecular orbitals are generally found to reflect the usually picture of bonding, i.e. they are localized between two nuclei, or in some cases, like diborane, extended over three nuclei. Although they indicate which atoms are bonded, they do not directly give any information about the strength of the bonds. Furthermore, localizing a set of MOs basicly corresponds to determining orbitals containing electron pairs. In structures with delocalized electrons (e.g. transition structures or conjugated systems) it may be difficult to achieve a proper localization of the MOs.

9.5 Natural Orbitals

The electron density calculated from a wave function is given as the square of the function, $|\Psi|^2 = \Psi * \Psi$. The reduced density matrix of order k, γ_k, is defined as[22]

$$\gamma_k(\mathbf{r}_1', \mathbf{r}_2', \ldots, \mathbf{r}_k', \mathbf{r}_1, \mathbf{r}_2, \ldots, \mathbf{r}_k)$$
$$= \binom{N}{k} \int \Psi^*(\mathbf{r}_1, \mathbf{r}_2, \ldots, \mathbf{r}_k, \mathbf{r}_{k+1}, \ldots, \mathbf{r}_N) \Psi(\mathbf{r}_1', \mathbf{r}_2', \ldots, \mathbf{r}_k', \mathbf{r}_{k+1}, \ldots, \mathbf{r}_N) \, d\mathbf{r}_{k+1} \ldots d\mathbf{r}_N$$

$$(9.29)$$

Note that the coordinates for Ψ^* and Ψ are different. Of special importance in electronic structure theory are the first- and second-order reduced density matrices, $\gamma_1(\mathbf{r}_1' \mathbf{r}_1)$ and $\gamma_2(\mathbf{r}_1', \mathbf{r}_2', \mathbf{r}_1, \mathbf{r}_2)$, since the Hamilton operator only contains one- and two-electron operators. Integrating the first-order density matrix over coordinate "1" yields the number of electrons, N, while the integral of the second-order density matrix over coordinates "1" and "2" is $N(N-1)/2$, i.e. the unique number of electron pairs. The first-order density matrix may be diagonalized, and the corresponding eigenvectors and -values are called *Natural Orbitals* (NO) and *Occupation Numbers*. For a single determinant RHF wave function the first-order density matrix is identical to the density matrix used in the formation of the Fock matrix (eq. (3.51)), and the natural orbitals have occupation numbers of either 0 or 2 (exactly). Since there are $N/2$ orbitals with degenerate eigenvalues of 2, the HF natural orbitals are not uniquely defined; they may be any linear combination of the canonical orbitals. For a multi-determinant wave function (MCSCF, CI, MP or CC) the occupation numbers may assume fractional values between 0 and 2. UHF wave functions (when different from RHF) will in general also give fractional occupations.

The original definition of natural orbitals was in terms of the density matrix from a full CI wave function, i.e. the best possible for a given basis set.[23] In that case the natural orbitals have the significance that they provide the fastest convergence. In order to obtain the lowest energy for a CI expansion using only a limited set of orbitals, the natural orbitals with the largest occupation numbers should be used.

When natural orbitals are determined from a wave function which only includes a limited amount of electron correlation (i.e. not full CI), the convergence property is not rigorously guaranteed, but since most practical methods recover 80–90% of the total electron correlation, the occupation numbers provide a good guideline for how important a given orbital is. This is the reason why natural orbitals are often used for evaluating which orbitals should be included in an MCSCF wave function (Section 4.6).

9.6 Natural Atomic Orbital and Natural Bond Orbital Analysis

The concept of natural orbitals may be used for distributing electrons into atomic and molecular orbitals, and thereby for deriving atomic charges and molecular bonds. The idea in the *Natural Atomic Orbital* (NAO) and *Natural Bond Orbital* (NBO) analysis developed by F. Weinholt and co-workers[24] is to use the one-electron density matrix for defining the shape of the atomic orbitals in the molecular environment, and derive molecular bonds from electron density between atoms.

Let us assume that the basis functions have been arranged so that all orbitals located on center A are before those on centre B, which are before those on centre C etc.

$$\chi_1^A, \chi_2^A, \chi_3^A, \ldots, \chi_k^B, \chi_{k+1}^B, \chi_{k+2}^B, \ldots, \chi_n^C, \chi_{n+1}^C, \chi_{n+1}^C, \ldots \qquad (9.30)$$

The density matrix can be written in terms of blocks of basis functions belonging to a specific centre as

$$\mathbf{D} = \begin{pmatrix} \mathbf{D}^{AA} & \mathbf{D}^{AB} & \mathbf{D}^{AC} & \cdots \\ \mathbf{D}^{AB} & \mathbf{D}^{BB} & \mathbf{D}^{BC} & \cdots \\ \mathbf{D}^{AC} & \mathbf{D}^{BC} & \mathbf{D}^{CC} & \cdots \\ \cdots\cdots\cdots\cdots\cdots\cdots\cdots\cdots\cdots \end{pmatrix} \qquad (9.31)$$

The Natural Atomic Orbitals for atom A in the molecular environment may be defined as those which diagonalize the $\mathbf{D}^{AA}$ block, NAOs for atom B as those which diagonalize the $\mathbf{D}^{BB}$ block etc. These NAOs will in general not be orthogonal, and the orbital occupation numbers will therefore not sum to the total number of electrons. To achieve a well-defined division of the electrons, the orbitals should be orthogonalized.

The NAOs will normally resemble the pure atomic orbitals (as calculated for an isolated atom), and may be divided into a "natural minimal basis" (corresponding to the occupied atomic orbitals for the isolated atom), and a remaining set of natural "Rydberg" orbitals based on the magnitude of the occupation numbers. The minimal set of NAOs will normally be strongly occupied (i.e. having occupation numbers significantly different from zero), while the Rydberg NAOs usually will be weakly occupied (i.e. having occupation numbers close zero). There are as many NAOs as the size of the atomic basis set, and the number of Rydberg NAOs thus increases as the basis set is enlarged. It is therefore desirable that the orthogonalization procedure preserves the form of the strongly occupied orbitals as much as possible, which is achieved by using an occupancy-weighted orthogonalizing matrix. If all orbital occupancies are identical to 2 or 0, the orthogonalization is identical to the Löwdin method (Eq. (9.8)). The procedure is as follows

(1) Each of the atomic blocks in the density matrix is diagonalized to produce a set of non-orthogonal NAOs, often denoted "pre-NAOs".
(2) The strongly occupied pre-NAOs for each centre are made orthogonal to all the strongly occupied pre-NAOs on the other centres by an occupancy-weighted procedure.
(3) The weakly occupied pre-NAOs on each centre are made orthogonal to the strongly occupied NAOs on the same centre by a standard Gram–Schmidt orthogonalization.
(4) The weakly occupied NAOs are made orthogonal to all the weakly occupied NAOs on the other centres by an occupancy-weighted procedure.

The final set of orthogonal orbitals are simply denoted NAOs, and the diagonal elements of the density matrix in this basis are the orbital populations. Summing all contributions from orbitals belonging to a specific centre produces the atomic charge. It is usually found that the natural minimal NAOs contribute $99 + \%$ of the electron density, and they form a very compact representation of the wave function in terms of atomic orbitals. The further advantage of the NAOs is that they are defined from the density matrix, guaranteeing that the electron occupation is between 0 and 2, and that they converge to well-defined values as the size of the basis set is increased. Furthermore, the analysis may also be performed for correlated wave functions.

Once the density matrix has been transformed to the NAO basis, bonds between atoms may be identified from the off-diagonal blocks. The procedure involves the following steps.

(1) NAOs for an atomic block in the density matrix which have occupation numbers very close to 2 (say >1.999) are identified as core orbitals. Their contributions to the density matrix are removed.

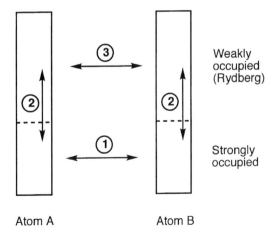

Figure 9.4 Illustration of the orthogonalization order in the NAO analysis

(2) NAOs for an atomic block in the density matrix which have large occupancy numbers (say >1.90) are identified as lone pair orbitals. Their contributions to the density matrix are also removed.

(3) Each pair of atoms (AB, AC, BC,...) is now considered, and the two-by-two sub-blocks of the density matrix (with the core and lone pair contributions removed) are diagonalized. Natural bond orbitals are identified as eigenvectors which have large eigenvalues (occupation numbers larger than say 1.90).

(4) If an insufficient number of NBOs are generated by the above procedure (sum of occupation numbers for core, lone pair and bond orbitals significantly less than the number of electrons), the criteria for accepting a NBO may be gradually lowered until a sufficiently large fraction of the electrons has been assigned to bonds. Alternatively, a search may be initiated for three-centre bonds. The contributions to the density matrix from all diatomic bonds are removed, and all three-by-three sub-blocks are diagonalized. Such three-centre bonds are quite rare, boron systems being the most notable exception.

Once NBOs have been identified, they may be written as linear combinations of the NAOs, forming a localized picture of the "atomic" orbitals involved in the bonding.

9.7 Computational Considerations

Population analyses based on basis functions (such as Mulliken or Löwdin) require insignificant computational time. The NAO analysis involves only matrix diagonalization of small subsets of the density matrix, and also requires a negligible amount of computer time, although it is more involved than a Mulliken or Löwdin analysis. The determination of ESP fitted charges requires an evaluation of the potential at many (often several thousand) points in space, and a subsequent solution of a matrix equation for minimizing the least squares expression. For large systems this is no longer completely trivial in terms of computer time. The AIM population analysis requires a complete topological analysis of the electron density surface, and a subsequent numerical integration of the atomic basins. For medium sized systems and quality wave functions, this analysis may be more time-consuming than determining the wave function itself. GAPT charges require calculation of the second derivative of the wave function, and this is computationally demanding, especially for large molecules and/or correlated wave functions. There is little doubt that these computational considerations partly explain the popularity of especially the Mulliken population analysis, despite its well-known shortcomings. For analysis purposes the NAO procedure is an attractive method, but for modelling purposes (i.e. force field charges) ESP charges are clearly the logical choice.

9.8 Examples

The tables below give some examples of atomic charges and bond orders calculated by various methods as a function of the basis set at the HF level of theory. It is evident that the Mulliken and Löwdin methods do not converge as the basis set is increased, and the values in general behave unpredictably. Especially the presence of diffuse functions lead to absurd behaviours, as the aug-cc-pVXZ basis sets illustrate for CH_4. Note also that

Table 9.1 Atomic charges for carbon in CH_4

Basis	Mulliken	Löwdin	ESP Fit	NAO	AIM
STO-3G	−0.26	−0.15	−0.38	−0.21	+0.25
3-21G	−0.80	−0.38	−0.45	−0.89	−0.01
6-31G(d,p)	−0.47	−0.43	−0.36	−0.88	+0.26
6-311G(2d,2p)	−0.14	−0.13	−0.36	−0.69	+0.19
6-311++G(2d,2p)	−0.18	−0.20	−0.36	−0.71	+0.19
cc-pVDZ	−0.13	−0.76	−0.31	−0.79	+0.32
cc-pVTZ	−0.37	−0.21	−0.35	−0.72	
cc-pVQZ	−0.27	−0.07	−0.36		
aug-cc-pVDZ	+0.63	−0.43	−0.35	−0.77	+0.33
aug-cc-pVTZ	−1.20	+0.05	−0.37	−0.72	

Table 9.2 Atomic charges for oxygen in H_2O

Basis	Mulliken	Löwdin	ESP Fit	NAO	AIM
STO-3G	−0.39	−0.27	−0.65	−0.41	−0.89
3-21G	−0.74	−0.46	−0.90	−0.87	−0.93
6-31G(d,p)	−0.67	−0.44	−0.81	−0.97	−1.24
6-311G(2d,2p)	−0.52	−0.00	−0.74	−0.91	−1.24
6-311++G(2d,2p)	−0.47	−0.12	−0.76	−0.93	−1.25
cc-pVDZ	−0.29	−0.58	−0.76	−0.91	−1.27
cc-pVTZ	−0.48	−0.11	−0.75	−0.92	
cc-pVQZ	−0.51	+0.23	−0.75		
aug-cc-pVDZ	−0.26	−0.39	−0.74	−0.96	−1.26
aug-cc-pVTZ	−0.41	+0.12	−0.74	−0.93	

Table 9.3 Bond orders for CH_4 and H_2O

Basis	CH_4, **DS**	CH_4, AIM	H_2O, **DS**	H_2O, AIM
STO-3G	0.99	1.00	0.95	0.81
3-21G	0.93	1.00	0.83	0.80
6-31G(d,p)	0.98	1.00	0.89	0.62
6-311G(2d,2p)	1.00	1.00	0.99	0.63
6-311++G(2d,2p)	1.00	1.00	0.96	0.62
cc-pVDZ	1.00	0.99	1.03	0.60
cc-pVTZ	0.98		1.01	
cc-pVQZ	0.98		1.01	
aug-cc-pVDZ	0.91	0.99	1.11	0.62
aug-cc-pVTZ	0.86		1.04	

The bond order denoted **DS** is calculated from eq. (9.11).

for sufficiently large basis sets, the charge on oxygen in H_2O can be calculated to be <u>less</u> than on carbon in CH_4! The ESP fitted charges, as well as those derived by the NAO and AIM procedures, attain well-defined values as the basis set is enlarged, and they are rather insensitive to the presence of diffuse functions. The charges assigned by these three methods, however, differ significantly, e.g. the carbon in CH_4 may be assigned charges between +0.2 (AIM) and −0.7 (NAO).

References

1. S. M. Bachrach, *Rev. Comput. Chem.*, **5** (1994), 171.
2. R. S. Mulliken, *J. Chem. Phys.*, **36** (1962), 3428.
3. P.-O. Löwdin, *Adv. Quantum. Chem.*, **5** (1970), 185.
4. L. Pauling, *J. Am. Chem. Soc.*, **69** (1947), 542.
5. I. Mayer, *Chem. Phys. Lett.*, **97** (1983), 270.
6. D. E. Williams, *Rev. Comput. Chem.*, **2** (1991), 219.
7. C. I. Bayly, P. Cieplak, W. D. Cornell and P. A. Kollman, *J. Phys. Chem.*, **97** (1993), 10269.
8. C. A. Reynolds, J. W. Essex and W. G. Richards, *J. Am. Chem. Soc.*, **114** (1992), 9075.
9. M. M. Francl, C. Carey, L. E. Chilian and D. M. Gange, *J. Comput. Chem.*, **17** (1996), 367.
10. C. Aleman, M. Orozro and F. J. Luque, *Chem. Phys.*, **189** (1994), 573.
11. U. Koch and E. Egert, *J. Comput. Chem.*, **16** (1995), 937.
12. A. J. Stone and M. Alderton, *Mol. Phys.*, **56** (1985), 1047.
13. R. F. W. Bader, *Atoms in Molecules*, Clarendon Press, Oxford 1990; R. F. W. Bader, *Chem. Rev.*, **91** (1991), 893.
14. C. L. Perrin, *J. Am. Chem. Soc.*, **113** (1991), 2865.
15. C. Mei, K. E. Edgecombe, V. H. Smith Jr, and A. Heilingbrunner, *Int. J. Quantum. Chem.*, **48** (1993), 287.
16. J. Cioslowski, *J. Am. Chem. Soc.*, **111** (1989), 8333.
17. J. Pipek and P. G. Mezey, *Chem. Phys.*, **90** (1989), 4916.
18. S. F. Boys, *Rev. Mod. Phys.*, **32** (1960), 296.
19. C. Edmiston and K. Ruedenberg, *J. Chem. Phys.*, **43** (1965), S97.
20. W. von Niessen, *J. Chem. Phys.*, **56** (1972), 4290.
21. R. C. Raffeneti, K. Ruedenberg, C. L. Jansen and H. F. Schaefer, *Theo. Chim. Acta*, **86** (1992), 149.
22. R. G. Parr and W. Yang *Density Functional Theory*, Oxford University Press, 1989.
23. P.-O. Löwdin, *Phys. Rev.*, **97** (1955), 1474.
24. A. E. Reed, L. A. Curtiss and F. Weinholt, *Chem. Rev.*, **88** (1988), 899.

10 Molecular Properties

The focus in Chapters 3 and 4 has been on determining the wave function and its energy at a given geometry in the absence of external fields (electric or magnetic). While relative energies are certainly of interest, there are many other molecular properties that can be calculated by electronic structure methods. Most properties may be defined as the response of a wave function, an energy or expectation value of an operator to a perturbation, where the perturbation may be any kind of operator not present in the Hamiltonian used for solving the Schrödinger equation. It may for example be terms arising in a relativistic treatment (e.g. spin–orbit interactions, see Section 8.2), which can be added as perturbations in non-relativistic theory. It may also be external fields (electric or magnetic) or an internal perturbation, such as a nuclear spin. If we also include "perturbations" like adding or removing an electron, electron affinities and ionization potentials can also be included in this definition. There are a few remaining properties which cannot easily be characterized as a response to a perturbation, most notably transition moments, which determine absorption intensities. These depend on matrix elements between two different wave functions.

We will here consider four types of perturbation: external electric ($\mathbf{F}$) or magnetic field ($\mathbf{B}$), nuclear magnetic moment (nuclear spin, $\mathbf{I}$) and a change in the nuclear geometry ($\mathbf{R}$). The first two, electric and magnetic fields, may either be *time independent*, leading to *static* properties, or *time dependent*, leading to *dynamic* properties. Time-dependent fields are usually associated with electromagnetic radiation characterized by a frequency. Static properties may be considered as the limiting case of dynamic properties when the frequency goes to zero. We will only consider the static case here, and again concentrate on properties of a single molecule for a fixed geometry. A direct comparison with (gas phase) experimental macroscopic quantities may be done by proper averaging over for example vibrational and rotational states. We will furthermore concentrate on the electronic contribution to properties, the corresponding nuclear contribution (if present) is normally trivial to calculate as it is independent of the wave function.[1]

There are three main methods for calculating the effect of a perturbation: derivative techniques, perturbation theory and propagator methods. The former two are closely related while propagator methods are somewhat different, and will be discussed separately.

The derivative formulation is perhaps the easiest to understand. In this case the energy is expanded in a Taylor series in the perturbation strength λ.

$$E(\lambda) = E(0) + \frac{\partial E}{\partial \lambda}\lambda + \frac{1}{2}\frac{\partial^2 E}{\partial \lambda^2}\lambda^2 + \frac{1}{6}\frac{\partial^3 E}{\partial \lambda^3}\lambda^3 + \ldots \qquad (10.1)$$

The nth-order property is the nth-order derivative of the energy, $\partial^n E/\partial \lambda^n$ (the factor $1/n!$ may or may not be included in the property). Note that the perturbation is usually a vector, and the first derivative is therefore also a vector, the second derivative a matrix, the third derivative a (third-order) tensor etc.

10.1 Examples

10.1.1 External Electric Field

The interaction of an electronic charge distribution $\rho(\mathbf{r})$ with an electric potential $V(\mathbf{r})$ gives an energy correction.

$$E = \int \rho(\mathbf{r})V(\mathbf{r})\,d\mathbf{r} \qquad (10.2)$$

Since the electric field $(\mathbf{F} = -\partial V/\partial \mathbf{r})$ normally is fairly uniform at the molecular level, it is useful to write E as a multipole expansion.

$$E = qV - \mathbf{\mu}\,\mathbf{F} - \tfrac{1}{2}\mathbf{Q}\mathbf{F}' - \ldots \qquad (10.3)$$

Here q is the net charge (monopole), $\mathbf{\mu}$ is the (electric) dipole moment, $\mathbf{Q}$ is the quadrupole moment, and $\mathbf{F}$ and $\mathbf{F}'$ are the field and field gradient $(\partial \mathbf{F}/\partial \mathbf{r})$, respectively. The dipole moment and electric field are vectors, and the $\mathbf{\mu}\mathbf{F}$ term should be interpreted as the dot product $(\mathbf{\mu} \cdot \mathbf{F} = \mu_x F_x + \mu_y F_y + \mu_z F_z)$. The quadrupole moment and field gradient are 3×3 matrices, and $\mathbf{Q}\mathbf{F}'$ is the sum of all product terms. For an external field it is rarely necessary to go beyond the quadrupole term, but for molecular interactions the octupole moment may also be important (it is for example the first non-vanishing moment for spherical molecules like CH_4).

In the absence of an external field, the unperturbed dipole and quadrupole moments may be calculated from the electronic wave function as simple expectation values.

$$\begin{aligned} \mathbf{\mu} &= \langle \Psi | \mathbf{r} | \Psi \rangle \\ \mathbf{Q} &= \langle \Psi | \mathbf{r}\mathbf{r}^t | \Psi \rangle \end{aligned} \qquad (10.4)$$

The superscript t denotes a transposition of the $\mathbf{r}$-vector, i.e. converting it from a column to a row vector. The $\mathbf{r}\mathbf{r}^t$ notation for the quadrupole moment therefore indicates a 3×3 matrix containing the products of the x-, y- and z-coordinates, e.g. the Q_{xy} component is calculated as the expectation value of xy.

The presence of a field influences the wave function, and leads to induced dipole, quadrupole etc. moments. For the dipole moment this may be written as

$$\mathbf{\mu} = \mathbf{\mu}_0 + \mathbf{\alpha}\mathbf{F} + \tfrac{1}{2}\mathbf{\beta}\mathbf{F}^2 + \tfrac{1}{6}\mathbf{\gamma}\mathbf{F}^3 + \ldots \qquad (10.5)$$

where $\mathbf{\mu}_0$ is the permanent dipole moment, $\mathbf{\alpha}$ is the (dipole) *polarizability*, $\mathbf{\beta}$ is the (first) *hyperpolarizability*, $\mathbf{\gamma}$ is the second hyperpolarizability etc. The quadrupole moment may similarly be expanded in the field by means of a quadrupole polarizability, hyperpolarizability etc.

For a homogeneous field (i.e. the field gradient and higher derivatives are zero), the total energy of a neutral molecule may be written as a Taylor expansion.

$$E(\mathbf{F}) = E(0) + \frac{\partial E}{\partial \mathbf{F}}\mathbf{F} + \frac{1}{2}\frac{\partial^2 E}{\partial \mathbf{F}^2}\mathbf{F}^2 + \frac{1}{6}\frac{\partial^3 E}{\partial \mathbf{F}^3}\mathbf{F}^3 + \frac{1}{24}\frac{\partial^4 E}{\partial \mathbf{F}^4}\mathbf{F}^4 + \cdots \qquad (19.6)$$

According to eq. (10.3) we also have $E = -\boldsymbol{\mu} \cdot \mathbf{F}$, where $\boldsymbol{\mu}$ is given by the expression in eq. (10.5). Carrying out the differentiation in eq. (10.6) shows that the first derivative is the (permanent) dipole moment $\boldsymbol{\mu}_0$, the second derivative is the polarizability $\boldsymbol{\alpha}$, the third derivative is the hyperpolarizability $\boldsymbol{\beta}$ etc.

$$E(\mathbf{F}) = E(0) - \boldsymbol{\mu}_0\mathbf{F} - \tfrac{1}{2}\boldsymbol{\alpha}\mathbf{F}^2 - \tfrac{1}{6}\boldsymbol{\beta}\mathbf{F}^3 - \tfrac{1}{24}\boldsymbol{\gamma}\mathbf{F}^4 - \cdots \qquad (10.7)$$

Note that the constant factor in front of the higher-order terms differs between eqs. (10.3) and (10.5)/(10.7).

10.1.2 *External Magnetic Field*

The interaction with a magnetic field may similarly be written in term of magnetic dipole, quadrupole etc. moments (there is no magnetic monopole, corresponding to electric charge). Since the magnetic interaction is substantially smaller in magnitude than the electric, only the dipole term is normally considered.

$$E = -\mathbf{mB} - \cdots \qquad (10.8)$$

The dipole moment $\mathbf{m}$ depends on the total angular momentum, which may be written in terms of the orbital angular moment operator $\mathbf{L}$ and the total electron spin $\mathbf{S}$.

$$\mathbf{m} = -\tfrac{1}{2}\langle \Psi | \mathbf{L} + g_e\mathbf{S} | \Psi \rangle$$
$$\mathbf{L} = (\mathbf{r} - \mathbf{R}_G) \times \mathbf{p} \qquad (10.9)$$

Here $\mathbf{R}_G$ is the gauge origin (discussed in Section 10.7), and the electronic g_e-factor is a constant approximately equal to 2.0023. The orbital part of the permanent magnetic dipole moment will be zero for all non-degenerate wave functions (i.e. belonging to A, B or Σ representations), since the $\mathbf{L}$ operator is purely imaginary ($\mathbf{p} = -i\mathbf{V}$) and the wave function in such cases is real. Similarly, only degenerate open-shell states (doublet, triplet, etc.) have the spin part of the magnetic dipole moment different from zero. Since the large majority of stable molecules are closed-shell singlets, it follows that permanent magnetic dipole moments ($\mathbf{m}_0$) are quite rare. The presence of a field, however, may induce a magnetic dipole moment, which again can be defined in terms of a Taylor expansion of the total energy (μ_0 is the vacuum permeability, equal to $4\pi/c^2$ in atomic units, Appendix D).

$$E(\mathbf{B}) = E(0) - \frac{\partial E}{\partial \mathbf{B}}\mathbf{B} - \frac{1}{2}\frac{\partial^2 E}{\partial \mathbf{B}^2}\mathbf{B}^2 - \cdots$$
$$E(\mathbf{B}) = E(0) - \mathbf{m}_0\mathbf{B} - \frac{1}{2\mu_0}\xi\mathbf{B}^2 - \cdots \qquad (10.10)$$

The second derivative is the *magnetizability* ξ (the corresponding macroscopic quantity is called the magnetic *susceptibility* χ).

10.1.3 Internal Magnetic Moment

The perturbation may also be an internal magnetic moment $\mathbf{I}$, arising from a nuclear spin ($\mathbf{I}$ is here taken to include the proportionality constants between spin and moment, i.e. it includes the $g_A\mu_N$ factor from Section 8.2).

$$E(\mathbf{I}_1, \mathbf{I}_2, \ldots) = E(\mathbf{0}) + \frac{\partial E}{\partial \mathbf{I}_1}\mathbf{I}_1 + \frac{1}{2}\frac{\partial^2 E}{\partial \mathbf{I}_1 \partial \mathbf{I}_2}\mathbf{I}_1\mathbf{I}_2 + \cdots$$

$$E(\mathbf{I}_1, \mathbf{I}_2, \ldots) = E(\mathbf{0}) + \mathbf{g}\mathbf{I}_1 + h\mathbf{J}\mathbf{I}_1\mathbf{I}_2 + \cdots \tag{10.11}$$

The first derivative is the hyperfine coupling constant $\mathbf{g}$ (as measured by ESR), the second derivative with respect to two different nuclear spins is the NMR coupling constant, $\mathbf{J}$ (Planck's constant appears owing to the convention of reporting coupling constants in Hertz, and the factor of 1/2 disappears since we implicitly only consider distinct pairs of nuclei).

10.1.4 Geometry Change

The change in energy for moving a nucleus can also be written as a Taylor expansion.

$$E(\mathbf{R}) = E(\mathbf{R}_0) + \frac{\partial E}{\partial \mathbf{R}}(\mathbf{R} - \mathbf{R}_0) + \frac{1}{2}\frac{\partial^2 E}{\partial \mathbf{R}^2}(\mathbf{R} - \mathbf{R}_0)^2 + \frac{1}{6}\frac{\partial^3 E}{\partial \mathbf{R}^3}(\mathbf{R} - \mathbf{R}_0)^3 + \cdots$$

$$E(\mathbf{R}) = E(\mathbf{R}_0) + \mathbf{g}(\mathbf{R} - \mathbf{R}_0) + \frac{1}{2}\mathbf{H}(\mathbf{R} - \mathbf{R}_0)^2 + \frac{1}{6}\mathbf{K}(\mathbf{R} - \mathbf{R}_0)^3 + \cdots$$

$$\tag{10.12}$$

The first derivative is the gradient $\mathbf{g}$, the second derivative is the force constant (Hessian) $\mathbf{H}$, the third derivative is the anharmonicity $\mathbf{K}$ etc. If the $\mathbf{R}_0$ geometry is a stationary point ($\mathbf{g} = \mathbf{0}$) the force constant matrix may be used for evaluating harmonic vibrational frequencies and normal coordinates, $\mathbf{q}$, as discussed in Section 13.1. If higher-order terms are included in the expansion, it is possible to determine also anharmonic frequencies and phenomena such as Fermi resonance.

10.1.5 Mixed Derivatives

Mixed derivatives refer to cross terms if the energy is expanded in more than one perturbation. There are many such mixed derivatives which translate into molecular properties, below are a few examples.

The change in the dipole moment with respect to a geometry displacement along a normal coordinate is approximately proportional to the intensity of an IR absorption. In the so-called double harmonic approximation (terminating the expansion at first order in the electric field and geometry), the intensity is (except for some constants)

$$\text{IR Intensity} \quad \propto \left(\frac{\partial \boldsymbol{\mu}}{\partial \mathbf{q}}\right)^2 \propto \left(\frac{\partial^2 E}{\partial \mathbf{R} \partial \mathbf{F}}\right)^2 \tag{10.13}$$

In the double harmonic approximation, only fundamental bands can have an intensity different from zero. Including higher-order terms in the expansion allows calculation of intensities of overtone bands, as well as adding contributions to the fundamental bands.

The intensity of a Raman band in the harmonic approximation is given by the derivative of the polarizability with respect to a normal coordinate.

$$\text{Raman Intensity} \quad \propto \left(\frac{\partial\boldsymbol{\alpha}}{\partial\mathbf{q}}\right)^2 \propto \left(\frac{\partial^3 E}{\partial\mathbf{R}\partial\mathbf{F}^2}\right)^2 \tag{10.14}$$

The mixed derivative of an external and an internal magnetic field (nuclear spin) is the NMR shielding constant, $\boldsymbol{\sigma}$.

$$\text{NMR Shielding} \quad \propto \left(\frac{\partial^2 E}{\partial\mathbf{B}\partial\mathbf{I}}\right) \tag{10.15}$$

Table 10.1 below gives some examples of properties which may be calculated from derivatives of a certain order with respect to the above four perturbations.

$$\text{Property} \quad \propto \frac{\partial^{n_F+n_B+n_I+n_R} E}{\partial\mathbf{F}^{n_F}\partial\mathbf{B}^{n_B}\partial\mathbf{I}^{n_I}\partial\mathbf{R}^{n_R}} \tag{10.16}$$

All of these properties can be calculated at different levels of sophistication (electron correlation and basis sets). It should be noted that dynamic properties, where one or more of the external electric and/or magnetic fields are time-dependent, may involve different frequencies.

Table 10.1 Properties which may be calculated from derivatives of the energy

n_F	n_B	n_I	n_R	Property
0	0	0	0	Energy
1	0	0	0	Electric dipole moment
0	1	0	0	Magnetic dipole moment
0	0	1	0	Hyperfine coupling constant
0	0	0	1	Energy gradient
2	0	0	0	Electric polarizability
0	2	0	0	Magnetizability
0	0	2	0	Spin–spin coupling (for different nuclei)
0	0	0	2	Harmonic vibrational frequencies
1	0	0	1	Infra-red absorption intensities
1	1	0	0	Circular dichroism
0	1	1	0	Nuclear magnetic shielding
3	0	0	0	(first) Electric hyperpolarizability
0	3	0	0	(first) Hypermagnetizability
0	0	0	3	(cubic) Anharmonic corrections to vibrational frequencies
2	0	0	1	Raman intensities
2	1	0	0	Magnetic circular dichroism (Faraday effect)
1	0	0	2	Infra-red intensities for overtone and combination bands
4	0	0	0	(second) Electric hyperpolarizability
0	4	0	0	(second) Hypermagnetizability
0	0	0	4	(quartic) Anharmonic corrections to vibrational frequencies
2	0	0	2	Raman intensities for overtone and combination bands
2	2	0	0	Cotton–Mutton effect

10.2 Perturbation Methods

Let us first look at some general features. The presence of a perturbation will give rise to extra terms in the Hamiltonian; we will in the following need to consider operators which are both linear and quadratic.

$$\mathbf{H} = \mathbf{H}_0 + \lambda\mathbf{P}_1 + \lambda^2\mathbf{P}_2 \qquad (10.17)$$

$\mathbf{H}_0$ is the normal electronic Hamilton operator, and the perturbations are described by the operators $\mathbf{P}_1$ and $\mathbf{P}_2$, with λ determining the strength. Based on an expansion in <u>exact</u> wave functions, Rayleigh–Schrödinger perturbation theory (section 4.8) gives the first- and second-order energy corrections.

$$W_1 = \lambda\langle\Psi_0|\mathbf{P}_1|\Psi_0\rangle$$

$$W_2 = \lambda^2\left[\langle\Psi_0|\mathbf{P}_2|\Psi_0\rangle + \sum_{i\neq0}\frac{\langle\Psi_0|\mathbf{P}_1|\Psi_i\rangle\langle\Psi_i|\mathbf{P}_1|\Psi_0\rangle}{E_0 - E_i}\right] \qquad (10.18)$$

The first-order term is identical to eq. (4.36), while the second-order equation corresponds to eq. (4.38) with an additional term involving the expectation value of $\mathbf{P}_2$ over the unperturbed wave function. The first-order energy correction is identified with the first-order property, the second-order correction with the second-order property, etc. Although these expressions only hold for exact wave functions, they may be used also for approximate wave functions. The methodology of how the general expressions can be reduced to formulas involving molecular integrals is analogous to that used in Section 3.3. The first-order term is simply the expectation value of the perturbation operator over the unperturbed wave function, and is easy to calculate. The second-order property, however, involves a sum over all excited states. In some cases, mainly associated with semi-empirical methods, second-order properties are evaluated directly from eq. (10.18), known as *Sum Over States* (SOS) methods. Since this involves a determination of all excited states, it is very inefficient for *ab initio* methods. A computationally efficient way of calculating such properties is by means of propagator methods (Section 10.9).

10.3 Derivative Techniques

Derivative techniques consider the energy in the presence of the perturbation, perform an analytical differentiation of the energy n times to derive a formula for the nth-order property, and let the perturbation strength go to zero.

Let us write the energy as

$$E(\lambda) = \langle\Psi(\lambda)|\mathbf{H}_0 + \lambda\mathbf{P}_1 + \lambda^2\mathbf{P}_2|\Psi(\lambda)\rangle \qquad (10.19)$$

which is strictly true for HF, MCSCF and CI wave functions, and can be generalized to MP and CC methods as shown in Section 10.4. The perturbation-dependent terms in the operator are written explicitly, while the wave function dependence is implicit, via the parameterization (MO and state coefficients) and possibly also the basis functions.

The first derivative of the energy can be written as

$$\frac{\partial E}{\partial\lambda} = \left\langle\frac{\partial\Psi}{\partial\lambda}\middle|\mathbf{H}_0+\lambda\mathbf{P}_1+\lambda^2\mathbf{P}_2\middle|\Psi\right\rangle + \langle\Psi|\mathbf{P}_1 + 2\lambda\mathbf{P}_2|\Psi\rangle + \left\langle\Psi\middle|\mathbf{H}_0+\lambda\mathbf{P}_1+\lambda^2\mathbf{P}_2\middle|\frac{\partial\Psi}{\partial\lambda}\right\rangle$$

$$(10.20)$$

For real wave functions the first and third terms are identical. Letting the perturbation strength go to zero yields

$$\frac{\partial E}{\partial \lambda}\bigg|_{\lambda=0} = \langle \Psi_0 | \mathbf{P}_1 | \Psi_0 \rangle + 2 \left\langle \frac{\partial \Psi_0}{\partial \lambda} \bigg| \mathbf{H}_0 \bigg| \Psi_0 \right\rangle \qquad (10.21)$$

The wave function depends on the perturbation indirectly, via parameters in the wave function ($\mathbf{C}$), and possibly also the basis functions (χ). The wave function parameters may be MO coefficients (HF), state coefficients (CI, MP, CC) or both (MCSCF).

$$\frac{\partial \Psi}{\partial \lambda} = \frac{\partial \Psi}{\partial \chi} \frac{\partial \chi}{\partial \lambda} + \frac{\partial \Psi}{\partial \mathbf{C}} \frac{\partial \mathbf{C}}{\partial \lambda} \qquad (10.22)$$

Assuming for the moment that the basis functions are independent of the perturbation ($\partial \chi / \partial \lambda = 0$), the derivative (10.21) may be written as

$$\frac{\partial E}{\partial \lambda}\bigg|_{\lambda=0} = \langle \Psi_0 | \mathbf{P}_1 | \Psi_0 \rangle + 2 \frac{\partial \mathbf{C}}{\partial \lambda} \left\langle \frac{\partial \Psi_0}{\partial \mathbf{C}} \bigg| \mathbf{H}_0 \bigg| \Psi_0 \right\rangle \qquad (10.23)$$

If the wave function is variationally optimized with respect to <u>all</u> parameters (HF or MCSCF, but <u>not</u> CI), the last term disappears since the energy is stationary with respect to a variation of the MO/state coefficients ($\mathbf{H}_0, \mathbf{P}_1$ and $\mathbf{P}_2$ do not depend on the parameters $\mathbf{C}$).

$$\frac{\partial E}{\partial \mathbf{C}} = \frac{\partial}{\partial \mathbf{C}} \langle \Psi | \mathbf{H}_0 + \lambda \mathbf{P}_1 + \lambda^2 \mathbf{P}_2 | \Psi \rangle$$

$$= 2 \left\langle \frac{\partial \Psi}{\partial \mathbf{C}} \bigg| \mathbf{H}_0 + \lambda \mathbf{P}_1 + \lambda^2 \mathbf{P}_2 \bigg| \Psi \right\rangle \qquad (10.24)$$

$$\frac{\partial E}{\partial \mathbf{C}}\bigg|_{\lambda=0} = 2 \left\langle \frac{\partial \Psi_0}{\partial \mathbf{C}} \bigg| \mathbf{H}_0 \bigg| \Psi_0 \right\rangle = 0$$

Variational wave functions thus obey the *Hellmann–Feynman* theorem.

$$\frac{\partial}{\partial \lambda} \langle \Psi | \mathbf{H} | \Psi \rangle = \left\langle \Psi \bigg| \frac{\partial \mathbf{H}}{\partial \lambda} \bigg| \Psi \right\rangle \qquad (10.25)$$

In such cases the expression from first-order perturbation theory (10.18) yields a result identical to the first derivative of the energy with respect to λ. For wave functions which are not completely optimized with respect to all parameters (CI, MP or CC), the Hellmann–Feynman theorem does not hold, and a first-order property calculated as an expectation value will not be identical to that obtained as an energy derivative. Since the Hellmann–Feynman theorem holds for an exact wave function, the difference between the two values becomes smaller as the quality of an approximate wave function increases; however, for practical applications the difference is not negligible. It has been argued that the derivative technique resembles the physical experiment more, and consequently formula (10.21) should be preferred over (10.18).

The second derivative of the energy can be written as

$$\frac{\partial^2 E}{\partial \lambda^2} = 2 \left\langle \frac{\partial^2 \Psi}{\partial \lambda^2} \bigg| \mathbf{H}_0 + \lambda \mathbf{P}_1 + \lambda^2 \mathbf{P}_2 \bigg| \Psi \right\rangle + 4 \left\langle \frac{\partial \Psi}{\partial \lambda} \bigg| \mathbf{P}_1 + 2\lambda \mathbf{P}_2 \bigg| \Psi \right\rangle$$
$$+ 2 \left\langle \frac{\partial \Psi}{\partial \lambda} \bigg| \mathbf{H}_0 + \lambda \mathbf{P}_1 + \lambda^2 \mathbf{P}_2 \bigg| \frac{\partial \Psi}{\partial \lambda} \right\rangle + 2 \langle \Psi | \mathbf{P}_2 | \Psi \rangle \qquad (10.26)$$

In the limit of the perturbation strength going to zero this reduces to

$$\frac{\partial^2 E}{\partial \lambda^2}\bigg|_{\lambda=0} = 2\left\langle \frac{\partial^2 \Psi_0}{\partial \lambda^2}\bigg|\mathbf{H}_0\bigg|\Psi_0\right\rangle + 4\left\langle \frac{\partial \Psi_0}{\partial \lambda}\bigg|\mathbf{P}_1\bigg|\Psi_0\right\rangle$$
$$+ 2\left\langle \frac{\partial \Psi_0}{\partial \lambda}\bigg|\mathbf{H}_0\bigg|\frac{\partial \Psi_0}{\partial \lambda}\right\rangle + 2\langle\Psi_0|\mathbf{P}_2|\Psi_0\rangle \tag{10.27}$$

The implicit wave function dependence on C allows the derivative to be written as

$$\frac{\partial^2 E}{\partial \lambda^2}\bigg|_{\lambda=0} = 2\frac{\partial^2 \mathbf{C}}{\partial \lambda^2}\left\langle \frac{\partial \Psi_0}{\partial \mathbf{C}}\bigg|\mathbf{H}_0\bigg|\Psi_0\right\rangle + 2\left(\frac{\partial \mathbf{C}}{\partial \lambda}\right)^2\left\langle \frac{\partial^2 \Psi_0}{\partial \mathbf{C}^2}\bigg|\mathbf{H}_0\bigg|\Psi_0\right\rangle$$
$$+ 4\left(\frac{\partial \mathbf{C}}{\partial \lambda}\right)\left\langle \frac{\partial \Psi_0}{\partial \mathbf{C}}\bigg|\mathbf{P}_1\bigg|\Psi_0\right\rangle + 2\left(\frac{\partial \mathbf{C}}{\partial \lambda}\right)^2\left\langle \frac{\partial \Psi_0}{\partial \mathbf{C}}\bigg|\mathbf{H}_0\bigg|\frac{\partial \Psi_0}{\partial \mathbf{C}}\right\rangle \tag{10.28}$$
$$+ 2\langle\Psi_0|\mathbf{P}_2|\Psi_0\rangle$$

For a variationally optimized wave function, the first term is again zero (eq. (10.24)). Furthermore, the second term, which involves calculation of the second derivative of the wave function with respect to the parameters, can be avoided. This can be seen by differentiating the stationary condition (10.24) with respect to the perturbation.

$$\frac{\partial}{\partial \lambda}\left\langle \frac{\partial \Psi}{\partial \mathbf{C}}\bigg|\mathbf{H}_0 + \lambda\mathbf{P}_1 + \lambda^2\mathbf{P}_2\bigg|\Psi\right\rangle\bigg|_{\lambda=0} =$$
$$\left(\frac{\partial \mathbf{C}}{\partial \lambda}\right)\left\langle \frac{\partial^2 \Psi_0}{\partial \mathbf{C}^2}\bigg|\mathbf{H}_0\bigg|\Psi_0\right\rangle + \left\langle \frac{\partial \Psi_0}{\partial \mathbf{C}}\bigg|\mathbf{P}_1\bigg|\Psi_0\right\rangle + \left(\frac{\partial \mathbf{C}}{\partial \lambda}\right)\left\langle \frac{\partial \Psi_0}{\partial \mathbf{C}}\bigg|\mathbf{H}_0\bigg|\frac{\partial \Psi_0}{\partial \mathbf{C}}\right\rangle = 0 \tag{10.29}$$

The second derivative in eq. (10.28) therefore reduces to

$$\frac{\partial^2 E}{\partial \lambda^2}\bigg|_{\lambda=0} = 2\left(\frac{\partial \mathbf{C}}{\partial \lambda}\right)\left\langle \frac{\partial \Psi_0}{\partial \mathbf{C}}\bigg|\mathbf{P}_1\bigg|\Psi_0\right\rangle + 2\langle\Psi_0|\mathbf{P}_2|\Psi_0\rangle \tag{10.30}$$

or in a more compact notation

$$\frac{\partial^2 E}{\partial \lambda^2}\bigg|_{\lambda=0} = 2\left\langle \frac{\partial \Psi_0}{\partial \lambda}\bigg|\mathbf{P}_1\bigg|\Psi_0\right\rangle + 2\langle\Psi_0|\mathbf{P}_2|\Psi_0\rangle \tag{10.31}$$

i.e. only the first-order change in the wave function is necessarily. For exact wave functions (10.31) becomes identical to the perturbation expression (10.18), since the first derivative of the wave function then may be expanded in a complete set of eigenfunctions (eq. (4.35)).

$$\frac{\partial \Psi_0}{\partial \lambda} = \sum_{i=1}^{\infty} a_i \Psi_i, \quad a_i = \frac{\langle\Psi_i|\mathbf{P}_1|\Psi_0\rangle}{E_0 - E_i} \tag{10.32}$$

10.4 Lagrangian Techniques

For variationally optimized wave functions (HF or MCSCF) there is a $2n + 1$ rule, analogous to the perturbational energy expression in Section 4.8 (eq. (4.34)): knowledge of the nth derivative (also called the response) of the wave function is sufficient for calculating a property to order $2n + 1$. For non-variational wave functions eq. (10.27) suggests that the nth-order wave function response is required for calculating the nth-order property. This may be avoided, however, by a technique first illustrated for CISD geometry derivatives by Handy and Schaefer,[2] often referred to as the *Z-vector* method. It has since been generalized to cover other types of wave functions and derivatives by formulating it in terms of a Lagrange function.[3]

The idea is to construct a Lagrange function which has the same energy as the non-variational wave function, but which is variational in all parameters. Consider for example a CI wave function, which is variational in the state coefficients (**a**) but not in the MO coefficients (**c**) (note that we employ lower case **c** for the MO coefficients, but capital **C** to denote all wave function parameters, i.e. **C** contains both **a** and **c**), since they are determined by the stationary condition for the HF wave function.

$$\frac{\partial E_{CI}}{\partial \mathbf{a}} = \frac{\partial}{\partial \mathbf{a}} \langle \Psi_{CI}(\mathbf{a}, \mathbf{c}) | \mathbf{H} | \Psi_{CI}(\mathbf{a}, \mathbf{c}) \rangle = 2 \left\langle \frac{\partial \Psi_{CI}}{\partial \mathbf{a}} \middle| \mathbf{H} \middle| \Psi_{CI} \right\rangle = 0$$

$$\frac{\partial E_{CI}}{\partial \mathbf{c}} = \frac{\partial}{\partial \mathbf{c}} \langle \Psi_{CI}(\mathbf{a}, \mathbf{c}) | \mathbf{H} | \Psi_{CI}(\mathbf{a}, \mathbf{c}) \rangle = 2 \left\langle \frac{\partial \Psi_{CI}}{\partial \mathbf{C}} \middle| \mathbf{H} \middle| \Psi_{CI} \right\rangle \neq 0 \qquad (10.33)$$

$$\frac{\partial E_{HF}}{\partial \mathbf{c}} = \frac{\partial}{\partial \mathbf{c}} \langle \Psi_{HF}(\mathbf{c}) | \mathbf{H} | \Psi_{HF}(\mathbf{c}) \rangle = 2 \left\langle \frac{\partial \Psi_{HF}}{\partial \mathbf{c}} \middle| \mathbf{H} \middle| \Psi_{HF} \right\rangle = 0$$

Consider now the Lagrange function

$$L_{CI} = E_{CI} + \kappa \frac{\partial E_{HF}}{\partial \mathbf{c}} \qquad (10.34)$$

where κ contains a set of Lagrange multipliers. The derivatives of the Lagrange function with respect to **a**, **c** and κ are given as

$$\frac{\partial L_{CI}}{\partial \mathbf{a}} = \frac{\partial E_{CI}}{\partial \mathbf{a}} = 0$$

$$\frac{\partial L_{CI}}{\partial \kappa} = \frac{\partial E_{HF}}{\partial \mathbf{c}} = 0 \qquad (10.35)$$

$$\frac{\partial L_{CI}}{\partial \mathbf{c}} = \frac{\partial E_{CI}}{\partial \mathbf{c}} + \kappa \frac{\partial^2 E_{HF}}{\partial \mathbf{c}^2} = 0$$

The first two derivatives are zero owing to the properties of the CI and HF wave functions, eq. (10.33). The last equation is zero by virtue of the Lagrange multipliers, i.e. we <u>choose</u> κ such that $\partial L_{CI}/\partial \mathbf{c} = 0$. It may be written more explicitly as

$$\frac{\partial L_{CI}}{\partial \mathbf{c}} = 2 \left\langle \frac{\partial \Psi_{CI}}{\partial \mathbf{c}} \middle| \mathbf{H} \middle| \Psi_{CI} \right\rangle + 2\kappa \left[\left\langle \frac{\partial^2 \Psi_{HF}}{\partial \mathbf{c}^2} \middle| \mathbf{H} \middle| \Psi_{HF} \right\rangle + \left\langle \frac{\partial \Psi_{HF}}{\partial \mathbf{c}} \middle| \mathbf{H} \middle| \frac{\partial \Psi_{HF}}{\partial \mathbf{c}} \right\rangle \right] = 0$$

$$(10.36)$$

Note that no new operators are involved, only derivatives of the CI or HF wave function with respect to the MO coefficients. The matrix elements can thus be calculated from the same integrals as the energy itself, as discussed in Sections 3.5 and 4.2.1.

The derivative with respect to a perturbation can now be written as

$$\frac{\partial L_{CI}}{\partial \lambda} = \frac{\partial E_{CI}}{\partial \lambda} + \kappa \frac{\partial}{\partial \lambda} \left(\frac{\partial E_{HF}}{\partial \mathbf{c}} \right) \qquad (10.37)$$

or more explicitly as

$$\frac{\partial L_{\mathrm{CI}}}{\partial \lambda} = \frac{\partial}{\partial \lambda} \langle \Psi_{\mathrm{CI}} | \mathbf{H} | \Psi_{\mathrm{CI}} \rangle + \kappa \frac{\partial}{\partial \lambda} \left\langle \frac{\partial \Psi_{\mathrm{HF}}}{\partial \mathbf{c}} \middle| \mathbf{H} \middle| \Psi_{\mathrm{HF}} \right\rangle$$

$$\frac{\partial L_{\mathrm{CI}}}{\partial \lambda} = \left\langle \Psi_{\mathrm{CI}} \middle| \frac{\partial \mathbf{H}}{\partial \lambda} \middle| \Psi_{\mathrm{CI}} \right\rangle + 2 \frac{\partial \mathbf{a}}{\partial \lambda} \left\langle \frac{\partial \Psi_{\mathrm{CI}}}{\partial \mathbf{a}} \middle| \mathbf{H} \middle| \Psi_{\mathrm{CI}} \right\rangle + 2 \frac{\partial \mathbf{c}}{\partial \lambda} \left\langle \frac{\partial \Psi_{\mathrm{CI}}}{\partial \mathbf{c}} \middle| \mathbf{H} \middle| \Psi_{\mathrm{CI}} \right\rangle$$

$$+ \kappa \left[\left\langle \frac{\partial \Psi_{\mathrm{HF}}}{\partial \mathbf{c}} \middle| \frac{\partial \mathbf{H}}{\partial \lambda} \middle| \Psi_{\mathrm{HF}} \right\rangle + 2 \frac{\partial \mathbf{c}}{\partial \lambda} \left(\left\langle \frac{\partial^2 \Psi_{\mathrm{HF}}}{\partial \mathbf{c}^2} \middle| \mathbf{H} \middle| \Psi_{\mathrm{HF}} \right\rangle + \left\langle \frac{\partial \Psi_{\mathrm{HF}}}{\partial \mathbf{c}} \middle| \mathbf{H} \middle| \frac{\partial \Psi_{\mathrm{HF}}}{\partial \mathbf{c}} \right\rangle \right) \right]$$

$$(10.38)$$

The second term disappears since the CI wave function is variational in the state coefficients, eq. (10.33). The three terms involving the derivative of the MO coefficients $(\partial \mathbf{c}/\partial \lambda)$ also disappear owing to our choice of the Lagrange multipliers, eq. (10.36). If we furthermore adapt the definition that $\partial \mathbf{H}/\partial \lambda = \mathbf{P}_1$ (eq. (10.17)), the final derivative may be written as

$$\frac{\partial L_{\mathrm{CI}}}{\partial \lambda} = \langle \Psi_{\mathrm{CI}} | \mathbf{P}_1 | \Psi_{\mathrm{CI}} \rangle + \kappa \left\langle \frac{\partial \Psi_{\mathrm{HF}}}{\partial \mathbf{c}} \middle| \mathbf{P}_1 \middle| \Psi_{\mathrm{HF}} \right\rangle \qquad (10.39)$$

where the Lagrange multipliers are determined from eq. (10.36).

What has been accomplished? The original expression (10.23) contains the derivative of the MO coefficients with respect to the perturbation $(\partial \mathbf{c}/\partial \lambda)$, which can be obtained by solving the CPHF equations (Section 10.5). For geometry derivatives, for example, there will be $3N$ different perturbations, i.e. we need to solve $3N$ sets of CPHF equations. The Lagrange expression (10.39), on the other hand, contains a set of Lagrange multipliers κ which are independent of the perturbation, i.e. we need only solve one equation for κ, (10.36). Furthermore, the CPHF equations involve derivatives of the basis functions, while the equation for κ only involves integrals of the same type as for calculating the energy itself.

The Lagrange technique may be generalized to other types of non-variational wave functions (MP and CC), and to higher-order derivatives. It is found that the $2n + 1$ rule is recovered, i.e. if the wave function response is known to order n, the $(2n + 1)$th-order property may be calculated for any type of wave function.

10.5 Coupled Perturbed Hartree–Fock

Although a calculation of the wave function response can be avoided for the first derivative, it is necessary for second (and higher) derivatives. Eq. (10.29) gives directly an equation for determining the (first-order) response, which is structurally the same as eq. (10.36). For an HF wave function, an equation of the change in the MO coefficients may also be formulated from the Hartree–Fock equation, eq. (3.50).

$$\mathbf{F}^{(0)} \mathbf{C}^{(0)} = \mathbf{S}^{(0)} \mathbf{C}^{(0)} \boldsymbol{\varepsilon}^{(0)} \qquad (10.40)$$

The superscript (0) here denotes the unperturbed system. The orthonormality of the molecular orbitals (eq. (3.20)) can be expressed as

$$\mathbf{C}^{\dagger(0)} \mathbf{S}^{(0)} \mathbf{C}^{(0)} = \mathbf{1} \qquad (10.41)$$

Expanding each of the $\mathbf{F}$, $\mathbf{C}$, $\mathbf{S}$ and $\boldsymbol{\varepsilon}$ matrices in terms of a perturbation parameter (e.g. $\mathbf{F} = \mathbf{F}^{(0)} + \lambda \mathbf{F}^{(1)} + \lambda^2 \mathbf{F}^{(2)} + \ldots$) and collecting all the first-order terms (analogous to the strategy used in Section 4.8) gives

$$\mathbf{F}^{(1)}\mathbf{C}^{(0)} + \mathbf{F}^{(0)}\mathbf{C}^{(1)} = \mathbf{S}^{(1)}\mathbf{C}^{(0)}\boldsymbol{\varepsilon}^{(0)} + \mathbf{S}^{(0)}\mathbf{C}^{(1)}\boldsymbol{\varepsilon}^{(0)} + \mathbf{S}^{(0)}\mathbf{C}^{(0)}\boldsymbol{\varepsilon}^{(1)}$$
$$(\mathbf{F}^{(0)} + \mathbf{S}^{(0)}\boldsymbol{\varepsilon}^{(0)})\mathbf{C}^{(1)} = (\mathbf{F}^{(1)} + \mathbf{S}^{(0)}\boldsymbol{\varepsilon}^{(1)} + \mathbf{S}^{(1)}\boldsymbol{\varepsilon}^{(0)})\mathbf{C}^{(0)}$$

$$(10.42)$$

while the orthonormality condition becomes

$$\mathbf{C}^{\dagger(1)}\mathbf{S}^{(0)}\mathbf{C}^{(0)} + \mathbf{C}^{\dagger(0)}\mathbf{S}^{(1)}\mathbf{C}^{(0)} + \mathbf{C}^{\dagger(0)}\mathbf{S}^{(0)}\mathbf{C}^{(1)} = 0 \qquad (10.43)$$

Equation (10.42) are the first-order *Coupled Perturbed Hartree–Fock* (CPHF) equations.[4] The perturbed MO coefficients are given in terms of unperturbed quantities and the first-order Fock, Lagrange ($\boldsymbol{\varepsilon}$) and overlap matrices. The $\mathbf{F}^{(1)}$ term is given as (eq. (3.52)).

$$\mathbf{F}^{(1)} = \mathbf{h}^{(1)} + \mathbf{G}^{(1)}\mathbf{D}^{(0)} + \mathbf{G}^{(0)}\mathbf{D}^{(1)} \qquad (10.44)$$

where $\mathbf{h}$ is the one-electron (core) matrix, $\mathbf{D}$ is the density matrix and $\mathbf{G}$ is the tensor containing the two-electron integrals. The density matrix is given as a product of MO coefficients (eq. (3.51)).

$$\mathbf{D}^{(0)} = \mathbf{C}^{\dagger(0)}\mathbf{C}^{(0)}$$
$$\mathbf{D}^{(1)} = \mathbf{C}^{\dagger(1)}\mathbf{C}^{(0)} + \mathbf{C}^{\dagger(0)}\mathbf{C}^{(1)}$$

$$(10.45)$$

The $\mathbf{S}^{(1)}$, $\mathbf{h}^{(1)}$ and $\mathbf{g}^{(1)}$ quantities are (first) derivatives of overlap, one- and two-electron integrals over basis functions.

$$S_{\alpha\beta}^{(1)} = \langle \chi_\alpha | \chi_\beta \rangle^{(1)} = \frac{\partial}{\partial\lambda}\langle \chi_\alpha | \chi_\beta \rangle$$

$$h_{\alpha\beta}^{(1)} = \langle \chi_\alpha | \mathbf{h} | \chi_\beta \rangle^{(1)} = \frac{\partial}{\partial\lambda}\langle \chi_\alpha | \mathbf{h} | \chi_\beta \rangle \qquad (10.46)$$

$$g_{\alpha\beta\gamma\delta}^{(1)} = \langle \chi_\alpha \chi_\beta | \mathbf{g} | \chi_\gamma \chi_\delta \rangle^{(1)} = \frac{\partial}{\partial\lambda}\langle \chi_\alpha \chi_\beta | \mathbf{g} | \chi_\gamma \chi_\delta \rangle$$

The derivatives over the integrals may involve derivatives of the basis functions or the operator, or both (see Section 10.8). Using eqs. (10.45) and (10.44) in eq. (10.42) gives a set of linear equations relating $\mathbf{C}^{(1)}$ to $\mathbf{S}^{(1)}$, $\mathbf{h}^{(1)}$, $\mathbf{g}^{(1)}$ and $\mathbf{C}^{(0)}$.

Just as the variational condition for an HF wave function can be formulated either as a matrix equation or in terms of orbital rotations (Sections 3.5 and 3.6), the CPHF may also be viewed as a rotation of the molecular orbitals. In the absence of a perturbation the molecular orbitals make the energy stationary, i.e. the derivatives of the energy with respect to a change in the MOs are zero. This is equivalent to the statement that the off-diagonal elements of the Fock matrix between the occupied and virtual MOs are zero.

$$\langle \phi_i | \mathbf{F} | \phi_a \rangle = 0$$
$$\langle \phi_i | \mathbf{h} | \phi_a \rangle + \sum_{k=1}^{N}(\langle \phi_i \phi_k | \mathbf{g} | \phi_a \phi_k \rangle - \langle \phi_i \phi_k | \mathbf{g} | \phi_k \phi_a \rangle) = 0$$

$$(10.47)$$

When a perturbation is introduced, the stationary condition means that the orbitals must change, which may be described as a mixing of the unperturbed MOs. In other words, the stationary orbitals in the presence of a perturbation is given by a unitary transformation of the unperturbed orbitals (see also Section 3.6).

$$\phi_i' = \sum_{j=1}^{M} U_{ji}\phi_j \tag{10.48}$$

The $\mathbf{U}$ matrix describes how the MOs change, i.e. it contains the derivatives of the MO coefficients. In the absence of a perturbation $\mathbf{U}$ is the identity matrix.

Let us now explicitly make $\mathbf{U}^{(1)}$ the matrix containing the first-order changes in the MO coefficients.

$$|\phi_i\rangle \rightarrow |\phi_i\rangle + \lambda \sum_{j=1}^{M} U_{ji}^{(1)}|\phi_j\rangle + \ldots \tag{10.49}$$

In terms of the matrix formulation in eqs. (10.42)–(10.43), the equivalent of (10.48) is

$$\mathbf{C}^{(1)} = \mathbf{U}^{(1)}\mathbf{C}^{(0)} \tag{10.50}$$

An equation for the $\mathbf{U}^{(1)}$ elements can be obtained from the condition that the Fock matrix is diagonal, and expanding all involved quantities to first-order.

$$\langle\phi_\alpha|\mathbf{h}|\phi_\beta\rangle \rightarrow \langle\phi_\alpha|\mathbf{h}|\phi_\beta\rangle)^{(0)} + \langle\phi_\alpha|\mathbf{h}|\phi_\beta\rangle^{(1)}$$
$$\langle\phi_\alpha\phi_\beta|\mathbf{g}|\phi_\gamma\phi_\delta\rangle \rightarrow \langle\phi_\alpha\phi_\beta|\mathbf{g}|\phi_\gamma\phi_\delta\rangle^{(0)} + \langle\phi_\alpha\phi_\beta|\mathbf{g}|\phi_\gamma\phi_\delta\rangle^{(1)} \tag{10.51}$$

The $\langle\phi_\alpha|\mathbf{h}|\phi_\beta\rangle^{(1)}$ and $\langle\phi_\alpha\phi_\beta|\mathbf{g}|\phi_\gamma\phi_\delta\rangle^{(1)}$ elements are integral derivatives with respect to the perturbation, analogous to eq. (10.46), but expressed in terms of molecular orbitals. Inserting these expansions into the $\langle\phi_i|\mathbf{F}|\phi_a\rangle = 0$ condition and collecting all terms which are first-order in λ gives a matrix equation which can be written as

$$\mathbf{A}^{(0)}\mathbf{U}^{(1)} = \mathbf{B}^{(1)} \tag{10.52}$$

The $\mathbf{A}^{(0)}$ matrix contains only unperturbed quantities ($\langle\phi_\alpha|\mathbf{h}|\phi_\beta\rangle^{(0)}$ and $\langle\phi_\alpha\phi_\beta|\mathbf{g}|\phi_\gamma\phi_\delta\rangle^{(0)}$), while the $\mathbf{B}^{(1)}$ matrix contains first derivatives ($\langle\phi_\alpha|\mathbf{h}|\phi_\beta\rangle^{(1)}$ and $\langle\phi_\alpha\phi_\beta|\mathbf{g}|\phi_\gamma\phi_\delta\rangle^{(1)}$).

Since the energy is independent of a rotation among the occupied or virtual orbitals, only the mixing of occupied and virtual orbitals is determined by requiring that the energy is stationary. The occupied–occupied and virtual–virtual mixing may be fixed from the orthonormality condition (eq. (10.43)), or equivalently, by requiring the perturbed Fock matrix to be diagonal also in the occupied–occupied and virtual–virtual blocks. Without these additional requirements the procedure is normally just called *Coupled Hartree–Fock* (CHF).

The CPHF equations are linear and can be determined by standard matrix operations. The size of the $\mathbf{U}$ matrix is the number of occupied orbitals times the number of virtual orbitals, which in general is quite large, and the CPHF equations are normally solved by iterative methods. Furthermore, as illustrated above, the CPHF equations may be formulated either in an atomic orbital or molecular orbital basis. Although the latter has computational advantages in certain cases, the former is more suitable for use in connection with direct methods (where the atomic integrals are calculated as required), as discussed in Section 3.8.5.

There will be one CPHF equation to be solved for each perturbation. If it is an electric or magnetic field, there will in general be three components (F_x, F_y, F_z), if it is a geometry perturbation there will be $3N$ (actually only $3N - 6$ independent) components, N being the number of atoms. Since the $\mathbf{A}^{(0)}$ matrix is independent of the nature of the perturbation, such multiple CPHF equations are often solved simultaneously.

The CPHF procedure may be generalized to higher order. Extending the expansion to second-order allows derivation of an equation for the second-order change in the MO coefficients, by solving a second-order CPHF equation etc.

For perturbation-dependent basis sets (e.g. geometry derivatives) the (first-order) CPHF equations involve (first) derivatives of the one- and two-electron integrals with respect to the perturbation. For basis functions which are independent of the perturbation (e.g. an electric field), these derivatives are zero. Typically the solution of each CPHF equation (for each perturbation) requires approximately 1/2 the time for solving the HF equations themselves. For basis set dependent perturbations, the first-order CPHF equations are only needed for calculating second (and higher) derivatives, which have terms involving second (and higher) derivatives of the integrals themselves, and solving the CPHF equations is usually not the bottleneck in these cases.

Without the Lagrange technique for non-variational wave functions (CI, MP and CC), the nth-order CPHF is needed for the nth-derivative. Consider for example the MP2 energy correction, eq. (4.45).

$$\text{MP2} = \sum_{i<j}^{\text{occ}} \sum_{a<b}^{\text{vir}} \frac{[\langle \phi_i \phi_j | \phi_a \phi_b \rangle - \langle \phi_i \phi_j | \phi_b \phi_a \rangle]^2}{\varepsilon_i + \varepsilon_j - \varepsilon_a - \varepsilon_b} \tag{10.53}$$

The derivative of a molecular integral is given by

$$\frac{\partial}{\partial \lambda} \langle \phi_i \phi_j | \phi_k \phi_l \rangle = \frac{\partial}{\partial \lambda} \sum_{\alpha\beta\gamma\delta} c_{i\alpha} c_{j\beta} c_{k\gamma} c_{l\delta} \langle \chi_\alpha \chi_\beta | \chi_\gamma \chi_\delta \rangle \tag{10.54}$$

i.e. this requires both the derivative of the MO coefficients and the two-electron integrals in the AO basis. The denominator will give rise to derivatives of the MO energies, which can be obtained by solving the CPHF equations. A straight forward differentiation of eq. (10.53) thus leads to a formula where the first-order response is required.

Let us exemplify some of the above generalizations for the case of a HF wave function.

10.6   Electric Field Perturbation

If the perturbation is a homogeneous electric field $\mathbf{F}$, the perturbation operator $\mathbf{P}_1$ (eq. (10.17)) is the position vector $\mathbf{r}$ and $\mathbf{P}_2$ is zero. Assuming that the basis functions are independent of the electric field (as is normally the case), the first-order HF property, the dipole moment, from the derivative formula (10.21) is given as (since an HF wave function obeys the Hellmann–Feynman theorem)

$$\frac{\partial E_{\text{HF}}}{\partial \mathbf{F}} = \langle \Psi_0 | \mathbf{r} | \Psi_0 \rangle \tag{10.55}$$

which is equivalent to the expression from first-order perturbation theory, (10.18). For non-variational wave functions the dipole moment calculated by the two approaches will be different, since the derivative of the wave function with respect to the field will not be zero.

The second-order property, the dipole polarizability, as given by the derivative formula eq. (10.31), is

$$\frac{\partial^2 E_{HF}}{\partial \mathbf{F}^2} = 2\left\langle \frac{\partial \Psi_0}{\partial \mathbf{F}} \left| \mathbf{r} \right| \Psi_0 \right\rangle \tag{10.56}$$

while second-order perturbation theory (eq. (10.18) yields

$$W_2 = \sum_{i \neq 0} \frac{|\langle \Psi_0 | \mathbf{r} | \Psi_i \rangle|^2}{E_0 - E_i} \tag{10.57}$$

10.7 Magnetic Field Perturbation

The situation is somewhat more complicated when the perturbation is a magnetic field. An electric field interacts directly with the charged particles (electron and nuclei), and adds a <u>potential</u> energy term to the Hamilton operator. A magnetic field, however, interacts with the magnetic moments generated by the <u>movement</u> of the charged particles (electrons), i.e. a magnetic perturbation changes the <u>kinetic</u> energy operator. The generalized (also called the canonical) momentum operator π is defined as

$$\pi = \mathbf{p} - q\mathbf{A} \tag{10.58}$$

where q is the charge and $\mathbf{A}$ is the *vector potential* associated with the magnetic field (more correctly, the *magnetic induction* or *flux* density, being different from the magnetic field by a factor of $4\pi \times 10^{-7}\,\mathrm{Hm}^{-1}$).

$$\mathbf{B} = \nabla \times \mathbf{A} \tag{10.59}$$

($\nabla\times$ is the curl operator). Only the kinetic energy of the electrons are considered within the Born–Oppenheimer approximation, and the generalized momentum becomes ($q = -1$)

$$\pi = \mathbf{p} + \mathbf{A} \tag{10.60}$$

The vector potential is not uniquely defined since the gradient of any scalar function may be added (the curl of a derivative is always zero). It is convention to select it as

$$\mathbf{A}(\mathbf{r}) = \tfrac{1}{2}\mathbf{B} \times (\mathbf{r} - \mathbf{R}_G) \tag{10.61}$$

where $\mathbf{R}_G$ is referred to as the *gauge origin*, i.e. the origin of the coordinate system.

10.7.1 External Magnetic Field

The Hamilton operator in the presence of a magnetic field is given as

$$\mathbf{H} = \tfrac{1}{2}\pi^2 + \mathbf{V} = \tfrac{1}{2}(\mathbf{p} + \mathbf{A})^2 + \mathbf{V} \tag{10.62}$$

where

$$(\mathbf{p} + \mathbf{A})^2 = \mathbf{p}^2 + \mathbf{p} \cdot \mathbf{A} + \mathbf{A} \cdot \mathbf{p} + \mathbf{A}^2 \tag{10.63}$$

Since $\mathbf{p} = -i\nabla$, the $\mathbf{p} \cdot \mathbf{A}$ term gives

$$(\mathbf{p} \cdot \mathbf{A})\Psi = -i(\nabla \cdot \mathbf{A})\Psi = -i\mathbf{A} \cdot (\nabla\Psi) - i\Psi(\nabla \cdot \mathbf{A}) \qquad (10.64)$$

Selecting the *Coulomb gauge*, where $\nabla \cdot \mathbf{A} = 0$, gives $\mathbf{p} \cdot \mathbf{A} = \mathbf{A} \cdot \mathbf{p}$, and thereby

$$\mathbf{H} = \tfrac{1}{2}\mathbf{p}^2 + \mathbf{A} \cdot \mathbf{p} + \tfrac{1}{2}\mathbf{A}^2 + \mathbf{V} \qquad (10.65)$$

The first term is identical to the usual kinetic energy operator. Inserting the expression for the vector potential (10.61) yields

$$
\begin{aligned}
\mathbf{A} \cdot \mathbf{p} &= \left(\frac{1}{2}\mathbf{B} \times (\mathbf{r} - \mathbf{R}_G)\right) \cdot \mathbf{p} \\
&= \frac{1}{2}\mathbf{B} \cdot (\mathbf{r} - \mathbf{R}_G) \times \mathbf{p} = \frac{1}{2}\mathbf{B} \cdot \mathbf{L} \\
\frac{1}{2}\mathbf{A}^2 &= \frac{1}{2}\left(\frac{1}{2}\mathbf{B} \times (\mathbf{r} - \mathbf{R}_G)\right) \cdot \left(\frac{1}{2}\mathbf{B} \times (\mathbf{r} - \mathbf{R}_G)\right) \\
&= \frac{1}{8}\{\mathbf{B}^2 \cdot (\mathbf{r} - \mathbf{R}_G)^2 - [\mathbf{B} \cdot (\mathbf{r} - \mathbf{R}_G)]^2\}
\end{aligned}
\qquad (10.66)
$$

where the vector identities $\mathbf{a} \times \mathbf{b} \cdot \mathbf{c} = \mathbf{a} \cdot \mathbf{b} \times \mathbf{c}$ and $(\mathbf{a} \times \mathbf{b}) \cdot (\mathbf{a} \times \mathbf{b}) = (\mathbf{a} \cdot \mathbf{a})(\mathbf{b} \cdot \mathbf{b}) - (\mathbf{a} \cdot \mathbf{b})(\mathbf{a} \cdot \mathbf{b})$ have been used, and the angular momentum operator $\mathbf{L}$ is as defined in eq. (10.9). The presence of a magnetic field introduces two new terms, being linear and quadratic in the field. The $\mathbf{B} \cdot \mathbf{L}$ term is the orbital analogue of the Zeeman effect for the electron spin, discussed in section 8.1 (eq. (8.19)). The second-order property is the magnetizability ξ, which according to eqs. (10.17) and (10.18)/(10.27) contains contributions from both linear and quadratic perturbation operators. The $\mathbf{P}_1^\xi$ operator is (half) the angular moment $\mathbf{L}$, while the $\mathbf{P}_2^\xi$ operator may be written as

$$\mathbf{P}_2^\xi = \tfrac{1}{8}[(\mathbf{r} - \mathbf{R}_G)^{\mathrm{t}}(\mathbf{r} - \mathbf{R}_G) - (\mathbf{r} - \mathbf{R}_G)(\mathbf{r} - \mathbf{R}_G)^{\mathrm{t}}] \qquad (10.67)$$

Here $(\mathbf{r} - \mathbf{R}_G)^{\mathrm{t}}(\mathbf{r} - \mathbf{R}_G)$ is the dot product times a unit matrix (i.e. $(\mathbf{r} - \mathbf{R}_G) \cdot (\mathbf{r} - \mathbf{R}_G)\mathbf{I}$) and $(\mathbf{r} - \mathbf{R}_G)(\mathbf{r} - \mathbf{R}_G)^{\mathrm{t}}$ is a 3×3 matrix containing the products of the x, y, z components, analogous to the quadrupole moment, eq. (10.4). Note that both the $\mathbf{L}$ and $\mathbf{P}_2^\xi$ operators are gauge dependent. When field-independent basis functions are used the first-order property, the HF magnetic dipole moment, is given as the expectation value over the unperturbed wave function (for a singlet state) eqs. (10.18)/(10.23).

$$\frac{\partial E_{\mathrm{HF}}}{\partial \mathbf{B}} = \frac{1}{2}\langle\Psi_0|\mathbf{L}|\Psi_0\rangle \qquad (10.68)$$

The second-order term, the magnetizability, has two components. The derivative expression (10.31) is

$$\frac{\partial^2 E_{\mathrm{HF}}}{\partial \mathbf{B}^2} = 2\langle\Psi_0|\mathbf{P}_2^\xi|\Psi_0\rangle + 2\left\langle\frac{\partial\Psi_0}{\partial\mathbf{B}}\middle|\mathbf{L}\middle|\Psi_0\right\rangle \qquad (10.69)$$

while second-order perturbation theory (eq. (10.18)) yields

$$W_2 = \langle\Psi_0|\mathbf{P}_2^\xi|\Psi_0\rangle + \sum_{i=1}^{\infty}\frac{|\langle\Psi_0|\mathbf{L}|\Psi_i\rangle|^2}{E_0 - E_i} \qquad (10.70)$$

The first term is referred to as the *diamagnetic* contribution, while the latter is the *paramagnetic* part of the magnetizability. Each of the two components depend on the selected gauge origin; however, for exact wave functions these cancel exactly. For approximate wave functions this is not guaranteed, and as a result the total property may depend on where the origin for the vector potential (eq. (10.61)) has been chosen.

10.7.2 Nuclear Spin

A nucleus with non-zero spin acts as a magnetic dipole, giving raise to a vector potential $\mathbf{A}_A$.

$$\mathbf{A}_A = \frac{\mu_0 \mathbf{I}_A \times (\mathbf{r} - \mathbf{R}_A)}{4\pi |\mathbf{r} - \mathbf{R}_A|^3} \tag{10.71}$$

Here $\mathbf{I}_A$ is the magnetic moment of nucleus $\mathbf{A}$ and $\mathbf{R}_A$ is the position (the nucleus is the natural Gauge origin). Adding this to the external vector potential in eq. (10.62) and expanding as in (10.63) gives

$$(\mathbf{p} + \mathbf{A} + \mathbf{A}_A)^2 = \mathbf{p}^2 + \mathbf{A}^2 + \mathbf{A}_A^2 + 2\mathbf{A} \cdot \mathbf{p} + 2\mathbf{A}_A \cdot \mathbf{p} + 2\mathbf{A} \cdot \mathbf{A}_A \tag{10.72}$$

The $\mathbf{p} \cdot \mathbf{A}_A, \mathbf{A}_A^2$ and $\mathbf{A} \cdot \mathbf{A}_A$ terms give the following operators, respectively.

$$\mathbf{H}^{PSO} = \frac{\mu_0}{4\pi} \frac{\mathbf{I}_A \cdot ((\mathbf{r} - \mathbf{R}_A) \times \mathbf{p})}{|\mathbf{r} - \mathbf{R}_A|^3} = \mathbf{I}_A \cdot \mathbf{P}_1^{PSO}$$

$$\mathbf{H}^{DSO} = -\frac{\mu_0^2}{32\pi^2} \frac{(\mathbf{I}_A \times (\mathbf{r} - \mathbf{R}_A)) \cdot (\mathbf{I}_B \times (\mathbf{r} - \mathbf{R}_B))}{|\mathbf{r} - \mathbf{R}_A|^3 |\mathbf{r} - \mathbf{R}_B|^3} = \mathbf{I}_A \cdot \mathbf{P}_2^{DSO} \cdot \mathbf{I}_B \tag{10.73}$$

$$\mathbf{H}^{\sigma} = \frac{\mu_0}{8\pi} \frac{(\mathbf{B} \times (\mathbf{r} - \mathbf{R}_G)) \cdot (\mathbf{I}_A \times (\mathbf{r} - \mathbf{R}_A))}{|\mathbf{r} - \mathbf{R}_A|^3} = \mathbf{B} \cdot \mathbf{P}_2^{\sigma} \cdot \mathbf{I}_A$$

$\mathbf{H}^{PSO}$ and $\mathbf{H}^{DSO}$ are called the *Paramagnetic* and *Diamagnetic Spin–Orbit* operators, while $\mathbf{H}^{\sigma}$ gives the diamagnetic part of the nuclear shielding, eq. (10.15). $\mathbf{H}^{PSO}$ is identical to that in eq. (8.32). $\mathbf{H}^{PSO}$ depends only on one perturbation (nuclear spin) and is a $\mathbf{P}_1$ operator in the above classification (eq. (10.17)), while $\mathbf{H}^{DSO}$ and $\mathbf{H}^{\sigma}$ refer to two perturbations, and are of the $\mathbf{P}_2$ type. They may be written as

$$\mathbf{P}_1^{PSO} = \frac{\mu_0}{4\pi} \frac{(\mathbf{r} - \mathbf{R}_A) \times \mathbf{p}}{|\mathbf{r} - \mathbf{R}_A|^3}$$

$$\mathbf{P}_2^{DSO} = \frac{\mu_0^2}{32\pi^2} \frac{(\mathbf{r} - \mathbf{R}_A)^t (\mathbf{r} - \mathbf{R}_B) - (\mathbf{r} - \mathbf{R}_A)(\mathbf{r} - \mathbf{R}_B)^t}{|\mathbf{r} - \mathbf{R}_A|^3 |\mathbf{r} - \mathbf{R}_B|^3} \tag{10.74}$$

$$\mathbf{P}_2^{\sigma} = \frac{\mu_0}{8\pi} \frac{(\mathbf{r} - \mathbf{R}_G)^t (\mathbf{r} - \mathbf{R}_A) - (\mathbf{r} - \mathbf{R}_G)(\mathbf{r} - \mathbf{R}_A)^t}{|\mathbf{r} - \mathbf{R}_A|^3}$$

where the notation for $\mathbf{P}_2^{DSO}$ and $\mathbf{P}_2^{\sigma}$ is analogous to that in eq. (10.67) and $\mathbf{R}_{A/B}$ denotes nuclear positions.

The NMR shielding, which is the mixed second derivative with respect to a nuclear spin and an external magnetic field, has in analogy with the magnetizability a diamagnetic and a paramagnetic part.[5] The diamagnetic part arises from $\mathbf{P}_2^{\sigma}$, while the paramagnetic contribution contains products of matrix elements from $\mathbf{P}_1^{PSO}$ (from the nuclear spin) and the angular moment operator $\mathbf{L}$ (from the external field). Written in terms of the perturbation formula (10.18), the expression for the nuclear shielding for atom A becomes

$$\boldsymbol{\sigma}_A = \langle \Psi_0 | \mathbf{P}_2^\sigma | \Psi_0 \rangle + \sum_{i=1}^{\infty} \frac{\langle \Psi_0 | \mathbf{P}_1^{PSO} | \Psi_i \rangle \langle \Psi_i | \mathbf{L} | \Psi_0 \rangle + \langle \Psi_0 | \mathbf{L} | \Psi_i \rangle \langle \Psi_i | \mathbf{P}_1^{PSO} | \Psi_0 \rangle}{E_0 - E_i}$$

(10.75)

All the operators $\mathbf{P}_2^\sigma$, $\mathbf{P}_1^{PSO}$ and $\mathbf{L}$ are gauge dependent, relating to the position of atom A via eqs. (10.9) and (10.74), and each of the dia- and para-magnetic terms depends on the chosen gauge.

In order to describe nuclear spin–spin coupling, we need to include electron and nuclear spins, which are not present in the non-relativistic Hamilton operator. A relativistic treatment, as shown in Section 8.2, gives a direct nuclear–nuclear coupling term (eq. (8.33)).

$$H_{nn}^{SS} = \frac{\mu_0}{8\pi} \left(\frac{\mathbf{I}_A \cdot \mathbf{I}_B}{|\mathbf{R}_A - \mathbf{R}_B|^3} - 3 \frac{(\mathbf{I}_A \cdot (\mathbf{R}_A - \mathbf{R}_B))((\mathbf{R}_A - \mathbf{R}_B) \cdot \mathbf{I}_B)}{|\mathbf{R}_A - \mathbf{R}_B|^5} \right)$$

(10.76)

For rapidly tumbling molecules (solution or gas phase) this contribution averages out to zero, but it is significant for solid state NMR. The $\mathbf{H}_{ne}^{SS}$ operator (eq. (8.32)) gives an indirect term, which is normally written as two separate operators.

$$\mathbf{H}^{FC} = \frac{g_e \mu_0}{3} (\mathbf{s} \cdot \mathbf{I}_A) \delta(\mathbf{r} - \mathbf{R}_A) = \mathbf{I}_A \cdot \mathbf{P}_1^{FC}$$

$$\mathbf{H}^{SD} = -\frac{g_e \mu_0}{8\pi} \left[\frac{\mathbf{s} \cdot \mathbf{I}_A}{|\mathbf{r} - \mathbf{R}_A|^3} - 3 \frac{(\mathbf{s} \cdot (\mathbf{r} - \mathbf{R}_A))((\mathbf{r} - \mathbf{R}_A) \cdot \mathbf{I}_A)}{|\mathbf{r} - \mathbf{R}_A|^5} \right] = \mathbf{I}_A \cdot \mathbf{P}_1^{SD}$$

(10.77)

$\mathbf{H}^{FC}$ is the *Fermi Contact* and $\mathbf{H}^{SD}$ is the *Spin–Dipolar* operator, where $\mathbf{s}$ is the (electron) spin operator. $\mathbf{H}^{FC}$ and $\mathbf{H}^{SD}$ depend only on one nucleus and are thus $\mathbf{P}_1$ type operators.

$$\mathbf{P}_1^{FC} = \frac{g_e \mu_0}{3} \delta(\mathbf{r} - \mathbf{R}_A) \mathbf{s}$$

$$\mathbf{P}_1^{SD} = -\frac{g_e \mu_0}{8\pi} \left[\frac{\mathbf{s}}{|\mathbf{r} - \mathbf{R}_A|^3} - 3 \frac{(\mathbf{s} \cdot (\mathbf{r} - \mathbf{R}_A))(\mathbf{r} - \mathbf{R}_A)}{|\mathbf{r} - \mathbf{R}_A|^5} \right]$$

(10.78)

The $\mathbf{H}^{FC}$ and $\mathbf{H}^{SD}$ operators determine the isotropic and anisotropic parts of the hyperfine coupling constant (eq. (10.11)), respectively. The latter contribution averages out for rapidly tumbling molecules (solution or gas phase), and the (isotropic) hyperfine coupling constant is therefore determined by the Fermi–Contact contribution, i.e. the electron density at the nucleus.

The indirect spin–spin coupling between nuclei A and B, which is the one observed in solution phase NMR, contains several contributions, as seen from eq. (10.18)

$$\mathbf{J}_{AB} = \langle \Psi_0 | \mathbf{P}_{2,AB}^{DSO} | \Psi_0 \rangle + \sum_{i \neq 0} \frac{\langle \Psi_0 | \mathbf{P}_{1,A} | \Psi_i \rangle \langle \Psi_i | \mathbf{P}_{1,B}' | \Psi_0 \rangle}{E_0 - E_i}$$

(10.79)

The first part can be evaluated as the expectation value of $\mathbf{P}_2^{DSO}$ (eq. (10.74). The second part contains six pieces, corresponding to all combinations of $\mathbf{P}_1^{PSO}, \mathbf{P}_1^{FC}$ and $\mathbf{P}_1^{SD}$ (eqs. (10.74) and (10.78)). $\mathbf{P}_1^{FC}$ and $\mathbf{P}_1^{SD}$ contain the electron spin operators and for a singlet ground state (as is usually the case), this means that the excited state Ψ_i in the summation must be a triplet state. Since $\mathbf{P}_1^{PSO}$ does not depend on eletronic spin, this means that the combinations of $\mathbf{P}_1^{PSO}$ with either $\mathbf{P}_1^{FC}$ or $\mathbf{P}_1^{SD}$ give zero contribution. For rapidly tumbling molecules, it can be shown the cross term between $\mathbf{P}_1^{FC}$ and $\mathbf{P}_1^{SD}$ averages out. For the trace (sum of the diagonal terms) of the 3×3 coupling matrix $\mathbf{J}$, which is the observed coupling constant, only the three "diagonal" terms $(\mathbf{P}_1 = \mathbf{P}_1')$ in (10.79) thus survive. It is often found that the Fermi–Contact term is the most important, especially for one-bond couplings $(^2\mathbf{J})$, followed by the Paramagnetic Spin–Orbit, while the Diamagnetic Spin–Orbit and Spin–Dipolar contributions are small.[6] It should be noted, however, that this is based on relatively few data.

10.7.3 Gauge Dependence of Magnetic Properties

There are two factors which make the calculation of magnetic properties somewhat more complicated than the corresponding electric properties. First, the angular moment operator $\mathbf{L}$ is imaginary (eq. (10.9)), implying that the wave function must be allowed to be complex. Second, the presence of the gauge origin in the operators means that the results may be origin dependent. An exact wave function will of course give origin-independent results, as will an HF wave function if a complete basis set is employed. In practice, however, a finite basis must be employed, and standard basis sets will yield results which depend on where the user has chosen the origin of the gauge to be. The centre of mass is often used, but this is by no means a unique choice. The gauge error depends on the distance between the wave function and the gauge origin, and some methods try to minimize the error by selecting separate gauges for each (localized) molecular orbital. Two such methods are known as *Individual Gauge for Localized Orbitals* (IGLO)[7] and *Localized Orbital/local oRiGin* (LORG).[8]

A more recent implementation, which completely eliminates the gauge dependence, is to make the basis functions explicitly dependent on the magnetic field by inclusion of a complex phase factor referring to the position of the basis function (usually the nucleus).

$$\mathbf{X}_A(\mathbf{r}, \mathbf{R}_A) = e^{-\frac{i}{c}\mathbf{A}_A \cdot \mathbf{r}}\chi_A(\mathbf{r}, \mathbf{R}_A)$$
$$\chi_A(\mathbf{r}, \mathbf{R}_A) = Ne^{-\alpha(\mathbf{r}-\mathbf{R}_A)^2} \tag{10.80}$$
$$\mathbf{A}_A = \tfrac{1}{2}\mathbf{B} \times (\mathbf{R}_A - \mathbf{R}_G)$$

Such orbitals are known as *London Atomic Orbitals* (LAO) or *Gauge Including/ Invariant Atomic Orbitals* (GIAO).[9] The effect is that matrix elements involving GIAOs only contain a <u>difference</u> in vector potentials, thereby removing the reference to an absolute gauge origin. For the overlap and potential energy it is straightforward to see that matrix elements become independent of the gauge origin.

$$\langle \mathbf{X}_A | \mathbf{X}_B \rangle = \left\langle \chi_A \left| e^{\frac{i}{c}(\mathbf{A}_A - \mathbf{A}_B) \cdot \mathbf{r}} \right| \chi_B \right\rangle$$
$$\langle \mathbf{X}_A | \mathbf{V} | \mathbf{X}_B \rangle = \langle \chi_A | e^{\frac{i}{c}(\mathbf{A}_A - \mathbf{A}_B) \cdot \mathbf{r}} \mathbf{V} | \chi_B \rangle \tag{10.81}$$
$$\mathbf{A}_A - \mathbf{A}_B = \tfrac{1}{2}\mathbf{B} \times (\mathbf{R}_A - \mathbf{R}_B)$$

The kinetic energy is slightly more complicated, but it can be shown that the following relation holds:

$$
\begin{aligned}
\langle \mathbf{X}_A | \boldsymbol{\pi}^2 | \mathbf{X}_B \rangle &= \langle \mathbf{X}_A | (\mathbf{p} + \tfrac{1}{2}\mathbf{B} \times (\mathbf{r} - \mathbf{R}_G))^2 | \mathbf{X}_B \rangle \\
&= \langle \chi_A | e^{\frac{i}{c}(\mathbf{A}_A - \mathbf{A}_B)\cdot\mathbf{r}} (\mathbf{p} + \tfrac{1}{2}\mathbf{B} \times (\mathbf{r} - \mathbf{R}_B))^2 | \chi_B \rangle
\end{aligned}
\tag{10.82}
$$

Note that $\mathbf{R}_G$ has been replaced by $\mathbf{R}_B$ in the last bracket. The use of GIAOs as basis functions makes all matrix elements, and therefore all properties, independent of the gauge origin. The wave function itself, however, is expressed in term of the basis functions, and therefore becomes origin dependent, by means of a complex phase factor. The use of perturbation-dependent basis functions has the further advantage of strongly reducing the need for high angular momentum functions, i.e. the property is typically calculated with an accuracy comparable to that of the unperturbed system.[10]

While LAO/GIAO had been proposed well before the advent of modern computational chemistry, it was only developments in calculating (geometrical) derivatives of the energy (and wave function) that made it practical to use field-dependent orbitals.[11]

10.8 Geometry Perturbations

If the perturbation is a change in geometry, the basis set is clearly perturbation dependent since the functions move along with the nuclei. Standard perturbation theory is therefore not suitable for calculating molecular gradients.

The general formula for the derivative is given by eq. (10.21).

$$
\frac{\partial E}{\partial \mathbf{R}} = \left\langle \Psi_0 \left| \frac{\partial \mathbf{H}}{\partial \mathbf{R}} \right| \Psi_0 \right\rangle + 2 \left\langle \frac{\partial \Psi_0}{\partial \mathbf{R}} \left| \mathbf{H}_0 \right| \Psi_0 \right\rangle
\tag{10.83}
$$

The first term is the Hellmann–Feynman force and the second is the wave function response. The latter now contains contributions both from a change in basis functions and MO coefficients.

$$
\frac{\partial \Psi}{\partial \mathbf{R}} = \frac{\partial \Psi}{\partial \chi}\frac{\partial \chi}{\partial \mathbf{R}} + \frac{\partial \Psi}{\partial \mathbf{c}}\frac{\partial \mathbf{c}}{\partial \mathbf{R}}
\tag{10.84}
$$

For an HF wave function the MO dependence disappears owing to the variational nature ($\partial \Psi / \partial \mathbf{c} = 0$).

Since geometry derivatives are important for optimizing geometries, it may be useful to look in more detail at the quantities involved in calculating first and second derivatives of an HF wave function. Such formulas are most easily derived directly from the HF energy expressed in terms of the atomic quantities (eq. (3.53).[12]

$$
E_{\mathrm{HF}} = \sum_{\alpha\beta}^{M} D_{\alpha\beta} h_{\alpha\beta} + \frac{1}{2}\sum_{\alpha\beta\gamma\delta}^{M} D_{\alpha\beta} D_{\gamma\delta}(\langle \chi_\alpha \chi_\gamma | \chi_\beta \chi_\delta \rangle - \langle \chi_\alpha \chi_\gamma | \chi_\delta \chi_\beta \rangle) + V_{\mathrm{nn}}
$$

$$
\tag{10.85}
$$

Differentiation (using λ as a general geometrical displacement of a nucleus) yields the expression

$$
\begin{aligned}
\frac{\partial E_{HF}}{\partial \lambda} =& \sum_{\alpha\beta}^{M} \left(\frac{\partial D_{\alpha\beta}}{\partial \lambda} h_{\alpha\beta} + D_{\alpha\beta} \frac{\partial h_{\alpha\beta}}{\partial \lambda} \right) \\
&+ \frac{1}{2} \sum_{\alpha\beta\gamma\delta}^{M} \frac{\partial D_{\alpha\beta}}{\partial \lambda} D_{\gamma\delta} (\langle \chi_\alpha \chi_\gamma | \chi_\beta \chi_\delta \rangle - \langle \chi_\alpha \chi_\gamma | \chi_\delta \chi_\beta \rangle) \\
&+ \frac{1}{2} \sum_{\alpha\beta\gamma\delta}^{M} D_{\alpha\beta} \frac{\partial D_{\gamma\delta}}{\partial \lambda} (\langle \chi_\alpha \chi_\gamma | \chi_\beta \chi_\delta \rangle - \langle \chi_\alpha \chi_\gamma | \chi_\delta \chi_\beta \rangle) \\
&+ \frac{1}{2} \sum_{\alpha\beta\gamma\delta}^{M} D_{\alpha\beta} D_{\gamma\delta} \frac{\partial}{\partial \lambda} (\langle \chi_\alpha \chi_\gamma | \chi_\beta \chi_\delta \rangle - \langle \chi_\alpha \chi_\gamma | \chi_\delta \chi_\beta \rangle) + \frac{\partial V_{nn}}{\partial \lambda}
\end{aligned}
\tag{10.86}
$$

The third and fourth terms are identical and may be collected to cancel the factor of 1/2. Rearranging the terms gives

$$
\begin{aligned}
\frac{\partial E_{HF}}{\partial \lambda} =& \sum_{\alpha\beta}^{M} D_{\alpha\beta} \frac{\partial h_{\alpha\beta}}{\partial \lambda} + \frac{1}{2} \sum_{\alpha\beta\gamma\delta}^{M} D_{\alpha\beta} D_{\gamma\delta} \frac{\partial}{\partial \lambda} (\langle \chi_\alpha \chi_\gamma | \chi_\beta \chi_\delta \rangle - \langle \chi_\alpha \chi_\gamma | \chi_\delta \chi_\beta \rangle) + \frac{\partial V_{nn}}{\partial \lambda} \\
&+ \sum_{\alpha\beta}^{M} \frac{\partial D_{\alpha\beta}}{\partial \lambda} h_{\alpha\beta} + \sum_{\alpha\beta\gamma\delta}^{M} \frac{\partial D_{\alpha\beta}}{\partial \lambda} D_{\gamma\delta} (\langle \chi_\alpha \chi_\gamma | \chi_\beta \chi_\delta \rangle - \langle \chi_\alpha \chi_\gamma | \chi_\delta \chi_\beta \rangle)
\end{aligned}
\tag{10.87}
$$

The first two terms involve products of the density matrix with derivatives of the atomic integrals, while the two last terms can be recognized as derivatives of the density matrix times the Fock matrix (eq. (3.51)).

$$
\begin{aligned}
\frac{\partial E_{HF}}{\partial \lambda} =& \sum_{\alpha\beta}^{M} D_{\alpha\beta} \frac{\partial h_{\alpha\beta}}{\partial \lambda} + \frac{1}{2} \sum_{\alpha\beta\gamma\delta}^{M} D_{\alpha\beta} D_{\gamma\delta} \frac{\partial}{\partial \lambda} (\langle \chi_\alpha \chi_\gamma | \chi_\beta \chi_\delta \rangle - \langle \chi_\alpha \chi_\gamma | \chi_\delta \chi_\beta \rangle) \\
&+ \frac{\partial V_{nn}}{\partial \lambda} + \sum_{\alpha\beta}^{M} \frac{\partial D_{\alpha\beta}}{\partial \lambda} F_{\alpha\beta}
\end{aligned}
\tag{10.88}
$$

The derivative in eq. (10.88) of the nuclear repulsion (third term) is trivial since it does not involve electron coordinates. The one-electron derivatives are given as

$$
\begin{aligned}
h_{\alpha\beta} &= \langle \chi_\alpha | \mathbf{h} | \chi_\beta \rangle \\
\frac{\partial h_{\alpha\beta}}{\partial \lambda} &= \left\langle \frac{\partial \chi_\alpha}{\partial \lambda} \middle| \mathbf{h} \middle| \chi_\beta \right\rangle + \left\langle \chi_\alpha \middle| \frac{\partial \mathbf{h}}{\partial \lambda} \middle| \chi_\beta \right\rangle + \left\langle \chi_\alpha \middle| \mathbf{h} \middle| \frac{\partial \chi_\beta}{\partial \lambda} \right\rangle
\end{aligned}
\tag{10.89}
$$

The central of these is recognized as the Hellmann–Feynman force. The two-electron derivatives in eq. (10.88) become

$$\langle \chi_\alpha \chi_\gamma | \chi_\beta \chi_\delta \rangle = \langle \chi_\alpha \chi_\gamma | \mathbf{g} | \chi_\beta \chi_\delta \rangle$$

$$\frac{\partial}{\partial \lambda} \langle \chi_\alpha \chi_\gamma | \chi_\beta \chi_\delta \rangle = \left\langle \frac{\partial \chi_\alpha}{\partial \lambda} \chi_\gamma \middle| \mathbf{g} \middle| \chi_\beta \chi_\delta \right\rangle + \left\langle \chi_\alpha \frac{\partial \chi_\gamma}{\partial \lambda} \middle| \mathbf{g} \middle| \chi_\beta \chi_\delta \right\rangle$$

$$+ \left\langle \chi_\alpha \chi_\gamma \middle| \frac{\partial \mathbf{g}}{\partial \lambda} \middle| \chi_\beta \chi_\delta \right\rangle + \left\langle \chi_\alpha \chi_\gamma \middle| \mathbf{g} \middle| \frac{\partial \chi_\beta}{\partial \lambda} \chi_\delta \right\rangle \qquad (10.90)$$

$$+ \left\langle \chi_\alpha \chi_\gamma \middle| \mathbf{g} \middle| \chi_\beta \frac{\partial \chi_\delta}{\partial \lambda} \right\rangle$$

The central term is again the Hellmann–Feynman force, which vanishes since the two-electron operator $\mathbf{g}$ is independent of the nuclear positions.

The last term in eq. (10.88) involves a change in the density matrix, i.e. the MO coefficients.

$$D_{\alpha\beta} = \sum_{i=1}^{N} n_i c_{\alpha i} c_{\beta i}$$

$$\frac{\partial D_{\alpha\beta}}{\partial \lambda} = \sum_{i=1}^{N} n_i \left(\frac{\partial c_{\alpha i}}{\partial \lambda} c_{\beta i} + c_{\alpha i} \frac{\partial c_{\beta i}}{\partial \lambda} \right) \qquad (10.91)$$

Since the HF wave function is variational, the explicit calculation of the density derivatives can be avoided. The last term in eq. (10.88) may with eq. (10.91) be written as

$$\sum_{\alpha\beta}^{M} \frac{\partial D_{\alpha\beta}}{\partial \lambda} F_{\alpha\beta} = \sum_{\alpha\beta}^{M} \sum_{i=1}^{N} n_i \left(\frac{\partial c_{\alpha i}}{\partial \lambda} F_{\alpha\beta} c_{\beta i} + \frac{\partial c_{\beta i}}{\partial \lambda} F_{\alpha\beta} c_{\alpha i} \right) \qquad (10.92)$$

By virtue of the HF condition ($\mathbf{FC} = \mathbf{SC\varepsilon}$), eq. (10.92) may be written in terms of overlap integrals and MO energies.

$$\sum_{\alpha\beta}^{M} \frac{\partial D_{\alpha\beta}}{\partial \lambda} F_{\alpha\beta} = \sum_{\alpha\beta}^{M} \sum_{i=1}^{N} n_i \left(\frac{\partial c_{\alpha i}}{\partial \lambda} F_{\alpha\beta} c_{\beta i} + \frac{\partial c_{\beta i}}{\partial \lambda} F_{\alpha\beta} c_{\alpha i} \right)$$

$$= \sum_{\alpha\beta}^{M} \sum_{i=1}^{N} n_i \left(\frac{\partial c_{\alpha i}}{\partial \lambda} S_{\alpha\beta} c_{\beta i} \varepsilon_i + \frac{\partial c_{\beta i}}{\partial \lambda} S_{\alpha\beta} c_{\alpha i} \varepsilon_i \right) \qquad (10.93)$$

Finally, since the MOs are orthonormal, the derivatives of the coefficients may be replaced by derivatives of the overlap matrix.

$$\langle \phi_i | \phi_j \rangle = \sum_{\alpha\beta}^{M} c_{\alpha i} c_{\beta i} \langle \chi_\alpha | \chi_\beta \rangle = \sum_{\alpha\beta}^{M} c_{\alpha i} c_{\beta i} S_{\alpha\beta} = \delta_{ij}$$

$$\frac{\partial}{\partial \lambda} \langle \phi_i | \phi_j \rangle = \left(\frac{\partial c_{\alpha i}}{\partial \lambda} S_{\alpha\beta} c_{\beta i} + c_{\alpha i} \frac{\partial S_{\alpha\beta}}{\partial \lambda} c_{\beta i} + c_{\alpha i} S_{\alpha\beta} \frac{\partial c_{\beta i}}{\partial \lambda} \right) = 0 \qquad (10.94)$$

$$2 \frac{\partial c_{\alpha i}}{\partial \lambda} S_{\alpha\beta} c_{\beta i} = -c_{\alpha i} \frac{\partial S_{\alpha\beta}}{\partial \lambda} c_{\beta i}$$

The final derivative of the energy may thus be written as

$$\frac{\partial E_{HF}}{\partial \lambda} = \sum_{\alpha\beta}^{M} D_{\alpha\beta} \frac{\partial h_{\alpha\beta}}{\partial \lambda} + \frac{1}{2} \sum_{\alpha\beta\gamma\delta}^{M} D_{\alpha\beta} D_{\gamma\delta} \frac{\partial}{\partial \lambda} (\langle \chi_\alpha \chi_\gamma | \chi_\beta \chi_\delta \rangle - \langle \chi_\alpha \chi_\gamma | \chi_\delta \chi_\beta \rangle)$$

$$+ \frac{\partial V_{nn}}{\partial \lambda} - \sum_{\alpha\beta}^{M} W_{\alpha\beta} \frac{\partial S_{\alpha\beta}}{\partial \lambda} \tag{10.95}$$

or alternatively as

$$\frac{\partial E_{HF}}{\partial \lambda} = \sum_{\alpha\beta}^{M} D_{\alpha\beta} \frac{\partial h_{\alpha\beta}}{\partial \lambda} + \frac{1}{2} \sum_{\alpha\beta\gamma\delta}^{M} (D_{\alpha\beta} D_{\gamma\delta} - D_{\alpha\delta} D_{\gamma\beta}) \frac{\partial \langle \chi_\alpha \chi_\gamma | \chi_\beta \chi_\delta \rangle}{\delta\lambda}$$

$$+ \frac{\partial V_{nn}}{\delta\lambda} - \sum_{\alpha\beta}^{M} W_{\alpha\beta} \frac{\partial S_{\alpha\beta}}{\partial \lambda} \tag{10.96}$$

where the energy-weighted density matrix $\mathbf{W}$ has been introduced.

$$W_{\alpha\beta} = \sum_{i=1}^{N} \varepsilon_i c_{\alpha i} c_{\beta i} \tag{10.97}$$

Consider now the case where the perturbation λ is a specific nuclear displacement, $X_k \rightarrow X_k + \Delta X_k$. The derivatives of the one- and two-electron integrals are of two types, those involving derivatives of the basis functions, and those involving derivatives of the operators. The latter are given as

$$\frac{\partial \mathbf{h}}{\partial X_k} = \frac{\partial}{\partial X_k} \left(-\frac{1}{2} \nabla_i^2 - \sum_a \frac{Z_a}{|\mathbf{R}_a - \mathbf{r}_i|} \right) = \frac{(X_k - x_i)Z_k}{|\mathbf{R}_k - \mathbf{r}_i|^3}$$

$$\frac{\partial \mathbf{g}}{\partial X_k} = 0 \tag{10.98}$$

$$\frac{\partial \mathbf{V}_{nn}}{\partial X_k} = \frac{\partial}{\partial X_k} \left(\sum_{a>b} \frac{Z_a Z_b}{|\mathbf{R}_a - \mathbf{R}_b|} \right) = -\sum_{b \neq k} \frac{(X_k - X_b)Z_k Z_b}{|\mathbf{R}_k - \mathbf{R}_b|^3}$$

The derivative of the core operator $\mathbf{h}$ is a one-electron operator similar to the nucleus-electron attraction required for the energy itself (eq. (3.55)). The two-electron part yields zero, and the $\mathbf{V}_{nn}$ term is independent of the electronic wave function. The remaining terms in eqs. (10.89), (10.90) and (10.95) all involve derivatives of the basis functions. When these are Gaussian functions (as is usually the case) the derivative can be written in terms of two other Gaussian functions, having one lower and one higher angular momentum.

$$\chi_\alpha(\mathbf{R}_k) = N(X_k - x)^l (Y_k - y)^m (Z_k - z)^n e^{-\alpha(\mathbf{r}-\mathbf{R}_k)^2}$$

$$\frac{\partial \chi_\alpha(\mathbf{R}_k)}{\partial X_k} = N(X_k - x)^{l-1} (Y_k - y)^m (Z_k - z)^n e^{-\alpha(\mathbf{r}-\mathbf{R}_k)^2}$$

$$- 2N\alpha(X_k - x)^{l+1} (Y_k - y)^m (Z_k - z)^n e^{-\alpha(\mathbf{r}-\mathbf{R}_k)^2} \tag{10.99}$$

The derivative of a p-function can thus be written in terms of an s- and a d-type Gaussian function. The one- and two-electron integrals involving derivatives of basis functions are therefore of the same type as those used in the energy expression itself, the only difference is the angular momentum, and the fact that there are ~ 3 times as many of these derivative integrals than for the energy itself. Of all the terms in eqs. (10.95) and (10.96), the only significant computational effort is the derivatives of the two-electron integrals. Note, however, that the density matrix elements are known at the time when these integrals are calculated, and screening procedures analogous to those used in direct SCF techniques (Section 3.8.5) can be used to avoid calculating integrals which make insignificant contributions to the final result.

The second derivative of the energy with respect to a geometry change can be written as

$$
\frac{\partial^2 E_{\mathrm{HF}}}{\partial \lambda^2} = \sum_{\alpha\beta}^{M} D_{\alpha\beta} \frac{\partial^2 h_{\alpha\beta}}{\partial \lambda^2} + \frac{1}{2} \sum_{\alpha\beta\gamma\delta}^{M} D_{\alpha\beta} D_{\gamma\delta} \frac{\partial^2}{\partial \lambda^2} (\langle \chi_\alpha \chi_\gamma | \chi_\beta \chi_\delta \rangle - \langle \chi_\alpha \chi_\gamma | \chi_\delta \chi_\beta \rangle)
$$

$$
+ \frac{\partial^2 V_{\mathrm{nn}}}{\partial \lambda^2} - \sum_{\alpha\beta}^{M} W_{\alpha\beta} \frac{\partial^2 S_{\alpha\beta}}{\partial \lambda^2} + \sum_{\alpha\beta}^{M} \frac{\partial D_{\alpha\beta}}{\partial \lambda} \frac{\partial h_{\alpha\beta}}{\partial \lambda}
$$

$$
+ \sum_{\alpha\beta\gamma\delta}^{M} \frac{\partial D_{\alpha\beta}}{\partial \lambda} D_{\gamma\delta} \frac{\partial}{\partial \lambda} (\langle \chi_\alpha \chi_\gamma | \chi_\beta \chi_\delta \rangle - \langle \chi_\alpha \chi_\gamma | \chi_\delta \chi_\beta \rangle)
$$

$$
- \sum_{\alpha\beta}^{M} \frac{\partial W_{\alpha\beta}}{\partial \lambda} \frac{\partial S_{\alpha\beta}}{\partial \lambda} \tag{10.100}
$$

The first four terms only involve derivatives of operators and AO integrals; however, for the last three terms we need the derivative of the density matrix and MO energies. These can be obtained by solving the first-order CPHF equations (Section 10.5).

10.9 Propagator Methods

The name *Propagator*, also known as a *Greens function*, arises from a time-dependent evolution of a given quantity.[13] For two time-dependent operators $\mathbf{P}(t)$ and $\mathbf{Q}(t)$, a propagator may be defined as

$$
\langle\langle \mathbf{P}(t); \mathbf{Q}(t') \rangle\rangle = -i\theta(t - t')\langle \Psi_0 | \mathbf{P}(t)\mathbf{Q}(t') | \Psi_0 \rangle \pm i\theta(t' - t)\langle \Psi_0 | \mathbf{Q}(t)\mathbf{P}(t') | \Psi_0 \rangle \tag{10.101}
$$

where the $\pm$ sign depends on whether $\mathbf{P}$ and $\mathbf{Q}$ are number conserving operators or not, and $\theta(x)$ is the Heaviside stepfunction ($\theta(x) = 0$ for $x<0$ and $\theta(x) = 1$ for $x>0$). The propagator may be Fourier transformed to an *energy representation*, also called a *spectral* or *frequency representation* (ω is the frequency).

$$
\langle\langle \mathbf{P}; \mathbf{Q} \rangle\rangle_\omega = \sum_{i \neq 0} \frac{\langle \Psi_0 | \mathbf{P} | \Psi_i \rangle \langle \Psi_i | \mathbf{Q} | \Psi_0 \rangle}{\omega - E_i + E_0 + i\eta} \pm \frac{\langle \Psi_0 | \mathbf{Q} | \Psi_i \rangle \langle \Psi_i | \mathbf{P} | \Psi_0 \rangle}{\omega + E_i - E_0 - i\eta} \tag{10.102}
$$

Here η is an infinitesimally small number which ensures that the transformation is valid also when $\omega = \pm(E_i - E_0)$.

If the **P/Q** operators correspond to removal or addition of an electron, the propagator is called an *electron propagator*. The *poles* of the propagator (where the denominator is zero) correspond to ionization potentials and electron affinities.

If the **P/Q** operators are number conserving operators, the propagator is called a *Polarization Propagator* (PP). It may be viewed as the response of property **P** to perturbation **Q**. For the case where **P** = **Q** = **r** (the position operator), the propagator describes the response of the dipole moment $\langle\Psi_0|\mathbf{r}|\Psi_0\rangle$ to a linear field $\mathbf{F} = F\mathbf{r}$.

$$\langle\langle\mathbf{r};\mathbf{r}\rangle\rangle_\omega = \sum_{i\neq 0}\frac{|\langle\Psi_0|\mathbf{r}|\Psi_i\rangle|^2}{\omega - E_i + E_0 + i\eta} \pm \frac{|\langle\Psi_0|\mathbf{r}|\Psi_i\rangle|^2}{\omega + E_i - E_0 - i\eta} \tag{10.103}$$

The poles correspond to excitation energies, and the residues (numerator at the poles) to transition moments between the reference and excited states. In the limit where $\omega \to 0$ (i.e. where the perturbation is time independent), the propagator is identical to the second-order perturbation formula for a constant electric field (eq. (10.57)), i.e. the $\langle\langle\mathbf{r};\mathbf{r}\rangle\rangle_0$ propagator determines the static polarizability.

Choosing a non-zero value for ω corresponds to a time-dependent field with a frequency ω, i.e. the $\langle\langle\mathbf{r};\mathbf{r}\rangle\rangle_\omega$ propagator determines the frequency-dependent polarizability corresponding to an electric field described by the perturbation operator $\mathbf{Q}(t) = \mathbf{r}\cos(\omega t)$. Propagator methods are therefore well suited for calculating dynamical properties, and by suitable choices for the **P** and **Q** operators, a whole variety of properties may be calculated.[14]

Although eq. (10.103) for the propagator appears to involve the same effort as the perturbation approach (sum over all excited states, eq. (10.18)), the actual calculation of the propagator is somewhat different. Returning to the time representation of the polarization propagator, it may be written in terms of a *commutator*.

$$\langle\langle\mathbf{P}(t);\mathbf{Q}(t')\rangle\rangle = -i\theta(t-t')\langle\Psi_0|[\mathbf{P}(t),\mathbf{Q}(t')]|\Psi_0\rangle$$
$$[\mathbf{P}(t),\mathbf{Q}(t')] = \mathbf{P}(t)\mathbf{Q}(t') - \mathbf{Q}(t)\mathbf{P}(t') \tag{10.104}$$

Using the Heisenberg equation of motion

$$i\frac{d\mathbf{P}(t)}{dt} = [\mathbf{P}(t),\mathbf{H}] \tag{10.105}$$

for the propagator yields

$$i\frac{d\langle\langle\mathbf{P}(t);\mathbf{Q}(t')\rangle\rangle}{dt} = \delta(t-t')\langle\Psi_0|[\mathbf{P}(t),\mathbf{Q}(t)]|\Psi_0\rangle + \langle\langle[\mathbf{P}(t),\mathbf{H}];\mathbf{Q}(t')\rangle\rangle \tag{10.106}$$

Moving back to the frequency representation, and using that $\langle\langle[\mathbf{P},\mathbf{H}];\mathbf{Q}\rangle\rangle = \langle\langle\mathbf{P};[\mathbf{H},\mathbf{Q}]\rangle\rangle$, allows eq. (10.106) to be written as

$$\omega\langle\langle\mathbf{P};\mathbf{Q}\rangle\rangle_\omega = \langle\Psi_0|[\mathbf{P},\mathbf{Q}]|\Psi_0\rangle + \langle\langle\mathbf{P};[\mathbf{H},\mathbf{Q}]\rangle\rangle_\omega \tag{10.107}$$

This shows that a propagator may be written as an expectation value of a commutator plus another propagator involving a commutator and the Hamiltonian. Applying this formula iteratively gives

$$\langle\langle\mathbf{P};\mathbf{Q}\rangle\rangle_\omega = \omega^{-1}\langle\Psi_0|[\mathbf{P},\mathbf{Q}]|\Psi_0\rangle + \omega^{-2}\langle\Psi_0|[\mathbf{P},[\mathbf{H},\mathbf{Q}]]|\Psi_0\rangle$$
$$+ \omega^{-3}\langle\Psi_0|[\mathbf{P},[\mathbf{H},[\mathbf{H},\mathbf{Q}]]]|\Psi_0\rangle + \ldots \tag{10.108}$$

The propagator may thus be written as an infinite series of expectation values of increasingly complex operators over the reference wave function.

We now define identity and Hamilton *superoperators* as

$$
\begin{aligned}
\hat{\mathbf{I}}\mathbf{Q} &= \mathbf{Q} \\
\hat{\mathbf{H}}\mathbf{Q} &= [\mathbf{H}, \mathbf{Q}] \\
\hat{\mathbf{H}}^2\mathbf{Q} &= [\mathbf{H}, [\mathbf{H}, \mathbf{Q}]] \\
\hat{\mathbf{H}}^3\mathbf{Q} &= [\mathbf{H}, [\mathbf{H}, [\mathbf{H}, \mathbf{Q}]]]
\end{aligned}
\tag{10.109}
$$

where the "super" reflects that the $^{\wedge}$-operators work on operators rather than functions. The binary product corresponding to a bracket is in superoperator space defined as

$$
(\mathbf{P}|\mathbf{Q}) = \langle \Psi_0 | [\mathbf{P}^\dagger, \mathbf{Q}] | \Psi_0 \rangle
\tag{10.110}
$$

The infinite sum eq. (10.108) can then be written as an inverse.

$$
\langle\langle \mathbf{P}; \mathbf{Q} \rangle\rangle_\omega = \langle \Psi_0 | [\mathbf{P}, (\omega\hat{\mathbf{I}} - \hat{\mathbf{H}})^{-1}\mathbf{Q}] | \Psi_0 \rangle
\tag{10.111}
$$

This may be further transformed by an "inner projection" onto a complete set of excitation and de-excitation operators, $\mathbf{h}$: this is equivalent to inserting a "resolution of the identity" in the operator space (remember that superoperators work on operators).

$$
\langle\langle \mathbf{P}; \mathbf{Q} \rangle\rangle_\omega = (\mathbf{P}|\mathbf{h})(\mathbf{h}|\omega\hat{\mathbf{I}} - \hat{\mathbf{H}}|\mathbf{h})^{-1}(\mathbf{h}|\mathbf{Q})
\tag{10.112}
$$

For the electron propagator we may write $\mathbf{h}$ as

$$
\mathbf{h} = \{\mathbf{h}_1, \mathbf{h}_3, \mathbf{h}_5, \ldots\}
\tag{10.113}
$$

where $\mathbf{h}_1$ corresponds to addition or removal of an electron, $\mathbf{h}_3$ to addition or removal of an electron while simultaneously generating a single excitation or de-excitation, $\mathbf{h}_5$ to addition or removal of an electron while simultaneously generating a double excitation or de-excitation, etc.

For the polarization propagator we may write $\mathbf{h}$ as

$$
\mathbf{h} = \{\mathbf{h}_2, \mathbf{h}_4, \mathbf{h}_6, \ldots\}
\tag{10.114}
$$

where $\mathbf{h}_2$ generates all single excitations and de-excitations, $\mathbf{h}_4$ all double excitations and de-excitations, etc.

So far everything is exact. A complete manifold of excitation operators, however, means that all excited states are considered, i.e. a "full CI" approach. Approximate versions of propagator methods may be generated by restricting the excitation level, i.e. truncating $\mathbf{h}$. A complete specification furthermore requires a selection of the reference, normally taken as either an HF or MCSCF wave function.

The simplest polarization propagator corresponds to choosing an HF reference and including only the $\mathbf{h}_2$ operator, known as the *Random Phase Approximation* (RPA). For the static case ($\omega = 0$) the resulting equations are identical to those obtained from a *Time-Dependent Hartree-Fock* (TDHF) analysis or Coupled Hartree–Fock approach, discussed in Section 10.5.

Splitting the $\mathbf{h}_2$ operator into an excitation and de-excitation part, $\mathbf{h}_2 = \{\mathbf{e}, \mathbf{d}\}$, allows the propagator to be written as two property vectors times an inverse matrix, often called the *principal propagator.*

$$\langle\langle\mathbf{P};\mathbf{Q}\rangle\rangle_\omega = ((\mathbf{P}|\mathbf{e}),(\mathbf{P}|\mathbf{d}))\begin{pmatrix}\omega\mathbf{1}-\mathbf{A} & -\mathbf{B} \\ -\mathbf{B} & -\omega\mathbf{1}-\mathbf{A}\end{pmatrix}^{-1}\begin{pmatrix}(\mathbf{e}\,|\,\mathbf{Q}) \\ (\mathbf{d}\,|\,\mathbf{Q})\end{pmatrix}\qquad(10.115)$$

The $\mathbf{A}$ and $\mathbf{B}$ matrices and $\mathbf{P}/\mathbf{Q}$ vectors are defined as

$$\mathbf{A} = (\mathbf{e}|\hat{\mathbf{H}}|\mathbf{e}) = \langle\Psi_0|[\mathbf{e}^\dagger,[\mathbf{H},\mathbf{e}]]|\Psi_0\rangle = \langle\Psi_0|[\mathbf{d},[\mathbf{H},\mathbf{e}]]|\Psi_0\rangle$$

$$\mathbf{B} = (\mathbf{d}|\hat{\mathbf{H}}|\mathbf{e}) = \langle\Psi_0|[\mathbf{d}^\dagger,[\mathbf{H},\mathbf{e}]]|\Psi_0\rangle = \langle\Psi_0|[\mathbf{e},[\mathbf{H},\mathbf{e}]]|\Psi_0\rangle \qquad(10.116)$$

$$(\mathbf{P}|\mathbf{e}) = \langle\Psi_0|[\mathbf{P}^\dagger,\mathbf{e}]|\Psi_0\rangle$$

The $\mathbf{A}$ matrix involves elements between singly excited states while $\mathbf{B}$ is given by matrix elements between doubly excited states and the reference. The $\mathbf{P}/\mathbf{Q}$ elements are matrix elements of the operator between the reference and a singly excited state. If $\mathbf{P} = \mathbf{r}$ this is a transition moment, and in the general case it is often denoted a "property gradient", in analogy with the case where the operator is the Hamiltonian (eq. (3.67)).

$$A_{ij}^{ab} = \langle\Psi_i^a|\mathbf{H}|\Psi_j^b\rangle - E_0\delta_{ij}\delta_{ab}$$

$$B_{ij}^{ab} = -\langle\Psi_0|\mathbf{H}|\Psi_{ij}^{ab}\rangle \qquad(10.117)$$

$$P_i^a = \langle\Psi_0|\mathbf{P}^\dagger|\Psi_i^a\rangle$$

The matrix elements may be reduced to orbital energies and two-electron integrals, as described in Section 4.2.1. Although not clear from this derivation, the principal propagator in eq. (10.115) is related to the $\mathbf{A}$ matrix in the CHF eq. (10.52), i.e. $(\mathbf{A}-\mathbf{B})$ in eq. (10.115) is the same as $\mathbf{A}$ in eq. (10.52).

Since the dimension of the principal propagator matrix may be large, it is impractical to calculate the inverse matrix in eq. (10.115) directly. In practice the propagator is therefore calculated in two steps, by first solving for an intermediate vector $\mathbf{X}$ (corresponding to $\mathbf{U}$ in eq. (10.50)).

$$\begin{pmatrix}\omega\mathbf{1}-\mathbf{A} & -\mathbf{B} \\ -\mathbf{B} & -\omega\mathbf{1}-\mathbf{A}\end{pmatrix}^{-1}\begin{pmatrix}(\mathbf{e}|\mathbf{Q}) \\ (\mathbf{d}|\mathbf{Q})\end{pmatrix} = \mathbf{X}$$

$$\begin{pmatrix}\omega\mathbf{1}-\mathbf{A} & -\mathbf{B} \\ -\mathbf{B} & -\omega\mathbf{1}-\mathbf{A}\end{pmatrix}\mathbf{X} = \begin{pmatrix}(\mathbf{e}|\mathbf{Q}) \\ (\mathbf{d}|\mathbf{Q})\end{pmatrix} \qquad(10.118)$$

and then multiplying it onto the property gradient

$$\langle\langle\mathbf{P};\mathbf{Q}\rangle\rangle_\omega = ((\mathbf{P}|\mathbf{e}),(\mathbf{P}|\mathbf{d}))\mathbf{X} \qquad(10.119)$$

The $\mathbf{X}$ vector in eq. (10.118) may be determined by iterative techniques, analogous to those used in direct CI (Section 4.2.4), i.e. the principal propagator matrix is never constructed explicitly. If the $\mathbf{Q}$ vector is set equal to zero in eq. (10.118), the equation corresponds to determining the poles of the principal propagator, i.e. the excitation energies. This is an eigenvalue problem, and finding the principal propagator for a CI wave function is equivalent to diagonalizing the CI Hamilton matrix (Section 4.2). For other types of reference wave functions (e.g. HF or MCSCF), the propagator formulation allows a generalization of calculating excitation energies.

The RPA method may be improved either by choosing an MCSCF reference wave function, leading to the MCRPA method, or by extending the operator manifold beyond

$\mathbf{h}_2$. By expanding the two parts of the propagator (property vector and principal propagator) as a function of the fluctuation potential (difference between the HF and exact electron–electron repulsion), it may be shown that RPA corresponds to terminating the expansion at first order. In the *Second-Order Polarization Propagator Approximation* (SOPPA), the expansion is carried out through second-order, which may be shown to require inclusion of the $\mathbf{h}_4$ operator. The *Higher RPA* (HRPA) method may be considered as an approximation to SOPPA, where the part involving the $\mathbf{h}_4$ operator is neglected. A full third-order propagator model has not yet been implemented, but two hybrid methods known as CCDPPA and CCSDPPA exist.[15] These models correspond to replacing the second-order perturbation coefficients with the corresponding coupled cluster amplitudes (eqs. (4.39) and (4.48)). They tend to perform somewhat better for cases where the HF reference contains significant multi-reference character.

Although the formal expression for the propagator and the second-order perturbation formula are identical, involving a sum over all excited states, the final expressions for the propagator refer only to the reference wave function, and the basic computational problem involves matrix elements between Slater determinants, and matrix manipulations. Modern implementations of propagator methods are computationally related to the derivative techniques discussed in Section 10.3. The significance is that propagator methods allow a calculation of a property directly, without having to construct all the excited states explicitly, i.e. avoiding the Sum Over States method. This also means that there are no excited wave functions directly associated with a given propagator method. The RPA method includes all singly and some doubly excited states, and typically generates results that are better than those from a CI calculation with single excitations only (if the $\mathbf{B}$ matrix is neglected in eq. (10.115), the results are identical those from to CIS), but not as good those from as CISD. Similarly, the SOPPA method involves an expansion through second-order, and typically gives results of MP2 quality, or slightly better.

A $\langle\langle\mathbf{P};\mathbf{Q}\rangle\rangle$ propagator is called a *linear* response function, since it measures the response of $\mathbf{P}$ to a perturbation $\mathbf{Q}$. The $\langle\langle\mathbf{r};\mathbf{r}\rangle\rangle$ propagator thus determines the polarizability, which is a second-order property. The concept may be generalized to higher orders, i.e. the *quadratic* response function, given as a $\langle\langle\mathbf{P};\mathbf{Q},\mathbf{R}\rangle\rangle$ propagator, determines third-order properties (the response of $\mathbf{P}$ to perturbations $\mathbf{Q}$ and $\mathbf{R}$), e.g. $\langle\langle\mathbf{r};\mathbf{r},\mathbf{r}\rangle\rangle$ gives the first hyperpolarizability. The *cubic* response function, given as a $\langle\langle\mathbf{P};\mathbf{Q},\mathbf{R},\mathbf{S}\rangle\rangle$ propagator, determines fourth-order properties, e.g. $\langle\langle\mathbf{r};\mathbf{r},\mathbf{r},\mathbf{r}\rangle\rangle$ gives the second hyperpolarizability etc. In the dynamic case, the higher-order properties may involve several different frequencies. The corresponding property may be written as for example $\boldsymbol{\beta}(-\omega;\omega_1,\omega_2)$, with $\omega=\omega_1+\omega_2$, where ω_1 and ω_2 are associated with the $\mathbf{Q}$ and $\mathbf{R}$ perturbations, respectively.

More recently *Equation Of Motion* (EOM) methods have been used in connection with other types of wave functions, most notably coupled cluster.[16] Such EOM methods are closely related to propagator methods, and give working equations which are similar to those encountered in propagator theory.

10.10 Property Basis Sets

The basis set requirements for obtaining a certain accuracy of a given molecular property are usually different from those required for a corresponding accuracy in energy. There is no analogy with the variational principle for properties, since the value in general is not bound. Basis sets for properties must therefore be tailored by adding functions until the desired accuracy is obtained. Given the nature of the perturbation, the specific need may be very different. An electric field, for example, essentially measures how easily the wave function distorts, i.e. it is primarily dependent on the most loosely bound electrons since they are the ones that are easiest polarized. The important part of the wave function is thus the "tail", necessitating diffuse functions in the basis set. Furthermore, an electric field polarizes the electron cloud, and polarization functions are therefore also important. For perturbation-independent basis functions, there is a "$2n + 1$" rule, i.e. if the unperturbed system is reasonably described by basis functions up to angular momentum L, then a basis set which includes functions up to angular momentum $L + n$ can predict properties up to order $2n + 1$. A minimum description of molecules containing first and second row atoms requires s- and p-functions, implying that d-functions are necessary for the polarizability and the first hyperpolarizability, and f-functions should be included for the second and third hyperpolarizability. A more realistic description, however, would include d-functions for the unperturbed system, necessitating f-functions for the polarizability.

A completely different type of property is for example spin–spin coupling constants, which contain interactions of electronic and nuclear spins. One of the operators is a delta function (Fermi–Contact, eq. (10.78)), which measures the quality of the wave function at a single point, the nuclear position. Since Gaussian functions have an incorrect behaviour at the nucleus (zero derivative compared with the "cusp" displayed by an exponential function), this requires addition of a number of very "tight" functions (large exponents) in order to predict coupling constants accurately.[17]

References

1. Y. Yamaguchi, Y. Osamura, J. D. Goddard and H. F. Schaefer III *A new dimension to quantum chemistry*, Oxford University Press, 1994; C. E. Dykstra, J. D. Augspurger, B. Kirtman and D. J. Malik, *Rev. Comput. Chem.*, **1** (1990), 83; D. B. Chesnut, *Rev. Comput. Chem.*, **8** (1996), 245; R. McWeeny *Methods of Molecular Quantum Mechanics*, Academic Press, 1992; J. Olsen and P. Jørgensen, *Modern Electronic Structure Theory*, Part II, ed. D. Yarkony, World Scientific, 1995, p. 857–990.
2. N. C. Handy and H. F. Schaefer III, *J. Chem. Phys.*, **81** (1984), 5031.
3. T. Helgaker and P. Jørgensen, *Theo. Chim. Acta*, **75** (1989), 111.
4. J. Gerratt and I. M. Mills, *J. Chem. Phys.*, **49** (1968), 1719.
5. J. Gauss, *J. Chem. Phys.*, **99** (1993), 3629.
6. S. A. Perera, M. Nooijen and R. J. Bartlett, *J. Chem. Phys.*, **104** (1996), 3290.
7. M. Schindler and W. Kutzelnigg, *J. Chem. Phys.*, **76** (1982), 1919).
8. A. E. Hansen and T. D. Bouman, *J. Chem. Phys.*, **82** (1985), 5035.
9. F. London, *J. Phys. Radium*, **8** (1937), 397; R. Ditchfield, *Mol. Phys.*, **27** (1974), 789.
10. K. Ruud, T. Helgaker, K. L. Bak, P. Jørgensen and H. J. Aa. Jensen, *J. Chem. Phys.*, **99** (1993), 3847; K. L. Bak, P. Jørgensen, T. Helgaker, K. Ruud and H. J. Aa. Jensen, *J. Chem. Phys.*, **100** (1994), 6620.
11. K. Wolinski, J. F. Hinton and P. Pulay, *J. Am. Chem. Soc.*, **112** (1990), 8251.
12. J. A. Pople, R. Krishnan, H. B. Schlegel and J. S. Binkley, *Int. J. Quantum Chem. Symp.*, **13** (1979), 225.

13. J. Oddershede, *Adv. Chem. Phys.*, **69** (1987), 201.
14. J. Oddershede and J. R. Sabin, *Int. J. Quantum Chem.*, **39** (1991), 371.
15. J. Oddershede, J. Geertsen and G. E. Scuseria, *J. Phys. Chem.*, **92** (1988), 3056.
16. J. F. Stanton and R. J. Bartlett, *J. Chem. Phys.*, **99** (1993), 5178; H. Koch and P. Jørgensen, *J. Chem. Phys.*, **93** (1990), 3333.
17. T. Helgaker, M. Jaszunski, K. Ruud, and A. Gorska, *Theor. Chem. Acta*, **99** (1998), 175.

11 Illustrating the Concepts

In this chapter we will illustrate some of the methods described in the previous sections. It is of course impossible to cover all types of bonding and geometries, but for highlighting the features we will look at the H_2O molecule. This is small enough that we can employ the full spectrum of methods and basis sets.

11.1 Geometry Convergence

The experimental geometry for H_2O has a bond length of 0.9578 Å and an angle of 104.48°.[1] Let us investigate how the calculated geometry change as a function of theoretical sophistication.

11.1.1 Ab Initio Methods

We need to look at the convergence as a function of basis set and amount of electron correlation (Figure 4.2). For the former we will use the correlation consistent basis sets of double, triple, quadruple, quintuple and, when possible, sextuple quality (Section 5.4.5), while the sensitivity to electron correlation will be sampled by the HF, MP2 and CCSD(T) methods (Sections 3.2, 4.8 and 4.9). Table 11.1 shows how the geometry changes as a function of basis set at the HF level of theory.

The HF results are clearly converged with the cc-pV5Z basis set, and the HF limit predicts a bond length which is too short, reminiscent of the incorrect dissociation of the single determinant wave function (Section 4.3). As a consequence, the bond angle becomes too large, owing to an overestimation of the repulsion between the two hydrogens. The underestimation of bond lengths at the HF level is quite general for covalent bonds, while the overestimation of bond angles is not. Although the increased repulsion/attraction between atom pairs in general is overestimated owing to too short bond lengths and too large a charge polarization, these factors may pull in different directions for a larger molecule, and bond angles may either be too large or too small. Note that the bond length decreases as the basis set is enlarged, thus for some systems a

Table 11.1. H_2O geometry as a function of basis set at the HF level of theory

Basis	R_{OH} (Å)	α_{HOH}
cc-pVDZ	0.9463	104.61
cc-pVTZ	0.9406	106.00
cc-pVQZ	0.9396	106.22
cc-pV5Z	0.9396	106.33
cc-pV6Z	0.9396	106.33

Table 11.2 H_2O geometry as a function of basis set at the MP2 level of theory

Basis	R_{OH} (Å)	α_{HOH}	ΔR_{OH} (Å)	$\Delta\alpha_{HOH}$
cc-pVDZ	0.9649	101.90	0.0186	−2.71
cc-pVTZ	0.9591	103.59	0.0185	−2.48
cc-pVQZ	0.9577	104.02	0.0181	−2.20
cc-pV5Z	0.9579	104.29	0.0184	−2.04
cc-pV6Z	0.9581	104.36	0.0185	−1.97

minimum or DZP type basis may give bond lengths which are longer than the experimental value. At the HF limit, however, covalent bond lengths will normally be too short.

The geometry variation at the MP2 level is shown in Table 11.2, with the change relative to the HF level given as Δ values. Including electron correlation at the MP2 level increases the bond length by about 0.018 Å, fairly independently of the basis set. As a consequence, the bond angle decreases, by about 2°. Note that the convergence in terms of basis set is much slower than at the HF level. From the observed behaviour the MP2 basis set limit may be estimated as 0.9582 ± 0.0001 Å and $104.40° \pm 0.04°$, which is already in good agreement with the experimental values. H_2O at the equilibrium geometry is a system where the HF is a good zero-order wave function, and perturbation methods should consequently converge fast. Indeed, the MP2 method recovers $\sim 95\%$ of the electron correlation energy, as shown below.

The variation at the CCSD(T) level is shown in Table 11.3, with the change relative to the MP2 level given as Δ values. Additional correlation with the CCSD(T) method gives only small changes relative to the MP2 level, and the effect of higher-order correlation diminishes as the basis set is enlarged. For H_2O the CCSD(T) method is virtually indistingable from CCSDT.[2]

The HF wave function contains equal amounts of ionic and covalent contributions (Section 4.3). For covalently bonded systems, like H_2O, the HF wave function is too ionic, and the effect of electron correlation is to increase the covalent contribution. Since the ionic dissociation limit is higher in energy than the covalent, the effect is that the equilibrium bond length increases when correlation methods are used. For dative bonds, such as metal–ligand compounds, the situation is reversed. In this case the HF wave function dissociates correctly, and bond lengths are normally too long. Inclusion of

Table 11.3 H_2O geometry as a function of basis set at the CCSD(T) level of theory

Basis	R_{OH} (Å)	α_{HOH}	ΔR_{OH} (Å)	$\Delta \alpha_{HOH}$
cc-pVDZ	0.9663	101.91	0.0014	0.01
cc-pVTZ	0.9594	103.58	0.0003	0.06
cc-pVQZ	0.9579	104.12	0.0002	0.10
cc-pV5Z	0.9580	104.38	0.0001	0.09

Table 11.4 H_2O geometry as a function of basis set at the MP2 level of theory including all electrons in the correlation

Basis	R_{OH} (Å)	α_{HOH}	ΔR_{OH} (Å)	$\Delta \alpha_{HOH}$
cc-pCVDZ	0.9643	101.91	-0.0005	0.04
cc-pCVTZ	0.9580	103.63	-0.0008	0.11
cc-pCVQZ	0.9569	104.14	-0.0009	0.12
cc-pCV5Z	0.9570	104.41	-0.0009	0.12

electron correlation adds attraction between ligands (dispersion interaction), which causes the metal–ligand bond lengths to contract.

The MP2 and CCSD(T) values in Tables 11.2 and 11.3 are for correlation of the valence electrons only, i.e. the frozen core approximation. In order to asses the effect of core-electron correlation, the basis set needs to be augmented with tight polarization functions. The corresponding MP2 results are shown in Table 11.4, where the Δ values refer to the change relative to the valence only MP2 with the same basis set. Essentially identical changes are found at the CCSD(T) level.

The effect of core-electron correlation is small, a small decrease in bond length and a corresponding small increase in bond angle. Addition of the CCSD(T)-MP2 changes (Table 11.3) and the MP2 core correlation correction to the estimated (valence) MP2 basis set limit results gives a bond length of 0.9574 Å and an angle of 104.61°, with error bars of perhaps ± 0.0004 Å and $\pm 0.06°$. Relativistic effects are small for this system, -0.00003 Å and $-0.07°$ as calculated at the Dirac–Fock level of theory (Table 8.1). Including these corrections allows a final predicted structure of 0.9574 Å and 104.54°, which are comparable to the experimental values of 0.9578 Å and 104.48°.

These results show that *ab initio* methods can give results of very high accuracy, provided that sufficiently large basis sets are used. Unfortunately, the combination of highly correlated methods, such as CCSD(T), and large basis sets means that such calculations are computationally expensive. Already for the H_2O system a CCSD(T) calculation with the cc-pV5Z basis is quite demanding. The results also show, however, that a quite respectable level of accuracy is reached at the MP2/cc-pVTZ level, which is applicable to a much larger variety of molecules. Furthermore, the errors at a given level are quite systematic, and relative values (comparing for example changes in geometries upon introduction of substituents) will be predicted with a substantially higher accuracy.

It should also be noted that the effect of electron correlation at the MP2 level (relative to HF) is largely independent of the basis set, but there is a significant coupling between

Table 11.5 H_2O bond distance (Å) as a function of basis set with different DFT functionals

Basis	SVWN	BLYP	BPW91	B3LYP	B3PW91
cc-pVDZ	0.9769	0.9799	0.9762	0.9687	0.9663
cc-pVTZ	0.9706	0.9716	0.9687	0.9613	0.9596
cc-pVQZ	0.9697	0.9703	0.9677	0.9602	0.9586
cc-pV5Z	0.9698	0.9703	0.9677	0.9602	0.9586

Table 11.6 H_2O bond angle as a function of basis set with different DFT functionals

Basis	SVWN	BLYP	BPW91	B3LYP	B3PW91
cc-pVDZ	102.47	101.81	101.78	102.74	102.68
cc-pVTZ	104.34	103.77	103.60	104.52	104.36
cc-pVQZ	104.71	104.21	103.97	104.89	104.68
cc-pV5Z	104.94	104.47	104.18	105.10	104.86

the basis set and the higher-order correlation (beyond MP2) effect. The importance of higher-order electron correlation decreases as the basis set is enlarged. This suggests that it is better to invest a given amount of computer time in performing a large basis set MP2 calculation than in performing a highly correlated calculation with a modest basis, at least when the HF is a good zero-order wave function.

11.1.2 DFT Methods

The two variable features in DFT methods are the basis set and the choice of the exchange–correlation potential. The performance for five popular functionals on the geometry for the cc-pVXZ basis sets is given in Tables 11.5 and 11.6. The grid size for the numerical integration of the exchange–correlation energy is sufficiently large that the error from incomplete grids can be neglected.

The geometry displays convergence characteristics similar to those obtained using the wave mechanics HF method (Table 11.1). A TZP type basis gives good results, and a QZP type is already close to the basis set limiting value. Compared with the experimental values of 0.9578 Å and 104.48°, the B3PW91 functional gives the best results, although the bond angle is slightly too large.

11.2 Total Energy Convergence

The total energy in *ab initio* theory is given relative to the separated particles, i.e. bare nuclei and electrons. The experimental value for an atom is the sum of all the ionization potentials; for a molecule there are additional contributions from the molecular bonds and associated zero-point energies. The experimental value for the total energy of H_2O is −76.480 a.u., and the estimated contribution from relativistic effects is −0.045 a.u. Including a mass correction of 0.0028 a.u. (a non-Born–Oppenheimer effect which accounts for the difference between finite and infinite nuclear masses) allows the "experimental" non-relativistic energy to be estimated at −76.438 ± 0.003 a.u.[3]

Table 11.7 % electron correlation recovered by different methods in the cc-pVDZ basis

Method	% EC
MP2	94.0
MP3	97.0
MP4	99.5
MP5	99.8
CCSD	98.3
CCSD(T)	99.7
CISD	94.5
CISDT	95.8
CISDTQ	99.9

For the cc-pVDZ basis set, the full CI result is available,[4] which allows an assessment of the performance of various approximate methods. The % of the electron correlation recovered by different methods is shown in Table 11.7. As already mentioned, the H_2O molecule is an easy system, for which the HF wave function provides a good reference. Furthermore, since there are only 10 electrons in H_2O, the effect of higher-order electron correlation is small. The intraorbital correlation between electron pairs dominates the correlation energy for such a small system, and the doubly excited configurations, which mainly describe the pair correlation, account for a large fraction of the total correlation energy. Consequently even the simple MP2 method performs exceedingly well, and the CCSD(T) result is for practical purposes identical to the full CI result. For such simple systems, the MP2 and MP3 % correlations are probably significantly higher than would be expected for a larger system.

The calculated E_{tot} as a function of basis set and electron correlation (valence electrons only) at the experimental geometry is given in Table 11.8. As the cc-pVXZ basis sets are fairly systematic in how they are extended from one level to the next, there is some justification for extrapolating the results to the "infinite" basis set limit (Section 5.4.5). The HF energy is expected to have an exponential behaviour, and a functional form of the type $A + B\exp(-Cn)$ with $n = 2$–6 yields an infinite basis set limit of -76.0676 a.u., in perfect agreement with the estimated HF limit of -76.0676 ± 0.0002 a.u.[5]

The correlation energy is expected to have an inverse power dependence once the basis set reaches a sufficient (large) size. Extrapolating the correlation contribution for $n = 3$–$5(6)$ with a function of the type $A + B(n+1)^{-C}$ yields the cc-pV∞Z values in Table 11.8. The extrapolated CCSD(T) energy is -76.376 a.u., yielding a valence correlation energy of -0.308 a.u.

The magnitude of the core correlation can be evaluated by including the oxygen 1s-electrons and using the cc-pCVXZ basis sets; the results are shown in Table 11.9. The extrapolated CCSD(T) correlation energy is -0.370 a.u. Assuming that the CCSD(T) method provides 99.7% of the full CI value, as indicated by Table 11.7, the extrapolated correlation energy becomes -0.371 a.u., well within the error limits on the estimated experimental value. The core (and core–valence) electron correlation is thus 0.063 a.u.,

Table 11.8 Total energy ($+76$ a.u.) as a function of basis set and electron correlation (valence only)

Method	cc-pVDZ	cc-pVTZ	cc-pVQZ	cc-pV5Z	cc-pV6Z	cc-pV∞Z
HF	− 0.02677	− 0.05713	− 0.06479	− 0.06704	− 0.06735	− 0.0676
MP2	− 0.22844	− 0.31863	− 0.34763	− 0.35860	− 0.36264	− 0.368
MP3	− 0.23544	− 0.32275	− 0.34939	− 0.35815	− 0.36094	− 0.364
MP4	− 0.24067	− 0.33302	− 0.36104	− 0.37051	− 0.37357	− 0.377
MP5	− 0.24120	− 0.33159				
CCSD	− 0.23801	− 0.32455	− 0.35080	− 0.35952		− 0.366
CCSD(T)	− 0.24104	− 0.33219	− 0.35979	− 0.36904		− 0.376
CISD	− 0.22997	− 0.31384	− 0.33922	− 0.34765		− 0.354

Table 11.9 Total energy ($+76$ a.u.) as a function of basis set and electron correlation (all electrons)

Method	cc-pCVDZ	cc-pCVTZ	cc-pCVQZ	cc-pCV5Z	cc-pCV∞Z (%EC)
HF	− 0.02718	− 0.05731	− 0.06490	− 0.06706	− 0.0677 (0.0)
MP2	− 0.26855	− 0.37486	− 0.40758	− 0.41939	− 0.430 (97.4)
MP3	− 0.27638	− 0.37984	− 0.41012	− 0.41978	− 0.430 (97.4)
MP4	− 0.28194	− 0.39079	− 0.42240	− 0.43268	− 0.440 (100.0)
MP5	− 0.28239	− 0.38907			
CCSD	− 0.27897	− 0.38154	− 0.41144	− 0.42104	− 0.428 (96.9)
CCSD(T)	− 0.28226	− 0.38978	− 0.42096	− 0.43105	− 0.438 (99.5)
CISD	− 0.26898	− 0.36799	− 0.39675	− 0.40599	− 0.412 (92.6)

which is comparable to the value for the valence electrons (i.e. 0.308 divided between four electron pairs is 0.077 a.u.).

The % of the total correlation energy is given in parenthesis in Table 11.9; in the infinite basis set limit the MP2 method recovers 97.4%. Notice, however, that while the perturbation series is smoothly convergent with the cc-pVDZ basis, it becomes oscillating with the larger basis set. With the cc-pVTZ basis, the MP5 result is higher in energy than MP4, and with the cc-pV5Z the MP3 result is higher than the MP2 value. This may be an indication that the perturbation series is actually divergent in a sufficiently large basis set. The extrapolated MP4 value is in perfect agreement with the experimental estimate, but this may be fortuitous. The CISD method performs rather poorly, yielding results which are worse than MP2, but at a cost similar to that of an MP4 calculation.

Since the CCSD(T) result essentially is equivalent to a full CI (Table 11.7), the data shows that the cc-pCVDZ basis is able to provide 70% of the <u>total</u> correlation energy. The corresponding values for the cc-pCVTZ, cc-pCVQZ and cc-pCV5Z basis sets are 90%, 96% and 99%, respectively. Slightly lower percentages have been found in other systems.[6] This illustrates the slow convergence of the correlation energy as a function of basis set. Each step up in basis set quality roughly doubles the number of functions. The cc-pCVDZ basis is capable of recovering 70% of the correlation energy, and improving the basis from cc-pCVDZ to cc-pCVTZ allows an additional 20% to be calculated. The next step up gives only 6% and the expansion from cc-pCVQZ to cc-pCV5Z only 3%. The last 5–10% of the correlation energy is therefore hard to get, requiring very large basis sets. This slow convergence is the principal limitation of traditional *ab initio* methods. The CCSD(T)/cc-pCV5Z total energy is still 4 kcal/mol off the experimentally derived non-relativistic value, the remaining error being distributed roughly equally between incomplete basis set and incomplete electron correlation effects. These errors are comparable to the non-Born–Oppenheimer correction of 2 kcal/mol, and substantially smaller than the relativistic correction of 28 kcal/mol. Calculating the total energy with an accuracy of ~ 1 kcal/mol is thus barely possible for this simple system.

Although the total energy calculated by DFT methods should in principle converge to the "experimental" value (-76.438 a.u.), there are no upper or lower bounds for the methods currently employed with approximate exchange–correlation functionals. Indeed, all the gradient based methods used here (BLYP, BPW91, B3LYP and B3PW91) give total energies well below the "experimental" value with the cc-pV5Z basis (-76.443 to -76.473 a.u.).

11.3 Dipole Moment Convergence

As examples of molecular properties we will look at how the dipole moment and harmonic vibrational frequencies converge as a function of level of theory.

11.3.1 Ab Initio Methods

The experimental value for the dipole moment is 1.847 D,[7] and the calculated value at different levels of theory is shown in Table 11.10. The dipole moment may be considered as the response of the wave function (and energy) to the presence of an external electric field, in the limit where the field strength is vanishingly small (Section 10.1.1). It is consequently sensitive to the representation of the wave function "tail", i.e. far from the nuclei, and diffuse functions are therefore expected to be important. Although the results with the regular cc-pVXZ basis sets may be converging, the rate of convergence is slow, as compared to the results for the basis sets augmented with diffuse

Table 11.10 H_2O dipole moment (Debye) as a function of theory (valence correlation only), experimental value is 1.847 D

Basis	HF	MP2	CCSD(T)
cc-pVDZ	2.057	1.964	1.936
cc-pVTZ	2.026	1.922	1.903
cc-pVQZ	2.008	1.904	1.890
cc-pV5Z	2.003	1.895	
cc-pV6Z	1.990		
aug-cc-pVDZ	2.000	1.867	1.848
aug-cc-pVTZ	1.984	1.852	1.839
aug-cc-pVQZ	1.982	1.858	1.848
aug-cc-pV5Z	1.982	1.861	

Table 11.11 H_2O dipole moment (Debye) as a function of theory (all electrons)

Basis	HF	MP2	CCSD(T)
aug-cc-pCVDZ	2.001	1.868	1.849
aug-cc-pCVTZ	1.983	1.857	1.843

Table 11.12 H_2O dipole moment (Debye) as a function of DFT functional and basis set; the experimental value is 1.847 D

Basis	SVWN	BLYP	BPW91	B3LYP	B3PW91
aug-cc-pVDZ	1.853	1.796	1.803	1.855	1.859
aug-cc-pVTZ	1.857	1.799	1.800	1.854	1.854
aug-cc-pVQZ	1.855	1.798	1.797	1.854	1.852
aug-cc-pV5Z	1.856	1.799	1.798	1.855	1.852

functions. This illustrates that care must be taken when calculating properties other than the total energy, as standard basis sets may not be able to describe important aspects of the wave function.

The HF dipole moment is too large, which is quite general, as the HF wave function overestimates the ionic contribution. The MP2 procedure recovers the large majority of the correlation effect, but the convergence with the aug-cc-pVXZ basis sets is not smooth, and does not readily allow an extrapolation. The CCSD(T) result with the aug-cc-pVQZ basis is very close to the experimental value, although remaining basis set effects and further correlation may change the value slightly. As expected for this property, the effect of core correlation is small, as shown by MP2 calculations in Table 11.11.

11.3.2 DFT Methods

Table 11.10 establishes that diffuse functions are mandatory for calculating dipole moments, and only results with the aug-cc-pVXZ basis sets are shown for the DFT methods in Table 11.12.

The calculated dipole moment is remarkably insensitive to the size of the basis set. Note that the SVWN value in this case is substantially better than BLYP and BPW91, i.e. this is a case where the theoretically "poorer" method provides better results than the more advanced gradient methods. Inclusion of "exact" exchange again improves the performance, and provides results very close to the experimental value, even with quite small basis sets.

11.4 Vibrational Frequencies' Convergence

The experimental values for the fundamental vibrational frequencies are $3756 \, \text{cm}^{-1}$, $3657 \, \text{cm}^{-1}$ and $1595 \, \text{cm}^{-1}$, while the corresponding harmonic values are $3943 \, \text{cm}^{-1}$, $3832 \, \text{cm}^{-1}$ and $1649 \, \text{cm}^{-1}$.[8] The differences due to anharmonicity are thus $187 \, \text{cm}^{-1}$, $175 \, \text{cm}^{-1}$ and $54 \, \text{cm}^{-1}$, i.e. 3–5% of the harmonic values.

Table 11.13 H_2O HF harmonic frequencies (cm^{-1}) as a function of basis set; experimental values are $3943\,cm^{-1}$, $3832\,cm^{-1}$ and $1649\,cm^{-1}$

Basis	ω_1	ω_2	ω_3
cc-pVDZ	4212	4114	1776
cc-pVTZ	4227	4127	1753
cc-pVQZ	4229	4130	1751
cc-pV5Z	4231	4131	1748

11.4.1 *Ab Initio Methods*

The calculated harmonic frequencies at the HF level are given in Table 11.13. Vibrational frequencies are examples of a slightly more complicated property. The frequencies are obtained from the force constant matrix (second derivative of the energy), evaluated at the equilibrium geometry (Section 13.1). Both the equilibrium geometry and the shape of the energy surface depend on the theoretical level. Part of the change in frequencies is due to changes in the geometry, since the force constant in general decreases with increasing bond length.

The HF vibrational frequencies are too high by about 7% relative to the experimental harmonic values, and by 10–13% relative to the anharmonic values. This overestimation is due to the incorrect dissociation and the corresponding bond lengths being too short (Table 11.1), and is consequently quite general. Vibrational frequencies at the HF level are therefore often scaled by ~ 0.9 to partly compensate for these systematic errors.

Inclusion of electron correlation normally lowers the force constants, since the correlation energy increases as a function of bond length. This usually means that vibrational frequencies decrease, although there are exceptions (vibrational frequencies also depend on off-diagonal force constants). The values calculated at the MP2 and CCSD(T) levels are shown in Tables 11.14 and 11.15.

The MP2 treatment recovers the majority of the correlation effect, and the CCSD(T) results with the cc-pVQZ basis sets are in good agreement with the experimental values. The remaining discrepancies of $9\,cm^{-1}$, $13\,cm^{-1}$ and $10\,cm^{-1}$ are mainly due to basis set inadequacies, as indicated by the MP2/cc-pV5Z results. The MP2 values are in respectable agreement with the experimental harmonic frequencies, but of course still overestimate the experimental fundamental ones by the anharmonicity. For this reason, calculated MP2 harmonic frequencies are often scaled by ~ 0.97 for comparison with experimental results.[9]

Table 11.14 H_2O MP2 harmonic frequencies (cm^{-1}) as a function of basis set (valence electrons only); experimental values are $3943\,cm^{-1}$, $3832\,cm^{-1}$ and $1649\,cm^{-1}$

Basis	ω_1	ω_2	ω_3
cc-pVDZ	3971	3852	1678
cc-pVTZ	3976	3855	1651
cc-pVQZ	3978	3855	1643
cc-pV5Z	3974	3849	1636

Table 11.15 H_2O CCSD(T) harmonic frequencies (cm^{-1}) as a function of basis set (valence electrons only)

Basis	ω_1	ω_2	ω_3
cc-pVDZ	3928	3822	1690
cc-pVTZ	3946	3841	1669
cc-pVQZ	3952	3845	1659

Table 11.16 H_2O MP2 harmonic frequencies (cm^{-1}) as a function of basis set (all electrons)

Basis	ω_1	ω_2	ω_3
cc-pCVDZ	3973	3853	1679
cc-pCVTZ	3976	3857	1651

The effect of core-electron correlation is small, as shown in Table 11.16. It should be noted that the valence and core correlation energy per electron pair is of the same magnitude, however, the core correlation is almost constant over the whole energy surface and consequently contributes very little to properties depending on relative energies, like vibrational frequencies. It should be noted that relativistic corrections for the frequencies are expected to be of the order of $1\,cm^{-1}$ or less.[10]

For comparison with experimental frequencies (which necessarily are anharmonic), there is normally little point in improving the theoretical level beyond MP2 with a TZ(2df,2pd) type basis set unless anharmonicity constants are calculated explicitly. Although anharmonicity can be approximately accounted for by scaling the harmonic frequencies by $\sim$0.97, the remaining errors in the harmonic force constants at this level are normally smaller than the corresponding errors due to variations in anharmonicity.

11.4.2 DFT Methods

The harmonic frequencies calculated with different DFT functionals as a function of basis set are shown in Tables 11.17–11.19. The convergence as a function of basis set is similar to that observed for the HF method. The B3PW91 functional again shows the best performance. With the cc-pV5Z basis set the deviations from the experimental harmonic frequencies are only $0\,cm^{-1}$, $9\,cm^{-1}$ and $17\,cm^{-1}$, substantially better that the

Table 11.17 H_2O highest harmonic frequency (cm^{-1}) as a function of basis set with different DFT functionals; the experimental value is $3943\,cm^{-1}$

Basis	SVWN	BLYP	BPW91	B3LYP	B3PW91
cc-pVDZ	3787	3691	3756	3852	3898
cc-pVTZ	3825	3753	3807	3900	3937
cc-pVQZ	3826	3762	3812	3906	3941
cc-pV5Z	3827	3767	3815	3909	3943

Table 11.18 H_2O second lowest harmonic frequency (cm^{-1}) as a function of basis set with different DFT functionals; the experimental value is $3832\,cm^{-1}$

Basis	SVWN	BLYP	BPW91	B3LYP	B3PW91
cc-pVDZ	3674	3589	3651	3750	3794
cc-pVTZ	3716	3654	3704	3800	3834
cc-pVQZ	3718	3663	3709	3806	3834
cc-pV5Z	3718	3666	3712	3808	3839

Table 11.19 H_2O lowest harmonic frequency (cm^{-1}) as a function of basis set with different DFT functionals; the experimental value is $1649\,cm^{-1}$

Basis	SVWN	BLYP	BPW91	B3LYP	B3PW91
cc-pVDZ	1581	1629	1632	1658	1660
cc-pVTZ	1561	1611	1613	1639	1640
cc-pVQZ	1556	1605	1607	1635	1636
cc-pV5Z	1551	1599	1603	1630	1632

results obtained with the CCSD(T)/cc-pVQZ and MP2/cc-pV5Z methods (Tables 11.14 and 11.15), at a cost similar to a wave mechanics HF calculation! It is also clear from Tables 11.17–11.19 that inclusion of "exact" exchange (B3LYP and B3PW91) substantially improves the performance. The "pure" DFT gradient methods, BLYP and BPW91, have errors $\sim150\,cm^{-1}$ for the stretching frequencies, and $\sim50\,cm^{-1}$ for the bend.

11.5 Bond Dissociation Curve

As can be seen in Table 11.9 it is very difficult to converge the total energy to a $\sim$ 1 kcal/ mol accuracy. The total energy, however, is in almost all cases irrelevant, the important quantity is the relative energy. Let us now examine how the shape of a potential energy surface depends on the theoretical level. We will look at two cases, stretching one of the O–H bonds in H_2O, and the HOH bending potential. The O–H dissociation curve is a case where the main change is associated with the difference in electron correlation between the two electrons in the bond being stretched. It should be noted that transition structures typically have bonds which are elongated by 0.5–0.8 Å, and the performance for the dissociation curve in this range will model the behaviour for describing bond breaking/forming reactions. The HOH bending energy, on the other hand, does not involve any bond breaking, and should therefore be less sensitive to the level of theory.

11.5.1 Basis Set Effect at the HF Level

Figure 11.1 shows the bond dissociation curve at the HF level with the STO-3G, 3-21G, 6-31G(d,p), cc-pVDZ and cc-pVQZ basis sets. The total energy drops considerably upon going from the STO-3G to the 3-21G and again to the 6-31G(d,p) basis. This is primarily due to the improved description of the oxygen 1s-orbital. The two different

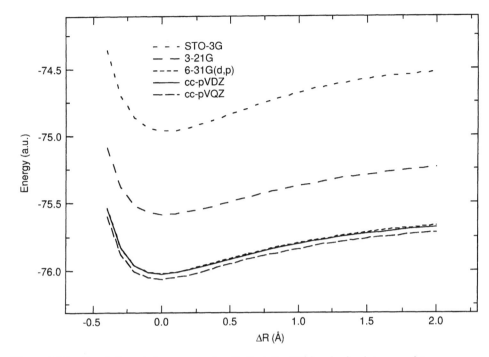

Figure 11.1   Bond dissociation curves for H_2O at the HF level, absolute energies

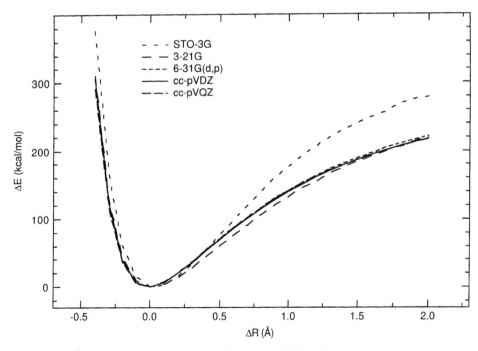

Figure 11.2   Bond dissociation curves for H_2O at the HF level, relative energies

types of DZP basis set, 6-31G(d,p) and cc-pVDZ, give very similar results, and the improvement upon going to the cc-pVQZ basis is relatively minor.

More important than the total energy is the shape of the curve, i.e. the energy relative to the equilibrium value, which is shown in Figure 11.2. The minimal STO-3G basis increases more steeply than the other basis sets, while the 3-21G is slightly too low in the $\Delta R = 0.3-1.3\,\text{Å}$ range. Considering that the STO-3G and 3-21G basis sets have the same number of primitive GTOs (Section 5.4.1), it is clear that uncontraction of the valence orbitals greatly improves the flexibility. The 6-31G(d,p), cc-pVDZ and cc-pVQZ basis sets give essentially identical curves, i.e. improvement of the basis set beyond DZP has a very minor effect at the HF level. Note also that the total energy for the 6–31 G(d,p) basis is $\sim$0.05 a.u. ($\sim$30 kcal/mol, Figure 11.1) above the HF limit, but this error is constant to within a few kcal/mol over the whole range.

11.5.2   *Performance of Different Types of Wave Function*

We will now look at how different types of wave functions behave when the O–H bond is stretched. The basis set used in all cases is the aug-cc-pVTZ, and the reference curve is taken as the [8, 8]-CASSCF result, which is slightly larger than a full-valence CI. As mentioned in Section 4.6, this allows a correct dissociation, and since all the valence electrons are correlated, it will generate a curve close to the full CI limit. The bond dissociation energy calculated at this level is 122.1 kcal/mol, which is comparable to the experimental value of 125.9 kcal/mol.

H_2O is a closed-shell singlet, and the HF wave function near the equilibrium geometry is of the RHF type. As one of the bonds is stretched, however, a UHF type will become lower in energy at some point (Section 4.4). Beyond this instability point, electron correlation methods may be based either on the RHF or UHF reference. The UHF wave function will be spin contaminated, which has some consequences as shown below. It should be noted that for open-shell species, one similarly has the option of using either an ROHF or a UHF reference wave function, but in such cases they will be different at all geometries, including near the equilibrium. In many cases, however, the UHF wave function is only slightly spin contaminated, and both approaches will then give similar results.

Figure 11.3 illustrates the behaviour of the single determinant wave functions, RHF, UHF and PUHF (Projected UHF, Section 4.4). The RHF energy continues to increase as the bond is stretched since it has the wrong dissociation limit, while the UHF converges to a value of 87 kcal/mol. At the equilibrium geometry the two electrons in the O–H bonding orbital are correlated, but this correlation energy disappears once the bond is broken. The UHF wave function correctly describes the dissociation limit in terms of energy, but does not recover any of the electron correlation at equilibrium (by definition, since UHF = RHF here). The difference between the UHF dissociation energy and the CASSCF value is therefore a measure of the amount of electron correlation in the O–H bond. With the present basis set this is 35 kcal/mol, a typical value for the correlation energy between two electrons in the same spatial MO.

At the dissociation limit the UHF wave function is essentially an equal mixture of a singlet and a triplet state, as discussed in Section 4.4. Removal of the triplet state by projection (PUHF) lowers the energy in the intermediate range, but has no effect when the bond is completely broken, since the singlet and triplet states are degenerate here.

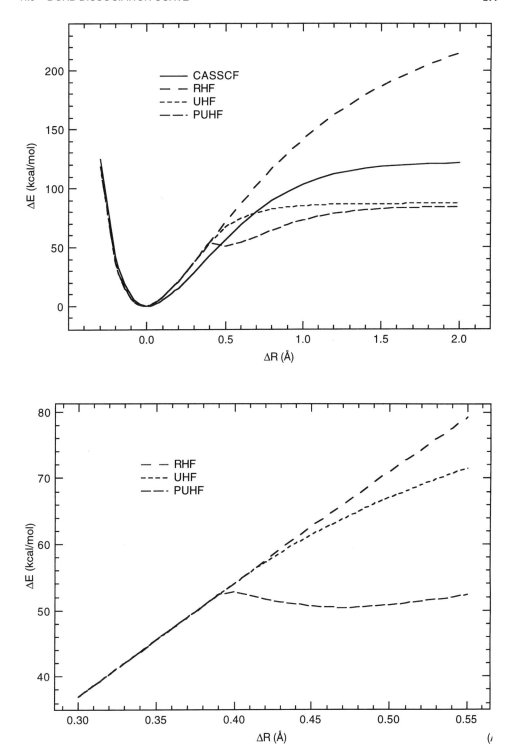

Figure 11.4 RHF, UHF and PUHF dissociation curves for H_2O near the instability point

The RHF/UHF instability point with this basis occurs when the bond is stretched by 0.40 Å. Figure 11.4 shows the behaviour of the energy curves in more detail in this region. It is seen that the PUHF has a discontinuous derivative at the instability point, and there is furthermore a shallow minimum right after the instability point, at an elongation of ~ 0.47 Å.

Since the RHF curve is too high in the transition structure region ($\Delta R \sim 0.5$–0.8), it is clear that RHF activation energies in general will be too large. UHF activation energies may either be too high or too low, but the PUHF value will essentially always be too low. Furthermore, the shape of a spin contaminated UHF energy surface will be too flat, and PUHF surfaces will be qualitatively wrong in the TS region. Spin contaminated UHF wave functions should consequently not be used for geometry optimizations.

The corresponding difference between restricted, unrestricted and projected unrestricted wave functions at the MP2 level is shown in Figure 11.5. The RMP2 raises too high, owing to the wrong dissociation limit of the underlying RHF. Both the UMP2 and PUMP2 dissociation energies are in reasonable agreement with the CASSCF value, but it is clear that the UMP2 energy is too high in the "intermediate" range owing to spin contamination. Figure 11.6 shows the curves in more detail near the RHF/UHF instability point. The UMP2 energy is <u>higher</u> than the RMP2, although the UHF energy is <u>lower</u> than the RHF. At the HF level, the UHF energy is lowest owing to a combination of spin contamination and inclusion of electron correlation (Section 4.8.2). Since the MP2 procedure recovers most of the electron correlation, only the energy raising effect due to spin contamination remains, and the UMP2 energy becomes higher

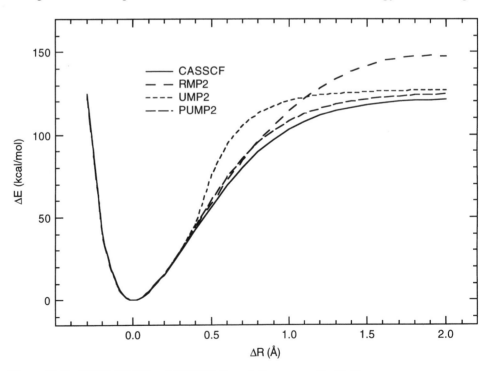

Figure 11.5 RMP2, UMP2 and PUMP2 dissociation curves for H_2O

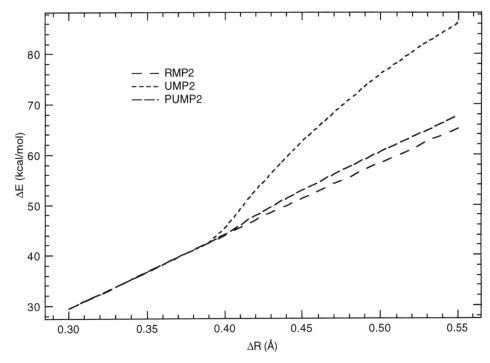

Figure 11.6 RMP2, UMP2 and PUMP2 dissociation curves for H_2O near the instability point

than RMP2. Removing the unwanted spin components makes the PUMP2 energy very similar to RMP2 for elongations less than ~ 1 Å, but it is significantly better at longer bond lengths owing to the correct dissociation in the UHF wave function. The RMP2 energy follows the "exact" curve closely to a ΔR of ~ 0.5 Å, and is in reasonable agreement out to ~ 1.0 Å. RMP2 activation energies are therefore often in quite reasonable agreement with experimental or higher-level theoretical values. It should also be noted that the discontinuity at the PUHF level essentially disappears when the projection is carried out on the MP2 wave function.

Figures 11.7 and 11.8 show the effect of extending the perturbation series at the RMP and UMP levels. Addition of more terms to the perturbation series improves the results, although the effect of MP3 over MP2 is minute. As the bond is stretched more than ~ 1.5 Å, the perturbation series breaks down owing to the RHF wave function becoming too poor a reference, and the energies start to decrease. The RMP4 method performs well out to an elongation of ~ 1.0 Å, and in the TS region where the bond is stretched $0.5-0.8$ Å, the MP4 error is less than 1 kcal/mol. Although real transition structures usually have more than one breaking/forming bond, and therefore are more sensitive to correlation effects, it is often found that the MP4 method with a suitably large basis can reproduce activation energies to within a few kcal/mol

The improvement brought about by extending the perturbation series beyond second order is very small when a UHF wave function is used as the reference, i.e. the higher-order terms do very little to reduce the spin contamination. In the dissociation limit the spin contamination is inconsequential, and the MP2, MP3 and MP4 results are all in

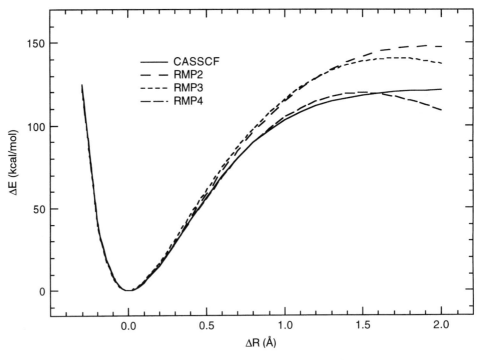

Figure 11.7   RMP2, RMP3 and RMP4 dissociation curves for H_2O

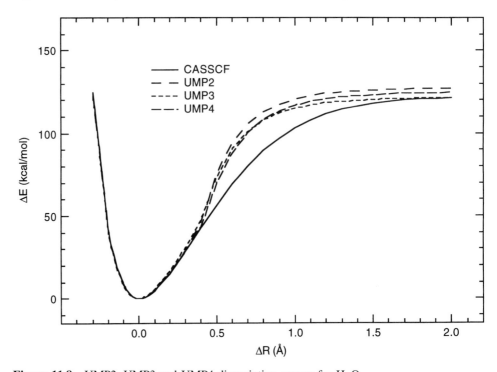

Figure 11.8   UMP2, UMP3 and UMP4 dissociation curves for H_2O

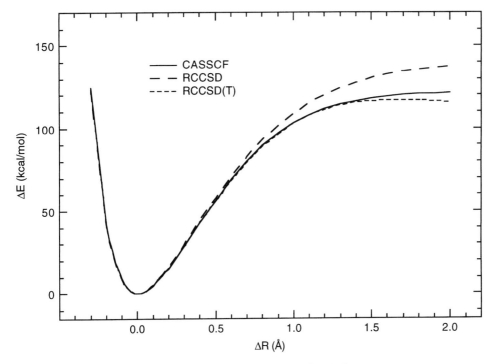

Figure 11.9   RCCSD and RCCSD(T) dissociation curves for H_2O

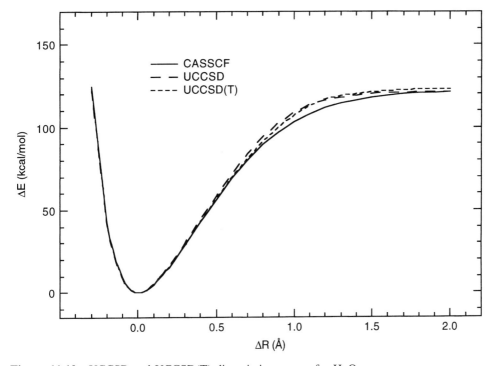

Figure 11.10   UCCSD and UCCSD(T) dissociation curves for H_2O

reasonable agreement with the "exact" CASSCF result (but too high compared with the experimental result owing to basis set limitations).

Figures 11.9 and 11.10 compare the performance of the CCSD and CCSD(T) methods, based on either an RHF or UHF reference wave function. Compared to the RMPn plot (Figure 11.7), it is seen that the infinite nature of coupled cluster causes it to perform somewhat better as the reference wave function becomes increasingly poor. While the RMP4 energy curve follows the "exact" out to an elongation of ~1.0 Å, the CCSD(T) has the same accuracy out to ~1.5 Å. Eventually, however, the wrong dissociation limit of the RHF wave also makes the coupled cluster methods break down, and the energy starts to decrease.

The spin contamination makes the UCC energy curves somewhat too high in the intermediate region, but the infinite nature of coupled cluster methods makes them significantly better at removing unwanted spin states as compared to UMPn methods (Figure 11.8).

The only generally applicable CI method is CISD, in which the singly and doubly excited configurations are treated variationally. These are also part of the MP4 method, which additionally has a term arising from disconnected quadruples, i.e. products of D-configurations, as well as a term due to (connected) triples. The CCSD method includes effects due to higher-order products of singles and doubles, i.e. sextuples, octuples etc. It is the inclusion of the product excitations that makes the MP and CC methods size extensive. Considering only the single and double excitations, and products thereof, allows a comparison between methods, and the performance of the CISD, MP4(SDQ) and CCSD models is shown in Figure 11.11.

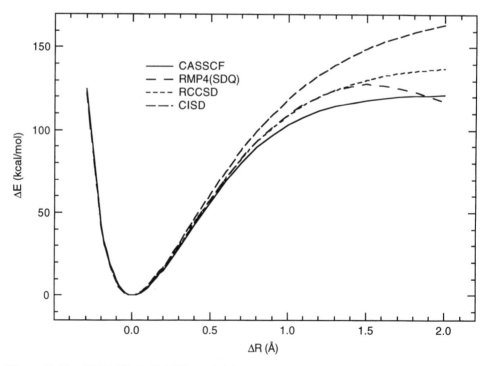

Figure 11.11   RMP4(SDQ), RCCSD and CISD dissociation curves for H_2O

It can clearly be seen that the CISD curve is worse than either of the other two, which are essentially identical out to a ΔR of 1.3 Å. The size inconsistency of the CISD method also has consequences for the energy curve when the bond is only half broken. Figure 11.11 illustrates why the use of CI methods has declined over the years, it normally gives less accurate results compared with MP or CC methods, but at a similar or higher computational cost. Furthermore, it is difficult to include the important triply excited configurations in CI methods (CISDT scales as M^8), but it is relatively easy to include them in MP or CC methods (MP4 and CCSD(T) scales as M^7).

11.5.3 DFT Methods

The performance of various DFT methods based on a restricted determinant is shown in Figure 11.12. The incorrect dissociation limit again makes the energy too high for large bond distances. In the intermediate region they all perform well, especially those including exact exchange (B3LYP and B3PW91), while the SVWN, BLYP and BPW91 are slightly too low.

The corresponding curves based on a UHF reference are shown in Figure 11.13. It is immediately clear that DFT methods do not have the "spin contamination" problem in the intermediate region, indeed spin contamination is not well defined in DFT.[11] Removing the "spin contamination" by projection methods gives rise to discontinuous derivatives and artificial minima, analogously to the PUHF case in Figures 11.3 and

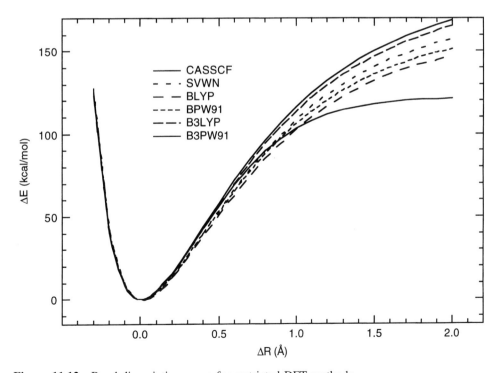

Figure 11.12 Bond dissociation curve for restricted DFT methods

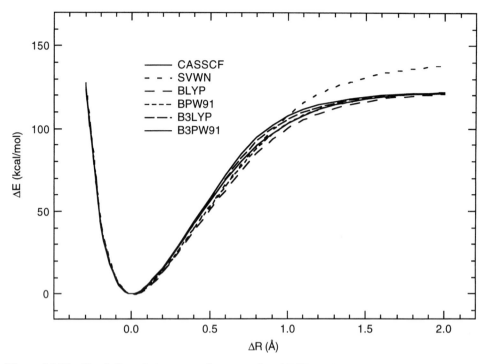

Figure 11.13 Bond dissociation curve for unrestricted DFT methods

11.4, and should consequently not be employed.[12] The local density method, SVWN, has a dissociation energy which is significantly too high (the LDA method suffers from overbinding, Section 6.1), and the gradient corrected BLYP and BPW91 are slightly too low in the intermediate region. The hybrid B3LYP and B3PW91 methods, however, follow the CASSCF closely over the whole range.

11.6 Angle Bending Curve

The angle bending in H_2O occurs without breaking any bonds, and the electron correlation energy is therefore relatively constant over the whole curve. The HF, MP2 and MP4 bending potentials are shown in Figure 11.14, where the reference curve is taken from a parametric fit to a large number of spectroscopic data.[13]

The HF method overestimates the barrier for linearity by 0.73 kcal/mol, while MP2 underestimates it by 0.76 kcal/mol. Furthermore, the HF curve increase slightly too steeply for small bond angles. The MP4 result, however, is within a few tenths of a kcal/ mol of the exact result over the whole curve. Compared to the bond dissociation discussed above, it is clear that relative energies of conformations which have similar bonding are fairly easy to calculate. While the HF and MP4 total energies with the aug-cc-pVTZ basis are ~260 kcal/mol and ~85 kcal/mol higher than the exact values at the equilibrium geometry (Table 11.8), these errors are essentially constant over the whole surface.

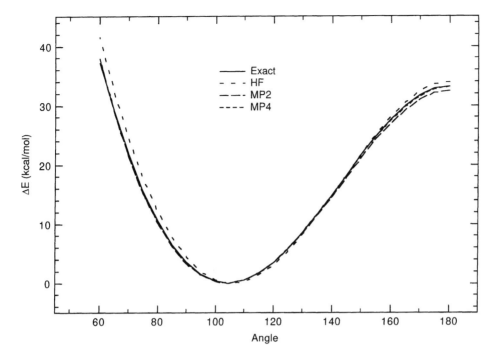

Figure 11.14 Angle bending curves for H_2O

11.7 Problematic Systems

The H_2O case is an example of a system where it is relatively easy to obtain good results. Nature is not always so kind; let us look at a couple of "theoretically difficult" cases.

11.7.1 The Geometry of FOOF

The FOOF molecule has an experimental geometry with an O–O bond length of 1.217 Å and a F–O bond length of 1.575 Å.[14] The calculated bond distances at different levels of theory are given in Table 11.20. Although the MP2/cc-pVDZ values are close to the experimental structure, the results with larger basis sets or inclusion of more electron correlation clearly show that this is accidental. Both the size of the basis set and the amount of electron correlation are important. The MP4(SDQ) geometry (1.278 Å and 1.495 Å) is very similar to CCSD, however, inclusion of the triply excited configurations in the full MP4(SDTQ) method has a huge effect. The O–O bond contracts and the F–O bonds are elongated to the point (>2.5 Å) where perturbation theory breaks down since the underlying RHF wave function becomes extremely poor. The MP4(SDTQ) model basically does not predict a stable FOOF molecule. The triples also have a large effect at the CCSD(T) level, but it is clear that the effect is wildly overestimated with the MP4 method. The best *ab initio* computational level reported is a CCSD(T) calculation with a TZ(2df) basis, where the O–O and the F–O distances are 1.218 Å and 1.589 Å, respectively, i.e. the effect of a set of f-functions is 0.002 Å and −0.025 Å. Even at this

Table 11.20 Bond distance (Å) in FOOF. Experimental values are 1.217 and 1.575 Å

		R_{OO}			R_{FO}	
	cc-pVDZ	DZP	TZ(2d)	cc-pVDZ	DZP	TZ(2d)
HF	1.304	1.308	1.301	1.368	1.362	1.361
MP2	1.210	1.266	1.140	1.581	1.521	1.728
MP3	1.302	1.320	1.301	1.455	1.449	1.450
CCSD	1.276	1.307	1.278	1.494	1.474	1.482
CCSD(T)	1.216	1.261	1.216	1.637	1.571	1.614
CISD	1.304	1.316	1.301	1.416	1.412	1.407
SVWN	1.202	1.222	1.186	1.556	1.536	1.573
BLYP	1.224	1.243	1.207	1.622	1.604	1.643
BPW91	1.211	1.231	1.119	1.612	1.589	1.623
B3LYP	1.240	1.264	1.222	1.523	1.502	1.540
B3PW91	1.229	1.254	1.217	1.517	1.491	1.524

Table 11.20 sourced from Ref. 15.

highly demanding level of theory is the F–O bond length still in error by 0.013 Å. For a simple system like H_2O, the same level of accuracy is reached already at the MP2 level with a DZP type basis.

The DFT methods are all well behaved, and perform surprisingly well for such a difficult system. The two hybrid methods, B3LYP and B3PW91, give results comparable to those for CCSD(T). The main problem is of course that there is no way of systematically improving the structure, or knowing beforehand whether DFT will be able to give a good description for the specific problem.

11.7.2 The Dipole Moment of CO

The experimental value for the dipole moment of CO is 0.122 D, with the polarity C^-O^+, for a bond length of 1.1281 Å.[16] Calculated values with the aug-cc-pVXZ basis sets[17] are given in Table 11.21. Some other results using other basis sets are shown in Table 11.22.

Table 11.21 Dipole moment (Debye) for CO; the experimental value is 0.122 D

	aug-cc-pVDZ	aug-cc-pVTZ	aug-cc-pVQZ	aug-cc-pV5Z
HF	− 0.255	− 0.263	− 0.265	− 0.265
MP2	0.296	0.280	0.275	0.273
MP3	0.076	0.047	0.036	0.032
MP4	0.220	0.222	0.216	0.214
CCSD	0.097	0.070	0.059	0.055
CCSD(T)	0.141	0.127	0.118	0.115
CISD	0.050	0.023	0.011	
SVWN	0.232	0.226	0.229	
BLYP	0.187	0.184	0.185	
BPW91	0.221	0.217	0.218	
B3LYP	0.091	0.086	0.087	
B3PW91	0.119	0.114	0.116	

Table 11.22 Dipole moment (Debye) for CO; the experimental value is 0.122 D

	aug-DZP	10s9p4d2f	ANO [4s3p2d1f]	ANO [7s6p5d3f2g1h]
HF	− 0.273	− 0.266		
MP2	0.303	0.282		
MP3	0.079	0.047		
MP4	0.223	0.235		
CCSD	0.100	0.071	0.067	0.075
CCSD(T)	0.142	0.130	0.107	0.110
CCSDT	0.140			

Table 11.22 sourced from Ref. 18.

The HF level as usual overestimates the polarity, in this case leading to an incorrect direction of the dipole moment. The MP perturbation series oscillates, and it is clear that the MP4 result is far from converged. The CCSD(T) method apparently recovers the most important part of the electron correlation, as compared to the full CCSDT result. However, even with the aug-cc-pV5Z basis sets, there is still a discrepancy of ~0.01 D relative to the experimental value.

The DFT methods are not particularly accurate, although for this specific problem the B3PW91 method gives a reasonably good result.

11.7.3 The Vibrational Frequencies of O_3

Ozone is an example of a molecule where the single reference RHF is quite poor, since there is considerably biradical character in the wave function (as illustrated in Figure 4.8). The harmonic vibrational frequencies derived from experiments are $1135 \, cm^{-1}$, $1089 \, cm^{-1}$ and $716 \, cm^{-1}$, where the band at $1089 \, cm^{-1}$ corresponds to an asymmetric stretch.[19] As this nuclear motion changes the relative weights of the ionic and biradical structures, the frequency is very sensitive to the quality of the wave function. Although the wave function is equally poor for all the frequencies, the two other vibrations (symmetric stretch and angle bending) conserve the C_{2v} symmetry, and thus benefit from a significant cancellation of errors.

The calculated frequencies at different levels of theory with the cc-pVTZ basis are given in Table 11.23, corresponding results with more sophisticated methods are shown in Table 11.24. The simple picture with ozone as a resonance structure between ionic and biradical forms would suggests that a two-configuration wave function should be able to give a qualitatively correct description. The [2,2]-CASSCF result, however, clearly shows that dynamical correlation is much more important. The poor RHF reference wave function is clearly seen in the MP results, the MP2 value being in error by a factor of 2 for the asymmetric stretch, and the MP4 result in error by ~$500 \, cm^{-1}$ for ω_2, despite reproducing ω_1 and ω_3 to within $30 \, cm^{-1}$. The coupled cluster methods are less sensitive to the quality of the HF wave function, and their results are in somewhat better agreement with the experimental values. The CCSD(T) results are within ~$20 \, cm^{-1}$ of the experimental values, but the results shown in Table 11.24 below suggest that this may partly be accidental. The DFT methods all perform very well, yielding results comparable to those at the CCSD or CCSD(T) levels, at a fraction of the

Table 11.23 Harmonic frequencies for O_3 with the cc-pVTZ basis

Method	ω_1	ω_2	ω_3
HF	1537	1418	867
MP2	1166	2241	743
MP3	1364	1713	798
MP4	1106	1592	695
CCSD	1278	1267	762
CCSD(T)	1154	1067	717
CISD	1407	1535	816
[2,2]-CASSCF	1189	1497	799
SVWN	1249	1148	744
BLYP	1130	980	683
BPW91	1177	1047	706
B3LYP	1252	1194	746
B3PW91	1288	1244	762
Experimental	1135	1089	716

Table 11.24 Harmonic frequencies for O_3 with other methods

Method/basis	ω_1	ω_2	ω_3
CCSD/DZP	1256	1240	748
CCSD(T)/DZP	1129	976	703
CCSDT/DZP	1141	1077	705
CCSD/ANO [5s4p3d2f]	1280	1262	766
CCSD(T)/ANO [5s4p3d2f]	1153	1053	718
CCSD/ANO [5s4p3d2f1g]	1292	1280	771
[12,9]-CASSCF/ANO [4s3p2d1f]	1100	1039	708
[12,9]-CASPT2/ANO [4s3p2d1f]	1087	998	691
Experimental	1135	1089	716

Table 11.24 sourced from Ref. 20.

computational cost. Even the local density functional, SVWN, gives acceptable results. It can be noted that the cc-pVTZ basis set is sufficiently for the DFT results to be essentially converged, and the results in Table 11.23 thus reflect the intrinsic accuracy of the different DFT methods.

The large difference between the CCSD, CCSD(T) and CCSDT results clearly indicates that triple excitations are important, and that the perturbative correction in CCSD(T) overestimates the effect. The CCSDT/DZP results are in good agreement with the experimental values, but the coupled cluster calculations with the larger ANO basis sets suggest that this is partly accidental. Note also that addition of g-functions has effects $\sim 10–20 \, \text{cm}^{-1}$. The [12,9]-CASSCF (full-valence space spanned by the atomic p-orbitals) with the ANO basis gives a good description, but inclusion of dynamical correlation by the CASPT2 method again spoils the agreement.

11.8   Relative Energies of C₄H₆ Isomers

The elaborate treatment for the H_2O system is only possible because of its small size. For larger systems, less rigorous methods must be employed. Let us as a more realistic example consider a determination of the relative stability of the C_4H_6 isomers shown in Figure 11.15. There are experimental values for the first eight structures,[21] which allows an evaluation of the performance of different methods. This in turn enables an estimate of how much trust should be put on the predicted values for **9**, **10** and **11**.

An investigation may start by optimizing the geometries by semi-empirical methods; this will give initial estimates of the energetics and provide reasonable starting geometries for higher-level *ab initio* calculations. Relative energies and associated errors relative to the experimental values for different semi-empirical methods are shown in Table 11.25.

MINDO/3 methods clearly have severe problems with some of the conjugated systems. The MNDO/AM1/PM3 family perform somewhat better, although none of them can predict the correct ordering. The SAM1 method is not an improvement for this case. The mean absolute deviation (MAD) for the predicted stabilities is ~10 kcal/mol, which is a typical accuracy for semi-empirical methods.

The next step up in terms of theory is *ab initio* HF with increasingly larger basis sets. Table 11.26 shows the results for various basis sets, where the geometries have been optimized with the STO-3G, 3-21G and 6-31G(d,p) basis sets, but with the latter used for the larger basis. The minimum STO-3G basis performs worse than the semi-

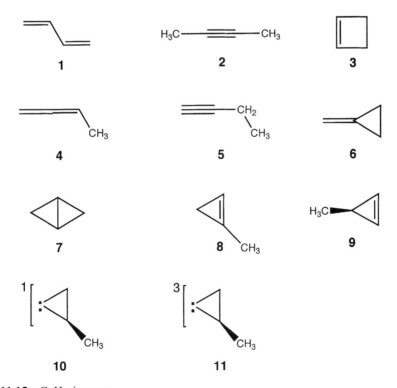

Figure 11.15   C₄H₆ isomers

Table 11.25 Energies (kcal/mol) relative to **1** calculated by semi-empirical methods

Isomer	MINDO/3	MNDO	AM1	PM3	SAM1	Exp.
2	− 19.9	− 4.1	2.1	− 1.3	− 5.5	8.6
3	1.2	2.1	15.8	6.6	13.8	11.2
4	− 3.4	4.6	7.2	7.0	2.0	12.4
5	− 3.8	7.2	7.6	4.7	3.7	13.2
6	1.9	8.9	17.7	13.5	15.1	21.7
7	17.8	35.1	48.2	38.2	44.1	25.6
8	10.1	24.7	34.7	26.3	28.2	31.9
9	20.0	31.6	37.8	29.2	34.2	
10	55.6	83.5	84.7	75.9	61.0	
11	54.2	84.5	90.3	80.3	80.4	
MAD	17.2	9.3	7.3	7.8	9.3	

Table 11.26 Energies relative to **1** calculated at the HF level with different basis sets

Isomer	STO-3G	3-21G	6-31G (d,p)	DZP	6-311G (2d,2p)	TZ (2d,2p)	Exp.
2	− 12.8	3.6	7.1	7.8	7.5	7.7	8.6
3	− 12.5	18.0	12.9	13.6	15.2	15.7	11.2
4	8.5	11.3	12.9	14.3	15.6	13.5	12.4
5	− 5.3	9.2	13.4	13.4	13.2	13.4	13.2
6	5.8	25.6	20.4	21.5	22.8	23.1	21.7
7	11.6	45.7	30.1	31.8	33.8	34.7	25.6
8	22.3	47.2	37.1	39.1	39.2	39.8	31.9
9	17.3	43.5	32.6	34.4	34.8	35.3	
10	51.7	83.5	72.7	72.4	74.6	75.3	
11	45.0	77.4	70.4	71.3	73.6	74.6	
MAD	15.3	8.0	2.1	2.7	3.6	3.6	

empirical methods, at a substantially higher computational cost. From experience, it is known that the geometry usually changes little beyond a DZP type basis, and relative energies change little beyond a TZ(2d,2p) basis. It is seen that the two different DZP type basis sets, 6-31G(d,p) and the Huzinaga contraction of the Dunning 9s5p primitive basis, give similar results, as do the 6-311G(2d,2p) and TZ(2d,2p). As the Pople style basis sets (6-31G and 6-311G) contain fewer primitive basis functions, they are computationally more efficient, and often preferred in practice. Notice also that the errors tend to increase as the basis set is enlarged, i.e. with a medium sized basis set such as 6-31G(d,p) there are some (fortuitous) cancellations of errors from incomplete basis set and neglect of electron correlation. The TZ(2d,2p) or 6-311G(2d,2p) results are stable to further increases in the basis set to within ~1 kcal/mol, but some isomers (**3**, **7** and **8**) are significantly underestimated in stability. With errors up to 9 kcal/mol, there is little point in including for example differences in zero-point energies, as these are expected to be at most 1–2 kcal/mol. Note also that the triplet carbene **11** is calculated to be lower in energy than the singlet **10** at the HF level of theory. Since **11** has one less electron pair than **10**, this stability will be reversed once correlation is taken into account.

Table 11.27 MP2/6-31G(d,p) energies relative to **1**, using either HF/6-31G(d,p) or MP2/6-31G(d,p) optimized geometries

Isomer	HF/6-31G(d,p) geometry	MP2/6-31G(d,p) geometry	Exp.
2	5.5	4.8	8.6
3	7.7	7.9	11.2
4	12.1	12.3	12.4
5	10.7	10.0	13.2
6	16.1	16.4	21.7
7	20.0	20.3	25.6
8	29.9	29.9	31.9
9	26.2	26.0	
10	81.3	81.9	
11	89.5	89.9	
MAD	3.2	3.3	

The next level up for improving the results would be to include electron correlation; the MP2 method clearly is an obvious first choice. Correlated calculations require polarization functions, and at least a DZP type basis is mandatory. The results at the MP2/6-31G(d,p) level using the HF/6-31G(d,p) optimized geometries, are shown in the first column in Table 11.27. The largest error is now reduced to less than 5 kcal/mol. Furthermore, as a good fraction of the correlation energy is recovered, the singlet carbene **10** is stabilized relative to **11** by ~10 kcal/mol.

In order to improve the results further, several features need to be addressed

(1) Optimizing the geometry at a correlated level, the MP2 method with a DZP type basis (6-31G(d,p) for example) would be a good starting points.
(2) Evaluation of frequencies for zero-point energy corrections.
(3) Improvement of the basis set at the MP2 level.
(4) Testing the sensitivity towards increases in electron correlation beyond MP2, for example by coupled cluster or MP4 calculations.

These effects are shown in Tables 11.27–11.30. The energetic changes due to optimization at the MP2 level, relative to the HF geometries, are less than 0.7 kcal/mol (Table 11.27).

The influence of zero-point energies is shown in Table 11.28. As the HF frequencies are systematically too high, the HF/6-31G(d,p) values are scaled by 0.92, and the MP2/6-31G(d,p) values are similarly scaled by 0.97.[22] The change in stabilities by zero-point energy corrections is less than 2 kcal/mol, and the difference between the HF and MP2 values is less than 0.7 kcal/mol.

Increasing the basis set to 6-311G(2d,2p) or TZ(2d,2p) gives changes of up to 5 kcal/mol, while addition of a set of f- and d-functions to form the 6-311G(2df,2pd) causes changes of ~3 kcal/mol. It is clear that further basis set extension may cause changes of a few kcal/mol.

Addition of electron correlation beyond MP2 is seen to have effects of up to 4–5 kcal/mol (Table 11.30). Although the difference between the CCSD(T) and MP2 results

Table 11.28 Zero-point energy corrections (kcal/mol)

Isomer	HF/6-31G(d,p)	MP2/6-31 G(d,p)
2	− 0.6	− 0.6
3	1.0	1.2
4	− 0.7	− 0.3
5	− 0.3	− 0.3
6	0.1	0.5
7	0.9	1.4
8	− 0.6	− 0.2
9	− 0.4	0.1
10	− 1.7	− 1.3
11	− 0.8	− 0.1

Table 11.29 Energies relative to **1** calculated at the MP2 level with different basis sets, using MP2/6-31G(d,p) optimized geometries

Isomer	6-31G (d,p)	6-311G (2d,2p)	TZ (2d,2p)	6-311G (2df,2pd)	Exp.
2	4.8	4.8	5.1	4.8	8.6
3	7.9	10.6	11.2	9.4	11.2
4	12.3	12.8	12.5	12.5	12.4
5	10.0	9.6	10.1	9.7	13.2
6	16.4	19.6	19.5	17.5	21.7
7	20.3	24.6	25.2	21.4	25.6
8	29.9	33.0	33.8	31.3	31.9
9	26.0	29.6	30.0	27.6	
10	81.9	84.0	84.8	83.3	
11	89.9	94.0	95.1	94.2	
r.m.s. error	3.3	1.8	1.6	2.6	

Table 11.30 Energies relative to **1** at different levels calculated with the 6-31G(d,p) basis sets at the MP2/6-31G(d,p) optimized geometry

Isomer	HF	MP2	MP3	MP4	CCSD	CCSD(T)	CISD	Exp.
2	7.9	4.8	9.1	7.4	8.5	9.2	7.5	8.6
3	12.6	7.9	8.9	10.1	10.1	10.6	9.1	11.2
4	12.7	12.3	12.3	12.2	12.2	12.5	12.1	12.4
5	14.2	10.0	14.0	12.5	13.5	14.1	13.2	13.2
6	20.0	16.4	17.6	18.5	18.4	19.1	17.2	21.7
7	29.8	20.3	23.7	25.2	25.8	26.5	23.8	25.6
8	37.2	29.9	32.2	32.8	33.3	33.7	32.9	31.9
9	32.6	26.0	28.5	29.0	29.6	30.1	28.7	
10	72.0	81.9	78.0	80.3	77.7	79.6	75.7	
11	70.1	89.9	88.2	91.2	88.3	90.6	82.4	
MAD	2.1	3.3	1.4	1.1	0.9	1.1	1.5	

Table 11.31 Energies relative to **1** by combining results from different calculations

Isomer	MP2 6-311G(2df,2pd)	Δ(CCSD(T)-MP2) 6-31G(d,p)	ΔZPE MP2/6-31G(d,p)	Sum	Exp.
2	4.8	4.4	− 0.6	8.6	8.6
3	9.4	2.7	1.2	13.3	11.2
4	12.5	0.2	− 0.3	12.4	12.4
5	9.7	4.1	− 0.3	13.5	13.2
6	17.5	2.7	0.5	20.7	21.7
7	21.4	6.2	1.4	29.0	25.6
8	31.3	3.8	− 0.2	34.9	31.9
9	27.6	4.1	0.1	31.8	
10	83.3	− 2.3	− 1.3	79.7	
11	94.2	0.7	− 0.1	94.8	
MAD	2.0			1.4	

Table 11.32 Energies relative to **1** calculated at DFT levels with the 6-311G(2d,2p) basis set, using MP2/6-31G(d,p) optimized geometries

Isomer	SVWN	BLYP	BPW91	B3LYP	B3PW91	Exp.
2	8.9	9.7	8.6	9.3	8.4	8.6
3	5.4	17.0	11.1	14.6	9.8	11.2
4	9.9	10.8	10.1	11.3	10.6	12.4
5	16.2	16.4	16.0	15.9	15.4	13.2
6	13.1	22.1	17.2	20.5	16.5	21.7
7	16.5	34.8	25.3	31.3	23.5	25.6
8	29.8	38.3	33.5	37.0	33.0	31.9
9	25.5	33.8	28.7	32.6	28.4	
10	81.0	84.2	80.6	82.0	79.1	
11	92.1	95.8	89.3	93.6	87.8	
MAD	4.5	4.0	1.7	2.8	2.0	

might decrease if larger basis sets were to be used, it shows that the MP2/6-311G (2df,2pd) level needs to be improved both with respect to additional electron correlation and larger basis set to provide results converged to ~1 kcal/mol. The CCSD(T) results are in excellent agreement with experiments, but the basis set effect shown in Table 11.29 and the neglect of zero-point energy corrections indicate that this is partly fortuitous. Note also that the CCSD results are slightly better than those for CCSD(T), despite the latter being the theoretically superior method, indicating cancellation of errors.

By assuming additivity in the style of the G2 procedure (Section 5.5), the CCSD(T)/ 6-31G(d,p) results may be combined with the changes due to basis set enlargement to 6-311G(2df,2pd) at the MP2 level and the zero-point energy corrections calculated at the MP2/6-31G(d,p) level. The results are shown in Table 11.31. From the observed accuracy of ±2 kcal/mol for structures **2–8**, the energetics of the species **9–11** may be assumed to be reliable to the same level of accuracy.

Results from various DFT methods are shown in Table 11.32. They in general perform well, yielding results of MP2 quality or better.

Table 11.33 Energies relative to **1** calculated by force field methods

Isomer	MM2	MM3	MMX	Exp.
2	10.9	10.6	11.3	8.6
3	12.8	12.6	11.5	11.2
4	13.6		12.6	12.4
5	14.6	14.6	14.9	13.2
6			21.7	21.7
7	26.7	27.0	23.9	25.6
8			33.9	31.9
9			31.8	
MAD	(1.5)	(1.6)	1.2	

Calculating relative energies of a series of hydrocarbons is of course well suited to force field methods, although a comparison of stabilities for isomers containing different numbers of "functional" groups (CH_3, CH_2, etc.) means that only force fields which are able to convert steric energies to heats of formation can be used (Section 2.2.9). Even for these relatively simple compounds, however, there are several "unusual" features for which adequate parameters are lacking. The straight MM2 and MM3 force fields lack parameters for the cyclopropenes **8** and **9**, the methylene-cyclopropane **6** and MM3 lacks parameters for the allene **4**. The carbenes **10** and **11** are of course outside the capabilities of force field methods. Table 11.33 compares the performances of the MM2 and MM3 methods, along with that of one of the modified MM2 models (MMX),[23] where parameters have been added to allow calculations on **4**, **6**, **8**, and **9**.

The performance is (as expected) very good. MMX provides relative (and absolute) stabilities with a MAD of only 1.2 kcal/mol, which is better than the estimates from the combined theoretical methods in Table 11.31. Considering that force field calculations require a factor of $\sim 10^5$ less computer time for these systems than the *ab initio* methods combined in Table 11.31, this clearly shows that knowledge of the strengths and weakness of different theoretical tools is important in selecting a proper model for answering a given question.

References

1. A. R. Hoy and P. R. Bunker, *J. Mol. Spect.*, **74** (1979), 1.
2. A. Halkier, P. Jørgensen, J. Gauss and T. Helgaker, *Chem. Phys. Lett.*, **274** (1997), 235.
3. A. Lüchow, J. B. Anderson and D. Feller, *J. Chem. Phys.*, **106** (1997), 7706.
4. J. Olsen, P. Jørgensen, H. Koch, A. Balkova and R. J. Bartlett, *J. Chem. Phys.*, **104** (1996), 8007.
5. T. Helgaker, W. Klopper, H. Koch and J. Noga, *J. Chem. Phys.*, **106** (1997), 9639.
6. A. K. Wilson and T. H. Dunning Jr, *J. Chem. Phys.*, **106** (1997), 8718.
7. S. A. Clough, Y. Beers, G. P. Klein and L. S. Rothman, *J. Chem. Phys.*, **59** (1973), 2254.
8. W. S. Benedict, N. Gailar and E. K. Plyler, *J. Chem. Phys.*, **24** (1956), 1139.
9. A. P. Scott and L. Radom, *J. Phys. Chem.*, **100** (1996), 16502.
10. J. M. L. Martin and T. J. Lee, *Chem. Phys. Lett.*, **225** (1994), 473.
11. J. A. Pople, P. M. W. Gill and N. C. Handy, *Int. J. Quantum. Chem.*, **56** (1995), 303.
12. J. M. Wittbrodt and H. B. Schlegel, *J. Chem. Phys.*, **105** (1996), 6574.

13. P. Jensen, *J. Mol. Spectrosc*, **133** (1989), 438.
14. R. H. Jackson, *J. Chem. Soc.* (1962), 4585.
15. G. E. Scuseria, *J. Chem. Phys.*, **94** (1991), 442; T. J. Lee, J. E. Rice, G. E. Scuseria and H. F. Schaefer III, *Theo. Chim. Acta*, **75** (1989), 81.
16. J. S. Muenter, *J. Mol. Spectrosc.*, **55** (1970), 490.
17. K. A. Peterson and T. H. Dunning Jr, *J. Mol. Struct.*, **400** (1997), 93.
18. L. A. Barnes, B. Liu and R. Lindh, *J. Chem. Phys.*, **98** (1993), 3972; G. E. Scuseria, M. D. Miller, F. Jensen and J. Geertsen, *J. Chem. Phys.*, **94** (1991), 6660.
19. A. Barbe, C. Secroun and P. Jouve, *J. Mol. Spectrosc.*, **49** (1974), 171.
20. J. D. Watts, J. F. Stanton and R. J. Bartlett, *Chem. Phys. Lett.*, **178** (1991), 471; K. A. Peterson, R. C. Mayrhofer, E. L. Sibert III and R. C. Woods, *J. Chem. Phys.*, **94** (1991), 414; T. J. Lee and G. E. Scuseria, *J. Chem. Phys.*, **93** (1990), 489; P. Borowski, K. Andersson, P.-Å. Malmqvist and B. O. Roos, *J. Chem. Phys.*, **97** (1992), 5568.
21. S. W. Benzon, F. R. Cruickshank, D. M. Golden, G. R. Haugen, H. E. O'Neal, A. S. Rodgers, R. Shaw and R. Walsh, *Chem. Rev.*, **69** (1969), 279.
22. A. P. Scott and L. Radom, *J. Phys. Chem.*, **100** (1996), 16502.
23. K. E. Gilbert, *PCMODEL*, Serena Software.

12 Transition State Theory and Statistical Mechanics

Consider a chemical reaction of the type $A + B \rightarrow C + D$. The rate of reaction may be written as

$$\frac{d[C]}{dt} = \frac{d[D]}{dt} = -\frac{d[A]}{dt} = -\frac{d[B]}{dt} = k[A]^n[B]^m \tag{12.1}$$

where k is the rate constant. If k is known, the concentration of the various species can be calculated at a given time from the initial concentrations. At the microscopic level, the rate constant is a function of the quantum states of A, B, C and D, e.g. the electronic, translational, rotational and vibrational quantum numbers. The macroscopic rate constant is an average over such "microscopic" rate constants, weighted by the probability of finding a molecule with a given set of quantum numbers. For systems in equilibrium, the probability of finding a molecule in a certain state depends on its energy by means of the Boltzman distribution. The macroscopic rate constant thereby becomes a function of temperature.

12.1 Transition State Theory

Transition State Theory (TST) assumes that a reaction proceeds from one energy minimum to another via an intermediate maximum.[1] The *Transition State* is the configuration which divides the reactant and product parts of the surface (i.e. a molecule which has reached the transition state will continue on to product), while the geometrical configuration of the energy maximum is called the *Transition Structure*. Within standard TST the transition state and transition structure are identical, but this is not necessarily the case for more refined models. Nevertheless, the two terms are often used interchangeably, and share the same acronym TS. The reaction proceeds via a "reaction coordinate", usually taken to be negative at the reactant, zero at the TS and positive for the product. The reaction coordinate leads from the reactant to the product along a path where the energy is as low as possible, and the TS is the point where the energy has a

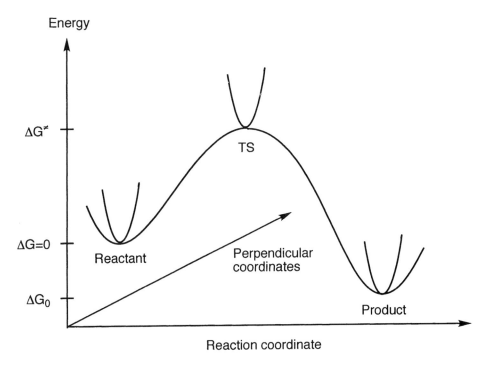

Figure 12.1   Schematic illustration of a reaction path

maximum. In the multidimensional case it is thus a first-order saddle point on the potential energy surface, a maximum in the reaction coordinate direction and a minimum along all other coordinates. TST is a semi-classical theory, where the dynamics along the reaction coordinate is treated classically, while the perpendicular directions take into account the quantization of for example the vibrational energy.

Transition state theory assumes an equilibrium energy distribution among all possible quantum states at all points along the reaction coordinate. The probability of finding a molecule in a given quantum state is proportional to $e^{-\Delta E/k_B T}$, which is a Boltzmann distribution. Assuming that the molecules at the TS are in equilibrium with the reactant, the macroscopic rate constant can be expressed as

$$k = \frac{k_B T}{h} e^{-\Delta G^{\neq}/RT} \tag{12.2}$$

$\Delta G^{\neq}$ is the Gibbs free energy difference between the TS and reactant, and k_B is Boltzmann's constant. Actually, the TST expression only holds if all molecules which pass from the reactant over the TS go on to product. To allow for "recrossings", where a molecule passes over the TS but is reflected back to the reactant side, a *transmission coefficient*, κ, is sometimes introduced. This factor also allows for the quantum mechanical phenomenon of tunnelling, i.e. molecules that have insufficient energy to pass <u>over</u> the TS may tunnel <u>through</u> the barrier and appear on the product side. The transmission coefficient is difficult to calculate, but is usually close to one, and rarely falls outside the range 0.5–2. At low temperatures the tunnelling contribution dominates, leading to $\kappa > 1$, while the recrossing effect is most important at high temperatures, giving $\kappa < 1$. For the majority of reactions the calculated accuracy in $\Delta G^{\neq}$ introduces errors much larger than a factor of 2, and the transmission coefficient is usually ignored.

From the TST expression (12.2) it is clear that if the free energy of the reactant and TS can be calculated, the reaction rate follows trivially. Similarly, the equilibrium constant for a reaction can be calculated from the free energy difference between the reactant(s) and product(s).

$$K_{eq} = e^{-\Delta G_0/RT} \tag{12.3}$$

The Gibbs free energy is given in terms of the enthalpy and entropy, $G = H - TS$. The enthalpy and entropy for a macroscopic ensemble of particles may be calculated from properties of the individual molecules by means of statistical mechanics.

12.2 Statistical Mechanics

The key feature in statistical mechanics is the *partition function*.[2] Just as the wave function is the corner-stone of quantum mechanics (from that everything else can be calculated by applying proper operators), the partition function allows calculation of all macroscopic functions in statistical mechanics. The partition function for a single molecule is usually denoted q and defined as a sum of exponential terms involving all possible quantum energy states; Q is the partition function for N molecules.

$$q = \sum_i^{\text{all states}} e^{-\varepsilon_i/k_B T}$$

$$Q = q^N \text{ (different particles)} \tag{12.4}$$

$$Q = \frac{q^N}{N!} \text{ (identical particles)}$$

The partition function may alternatively be written as a sum over all distinct energy levels, times a degeneracy factor g_i.

$$q = \sum_i^{\text{all levels}} g_i e^{-\varepsilon_i/k_B T} \tag{12.5}$$

Once the partition function is known, thermodynamic functions such as the internal energy U and Helmholtz free energy A may be calculated according to

$$U = k_B T^2 \left(\frac{\partial \ln Q}{\partial T}\right)_V \tag{12.6}$$

$$A = -k_B T \ln Q$$

Macroscopic observables, such as pressure P or heat capacity at constant volume C_V, may be calculated as derivatives of thermodynamic functions.

$$P = -\left(\frac{\partial A}{\partial V}\right)_T = k_B T \left(\frac{\partial \ln Q}{\partial V}\right)_T$$

$$C_V = \left(\frac{\partial U}{\partial T}\right)_V = 2 k_B T \left(\frac{\partial \ln Q}{\partial T}\right)_V + k_B T^2 \left(\frac{\partial^2 \ln Q}{\partial T^2}\right)_V \tag{12.7}$$

Other thermodynamical functions, such as the enthalpy H, the entropy S and Gibbs free energy G, may be constructed from these relations.

$$H = U + PV = k_B T^2 \left(\frac{\partial \ln Q}{\partial T}\right)_V + k_B TV \left(\frac{\partial \ln Q}{\partial V}\right)_T$$

$$S = \frac{U - A}{T} = k_B T \left(\frac{\partial \ln Q}{\partial T}\right)_V + k_B \ln Q \tag{12.8}$$

$$G = H - TS = k_B TV \left(\frac{\partial \ln Q}{\partial V}\right)_T - k_B T \ln Q$$

In order to calculate $q(Q)$ all possible quantum states are needed. It is usually assumed that the energy of a molecule can be approximated as a sum of terms involving translational, rotational, vibrational and electronical states. Except for a few cases this is a good approximation. For linear, "floppy" (soft bending potential), molecules the separation of the rotational and vibrational modes may be problematic. If two energy surfaces come close together (avoided crossing), the separability of the electronic and vibrational modes may be a poor approximation (breakdown of the Born–Oppenheimer approximation, Section 3.1).

There are in principle also energy levels associated with nuclear spins. In the absence of an external magnetic field, these are degenerate and consequently contribute a constant term to the partition function. As nuclear spins do not change during chemical reactions, we will ignore this contribution.

The assumption that the energy can be written as a sum of terms implies that the partition function can be written as a product of terms. As the enthalpy and entropy contributions involve taking the logarithm of q, the product thus transforms into sums of enthalpy and entropy contributions.

$$\varepsilon_{tot} = \varepsilon_{trans} + \varepsilon_{rot} + \varepsilon_{vib} + \varepsilon_{elec}$$

$$q_{tot} = q_{trans}\, q_{rot}\, q_{vib}\, q_{elec}$$

$$H_{tot} = H_{trans} + H_{rot} + H_{vib} + H_{elec} \tag{12.9}$$

$$S_{tot} = S_{trans} + S_{rot} + S_{vib} + S_{elec}$$

For each of the partition functions the sum over allowed quantum states runs to infinity; however, since the energies become larger, the partition functions are finite. Let us examine each of the q-factors in a little more detail.

12.2.1 q_{trans}

The allowed quantum states for the translational energy are determined by placing the molecule in a "box", i.e. the potential is zero inside the box but infinite outside. The only purpose of the box is to allow normalization of the translational wave function, i.e. the exact size is not important. The solutions to the Schrödinger equation for such a "particle in a box" are standing waves, cosine and sine functions, and the energy levels are very close together. The summation in the partition function can therefore be replaced by an integral (an integral is just a sum in the limit of infinitely-small contributions). The translational partition function becomes

$$q_{\text{trans}} = \left(\frac{2\pi M k_B T}{h^2}\right)^{3/2} V \tag{12.10}$$

The only molecular parameter which enters is the total molecular mass M. The volume depends on the number of particles. It is customary to work on a molar scale, in which case V is the volume of one mole of (ideal) gas.

12.2.2 q_{rot}

In the lowest approximation the rotation of a molecule is assumed to occur with a geometry that is independent of the rotational quantum number. A more refined treatment allows the geometry to "stretch" with rotational energy, this may be described by adding a "centrifugal" correction. Such corrections are typically of the order of a few %. The energy levels calculated from the Schrödinger equation for a diatomic "rigid rotor" are given by

$$\varepsilon_{\text{rot}} = J(J+1)\frac{h^2}{8\pi^2 I} \tag{12.11}$$

where J is a quantum number running from zero to infinity and I is the moment of inertia given by

$$I = m_1 r_1^2 + m_2 r_2^2 \tag{12.12}$$

with r_i being a coordinate relative to the centre of mass. For all molecules, except very light species such as H_2 and LiH, the moment of inertia is so large that the spacing between the rotational energy levels is much smaller than $k_B T$ at ambient temperatures. As for q_{trans}, this means that the summation in eq. (12.4) can be replaced by an integral, yielding

$$q_{\text{rot}} = \frac{8\pi^2 I k_B T}{h^2 \sigma} \tag{12.13}$$

The symmetry index σ is 2 for a homonuclear and 1 for a heteronuclear diatomic molecule.

For a polyatomic molecule the equivalent of eq. (12.12) is a 3×3 matrix.

$$\mathbf{I} = \begin{pmatrix} \sum_i m_i(y_i^2 + z_i^2) & -\sum_i m_i x_i y_i & -\sum_i m_i x z_i \\ -\sum_i m_i x_i y_i & \sum_i m_i(x_i^2 + z_i^2) & -\sum_i m_i y_i z_i \\ -\sum_i m_i x_i z_i & -\sum_i m_i y_i z_i & \sum_i m_i(x_i^2 + y_i^2) \end{pmatrix} \tag{12.14}$$

where the coordinates again are relative to the centre of mass. By choosing a suitable coordinate transformation this matrix may be diagonalized (Chapter 13), with the eigenvalues being the *moments of inertia* and the eigenvectors called *principal axes of inertia*. For a general polyatomic molecule the rotational energy levels cannot be written in a simple form. A good approximation, however, can be obtained from classical mechanics, resulting in the following partition function.

$$q_{\text{rot}} = \frac{\sqrt{\pi}}{\sigma} \left(\frac{8\pi^2 k_B T}{h^2}\right)^{3/2} \sqrt{I_1 I_2 I_3} \tag{12.15}$$

Here I_i are the three moments of inertia. The symmetry index σ is the order of the rotational subgroup in the molecular point group (i.e. the number of proper symmetry operations), for H_2O it is 2, for NH_3 it is 3, for benzene it is 12 etc. The rotational partition function requires only information about the atomic masses and positions (eq. (12.14)), i.e. the molecular geometry.

12.2.3 q_{vib}

In the lowest approximation the molecular vibrations may be described as those of a harmonic oscillator. These can be derived by expanding the energy as a function of the nuclear coordinates in a Taylor series around the equilibrium geometry. For a diatomic molecule this is the internuclear distance R.

$$E(R) = E(R_0) + \frac{dE}{dR}(R - R_0) + \frac{1}{2}\frac{d^2E}{dR^2}(R - R_0)^2 + \frac{1}{6}\frac{d^3E}{dR^3}(R - R_0)^3 + \cdots \quad (12.16)$$

The first term may be taken as zero, this is just the zero point for the energy. The second term (the gradient) vanishes since the expansion is around the equilibrium geometry. Keeping only the lowest non-zero term results in the harmonic approximation, with k being the force constant.

$$E(\Delta R) \simeq \frac{1}{2}\frac{d^2E}{dR^2}\Delta R^2 = \frac{1}{2}k\Delta R^2 \quad (12.17)$$

Including higher-order terms leads to anharmonic corrections to the vibration, such effects are typically of the order of a few %. The energy levels obtained from the Schrödinger equation for a one-dimensional harmonic oscillator (diatomic system) are given by

$$\varepsilon_{vib} = \left(n + \frac{1}{2}\right)h\nu$$

$$\nu = \frac{1}{2\pi}\sqrt{\frac{k}{\mu}} \quad (12.18)$$

where n is a quantum number running from zero to infinity and ν is the vibrational frequency given in terms of the force constant $k\,(\partial^2 E/\partial R^2)$ and the reduced mass $\mu = m_1 m_2/(m_1 + m_2)$. Contrary to the translational and rotational energies, the difference between vibrational energy levels is not small compared to $k_B T$, it is typically of the same order of magnitude for temperatures around 300 K. The summation for q_{vib} (eq. (12.4)) can therefore not be replaced by an integral. Owing to the regular spacing of the energy level, however, the infinite summation can be written in a closed form.

$$q_{vib} = \sum_{n=0}^{\infty} e^{-\left(n+\frac{1}{2}\right)\frac{h\nu}{k_B T}} = e^{-\frac{h\nu}{2k_B T}}\sum_{n=0}^{\infty} e^{-\frac{nh\nu}{k_B T}}$$

$$q_{vib} = e^{-\frac{h\nu}{2k_B T}}\left(1 + e^{-\frac{h\nu}{k_B T}} + e^{-\frac{2h\nu}{k_B T}} + e^{-\frac{3h\nu}{k_B T}} + \ldots\right)$$

$$\left(1 + e^{-\frac{h\nu}{k_B T}} + e^{-\frac{2h\nu}{k_B T}} + e^{-\frac{3h\nu}{k_B T}} + \ldots\right) = \frac{1}{1 - e^{-\frac{h\nu}{k_B T}}} \quad (12.19)$$

$$q_{vib} = \frac{e^{-\frac{h\nu}{2k_B T}}}{1 - e^{-\frac{h\nu}{k_B T}}}$$

In the infinite sum each successive term is smaller than the previous by a constant factor $(e^{-h\nu/kT}$, which is $<1)$, and can therefore be expressed in a closed form. Only the vibrational frequency is needed for calculating the vibrational partition function for a harmonic oscillator, i.e. only the force constant and the atomic masses are required.

For a polynuclear molecule the force constant k is replaced by a $3N \times 3N$ matrix (N being the number of atoms in the molecule) containing all the second derivatives of the energy with respect to the coordinates. By a mass-weighting and transformation to a new coordinate system called the vibrational normal coordinates, this may be brought to a diagonal form, see Section 13.1 for details. In the vibrational normal coordinates, the $3N$-dimensional Schrödinger equation can be separated into $3N$ one-dimensional equations, each having the form of a harmonic oscillator. Of these three describe the overall translation and three (two for a linear molecule) describe the overall rotation, leaving $3N - 6(5)$ vibrations. If the stationary point is a minimum on the energy surface, the eigenvalues of the force constant matrix are all positive. If, however, the stationary point is a TS, one (and only one) of the eigenvalues is negative. This corresponds to the energy being a maximum in one direction and a minimum in all other directions. The "frequency" for the "vibration" along the eigenvector with a negative force constant will formally be imaginary, as it is the square root of a negative number (eq. (12.18)). For a TS there are thus only $3N - 7$ vibrations.

For a polyatomic molecule the total vibrational energy may be written as a sum of energies for each vibration, and the partition function as a product of partition functions.

$$q_{\text{vib}} = \prod_{i=1}^{3N-6(7)} \frac{e^{-\frac{h\nu_i}{2k_BT}}}{1 - e^{\frac{-h\nu_i}{k_BT}}} \qquad (12.20)$$

Only the vibrational frequencies are needed, which can be calculated from the force constant matrix and atomic masses.

12.2.4 q_{elec}

The electronic partition function involves a sum over electronic quantum states. These are the solutions to the electronic Schrödinger equation, i.e. the lowest (ground) state and all possible excited states. The energy difference between the ground and excited states is usually much larger than k_BT, which means that only the first term in the partition function summation (eq. (12.5)) is important. Defining the zero point for the energy as the electronic energy of the reactant, the electronic partition functions for the reactant and TS become

$$q_{\text{elec}}^{\text{reactant}} = g$$
$$q_{\text{elec}}^{\text{TS}} = g\,e^{-\frac{\Delta E^{\neq}}{k_BT}} \qquad (12.21)$$

The $\Delta E^{\neq}$ term is the difference in electronic energy between the reactant and TS, and g is the electronic degeneracy of the wave function. The degeneracy may be either in the spin part ($g = 1$ for a singlet, 2 for a doublet, 3 for a triplet etc.) or in the spatial part ($g = 1$ for wave functions belonging to an A, B or Σ representation in the point group, 2 for an E, π or Δ representation, 3 for a T representation etc.). The large majority of stable molecules have non-degenerate ground-state wave functions, and consequently $g = 1$.

12.3   Enthalpy and Entropy Contributions

Given the partition functions, the enthalpy and entropy terms may be calculated by carrying out the required differentiations in eq. (12.8). For one mole of molecules, the results for a non-linear system are (R being the gas constant)

$$H_{\text{trans}} = \frac{5}{2}RT$$

$$S_{\text{trans}} = \frac{5}{2}R + R\ln\left(\frac{V}{N_A}\left(\frac{2\pi M k_B T}{h^2}\right)^{3/2}\right)$$

$$H_{\text{rot}} = \frac{3}{2}RT$$

$$S_{\text{rot}} = \frac{1}{2}R\left[3 + \ln\left(\frac{\sqrt{\pi}}{\sigma}\left(\frac{8\pi^2 k_B T}{h^2}\right)^{3/2}\sqrt{I_1 I_2 I_3}\right)\right]$$

$$H_{\text{vib}} = R\sum_{i=1}^{3N-6(7)}\left(\frac{h\nu_i}{2k_B} + \frac{h\nu_i}{k_B}\frac{1}{e^{h\nu_i/k_B T}-1}\right)$$ (12.22)

$$S_{\text{vib}} = R\sum_{i=1}^{3N-6(7)}\left(\frac{h\nu_i}{k_B T}\frac{1}{e^{h\nu_i/k_B T}-1} - \ln(1 - e^{-h\nu_i/k_B T})\right)$$

$$H_{\text{elec}}^{\text{reactant}} = 0$$

$$H_{\text{elec}}^{\text{TS}} = \Delta E^{\neq}$$

$$S_{\text{elec}}^{\text{reactant}} = S_{\text{elec}}^{\text{TS}} = R\ln(g)$$

The rotational terms are slightly different for a linear molecule, and the vibrational terms will contain one more vibrational contribution.

$$H_{\text{rot}}(\text{linear}) = RT$$

$$S_{\text{rot}}(\text{linear}) = R\left[1 + \ln\left(\frac{8\pi^2 I k_B T}{\sigma h^2}\right)\right]$$ (12.23)

The vibrational enthalpy consists of two parts, the first is a sum of $h\nu/2$ contributions, this is the zero-point energies. The second part depends on temperature, and is a contribution from molecules which are not in the vibrational ground state. This contribution goes toward zero as the temperature goes to zero when all molecules are in the ground state. Note also that the sum over vibrational frequencies runs over $3N - 6$ for the reactant(s), but only $3N - 7$ for the TS. At the TS, one of the normal vibrations has been transformed into the reaction coordinate, which formally has an imaginary frequency.

To calculate $\Delta G^{\neq} = G_{\text{TS}} - G_{\text{reactant}}$, we need $\Delta H^{\neq}$ and $\Delta S^{\neq}$. $\Delta H_{\text{elec}}^{\neq}$ is directly the difference in electronic energy between the TS and the reactant. Except for complicated reactions involving several electronic states of different degeneracy (e.g. singlet molecules reacting via a triplet TS), $\Delta S_{\text{elec}}^{\neq}$ is zero.

For <u>unimolecular</u> reactions $\Delta H_{\text{trans}}^{\neq}$, $\Delta H_{\text{rot}}^{\neq}$, and $\Delta S_{\text{trans}}^{\neq}$ are zero, while $\Delta S_{\text{rot}}^{\neq}$ may be slightly different from zero due to a change in geometry (thereby changing the moments of inertia). The vibrational contribution to $\Delta H^{\neq}$ is usually a few kcal/mol negative, as there is one less vibration at the TS (lack of zero-point energy). The TS is normally somewhat more ordered than the reactant, typically giving a slightly negative $\Delta S^{\neq}_{\text{vib}}$.

For <u>bimolecular</u> reactions (i.e. where the reactant is two separate molecules) $\Delta H^{\neq}_{\text{trans}}$ and $\Delta H^{\neq}_{\text{rot}}$ contribute a constant -4 RT. The translational and rotational entropy changes are substantially negative, -30 to -50 e.u., due to the fact that there are six translational and six rotational modes in the reactants but only three of each at the TS. The six remaining degrees of freedom are transformed into the reaction coordinate and five new vibrations at the TS. These additional vibrations usually make $\Delta H^{\neq}_{\text{vib}}$ a few kcal/mol positive, and $\Delta S^{\neq}_{\text{vib}}$ positive by 5–10 e.u. For bimolecular reactions the entropy typically raises the free energy barrier by 10–15 kcal/mol, relative to the electronic energy alone.

Similarly, in order to calculate $\Delta G_0 = G_{\text{product}} - G_{\text{reactant}}$ we need ΔH_0 and ΔS_0. The generalization for the electronic, translational and rotational contributions to $\Delta H^{\neq}$ and $\Delta S^{\neq}$ given above also holds for ΔH_0 and ΔS_0. The considerations for a unimolecular reaction hold for reactions where the number of reactant and product molecules is the same, while the generalizations for a bimolecular reaction correspond to an addition where two reactants form a single product molecule (the reverse process being a fragmentation). The vibrational contribution to ΔH_0 and ΔS_0 for a "number conserving" reaction is usually small, since there is the same number of vibrational modes in the reactant and product. For an addition reaction the number of vibrational modes increases by six, and the contributions to ΔH_0 and ΔS_0 are again slightly positive, typically by a few kcal/mol and 5–10 e.u.

Tables 12.1–12.3 below give some examples of the magnitude of each term for two bimolecular reactions (Diels–Alder and S_N2 reactions, forming either one or two molecules as the product) and a unimolecular rearrangement (Claisen reaction). All values have been calculated at the MP2 level with the 6-31G(d) basis for the Diels–Alder and Claisen reactions, and the 6-31+G(d) basis for the S_N2 reaction. The values are given in kcal/mol at a temperature of 300 K ($RT = 0.60$ kcal/mol).

It should be noted that the experimental activation enthalpy for the Diels–Alder reaction is ~ 33 kcal/mol (estimated from the reverse reaction and the experimental reaction energy[3] i.e. the MP2/6-31G(d) value is ~ 14 kcal/mol too low. Similarly, the calculated reaction energy of -47 kcal/mol is in rather poor agreement with the

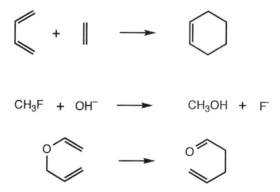

Figure 12.2   The Diels–Alder, S_N2 and Claisen reactions

Table 12.1 Diels–Alder reaction of butadiene and ethylene to form cyclohexene

	$\Delta H^{\neq}$	$\Delta S^{\neq}$	$-T\Delta S^{\neq}$	ΔH_0	ΔS_0	$-T\Delta S_0$
electronic	17.9	0	0	-52.5	0	0
vibrational	3.4	5.3	-1.6	7.6	1.5	-0.5
rotational	-0.9	-11.4	3.4	-0.9	-13.3	4.0
translational	-1.5	-34.7	10.4	-1.5	-34.7	10.4
total	18.9	-40.8	12.2	-47.3	-46.5	13.9
Exp.	~ 33	-41	12.3	-39.7	-44.8	13.4

Table 12.2 S_N2 reaction of OH^- with CH_3F to form CH_3OH and F^-

	$\Delta H^{\neq}$	$\Delta S^{\neq}$	$-T\Delta S^{\neq}$	ΔH_0	ΔS_0	$-T\Delta S_0$
electronic	5.1	0	0	-25.0	0	0
vibrational	2.1	6.8	-2.1	2.5	1.3	-0.4
rotational	-0.6	-0.4	0.1	-0.6	-4.3	1.3
translational	-1.5	-27.1	8.1	0	0	0
total	5.1	-20.7	6.1	-23.1	-3.0	0.9

Table 12.3 Claisen rearangement of allyl vinyl ether to form 5-hexenal

	$\Delta H^{\neq}$	$\Delta S^{\neq}$	$-T\Delta S^{\neq}$	ΔH_0	ΔS_0	$-T\Delta S_0$
electronic	23.4	0	0	-23.5	0	0
vibrational	-1.3	-8.6	2.6	0.3	-0.2	0.1
rotational	0	-0.3	0.1	0	0.4	-0.1
translational	0	0	0	0	0	0
total	22.1	-8.9	2.7	-23.2	0.2	0.0
Exp.	29.8	-7.7	2.3			

experimental value of -40 kcal/mol. The S_N2 reaction refers to the situation in the gas phase where the reagents initially form an ion–dipole complex, pass over the TS and form another ion–dipole complex. The energies given above are relative to the isolated reactants, which is the reason for the low activation energy. Note also that the rotational contribution to the reaction enthalpy is not zero, this is because one of the reactants is a diatomic molecule, while one of the products is an atom (which has no rotational term). The MP2/6–31G(d) activation enthalpy for the Claisen reaction is again somewhat lower than the experimental value of 29.8 kcal/mol[4] while the calculated activation entropy is in good agreement with the experimental value.

Summarizing, in order to calculate rate and equilibrium constants, we need to calculate $\Delta G^{\neq}$ and ΔG_0. This can be done if the geometry, energy and force constants are known for the reactant, TS and product. The translational and rotational contributions are trivial to calculate, while the vibrational frequencies require the full force constant matrix (i.e. all energy second derivatives), which may involve a significant computational effort.

The above treatment has made some assumptions, such as harmonic frequencies and "sufficiently small" energy spacing between the rotational levels. If a more elaborate treatment is required, the summation for the partition functions must be carried out explicitly. Many molecules also have internal rotations with quite small barriers. In the above they are assumed to be described by simple harmonic vibrations, which may be a poor approximation. Calculating the energy levels for a hindered rotor is somewhat complicated,[5] and is rarely done. If the barrier is very low, the motion may be treated as a free rotor, in which case it contributes a constant factor of RT to the enthalpy and $R/2$ to the entropy.

As can be seen from Tables 12.1–12.3, the electronic energy difference between the reactant/TS and reactant/product is the most important contribution to $\Delta G^{\neq}$ and ΔG_0. The electronic energy is furthermore the most difficult to calculate accurately. Let us consider three cases.

(A) The error in $\Delta E^{\neq}/\Delta E_0$ is $\sim 10\,$kcal/mol. It is clear that spending significant amounts of computer time in order to include vibrational, rotational and translational corrections has little meaning.
(B) The error in $\Delta E^{\neq}/\Delta E_0$ is $\sim 1\,$kcal/mol. The corrections from vibrations, rotations and translation now become important, and should be included. However, sophisticated treatments like anharmonic vibrations are unimportant.
(C) The error in $\Delta E^{\neq}/\Delta E_0$ is $\sim 0.1\,$kcal/mol. Corrections from vibrations, rotations and translation are clearly necessary. Explicit calculation of the partition functions for anharmonic vibrations and internal rotations may be considered. However, at this point other factors also become important for the activation energy. These include for example:

  (i) The position of the TS has been assumed to be at the maximum on the electronic energy surface. Actually it should be at the maximum on the ΔG surface. This would include entropy effects, and thus allow the position of the TS to depend on temperature. Such treatments are referred to as *Variational Transition State Theory* (VTST)[6] and are important for reactions with small (or zero) enthalpic barriers, such as recombinations of radicals.
  (ii) The possibility of recrossings and tunnelling (which requires a quantum description of the nuclear motion) should be included in order to produce a transmission coefficient.

Calculating the electronic barrier with an accuracy of $\sim 0.1\,$kcal/mol is only possible for very simple systems. An accuracy of $\sim 1\,$kcal/mol is usually considered a good, but hard to get, level of accuracy. The situation is slightly better for relative energies of stable species, but a $\sim 1\,$kcal/mol accuracy still requires a significant computational effort. Thermodynamic corrections beyond the rigid rotor/harmonic vibrations approximation are therefore rarely performed.

A prediction of $\Delta E^{\neq}/\Delta E_0$ to within $\sim 0.1\,$kcal/mol may produce a $\Delta G^{\neq}/\Delta G_0$ accurate to maybe $\sim 0.2\,$kcal/mol. This corresponds to a factor of ~ 1.4 error (at $T = 300\,$K) in the rate/equilibrium constant, which is poor compared to what is routinely obtained by experimental techniques. Calculating $\Delta G^{\neq}/\Delta G_0$ to within $\sim 1\,$kcal/mol is still only possible for fairly small systems. This corresponds to predicting the absolute rate constant, or the equilibrium distribution, to within a factor of

10. Theoretical calculations are therefore not very useful for predicting absolute rate or equilibrium constants. Relative rates, however, are somewhat easier. Often the interest is not in how fast a certain product is formed, but rather on what the rate difference is between two reactions. The absolute rate (only) influences how long the total reaction time will be, or how high the temperature should be. Rate differences, however, determine what the ratio between products is. When comparing calculated activation parameters for similar reactions, one can always hope for some "cancellation of errors". Theoretical methods are most useful for predicting and rationalizing different reaction pathways, not for predicting absolute rates.

The activation enthalpies and entropies are in principle dependent on temperature (eq. 12.22)), but only weakly so. For a limited temperature range they may be treated as constants. Obtaining these quantities experimentally is possible by measuring the reaction rate as a function of temperature, and plotting $\ln(k/T)$ against T^{-1} (eq. 12.24).

$$\ln\left(\frac{k}{T}\right) = \ln\left(\frac{k_B}{h}\right) + \frac{\Delta S^{\neq}}{R} - \frac{\Delta H^{\neq}}{RT} \tag{12.24}$$

Such plots should produce a straight line with the slope being equal to $-\Delta H^{\neq}/R$ and the intercept equal to $\ln(k_B/h) + \Delta S^{\neq}/R$. As the available temperature range typically is $\sim 100°$, the error in $\Delta H^{\neq}$ will typically be 0.1–0.5 kcal/mol. The activation entropy is determined by extrapolating outside the data points to $T = \infty$ ($1/T = 0$), and is usually somewhat less well defined, a typical error may be 5 e.u.

Experimentalists often analyse their data in terms of an Arrhenius expression instead of the TST expression eq. (12.2) by plotting $\ln(k)$ against T^{-1}.

$$k = Ae^{-\Delta E^{\neq}/RT}$$

$$\ln(k) = \ln(A) - \frac{\Delta E^{\neq}}{RT} \tag{12.25}$$

The connection with the TST expression (12.2) may be established from the definition (12.26) of the activation energy.

$$\Delta E^{\neq} \equiv RT^2\left(\frac{\partial \ln k}{\partial T}\right)_V \tag{12.26}$$

which results in the following transformations.

$$\Delta H^{\neq} = \Delta E^{\neq} - (1 - \Delta n)RT$$

$$\Delta S^{\neq} = R\left[\ln(A) - \ln\left(\frac{k_B T}{h}\right) - (1 - \Delta n)\right] \tag{12.27}$$

Here Δn is the change in the number of molecules from the reactant to the TS, i.e. $\Delta n = 0$ for a unimolecular reaction, -1 for a bimolecular reaction etc. For a solution phase reaction Δn is approximately 0.

References

1. H. Eyring, *J. Chem. Phys.*, **3** (1934), 107.
2. I. N. Levine, *Physical Chemistry*, McGraw-Hill, 1983; K. Lucas, *Applied Statistical Thermodynamics*, Springer-Verlag, 1991.

3. K. N. Houk, Y.-T. Lin and F. K. Brown, *J. Am. Chem. Soc.*, **108** (1986), 554.

4. F. W. Schuler and G. W. Murphy, *J. Am. Chem. Soc.*, **72** (1950), 3155.

5. W. Witschel and C. Hartwigsen, *Chem. Phys. Lett.*, **273** (1997), 304; D. G. Truhlar, *J. Comput. Chem.*, **12** (1990), 266.

6. M. S. Gordon and D. G. Truhlar, *Science*, **249** (1990), 491.

13 Change of Coordinate System

Many problems simplify significantly by choosing a suitable coordinate system. At the heart of these transformations is the "separability" theorem. If a Hamilton operator depending on N coordinates can be written as a sum of operators which only depend on one coordinate, the corresponding N coordinate wave function can be written as a product of one-coordinate functions, and the total energy as a sum of energies.

$$\mathbf{H}(x_1, x_2, x_3, \ldots) \, \Psi(x_1, x_2, x_3, \ldots) = E_{\text{tot}} \, \Psi(x_1, x_2, x_3, \ldots)$$

$$\mathbf{H}(x_1, x_2, x_3, \ldots) = \sum_i \mathbf{h}_i(x_i)$$

$$\mathbf{h}_i(x_i)\phi_i(x_i) = \varepsilon_i \phi_i(x_i) \tag{13.1}$$

$$E_{\text{tot}} = \sum_i \varepsilon_i$$

$$\Psi(x_1, x_2, x_3, \ldots) = \prod_i \phi_i(x_i)$$

Instead of solving one equation with N variables, the problem is transformed into solving N equations with only one variable.

Some coordinate transformations are non-linear, like transforming Cartesian to polar coordinates, where the polar coordinates are given in terms of square root and trigonometric functions of the Cartesian coordinates. This for example allows the Schrödinger equation for the hydrogen atom to be solved. Other transformations are linear, i.e. the new coordinate axes are linear combinations of the old coordinates. Such transformations can be used for reducing a matrix representation of an operator to a diagonal form. In the new coordinate system, the many-dimensional operator can be written as a sum of one-dimensional operators.

Consider a matrix $\mathbf{A}$ expressed in a coordinate system $\{x_1, x_2, x_3, \ldots, x_N\}$. The coordinate axes are the x_i vectors, these may be simple Cartesian axes, or one-variable functions, or many-variable functions. The matrix $\mathbf{A}$ is typically defined by an operator working on the coordinates. Some examples are:

(1) The force constant matrix in Cartesian coordinates.
(2) The Fock matrix in basis functions (atomic orbitals).
(3) The CI matrix in Slater determinants.

Finding the coordinates where these matrices are diagonal corresponds to finding:

(1) The vibrational normal coordinates.
(2) The molecular orbitals.
(3) The state coefficients, i.e. CI wave function(s).

The coordinate axis are usually orthonormal, but this is not necessary.

A linear coordinate transformation may be illustrated by a simple two-dimensional example. The new coordinate system is defined in term of the old by means of a rotation matrix, $\mathbf{U}$. In the general case the $\mathbf{U}$ matrix is *unitary* (complex elements), although for most applications it may be chosen to be *orthogonal* (real elements). This means that the matrix inverse is given by transposing the complex conjugate, $\mathbf{U}^{-1} = \mathbf{U}^{\dagger}$, or in the orthogonal case, simply as the transpose, $\mathbf{U}^{-1} = \mathbf{U}^{t}$. In the general N-dimensional case, $\mathbf{U}$ is an $N \times N$ matrix, but for illustration purposes we will use a two-dimensional example.

$$\mathbf{x}' = \mathbf{U}\mathbf{x}$$

$$\begin{pmatrix} x_1' \\ x_2' \end{pmatrix} = \begin{pmatrix} \cos\alpha \; \sin\alpha \\ -\sin\alpha \; \cos\alpha \end{pmatrix} \begin{pmatrix} x_1 \\ x_2 \end{pmatrix} \tag{13.2}$$

Consider now a function expanded to second order in the x-coordinate system

$$f(\mathbf{x}) = f(\mathbf{x}_0) + \mathbf{g}^{t}(\mathbf{x} - \mathbf{x}_0) + \tfrac{1}{2}(\mathbf{x} - \mathbf{x}_0)^{t}\mathbf{H}(\mathbf{x} - \mathbf{x}_0) \tag{13.3}$$

where $\mathbf{x}$ is a vector containing the coordinates, $\mathbf{g}$ is a vector containing the first derivatives of f with respect to $\mathbf{x}$, and $\mathbf{H}$ is the matrix of second derivatives. The superscript t denotes a transposition of a vector, i.e. converting it from a column to a row vector. Inserting unit matrices written in the form $\mathbf{U}^{-1}\mathbf{U} = \mathbf{U}^{t}\mathbf{U}$ gives

$$f(\mathbf{x}) = f(\mathbf{x}_0) + \mathbf{g}^{t}(\mathbf{U}^{t}\mathbf{U})(\mathbf{x} - \mathbf{x}_0) + \tfrac{1}{2}(\mathbf{x} - \mathbf{x}_0)^{t}(\mathbf{U}^{t}\mathbf{U})\mathbf{H}(\mathbf{U}^{t}\mathbf{U})(\mathbf{x} - \mathbf{x}_0)$$

$$f(\mathbf{x}) = f(\mathbf{x}_0) + [\mathbf{U}\mathbf{g}]^{t}[\mathbf{U}(\mathbf{x} - \mathbf{x}_0)] + \tfrac{1}{2}[\mathbf{U}(\mathbf{x} - \mathbf{x}_0)]^{t}[\mathbf{U}\mathbf{H}\mathbf{U}^{t}][\mathbf{U}(\mathbf{x} - \mathbf{x}_0)] \tag{13.4}$$

$$f'(\mathbf{x}') = f'(\mathbf{x}_0') + \mathbf{g}'^{t}(\mathbf{x}' - \mathbf{x}_0') + \tfrac{1}{2}(\mathbf{x}' - \mathbf{x}_0')^{t}\mathbf{H}'(\mathbf{x}' - \mathbf{x}_0')$$

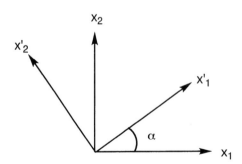

Figure 13.1 Rotation of a coordinate system

Assuming that the **H** matrix is symmetric ($H_{12} = H_{21}$), the elements in the **H**′ matrix are (using **U** from eq. (13.2))

$$H'_{11} = H_{11}\cos^2\alpha + H_{22}\sin^2\alpha + 2H_{12}(\cos\alpha)(\sin\alpha)$$
$$H'_{12} = H_{12}(\cos^2\alpha - \sin^2\alpha) + (H_{22} - H_{11})(\cos\alpha)(\sin\alpha) \qquad (13.5)$$
$$H'_{22} = H_{22}\cos^2\alpha + H_{11}\sin^2\alpha - 2H_{12}(\cos\alpha)(\sin\alpha)$$

Using trigonometric identities, the off-diagonal elements may be written as

$$2H'_{12} = 2H_{12}\cos(2\alpha) + (H_{22} - H_{11})\sin(2\alpha) \qquad (13.6)$$

These elements can be reduced to zero by choosing a proper rotational angle.

$$\tan(2\alpha) = \frac{2H_{12}}{H_{11} - H_{22}} \qquad (13.7)$$

In the rotated coordinate system the **H**′ matrix is diagonal, and the function variation in each of the two directions is thus <u>independent</u> of the other variable.

Diagonalizing a matrix is equivalent to determining its eigenvalues and -vectors. This may be seen from the definition of an eigenvalue ε and -vector **e** belonging to a matrix **A**.

$$\mathbf{A}\mathbf{e} = \varepsilon\mathbf{e} \qquad (13.8)$$

Consider the unitary transformation **U** which makes **A** diagonal.

$$[\mathbf{U}\mathbf{A}\mathbf{U}^t][\mathbf{U}\mathbf{e}] = \varepsilon[\mathbf{U}\mathbf{e}]$$
$$\mathbf{A}'\mathbf{e}' = \varepsilon\mathbf{e}' \qquad (13.9)$$

In the new coordinate system **A**′ is a diagonal matrix, and the (normalized) **e**′ vector is a new coordinate axis. The diagonal elements in **A**′ are therefore directly the eigenvalues, and since $\mathbf{e} = \mathbf{U}^t\mathbf{e}'$, the columns in the $\mathbf{U}^t$ matrix are the eigenvectors.

How does one determine this unique **U** matrix? In practice it is simple! There are many widely available computer routines which will take a matrix to its diagonal form. Given a matrix **A**, a call is made to a routine which returns the eigenvalues as a vector and the eigenvectors as a matrix, the first column being the eigenvector for the first eigenvalue, etc. Note also that diagonalization of a matrix readily allows a calculation of functions of matrices, like $\mathbf{A}^{-1/2}$ or $\exp(\mathbf{A})$. First make a unitary transformation which diagonalizes **A**, then replace each of the diagonal elements by the function of the element, and backtransform to the original coordinate system by $\mathbf{U}^t$ as illustrated in Figure 13.2. Below we give some examples of the above three cases.

$$\mathbf{A} = \begin{pmatrix} a_{11} & a_{12} & \cdots \\ a_{21} & a_{22} & \cdots \\ \cdots & \cdots & \cdots \end{pmatrix} \xrightarrow{\mathbf{U}} \begin{pmatrix} \varepsilon_1 & 0 & \cdots \\ 0 & \varepsilon_2 & \cdots \\ \cdots & \cdots & \cdots \end{pmatrix} \xrightarrow{f} \begin{pmatrix} f(\varepsilon_1) & 0 & \cdots \\ 0 & f(\varepsilon_2) & \cdots \\ \cdots & \cdots & \cdots \end{pmatrix} \xrightarrow{\mathbf{U}^t} f(\mathbf{A})$$

Figure 13.2 Constructing functions of matrices

13.1 Vibrational Normal Coordinates

The potential energy is approximated by a second-order Taylor expansion around the stationary geometry.

$$V(\mathbf{x}) \simeq V(\mathbf{x}_0) + \left(\frac{\mathrm{d}V}{\mathrm{d}\mathbf{x}}\right)^{\mathrm{t}} (\mathbf{x} - \mathbf{x}_0) + \frac{1}{2}(\mathbf{x} - \mathbf{x}_0)^{\mathrm{t}} \left(\frac{\mathrm{d}^2 V}{\mathrm{d}\mathbf{x}^2}\right)(\mathbf{x} - \mathbf{x}_0) \tag{13.10}$$

The energy for the expansion point, $V(\mathbf{x}_0)$, may be chosen as zero, and the first derivative is zero since $\mathbf{x}_0$ is a stationary point, i.e.

$$V(\Delta\mathbf{x}) = \tfrac{1}{2}\Delta\mathbf{x}^{\mathrm{t}}\mathbf{F}\,\Delta\mathbf{x} \tag{13.11}$$

where $\mathbf{F}$ is a $3N \times 3N$ (N being the number of atoms in the molecule) matrix containing the second derivatives of the energy with respect to the coordinates (the force constant matrix). The nuclear Schrödinger equation for an N-atom system then becomes

$$\left\{ -\sum_{i=1}^{3N} \frac{1}{2m_i} \frac{\partial^2}{\partial x_i^2} + \frac{1}{2}\Delta\mathbf{x}^{\mathrm{t}}\mathbf{F}\,\Delta\mathbf{x} \right\}\Psi = E\Psi \tag{13.12}$$

It is first transformed to mass-dependent coordinates by a $\mathbf{G}$ matrix containing the inverse square root of atomic masses (note that <u>atomic</u>, not <u>nuclear</u>, masses are used, this is in line with the Born–Oppenheimer approximation that the electrons follow the nucleus).

$$y_i = \sqrt{m_i}\,\Delta x_i$$

$$\frac{\partial^2}{\partial y_i^2} = \frac{1}{m_i}\frac{\partial^2}{\partial x_i^2}$$

$$G_{ij} = \frac{1}{\sqrt{m_i m_j}} \tag{13.13}$$

$$\left\{ -\sum_{i=1}^{3N} \frac{1}{2}\frac{\partial^2}{\partial y_i^2} + \frac{1}{2}\mathbf{y}^{\mathrm{t}}(\mathbf{FG})\mathbf{y} \right\}\Psi = E\,\Psi$$

A unitary transformation is then introduced which diagonalizes the $\mathbf{FG}$ matrix, yielding eigenvalues ε_i and eigenvectors $\mathbf{q}_i$. The kinetic energy operator is still diagonal in these coordinates.

$$\mathbf{q} = \mathbf{Uy}$$

$$\left\{ -\sum_{i=1}^{3N} \frac{1}{2}\frac{\partial^2}{\partial \mathbf{q}_i^2} + \frac{1}{2}\mathbf{q}^{\mathrm{t}}(\mathbf{U}(\mathbf{FG})\mathbf{U}^{\mathrm{t}})\mathbf{q} \right\}\Psi = E\,\Psi$$

$$-\sum_{i=1}^{3N} \left\{ \frac{1}{2}\frac{\partial^2}{\partial \mathbf{q}_i^2} + \frac{1}{2}\varepsilon_i \mathbf{q}_i^2 \right\}\Psi = E\,\Psi \tag{13.14}$$

$$\sum_{i=1}^{3N} \{\mathbf{h}_i\}\Psi = E\,\Psi$$

In the $\mathbf{q}$-coordinate system, the *vibrational normal coordinates*, the $3N$-dimensional Schrödinger equation can be separated into $3N$ one-dimensional Schrödinger equations, which are in the form of a standard harmonic oscillator. The eigenvectors of the **FG** matrix are the (mass-weighted) vibrational normal coordinates, and the eigenvalues ε_i are related to the vibrational frequencies as (analogously to eq. (12.18))

$$\nu_i = \frac{1}{2\pi} \sqrt{\varepsilon_i} \qquad (13.15)$$

When this procedure is carried out in Cartesian coordinates, there should be six (five for a linear molecule) eigenvalues of the **FG** matrix being exactly zero, corresponding to the translational and rotational modes. In real calculations, however, these values are not exactly zero. The three translational modes usually have "frequencies" very close to zero, typically less than 0.01 cm^{-1}. The deviation from zero is due to the fact that numerical operations are only carried out with a finite precision, and the accumulation of errors will typical give inaccuracies in ν of this magnitude. The residual "frequencies" for the rotational modes, however, may often be as large as 10–50 cm^{-1}. This is due to the fact that the geometry cannot be optimized to a gradient of exactly zero, again owing to numerical considerations. Typically the geometry optimization is considered converged if the root mean square (R.M.S.) gradient is less than $\sim 10^{-4}$–10^{-5} a.u., corresponding to the energy being converged to $\sim 10^{-6}$ a.u. The residual gradient shows up as vibrational frequencies for the rotations of the above magnitude.

If there are real frequencies of the same magnitude as the "rotational frequencies", mixing may occur and result in inaccurate values for the "true" vibrations. For this reason the translational and rotational degrees of freedom are normally removed from the force constant matrix before diagonalization. This may be accomplished by *projecting* the modes out.[1] Consider for example the following (normalized) vector describing a translation in the x-direction.

$$\mathbf{t}_x^t = \frac{1}{\sqrt{N}} \{1, 0, 0, 1, 0, 0, 1, \ldots, 0\} \qquad (13.16)$$

(the superscript t denotes a transposed vector, i.e. $\mathbf{t}_x$ is a column vector and $\mathbf{t}_x^t$ is a row vector). The $\mathbf{T}_x$ matrix removes the direction corresponding to translation in the x-direction.

$$\mathbf{T}_x = \mathbf{I} - \mathbf{t}_x \mathbf{t}_x^t \qquad (13.17)$$

By extending this to include vectors for all three translation and rotational modes, a projection matrix is obtained.

$$\mathbf{P} = \mathbf{I} - \mathbf{t}_x \mathbf{t}_x^t - \mathbf{t}_y \mathbf{t}_y^t - \mathbf{t}_z \mathbf{t}_z^t - \mathbf{r}_a \mathbf{r}_a^t - \mathbf{r}_b \mathbf{r}_b^t - \mathbf{r}_c \mathbf{r}_c^t \qquad (13.18)$$

The **r** vectors are the principal axes of inertia determined by diagonalization of the matrix of inertia (eq. (12.14)). By forming the matrix product $\mathbf{P}^t\mathbf{FP}$, the translation and rotational directions are removed from the force constant matrix, and consequently the six (five) trivial vibrations become exactly zero (within the numerical accuracy of the machine).

If the stationary point is a minimum on the energy surface, the eigenvalues of the **F** and **FG** matrices are all positive. If, however, the stationary point is a TS, one (and only one) of the eigenvalues is negative. This corresponds to the energy being a maximum in one direction and a minimum in all other directions. The "frequency" for the "vibration" along the eigenvector with a negative eigenvalue will formally be imaginary, as it is the square root of a negative number (eq. (13.15)). The corresponding eigenvector is the direction leading downhill from the TS towards the reactant and product. At the TS, the eigenvector for the imaginary frequency is the reaction coordinate. The whole reaction path may be calculated by sliding downhill to each side from the TS. This can be performed by taking a small step along the TS eigenvector, calculating the gradient and taking a small step in the negative gradient direction. The negative of the gradient always points downhill, and by taking a sufficiently large number of such steps an energy minimum is eventually reached. This is equivalent to a steepest descent minimization, but in practice more efficient methods are available, see Section 14.8 for details. The reaction path in mass-weighted coordinates is called the *Intrinsic Reaction Coordinate* (IRC).

It should be noted that the force constant matrix can be calculated at any geometry, but the transformation to normal coordinates is only valid at a stationary point, i.e. where the first derivative is zero. At a non-stationary geometry, a set of $3N-7$ *generalized frequencies* may be defined by removing the gradient direction from the force constant matrix (for example by projection techniques, eq. (13.17)) before transformation to normal coordinates.

13.2 Energy of a Slater Determinant

The variational problem is to minimize the energy of a single Slater determinant by choosing suitable values for the MO coefficients, under the constraint that the MOs remain orthonormal. With ϕ being an MO written as a linear combination of the basis functions (atomic orbitals) χ, this leads to a set of secular equations, $\mathbf{F}$ being the Fock matrix, $\mathbf{S}$ the overlap matrix and $\mathbf{C}$ containing the MO coefficients (Section 3.5).

$$\phi_i = \sum_\alpha c_{\alpha i}\, \chi_\alpha$$

$$F_{\alpha\beta} = \langle \chi_\alpha | \mathbf{F} | \chi_\beta \rangle \qquad (13.19)$$

$$S_{\alpha\beta} = \langle \chi_\alpha | \chi_\beta \rangle$$

$$\mathbf{FC} = \mathbf{S}\mathbf{C}\varepsilon$$

In this case the basis functions (coordinate system) are non-orthogonal, the overlaps are contained in the $\mathbf{S}$ matrix. By multiplying from the left by $\mathbf{S}^{-1/2}$ and inserting a unit matrix written in the form $\mathbf{S}^{-1/2}\mathbf{S}^{1/2}$, (13.19) may be reformulated as

$$\{\mathbf{S}^{-1/2}\mathbf{F}\mathbf{S}^{-1/2}\}\{\mathbf{S}^{1/2}\mathbf{C}\} = \{\mathbf{S}^{-1/2}\mathbf{S}^{1/2}\}\{\mathbf{S}^{1/2}\mathbf{C}\}\varepsilon$$
$$\mathbf{F}'\mathbf{C}' = \mathbf{C}'\varepsilon \qquad (13.20)$$

The latter equation is now in a standard form for determining the eigenvalues of the $\mathbf{F}'$ matrix. The eigenvectors contained in $\mathbf{C}'$ can then be backtransformed to the original coordinate system ($\mathbf{C} = \mathbf{S}^{-1/2}\mathbf{C}'$).

Equation (13.20) corresponds to a *symmetrical orthogonalization* of the basis. The initial coordinate system, (the basis functions χ) is non-orthogonal, but by multiplying with a matrix such as $\mathbf{S}^{-1/2}$, the new coordinate system has orthogonal axes.

$$\chi' = \mathbf{S}^{-1/2}\chi \qquad (13.21)$$

There are choice other than $\mathbf{S}^{-1/2}$; for such an orthogonalizing transformation, any matrix which has the property $\mathbf{X}^\dagger \mathbf{S} \mathbf{X} = \mathbf{I}$ can be used.

13.3 Energy of a CI Wave Function

The variational problem may again be formulated as a secular equation, where the coordinate axes are many-electron functions (Slater determinants), Φ, which are orthogonal (Section 4.2).

$$\Psi_{CI} = \sum_i a_i \Phi_i$$

$$H_{ij} = \langle \Phi_i | \mathbf{H} | \Phi_j \rangle \qquad (13.18)$$

$$\mathbf{Ha} = E\mathbf{a}$$

The **a** matrix contains the coefficients of the CI wave function. This problem may again be considered as selecting a basis where the Hamilton operator is diagonal (eq. (4.6) and Figure 4.4). In the initial coordinate system, the Hamilton matrix will have many off-diagonal elements. By a suitable unitary transformation it can be diagonalized. The diagonal elements are energies of many-electron CI wave functions, the ground and exited states. The corresponding eigenvectors contain the expansion coefficients a_i.

Finally some practical considerations. The time required for diagonalizing a matrix grows as the cube of the size of the matrix, and the amount of computer memory necessary for storing the matrix grows as the square of the size. Diagonalizing matrices up to $\sim 100 \times 100$ takes insignificant amounts of time, unless there are extraordinarily many such matrices. Matrices up to $\sim 1000 \times 1000$ pose no particular problems, although some considerations should be made of whether the time required for diagonalization is significant relative to other operations. Matrices larger than this require consideration. For example, just storing all the elements in a $10\,000 \times 10\,000$ matrix takes ~ 1Gb of memory (or disk space) on a computer. Determining all eigenvalues and -vectors of such a matrix takes a long time. For large-scale problems in quantum chemistry, however, one is usually not interested in all the eigenvalues and -vectors. In solving the CI matrix equation, for example, typically only the lowest eigenvalue and -vector are of interest; this is the ground-state energy and wave function. Large-scale diagonalizations are therefore normally solved by special iterative schemes which extract a few selected roots and eigenvectors.

Reference

1. W. H. Miller, N. C. Handy and J. E. Adams, *J. Chem. Phys.*, **72** (1980), 99.

14 Optimization Techniques

Many problems in computational chemistry can be formulated as an optimization of a multidimensional function.[1] Optimization is a general term for finding stationary points of a function, i.e. points where the first derivative is zero. In the majority of cases the desired stationary point is a minimum, i.e. all the second derivatives should be positive. In some cases the desired point is a first-order saddle point, i.e. the second derivative is negative in one, and positive in all other, directions. Some examples:

(1) The energy as a function of nuclear coordinates. Both minima and first-order saddle points (transition structures) are of interest. The energy function may be of the force field type, or from solving the electronic Schrödinger equation.
(2) An error function depending on parameters. Only minima are of interest, and the global minimum is usually (but not always) desired. This may for example be determination of parameters in a force field, a set of atomic charges, or a set of localized Molecular Orbitals.
(3) The energy of a wave function containing variational parameters, like an HF (one Slater determinant) or MCSCF (many Slater determinants) wave function. Parameters are typically the MO and state coefficients, but may also be for example basis function exponents. Usually only minima are desired, although in some cases saddle points may also be of interest (excited states).

The simple-minded approach for minimizing a function is to step one variable at a time until the function has reached a minimum, and then switch to another variable. This requires only the ability to calculate the function value for a given set of variables. However, as the variables are not independent, several cycles through the whole set are necessary for finding a minimum. This is impractical for more than 5–10 variables, and may not work anyway. Essentially all optimization methods used in computational chemistry thus assume that at least the first derivative of the function with respect to all variables, the gradient $\mathbf{g}$, can be calculated analytically (i.e. directly, and not as a numerical differentiation by stepping the variables). Some methods also assume that the second derivative matrix, the Hessian $\mathbf{H}$, can be calculated.

The function to be optimized, and its derivative(s), are calculated with a finite precision, which depends on the computational implementation. A stationary point can therefore not be located exactly, the gradient can only be reduced to a certain value. Below this value the numerical inaccuracies due to the finite precision will swamp the "true" functional behaviour. In practice the optimization is considered converged if the gradient is reduced below a suitable "cut-off" value. It should be noted that this in some cases may lead to problems, as a function with a very flat surface may meet the criteria without containing a stationary point.

There are three classes of commonly used optimization methods for finding minima, each with its advantages and disadvantages.

14.1 Steepest Descent

The gradient vector points in the direction where the function increases most, i.e. the function value can always be lowered by stepping in the opposite direction. In the *Steepest Descent* (SD) method, a series of function evaluations are performed in the negative gradient direction, i.e. along a search direction defined as $\mathbf{d} = -\mathbf{g}$. Once the function starts to increase, an approximate minimum may be determined for example by interpolation between the last three points. At this interpolated point a new gradient is calculated and used for the next line search. The steepest descent method is sure-fire; if the line minimization is carried out sufficiently accurately it will always lower the function value, and is therefore guaranteed to approach a minimum. It has, however, two main problems. Two subsequent line searches are necessarily perpendicular to each other. If there were a gradient component along the previous search direction the energy could be further lowered in this direction. There is therefore a tendency for each line search to partly spoil the function lowering obtained by the previous search. The steepest descent path oscillates around the minimum path, as illustrated in Figure 14.1. Furthermore, as the minimum is approached, the rate of convergence slows down. The steepest descent will actually never reach the minimum, it will crawl toward it at an ever decreasing speed.

An accurate line search will require several function evaluations along each search direction. Often the minimization along the line is only carried out fairly crudely, or a

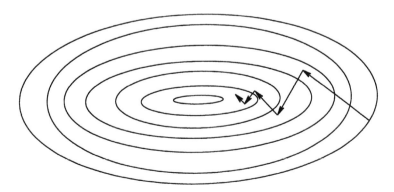

Figure 14.1 Steepest descent minimization

single step is simply taken along the negative gradient direction. The step size may be a fixed quantity or taken to depend on the magnitude of the gradient. It should be noted that the guarantee for lowering of the function value is lost when approximate line searches are used, e.g. simply taking a fixed step along the negative gradient direction will normally end up in an oscillatory state.

By nature the steepest descent method can only locate function minima. The advantage is that the algorithm is very simple, and requires only storage of the gradient vector. It is furthermore one of the few methods guaranteed to lower the function value. Its main use is to quickly relax a poor starting point, before some of the more advanced algorithms take over, or as a "backup" algorithm if the more sophisticated methods are unable to lower the function value.

14.2 Conjugate Gradient Methods

The main problem with the steepest descent method is the partial "undoing" of the previous step. The *Conjugate Gradient* (CG) method tries to repair this by performing each line search not along the current gradient, but along a line which is constructed so that it is "conjugate" to the previous search direction(s). If the surface is purely quadratic, each successive minimization will not generate gradient components along any of the previous directions. The first step is equivalent to a steepest descent step, but subsequent searches are performed along a line which is a mixture of the current negative gradient and the previous search direction.

$$\mathbf{d}_i = -\mathbf{g}_i + \beta_i \mathbf{d}_{i-1} \qquad (14.1)$$

There are several ways of choosing the β value. Some of the names associated with these methods are *Fletcher–Reeves, Polak–Ribiere* and *Hestenes–Stiefel.* Their definitions of β are

$$
\begin{aligned}
\beta_i^{\text{FR}} &= \frac{\mathbf{g}_i^{\text{t}} \mathbf{g}_i}{\mathbf{g}_{i-1}^{\text{t}} \mathbf{g}_{i-1}} \\[2mm]
\beta_i^{\text{PR}} &= \frac{\mathbf{g}_i^{\text{t}} (\mathbf{g}_i - \mathbf{g}_{i-1})}{\mathbf{g}_{i-1}^{\text{t}} \mathbf{g}_{i-1}} \\[2mm]
\beta_i^{\text{HS}} &= \frac{\mathbf{g}_i^{\text{t}} (\mathbf{g}_i - \mathbf{g}_{i-1})}{\mathbf{d}_{i-1}^{\text{t}} (\mathbf{g}_i - \mathbf{g}_{i-1})}
\end{aligned}
\qquad (14.2)
$$

The Polak–Ribiere prescription is usually preferred in practice. Conjugate gradient methods have much better convergence characteristics than the steepest descent, but they are again only able to locate minima. They do require slightly more storage than the steepest descent, since the previous gradient also must be saved.

14.3 Newton–Raphson Methods

Newton–Raphson (NR) methods expand the true function to second order around the current point $\mathbf{x}_0$.

$$f(\mathbf{x}) \simeq f(\mathbf{x}_0) + \mathbf{g}^{\text{t}} (\mathbf{x} - \mathbf{x}_0) + \tfrac{1}{2} (\mathbf{x} - \mathbf{x}_0)^{\text{t}} \mathbf{H} (\mathbf{x} - \mathbf{x}_0) \qquad (14.3)$$

Requiring the gradient of the second-order approximation (14.3) to be zero produces the following step.

$$(\mathbf{x} - \mathbf{x}_0) = -\frac{\mathbf{g}}{\mathbf{H}} = -\mathbf{H}^{-1}\mathbf{g} \tag{14.4}$$

In the coordinate system $(\mathbf{x}')$ where the Hessian is diagonal (i.e. performing a unitary transformation, see Chapter 13), the NR step may be written as

$$\Delta\mathbf{x}' = \sum_i \Delta x_i', \quad \Delta x_i' = -\frac{f_i}{\varepsilon_i} \tag{14.5}$$

where f_i is the projection of the gradient along the Hessian eigenvector with eigenvalue ε_i (the gradient component pointing in the direction of the ith eigenvector).

As the real function contains terms beyond second-order, the NR formula may be used iteratively for stepping towards a stationary point. Near a minimum, all the Hessian eigenvalues are positive (by definition), and the step direction is opposite to the gradient direction, as it should be. If, however, one of the Hessian eigenvalues is negative, the step in this direction will be <u>along</u> the gradient component, and thus <u>increase</u> the function value. In this case the optimization may end up at a stationary point with one negative Hessian eigenvalue, a first-order saddle point. In general, the NR method will attempt to converge on the "nearest" stationary point. Another problem is the use of the inverse Hessian for determining the step size. If one of the Hessian eigenvalues becomes close to zero, the step size goes towards infinity (except if the corresponding gradient component f_i is exactly zero). The NR step is thus without bound, it may take the variables far outside the region where the second-order Taylor expansion is valid. The latter region is often described by a *"Trust Radius"*. In some cases the NR step is taken as a search direction along which the function is minimized, analogously to the steepest descent and conjugate gradient methods. The augmented Hessian methods described below are normally more efficient.

The advantage of the NR method is that the convergence is second-order near a stationary point. If the function only contains terms up to second-order, the NR step will go to the stationary point in only one iteration. In general the function contains higher-order terms, but the second-order approximation becomes better and better as the stationary point is approached. Sufficiently close to the stationary point, the gradient is reduced <u>quadratically</u>. This means that if the gradient norm is reduced by a factor of 10 between two iterations, it will go down by a factor of 100 in the next iteration, and a factor of 10 000 in the next!

Besides the above-mentioned problems with step control, there are also other computational aspects which tend to make the straightforward NR problematic for many problem types. The true NR method requires calculation of the full second derivative matrix, which must be stored and inverted (diagonalized). For some function types a calculation of the Hessian is computationally demanding. For other cases, the number of variables is so large that manipulating a matrix the size of the number of variables squared is impossible. Let us address some solutions to these problems.

14.3.1 Step Control

There are two aspects in this. One is controlling the total length of the step, such that it does not exceed the region in which the second-order Taylor expansion is valid. The second is the problem of step direction. If the optimization is towards a minimum, the Hessian should have all positive eigenvalues in order for the step to be in the correct direction. If, however, the starting point is in a region where the Hessian has negative eigenvalues, the NR step will take it towards a saddle point or maximum. Both these problems can be solved by introducing a shift parameter λ (compare with eq. (14.5)).

$$\Delta \mathbf{x}' = \sum_i \Delta x_i', \quad \Delta x_i' = -\frac{f_i}{\varepsilon_i - \lambda} \tag{14.6}$$

If λ is chosen to be below the lowest Hessian eigenvalue, the denominator is always positive, and the step direction will thus be correct. Furthermore, if λ goes towards $-\infty$, the step size goes towards zero, i.e. the step size can be made arbitrarily small. Methods which modify the nature of the Hessian matrix by a shift parameter are known by names such as "*augmented Hessian*", "*level-shifted Newton–Raphson*", "*norm-extended Hessian*" or "*Eigenvector Following*" (EF), depending on how λ is chosen. We will here mention two popular methods for choosing λ.

The *Rational Function Optimization* (RFO)[2] expands the function in terms of a rational approximation instead of a straight second-order Taylor series (eq. (14.3)).

$$f(\mathbf{x}) \simeq \frac{f(\mathbf{x}_0) + \mathbf{g}^t(\mathbf{x} - \mathbf{x}_0) + \frac{1}{2}(\mathbf{x} - \mathbf{x}_0)^t \mathbf{H}(\mathbf{x} - \mathbf{x}_0)}{1 + \frac{1}{2}(\mathbf{x} - \mathbf{x}_0)^t \mathbf{S}(\mathbf{x} - \mathbf{x}_0)} \tag{14.7}$$

The $\mathbf{S}$ matrix is eventually set equal to a unit matrix which leads to the following equation for λ.

$$\sum_i \frac{f_i^2}{\varepsilon_i - \lambda} = \lambda \tag{14.8}$$

This is a one-dimensional equation in λ, which can be solved by standard (iterative) methods. There will in general be one more solution than there are degrees of freedom, but by choosing the lowest λ solution, it is ensured that the resulting step will be towards a minimum. The RFO step calculated from (14.6) will always be shorter than the pure NR step (14.5), but there is no guarantee that it will be within the trust radius. If the RFO step is too long, it may be scaled down by a simple multiplicative factor, however, if the factor is much smaller than 1 it follows that the resulting step may not be the optimum for the given trust radius.

Another way of choosing λ is to require that the step length be equal to the trust radius R, this is in essence the best step on a hypersphere with radius R. This is known as the *Quadratic Approximation* (QA) method.[3]

$$|\Delta \mathbf{x}'|^2 = \sum_i \left(\frac{f_i}{\varepsilon_i - \lambda}\right)^2 = R^2 \tag{14.9}$$

This may again have multiple solutions, but by choosing the lowest λ value the minimization step is selected. The maximum step size R may be taken as a fixed value, or allowed to change dynamically during the optimization. If for example the actual energy change between two steps agrees well with that predicted from the second-order Taylor expansion, the trust radius for the next step may be increased, and vice versa.

14.3.2   Obtaining the Hessian

The second problem, the computational aspect of calculating the Hessian, is often encountered in electronic structure calculations. Here the calculation of the second derivative matrix can be a factor of 10 more demanding than calculating the gradient. In such cases, an *updating* scheme may be used instead. The idea is to start off with an approximation to the Hessian, maybe just a unit matrix. The initial step will thus resemble a steepest descent step. As the optimization proceeds, the gradients at the previous points are used to make the Hessian a better approximation for the actual system. After two steps, the updated Hessian is a rather good approximation to the exact Hessian in the direction defined by these two points (but not in the other directions). There are many such updating schemes, some of the commonly used are associated with the names *Davidon–Fletcher–Powell* (DFP), *Broyden–Fletcher–Goldfarb–Shanno* (BFGS) and *Powell*. NR methods with such approximate Hessians are known as *pseudo–Newton–Raphson* or *variable metric* methods. It is clear that they do not converge as fast as true NR methods, where the exact Hessian is calculated in each step, but if five or ten steps may be taken for the same computational cost as one true NR step, the overall computational effort may be less. True NR methods converge quadratically near a stationary point, while pseudo-NR methods display a linear convergence. Far from a stationary point, however, the true NR method will typically also display only linear convergence.

Pseudo-NR methods are usually the best choice in geometry optimizations using an energy function calculated by electronic structure methods. The quality of the initial Hessian of course affects the convergence when an updating scheme is used. The best choice is usually an exact Hessian at the first point, however, this may not be the most cost-efficient strategy. In many cases a quite reasonable Hessian for a minimum search may be generated by simple rules connecting for example bond lengths and force constants.[4] Alternatively the initial Hessian may be taken from a calculation at a lower level of theory. As an initial exploration of an energy surface often is carried out at a low level of theory, followed by frequency calculations to establish the nature of the stationary points, the resulting force constants can be used to start an optimization at higher levels. This is especially useful for transition structure searches which require a quite accurate Hessian. The success of this strategy relies on the fact that the qualitative structure of an energy surface often is fairly insensitive to the level of theory, although there certainly are many examples where this is not the case.

14.3.3   Storing and Diagonalizing the Hessian

The last potential problem of all NR based methods is the storage and handling of the Hessian matrix. For methods where the calculation of the Hessian is easy, but the number of variables is very large, this may be a problem. A prime example here is geometry optimizations using a force field energy function. The computational effort for calculating the Hessian goes up roughly as the square of the number of atoms. Diagonalization of the Hessian matrix required for the NR optimization, however, depends on the cube of the matrix size, i.e. it goes up as the cube of the number of atoms. Since matrix diagonalization becomes a significant factor around a size of 1000×1000, it is clear that NR methods should not be used for force field optimizations beyond a few hundred atoms. For large systems the computational effort for predicting the geometry step will completely overwhelm the calculation of the energy, gradient and Hessian. The conjugate gradient method avoids handling of the Hessian and requires only storage of two gradient vectors. It is therefore usually the method of choice for force field optimizations.

For large systems many of the off-diagonal elements in the Hessian are very small, essentially zero (the coupling between distant atoms is very small). The Hessian for large systems is therefore a *sparse matrix*. NR methods which take advantage of this fact by neglecting off-diagonal blocks are denoted *truncated* NR. Some force field programs use an extreme example of this where only the 3×3 submatrices along the diagonal are retained. These 3×3 matrices contain the coupling elements between the x, y and z coordinates for a single atom. The task of inverting say a 3000×3000 matrix is thereby replaced by inverting 1000 3×3 matrices, reducing the computational cost for the diagonalization by a factor of 10^6. If the NR step is not taken directly, but rather used as a direction along which the function is minimized, truncated NR methods start to resemble the conjugate gradient method, although they are somewhat more complicated to implement.

14.4 Choice of Coordinates

Naively one may think that any set of coordinates which uniquely describes the function is equally good. This is not the case! A "good" set of coordinates may transform a divergent optimization into a convergent, or increase the rate of convergence. We will look specifically at the problem of optimizing a geometry given an energy function depending on nuclear coordinates, but the same considerations hold equally well for other types of optimization. We will furthermore use the straight Newton–Raphson formula (14.4) to illustrate the concepts. Given the first and second derivatives, the NR formula calculates the geometry step as the inverse of the Hessian times the gradient.

$$(\mathbf{x} - \mathbf{x}_0) = -\mathbf{H}^{-1}\mathbf{g} \tag{14.10}$$

In the coordinate system $(\mathbf{x}')$ where the Hessian is diagonal the step may be written as

$$\Delta\mathbf{x}' = \sum_i \Delta x_i', \quad \Delta x_i' = -\frac{f_i}{\varepsilon_i} \tag{14.11}$$

Essentially all computational programs calculate the fundamental properties, the energy and derivatives, in Cartesian coordinates. The Cartesian Hessian matrix has the dimension $3N \times 3N$, where N is the number of atoms. Of these, three describe the overall translation of the molecule, and three describe the overall rotation. In the molecular coordinate system there are only $3N - 6$ coordinates needed to uniquely describe the nuclear positions. Moving all the atoms in say the x-direction by the same amount does not change the energy, and the corresponding gradient component (and all higher derivatives) is zero. The Hessian matrix should therefore have six eigenvalues identical to zero, and the corresponding gradient components, f_i, should also be identical zero. In actual calculations, however, these values are certainly small, but not exactly zero. Numerical inaccuracies may introduce errors of perhaps 10^{-14}–10^{-16}. This can have rather drastic consequences. Consider for example a case where the gradient in the

x-translation direction is calculated to be 10^{-14}, while the corresponding Hessian eigenvalue is 10^{-16}. The NR step in this direction is then 100! This illustrates that care should be taken if redundant coordinates (i.e. more than are necessary to uniquely describe the system) are used in the optimization. In the case of Cartesian geometry optimization, the six translational and rotational degrees of freedom may be removed by projecting these components out of the Hessian prior to formation of the NR step (Section 13.1). The calculated "steps" in the zero eigenvalue directions are then simply neglected.

Another way of removing the six translational and rotational degrees of freedom is to use a set of internal coordinates. For a simple acyclic system these may be chosen as $3N - 1$ distances, $3N - 2$ angles and $3N - 3$ torsional angles, as illustrated in the construction of Z-matrices in Appendix E. In internal coordinates the six translational and rotational modes are automatically removed (since only $3N - 6$ coordinates are defined), and the NR step can be formed straightforwardly. For cyclic systems a choice of $3N - 6$ internal variables which span the whole optimization space may be somewhat more problematic, especially if symmetry is present.

Diagonalization of the Hessian is an example of a linear transformation; the eigenvectors are just linear combinations of the original coordinates. A linear transformation does not change the convergence/divergence properties, or the rate of convergence. We can form the NR step directly in Cartesian coordinates by inverting the Hessian and multiplying it with the gradient vector (eq. (14.10)), or we can transform the coordinates to a system where the Hessian is diagonal, form the ratios $-f_i/\varepsilon_i$ (eq. (14.11)) and backtransform to the original system. Both methods generate the exact same (except for round-off errors) NR step. Since we need to pay consideration to the six translational and rotational modes, however, the diagonal representation is advantageous.

The transformation from a set of Cartesian coordinates to a set of internal coordinates, which may for example be distances, angles and torsional angles, is an example of a non-linear transformation. The internal coordinates are connected with the Cartesian coordinates by means of square root and trigonometric functions, not simple linear combinations. A non-linear transformation will affect the convergence properties. This may be illustrate by considering a minimization of a Morse type function (eq. (2.5)) with $D = \alpha = 1$ and $x = \Delta R$.

$$E_{\text{Morse}}(x) = [1 - e^{-x}]^2 \tag{14.12}$$

We will consider two other variables obtained by a non-linear transformation: $y = e^{-x}$ and $z = e^x$. The minimum energy is at $x = 0$, corresponding to $y = z = 1$. Consider now an NR optimization starting at $x = -0.5$, corresponding to $y = 1.6487$ and $z = 0.6065$. Table 14.1 shows that the NR procedure in the x-variable requires four iterations before x is less than 10^{-4}. In the y-variable the optimization only requires one step to reach the $y = 1$ minimum exactly! The optimization in the z-variable takes six iterations before the value is within 10^{-4} of the minimum.

Consider now the same system starting from $x = 1 (y = 0.3679$ and $z = 2.7183)$. The x-variable optimization now diverges toward the $x = \infty$ limit, the y-variable optimization again converges (exactly) in one step, and the z-variable optimization also diverges toward the $z = \infty$ limit. The reason for this behaviour is seen when plotting the three functional forms as shown in Figures 14.2–14.4. The horizontal axis

Table 14.1 Convergence for different choices of variable

Iteration	x	y	z
0	− 0.5000	1.6487	0.6065
1	− 0.2176	1.0000	0.7401
2	− 0.0541		0.8667
3	− 0.0041		0.9579
4	0.0000		0.9951
5			0.9999
6			1.0000

Table 14.2 Convergence for different choices of variable

Iteration	x	y	z
0	1.0000	0.3679	2.7183
1	3.3922	1.0000	4.6352
2	4.4283		7.3225
3	5.4405		11.2981

covers the same range of x-variables for all three figures. In the x-variable space the second derivative is negative beyond $x = \ln 2 \, (= 0.69)$, and if the optimization is started at larger x-values, the optimization is no longer a minimization, but a maximization toward the $x = \infty$ asymptote. The function in the y-variable is a parabola, and the second-order expansion of the NR method is <u>exact</u>. <u>All</u> starting points consequently converge to the minimum in a single step. The transformation to the z-variable introduces a singularity at $z = 0$, and it can be seen from Figure 14.4 that the curve shape is much less quadratic than that of the original function. Using y as a variable is an example of a "good" non-linear transformation, while z is an example of a "poor" non-linear transformation.

These examples show that non-linear transformations may strongly affect the convergence properties of an optimization. The more "harmonic" the energy function is, the faster the convergence. One should therefore try to chose a set of coordinates where the third- and higher-order derivatives are as small as possible. Cartesian coordinates are not particularly good in this respect, but have the advantage that convergence properties are fairly uniform for different systems. A "good" set of internal coordinates may speed up the convergence, but a "bad" set of coordinates may slow it down or cause divergence. For acyclic systems the above-mentioned internal coordinates consisting of $3N - 1$ distances, $3N - 2$ angles and $3N - 3$ torsional angles are normally better than Cartesian coordinates. Cyclic systems, however, are notoriously difficult to chose a good set of internal coordinates for. Cyclopropane, for example, has three C–C bonds and three CCC angles, but only three independent variables (not counting the hydrogens). Choosing say two distances and one angle, introduces a strong coupling between the angle and distances owing to the "remote" C–C bond, which is described by the other three variables. Cartesian coordinates may display better convergence characteristics in such systems. Another problem is when very soft modes are present. A prototypical example is rotation of a methyl or hydroxyl group. Near the

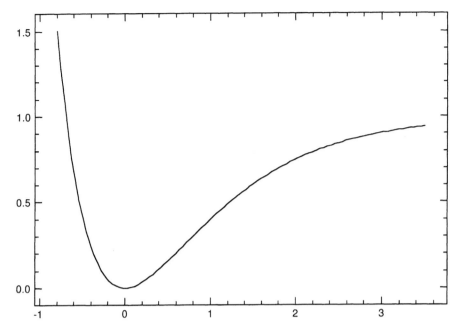

Figure 14.2   Morse curve as a function of x

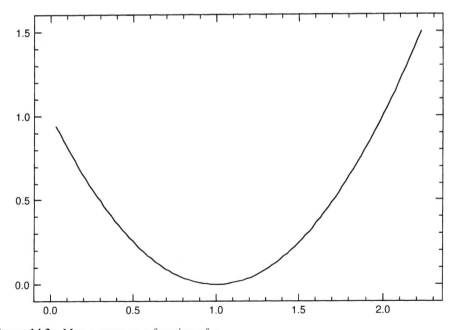

Figure 14.3   Morse curve as a function of y

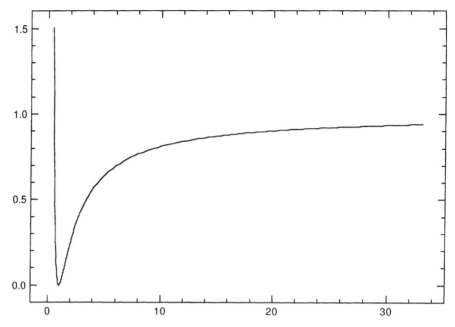

Figure 14.4 Morse curve as a function of z

minimum the energy changes very little as a function of the torsional angle, i.e. the corresponding Hessian eigenvalue is small. Consequently, even a small gradient may produce a large change in geometry. The potential is not very harmonic, and the result is that the optimization spends many iterations flopping from side to side. A similar problem is encountered in optimization of molecular clusters where the optimum structure is governed by weak van der Waals type interactions.

The problem of an "optimum" choice of coordinates has been addressed by Pulay and co-workers, who suggested using *natural internal coordinates*.[5] The atoms are first classified into three types: "terminal" (having only one bond), "ring" (part of a ring) or "internal". All distances between bonded atoms are used as variables. The ring and internal atoms are assumed to have a local symmetry depending on the number of terminal atoms attached to it, i.e. an internal atom with three terminal bonds has local C_{3v} symmetry, one with two terminal bonds has C_{2v} symmetry, a ring has locally D_{nh} symmetry etc. Suitable linear combinations of bending and torsional angles are then formed so that the coupling between these coordinates is exactly zero if the local symmetry is the exact symmetry. This will usually not be the case, but the local symmetry coordinates tend to minimize the coupling, and thus the magnitude of third and higher derivatives, thereby improving the NR performance. Natural internal coordinates appear to be a good choice for optimization to minima on an energy surface, since the bonding pattern is usually well defined for stable molecules. For locating transition structures, however, it is much less clear whether natural internal coordinates offer any special advantage. The bonding pattern is not as well defined for TSs, and a "good" set of coordinates at the starting geometry may become ill behaved during the optimization.

In the original formulation, a set of $3N - 6$ independent natural internal coordinates was chosen. It was later discovered that the same optimization characteristics could be obtained by using all distances, bending and torsional angles between atoms within bonding distance as variables.[6] Such a set of coordinates will in general be redundant (i.e. the number of coordinates is larger than $3N - 6$), and special care must be taken to handle this. More recently it has been argued that an "optimum" set of non-redundant coordinates may be extracted from a large set of (redundant) internal coordinates by selecting the eigenvectors corresponding to non-zero eigenvalues of the square of the matrix defining the transformation from Cartesian to internal coordinates. These linear combinations have been denoted *delocalized internal coordinates*,[7] and are in a sense a generalization of the natural internal coordinates.

Summarizing, the efficiency of Newton–Raphson based optimizations depends on the following factors:

(1) Hessian quality (exact or updated).
(2) Step control (augmented Hessian, choice of shift parameter(s)).
(3) Coordinates (cartesian, internal).
(4) Trust radius update (maximum step size allowed).

A comparison of different combinations of these can be found in ref. 8.

14.5 Transition Structure Optimization

Locating minima for functions is fairly easy. If everything else fails, the steepest descent is guaranteed to lower the function value. Finding first-order saddle points, transition structures, is much more difficult. There are <u>no</u> general methods which are guaranteed to work! Many different strategies have been proposed, the majority of which can be divided into two general categories, those based on interpolation between two minima, and those using only local information. Once the TS has been found, the whole reaction path may be located by tracing the Intrinsic Reaction Coordinate (Section 14.8), which corresponds to a steepest descent path in mass-weighted coordinates, from the TS to the reactant and product. Below is a description of some commonly used methods for locating TSs.

14.5.1 Methods Based on Interpolation Between Reactant and Product

These methods assume that the reactant and product geometries are known, and that a TS is located somewhere "between" these two end-points. They differ in how the interpolation is performed. It should be noted that many of the methods in this group do not actually locate the TS, they only locate a point close to it. The geometry can then be further refined by some of the "local" methods given in Section 14.5.9.

14.5.2 Linear and Quadratic Synchronous Transit

The *Linear Synchronous Transit* (LST) method[9] forms the geometry difference vector between the reactant and product, and locates the highest energy structure along this line. The assumption is that all variables change at the same rate along the reaction path.

This is in general not a good approximation, and only in simple systems does LST lead to a reasonable estimate of a TS. The *Quadratic Synchronous Transit* (QST) method approximates the reaction path by a parabola instead of a straight line. After the maximum on the LST is found, the QST is generated by minimizing the energy in the directions perpendicular to the LST path. The QST path may then be searched for an energy maximum. These methods are illustrated in Figure 14.5.

Bell and Crighton[10] refined the method by performing the minimization from the LST maximum in the directions <u>conjugate</u> to the LST instead of <u>orthogonal</u> as in the original formulation. A more recent variation of QST, called *Synchronous Transit-Guided Quasi-Newton* (STQN), uses a circle arc instead of a parabola for the interpolation, and uses the tangent to the circle for guiding the search towards the TS region.[11] Once the TS region is located, the optimization is switched to a quasi-Newton-Raphson (Section 14.5.9).

It should be noted that the success or failure of LST/QST, as with all optimizations, depends on the coordinates used in the interpolation. Consider for example the HNC to HCN rearrangement. In Cartesian coordinates the LST path preserves the linearity of the reactant and product, and thus predicts that the hydrogen moves <u>through</u> the nitrogen and carbon atoms. In internal coordinates, however, the angle changes from 0 ° to 180 °, and the LST will in this case locate a much more reasonable point with the hydrogen moving around the C–N moiety.

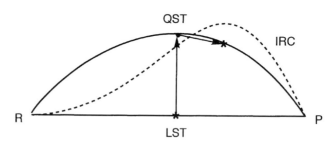

Figure 14.5 Illustration of the linear and quadratic synchronous transit methods. Energy maxima and minima are denoted by ∗ and •, respectively

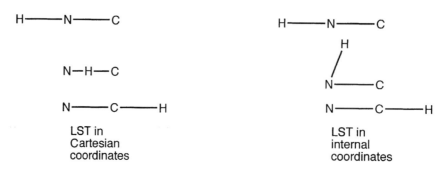

Figure 14.6 LST path in Cartesian and internal coordinates

14.5.3   "Saddle" Optimization Method

In the "*Saddle*" algorithm[12] the lowest of the two energy minima, reactant and product, is first identified. A trial structure is generated by displacing the geometry of the low energy species a fraction (for example 0.05) towards the high energy minimum. The trial structure is then optimized, subject to the constraint that the distance to the high energy minimum is constant. The lowest energy structure on the hypersphere becomes the new interpolation end-point, and the procedure is repeated. The two geometries will (hopefully) gradually converge on a low energy structure intermediate between the original two minima, as illustrated in Figure 14.7.

A related idea is used in the "*Line Then Plane*" (LTP) algorithm[13] where the constrained optimization is done in the hyperplane perpendicular to the interpolation line between the two end-points, rather than on a hypersphere.

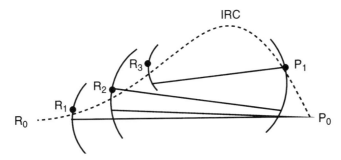

Figure 14.7   Illustration of the saddle method. Energy minima on the hyperspheres are denoted by •

14.5.4   The Chain Method

The *Chain method*[14] initially calculates the energy at a series of points placed at regular intervals (d_{max}) along a suitable reaction coordinate. This may either be a straight line, or involve an intermediate geometry to guide the search in a certain direction. The highest energy point is allowed to relax by a maximum step size of d_{max} along a direction defined by the gradient component orthogonal to the line between by the two neighbouring points. This process is repeated with the new highest energy point until the gradient becomes tangential to the path (within a specified threshold). When this happens, the current highest energy point cannot be further relaxed, and is instead moved to a maximum along the path. During the relaxation the chain may form loops, in which case intermediate point(s) is (are) discarded. Similarly, it may be necessary to add points to keep the distance between neighbours below d_{max}.

The *Locally Updated Planes* (LUP) minimization[15] is related to the chain method, but the relaxation is here done in the hyperplane perpendicular to the reaction coordinate, rather than along a line defined by the gradient. Furthermore, all the points are moved in each iteration, rather than one at a time.

The *Conjugate Peak Refinement* (CPR) method[16] may be considered as a dynamical version of the chain method, where points are added or removed based on a sequence of maximizations along line segments and minimizations along the conjugate directions.

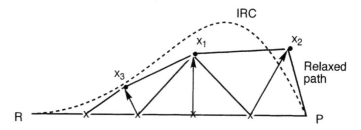

Figure 14.8 Illustration of the chain method. Initial points along the path are denoted by x, and relaxed points are denoted by •

The first cycle is analogous to the Bell and Crighton version of the QST: location of an energy maximum along a line between the reactant and product, followed by a sequential minimization in the conjugate directions. The corresponding point becomes a new path point, and an attempt is made to locate an LST maximum between the reactant and midpoint, and between the midpoint and product. If such a maximum is found, it is followed by a new conjugate minimization, which then defines a new intermediate point etc. The advantage over the chain and LUP methods is that points tend to be distributed in the important region near the TS, rather than uniformly over the whole reaction path.

In practice it may not be possible to minimize the energy in all the conjugate directions, since the energy surface in general is not quadratic. Once the gradient component along the LST path between two neighbouring points exceeds a suitable tolerance during the sequential line minimizations, the optimization is terminated, and the geometry becomes a new interpolation point. It may also happen that one of the interpolation points has the highest energy along the path, without being sufficiently close to a TS (as measured by the magnitude of the gradient), in which case the point is removed and a new interpolation is performed.

14.5.5 The Self Penalty Walk Method

In the *Self Penalty Walk* (SPW) method[17] the whole reaction path is approximated by minimizing the average energy along the path, given as a line integral between the reactant and product geometries (**R** and **P**).

$$S(\mathbf{R}, \mathbf{P}) = \frac{1}{L} \int_{\mathbf{R}}^{\mathbf{P}} E(\mathbf{x}) \, \mathbf{d}l\,(\mathbf{x}) \qquad (14.13)$$

The line element $\mathbf{d}l(\mathbf{x})$ belongs to the reaction path, which has a total length of L. In practice the line integral is approximated as a finite sum of M points, where M typically is of the order of 10–20.

$$S(\mathbf{R}, \mathbf{P}) \simeq \frac{1}{L} \sum_{i=1}^{M} E(\mathbf{x}_i) \Delta l_i \qquad (14.14)$$

In order to avoid all points aggregating near the minima (reactant and product), constraints are imposed for keeping the distance between two neighbouring points close

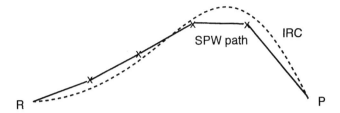

Figure 14.9   Illustration of the SPW method. Optimized path points are denoted by x

to the average distance. Furthermore, repulsion terms between all points are also added to keep the reaction path from forming loops. The resulting target function $T(\mathbf{R},\mathbf{P})$ may then be minimized by for example a conjugate gradient method.

$$T(\mathbf{R},\mathbf{P}) = \frac{1}{L}\sum_{i=1}^{M} E(\mathbf{x}_i)\Delta l_i + \gamma\sum_{i=0}^{M}\left(d_{i,i+1} - \bar{d}\right)^2 + \rho\sum_{i\geq j+1}^{M+1} \exp\left(-\frac{d_{i,j}}{\lambda\bar{d}}\right)$$

$$d_{ij} = |\mathbf{x}_i - \mathbf{x}_j| \tag{14.15}$$

$$\bar{d} = \sqrt{\frac{1}{M+1}\sum_{i=0}^{M} d_{i,i+1}^2}$$

The γ, ρ and λ parameters are suitable constants for weighting the distance and repulsion constraints relative to the average path energy. The initial reaction path may be taken simply as a straight line between the reactant and product, or one or more intermediate points may be added to guide the search in a certain direction. The grid point with the highest energy after minimization of the target function is the best estimate of the TS, and the sequence of line segments is an approximation to the IRC. Ayala and Schlegel have implemented a version[18] where one of the points is optimized directly to the TS and the remaining points form an approximation to the IRC path.

14.5.6   The Sphere Optimization Technique

The *Sphere* optimization technique[19] is related to the saddle method described in Section 14.5.3, and involves a sequence of constrained optimizations on hyperspheres with increasingly larger radii, using the reactant (or product) geometry as a constant expansion point. The lowest energy point on each successive hypersphere thus traces out a low energy path on the energy surface as illustrated in Figure 14.10.

Interpolation methods have the following characteristics.

(1) There may not be a TS connecting two minima directly. The algorithm may then find an intermediate geometry having a gradient substantially different from zero, i.e. no nearby stationary point.
(2) The TS found is not necessarily one which connects the two minima used in the interpolation. A calculation of the IRC path may reveal that it is a TS for a different reaction.

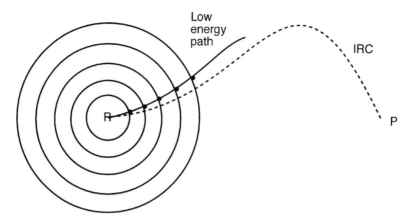

Figure 14.10   Illustration of the sphere method. Energy minima on the hyperspheres are denoted by •, while R indicates a (local) minimum in the full variable space

(3) There may be several TSs (and therefore at least one minimum) between the two selected end-points. Some algorithms may find one of these, and the two connecting minima can then be found by tracing the IRC path, or all the TSs and intermediate minima may be located. Methods which generate a sequence of points along the whole reaction path (e.g. CPR and SPW) are examples of the latter behaviour.

(4) The reaction path formed by a sequence of points generated by constrained optimizations may be discontinuous. For methods where two points are gradually moved from the reactant and product sides (e.g. saddle and LTP), this means that the distance between end-points does not converge towards zero.

(5) There may be more than one TS connecting two minima. As many of the interpolation methods start off by assuming a linear reaction coordinate between the reactant and product, the user needs to guide the initial search (for example by adding different intermediate structures) to find more than one TS.

(6) A significant advantage is that the constrained optimization can usually be carried out using only the first derivative of the energy. This avoids an explicit, and computationally expensive, calculation of the second derivative matrix, as is normally required by Newton–Raphson techniques.

(7) Each successive refinement of the TS estimate requires either location of an energy maximum or minimum along a one-dimensional path (typically a line), or a constrained optimization on an $N - 1$ dimensional hypersphere or -plane, or both (Table 14.3). A path minimization or maximization will normally involve several function evaluations, while a multidimensional minimization will require several gradient calculations. Geometry changes are often quite small in the end-game, but each step may still require a significant computational effort involving many function and/or gradient calculations. In such cases it is often advantageous to switch to one of the Newton–Raphson methods described in Section 14.5.9. The SPW method is slightly different as it transforms a constrained optimization into a global minimization, at the cost of increasing the number of variables by a factor of M, where M is the number of grid points.

Table 14.3 Optimization requirements in each refinement step

Method	Path minimization	Path maximization	Hyperplane minimization	Hypersphere minimization	Global minimization	Points moved
LST		×				1
QST		×	×			1
Saddle				×		1
LTP			×			2
Chain	×	(×)				1
LUP			×			M
CPR	×	×				1
SPW					×	M
Sphere				×		1

14.5.7   *Methods Based on Local Information*

Methods of the "local" type propagate the geometry by using only information about the function and its derivatives at the current point, i.e. they require no knowledge of the reactant and/or product geometries.

14.5.8   *Gradient Norm Minimizations*

Since transition structures are points where the gradient is zero, they may in principle be located by minimizing the gradient norm. This is in general not a good approach for two reasons: (1) there are typically many points where the gradient norm has a minimum without being zero, (2) any stationary point has a gradient norm of zero, thus all types of saddle points and minima/maxima may be found, not just TSs. Figure 14.11 shows an example of a one-dimensional function and its associated gradient norm. It is clear that a gradient norm minimization will only locate one of the two stationary points if started near $x = 1$ or $x = 9$. Most other starting points will converge on the shallow part of the function near $x = 5$. The often very small convergence radius makes gradient norm minimizations impractical for routine use.

14.5.9   *Newton–Raphson Methods*

By far the most common methods are based on some form of augmented Hessian Newton–Raphson approach. Sufficiently close to the TS, the standard NR formula will locate the TS rapidly. Sufficiently close means that the Hessian should have exactly one negative eigenvalue, and the eigenvector should be in the correct direction, along the "reaction coordinate". Furthermore, the NR step should be inside the trust radius. By using augmented Hessian techniques, the convergence radius may be enlarged over the straight NR approach, and first-order saddle points may be located even when started in a region where the Hessian does not have the correct structure, as long as the lowest eigenvector is in the "correct" direction.

Near a first-order saddle point the NR step maximizes the energy in one direction (along the Hessian TS eigenvector) and minimizes the energy along all other directions. Such a step may be enforced by choosing suitable shift parameters in the augmented

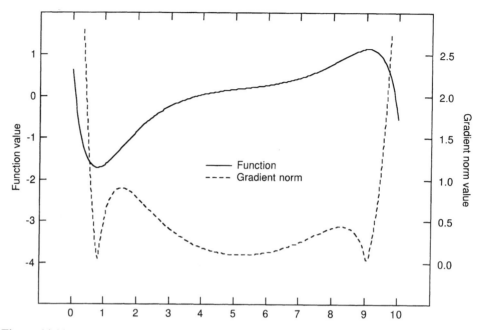

Figure 14.11   An example of a function and the associated gradient norm

Hessian method, i.e. the step is parameterized as in eq. (14.6). The minimization step is similar to that described in Section 14.3.1 for locating minima, the only difference is for the unique TS mode.

In the *Partitioned Rational Function Optimization* (P-RFO),[2] two shift parameters are employed.

$$\sum_{i \neq TS} \frac{f_i^2}{\varepsilon_i - \lambda} = \lambda, \quad \frac{f_{TS}^2}{\varepsilon_{TS} - \lambda_{TS}} = \lambda_{TS} \tag{14.16}$$

The λ for the minimization modes is determined as for the RFO method, eq. (14.8). The equation for λ_{TS} is quadratic, and by choosing the solution which is larger than ε_{TS} it is guaranteed that the step component in this direction is along the gradient, i.e. a maximization. As for the RFO step, there is no guarantee that the total step length will be within the trust radius.

The QA method[3] uses only one shift parameter, requiring that $\lambda_{TS} = -\lambda$, and restricts the total step length to the trust radius (compare with eq. (14.9)).

$$|\Delta \mathbf{x}'|^2 = \sum_{i \neq TS} \left(\frac{f_i}{\varepsilon_i - \lambda} \right)^2 + \left(\frac{f_{TS}}{\varepsilon_{TS} + \lambda} \right)^2 = R^2 \tag{14.17}$$

The exact same formula may be derived using the concept of an "image potential" (obtained by inverting the sign of f_{TS} and ε_{TS}), and the QA name is often used together with the TRIM acronym, which stands for *Trust Radius Image Minimization*.[20]

The ability of augmented Hessian methods for generating a search toward a first-order saddle point, even when started in a region where the Hessian has all positive eigenvalues, suggests that it may be possible to start directly from a minimum and "walk" to the TS by following a selected Hessian eigenvector uphill. Such mode followings, however, are only possible if the eigenvector being followed is only weakly coupled to the other eigenvectors (i.e. third and higher derivatives are small). All NR based methods assume that one of the Hessian eigenvectors points in the general direction of the TS, but this is only strictly true when the higher-order derivatives are small. If this is not the case, NR based methods may fail to converge even when started from a "good" geometry, where the Hessian has one negative eigenvalue. Note also that the magnitude of the higher derivatives depends on the choice of coordinates, i.e. a "good" choice of coordinates may transform a divergent optimization into a convergent one.

Pseudo-Newton–Raphson methods have traditionally been the preferred algorithms with ab *initio* wave function. The interpolation methods tend to have a somewhat poor convergence characteristic, requiring many function and gradient evaluations, and have consequently primarily been used in connection with semi-empirical and force field methods.

Newton–Raphson methods can be combined with extrapolation procedures, the best known of these is perhaps the GDIIS *(Geometry Direct Inversion in the Iterative Subspace)*,[21] which is directly analogous to the DIIS for electronic wave functions described in Section 3.8.1. In the GDIIS method the NR step is not taken from the last geometry, but from an interpolated point with a corresponding interpolated gradient.

$$\mathbf{x}_n^* = \sum_i^n c_i \mathbf{x}_i$$
$$\mathbf{g}_n^* = \sum_i^n c_i \mathbf{g}_i$$
(14.18)

The interpolated geometry and gradient are generated by requiring that the norm of an error vector is minimum, subject to a normalization condition.

$$\mathrm{ErrF}(\mathbf{c}) = \left| \sum_i c_i \mathbf{e}_i \right|$$
$$\sum_i c_i = 1$$
(14.19)

There are two common choices for the error vector; it can either be a "geometry" or a "gradient" vector, the latter being preferred in more recent work.[22]

$$\mathbf{e}_i = -\mathbf{H}_n^{-1}\mathbf{g} \quad \text{or} \quad \mathbf{e}_i = -\mathbf{g}_i$$
(14.20)

All NR methods assume that a reasonable guess of the TS geometry is available. Generating this guess is the whole trick. Some of the interpolating schemes described in Sections 14.5.1–14.5.6 may be useful in this respect. Other commonly used methods for generating TS guesses are based on performing a series of constrained optimizations, also known as "*coordinate driving*". One, or a few, internal coordinates are selected as "reaction variables" and fixed at certain values, while the remaining variables are optimized. The goal is to find a geometry where the residual gradients for the fixed variables are "sufficiently" small. The success of this method depends on whether a good choice of reaction variables has been made. They should normally be chosen among those which change substantially between reactant and product. A good choice has been made if the reaction coordinate at the TS contains large coefficients for these variables. The reaction coordinate at the TS, however, is only known after the TS has actually been found. If only one or two variables change significantly between reactant and product, this method usually works well. A couple of typical examples are the HNC to HCN rearrangement (reaction variable is the HCN angle) and S_N2 reactions of the type $X + CH_3Y \rightarrow XCH_3 + Y$ (reaction variables are the XC and CY distances). A mapping of more than two reaction variables becomes cumbersome, and rarely leads anywhere.

If a poor choice of reaction variables has been made, "hysteresis" is often observed. This is the term used when the optimization suddenly changes the geometry drastically for a small change in the fixed variables. Furthermore a series of optimizations made by increasing the fixed variable(s) to a given value may produce a result different from that produced when decreasing the fixed variable(s) to the same point. This indicates that the chosen reaction variable(s) do not contribute strongly to the actual reaction coordinate at the TS. Some TSs have reaction vectors which are not dominated by a few internal variables, and such TSs are difficult to find by constrained optimization methods. In some cases another set of (internal) coordinates may alleviate the problem, but finding these is part of the "black magic" involved in locating TSs.

If, on the other hand, the TS reaction vector essentially contains only a single internal variable, the coordinate driving works very well, and the constrained optimized geometry with the smallest residual gradient is a good approximation to the TS. A prototypical example of this is the rotation of a methyl group. All bond lengths and angles are essentially unchanged during the reaction, and the reaction coordinate is for all practical purposes the torsional angle. Good approximations to many conformational TSs can be generated by "driving" a selected torsional angle, and this is often the basis for conformational analysis using force field energy functions. It should be stressed that the highest energy structure along such a torsional driving profile is not exactly the TS, but it is usually a very good approximation to it.

There are two main problems with all NR based methods. One is the already mentioned need for a good starting geometry. The other is the requirement of a Hessian, which is quite expensive in terms of computer time for electronic structure methods. Contrary to minimizations, TS optimizations cannot start with a diagonal matrix and update it as the optimization proceeds. An NR TS search requires definition of a direction along which to maximize the energy, the reaction vector; i.e. the start Hessian should preferably have one negative eigenvalue. Normally the Hessian needs to be calculated explicitly at the first step, at subsequent steps the Hessian may be updated. If the geometry changes substantially during the optimization, however, it may be necessary to recalculate the Hessian at certain intervals. Finally, the choice of a "good" set of coordinates is even more critical in TS optimizations than in minimizations. A good set of coordinates enlarges the convergence region, relaxing the requirement of a good starting geometry. On the other hand, a poor set of coordinates decreases the convergence radius, forcing the user to generate a starting point very close to the actual TS in order for NR methods to work. Finally, NR methods are best suited for relatively "stiff" systems; large flexible systems with many small eigenvalues in the Hessian are better handled by some of the interpolation methods, such as SPW.

Mapping out whole reaction pathways by locating minima and connecting TSs is often computationally demanding. The (approximate) geometries of many of the important minima are often known in advance, and as mentioned above, energy minimizations are fairly uncomplicated. Locating TSs is much more involved. On a multidimensional energy surface, there will in general not be TSs connecting all pairs of minima. It is, however, essentially impossible to prove that a TS does not exist.

Symmetry can sometimes be used to facilitate the location of TSs. For some reactions, especially those where the reactant and product are identical, the TS will have a symmetry different from the reactant/product. The reaction vector will belong to one of the non-totally symmetric representations in the point group. The TS therefore can be located by constraining the geometry to a certain symmetry, and <u>minimizing</u> the energy. Consider for example the S_N2 reaction of Cl^- with CH_3Cl. The reactant and products have C_{3v} symmetry, but the TS has D_{3h} symmetry. Minimizing the energy under the constraint that the geometry should have D_{3h} symmetry will produce the lowest energy structure within this symmetry, which is the TS.

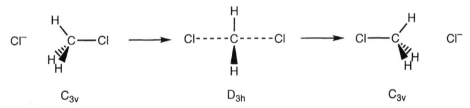

Figure 14.12   The TS for an identity S_N2 reaction has a higher symmetry than the reactant/product

For non-identity reactions it is often useful to start a search for stationary points by minimizing high symmetry geometries. A subsequent frequency calculation on the symmetry constrained (and minimized) structure will reveal the nature of the stationary point. If it is a minimum or TS we have already obtained useful information. If it turns out to be a higher-order saddle point, the normal coordinates associated with the imaginary frequencies show how the symmetry should be lowered to produce lower energy species, which may be either minima or TSs. As calculations on highly symmetric geometries are computationally less expensive than on non-symmetric structures, it is often a quite efficient strategy to start the investigation by concentrating on structures with symmetry.

14.5.10   *Gradient Extremal Methods*

Since there are no methods which are guaranteed to locate all TSs (short of mapping the whole surface, which is impossible for more than three or four variables), it is essentially impossible to prove that a TS does not exist. The failure to locate a TS connecting two minima may simply be due to the inability to generate a sufficiently good trial structure for NR methods, or interpolation methods converging to a TS not connecting the two desired minima.

An alternative method of locating stationary points is based on following *Gradient Extremal* (GE) paths[23]. A GE is a one-dimensional curve consisting of points where the derivative of the gradient norm is zero, subject to the constraint that the energy is constant. This may be shown to be equivalent to the condition that the gradient is an eigenvector of the Hessian.

$$\frac{\partial |\mathbf{g}|}{\partial \mathbf{x}} = 0, \qquad E(\mathbf{x}) = \text{constant}$$

$$\Updownarrow \qquad\qquad (14.21)$$

$$\mathbf{Hg} = \varepsilon \mathbf{g}$$

At stationary points there are GE curves leading along the direction of all Hessian eigenvectors, and by tracing out GE paths it is possible to locate many stationary points on a given surface. Although it is difficult to prove, it would appear that a complete mapping of all GE paths would locate all stationary points. It should be noted, however, that all types of stationary points are found, not only minima and TSs, but also higher-order saddle points. Furthermore, a GE path is in general not connected with the IRC, and may display features such as bifurcation and turning points. GEs often have complicated behaviours and tend to wander over a significant portion of the PES before ending up at another stationary point or leading to dissociation. Existing algorithms for following GEs require one component of the third derivative of the energy in each step, making it computationally expensive, and therefore limiting an exhaustive search to small systems. Nevertheless, GE following is one of the few methods for rigorously establishing whether or not a TS exists, and may be considered as a geometry equivalent of a "full-CI" calculation.

14.6 Constrained Optimization Problems

In some cases there are restrictions on the variables used to describe the function, for example:

(1) Certain geometrical restraints may be imposed, experimental data, for example, may indicate that some atom pairs are within a certain distance of each other.
(2) Fitting atomic charges to give a best match to a calculated electrostatic potential. The constraint is that the sum of atomic charges should equal the net charge of the molecule.
(3) A variation of wave function coefficients is subject to constraints like maintaining orthogonality of the MOs, and normalization of the MOs and the total wave function.

There are two main methods for enforcing such constraints. One is the *Penalty Function* approach, the other the method of *Lagrange Undetermined Multipliers*.

The penalty function approach adds a term of the type $k(r - r_0)^2$ to the function to be optimized. The variable r is constrained to be near the target value r_0, and the "force constant" k describes how important the constraint is compared with the unconstrained optimization. By making k arbitrary large, the constraint may be fulfilled to any given accuracy. It cannot, however, make the constraint variable exactly equal to r_0. This would require the constant k to go towards infinity, and in practice cause numerical problems when it becomes sufficiently large compared to the other terms. The penalty function approach is often used for restricting geometrical variables, like distances or angles, during geometry optimizations with force field methods. It may also be used for "driving" a selected variable, like a torsional angle. In certain cases the constraint is not to limit a variable to a single value, but rather keeping it between lower and upper limits. This is typically the situation for refining a force field structure subject to restraints imposed by experimental NOE data. In such cases the penalty function may be taken as a "flat bottom" potential, i.e. the penalty term is zero within the limits and increases harmonically outside the limits.

A more elegant method of enforcing constraints is the Lagrange method. The function to be optimized depends on a number of variables, $f(x_1, x_2, \ldots x_N)$, and the constraint condition can always be written as another function, $g(x_1, x_2, \ldots x_N) = 0$. Now define a Lagrange function as the original function minus a constant times the constraint function.

$$L(x_1, x_2, \ldots x_N, \lambda) = f(x_1, x_2, \ldots x_N) - \lambda g(x_1, x_2, \ldots x_N) \qquad (14.22)$$

If there is more than one constraint, one additional multiplier term is added for each constraint. The optimization is then performed on the Lagrange function by requiring that the gradient with respect to the x- and λ-variable(s) is equal to zero. In many cases the multipliers λ can be given a physical interpretation at the end. In the variational treatment of an HF wave function (Section 3.3), the MO orthogonality constraints turn out to be MO energies, and the multiplier associated with normalization of the total CI wave function (Section 4.2) becomes the total energy.

14.7   Locating the Global Minimum and Conformational Sampling

The methods described in sections 14.1–14.3 can only locate the "nearest" minimum, which is normally a *local* minimum, when starting from a given set of variables. In some cases the interest is in the lowest of all such minima, the *global* minimum, in other cases it is important to sample a large (preferably representative) set of local minima. Considering that the number of minima typically grows exponentially with the number of variables, the global optimization problem is an extremely difficult task for a multidimensional function.[24] It is often referred to as the *multiple minima* or *combinatorial explosion* problem in the literature. Consider for example a determination of the lowest energy conformation of linear alkanes, $CH_3(CH_2)_{n+1}CH_3$, by a force field method. There will in general be three possible energy minima for a rotation around each C–C bond. For butane there are thus three conformations, one *anti* and two *gauche* (which are symmetry equivalent). These minima may be generated by starting optimizations from three torsional angles separated by $120°$. In the $CH_3(CH_2)_{n+1}CH_3$ case there are n such rotatable bonds, for a possible 3^n different conformations. In order to find the global minimum we need to calculate the energy of them all. Assume for the moment that each optimization of a conformation takes one second of computer time. Table 14.4 gives the number of possible conformations, and the time required for optimizing them all.

Table 14.4 Possible conformation for linear alkanes, $CH_3(CH_2)_{n+1}CH_3$

n	Number of possible conformations (3^n)	Time 1 conformation $= 1$ s
1	3	3 s
5	243	4 min
10	59 049	16 h
15	14 348 907	166 d

It is clear that if the degrees of freedom exceed $\sim 10-12$, a systematic search is impossible. For the linear alkanes, it is known in advance that *anti* conformations in general are favoured over *gauche*, thus we may put some restrictions on the search, like having a maximum of three *gauche* interactions total. For most systems, however, there are no good guidelines for such a *priori* selections. Furthermore, for some cases the sampling interval must be less than 120 °; in ring systems it may be more like 60° increasing the potential number of conformations to 6^n. Cycloheptadecane is a frequently used test case for conformational searching, and various methods have established that there are 262 different conformations within 3 kcal/mol of the global minimum with the MM2 force field.[25] This system is close to the limit for being able to establish the global minimum.

Finding "reasonable" minima for large biomolecular systems is heavily dependent on selecting a "good" starting geometry. One way of attempting this is by "building up" the structure. A protein, for example, may be built from amino acids, which have been optimized to their global minimum, and/or smaller fragments of the whole structure may be subjected to a global minimum search. By combining such preoptimized fragments it is hoped that the starting geometry for the whole protein also will be "near" the global minimum for the full system.

The systematic, or grid, search is only possible for small molecules. For larger systems there are methods which can be used for perturbing a geometry from a local minimum to another minimum. Some commonly used methods for conformational sampling are:

(1) Stochastic and Monte Carlo Methods
(2) Molecular Dynamics
(3) Simulated Annealing
(4) Genetic Algorithms
(5) Diffusion Methods
(6) Distance Geometry Methods

None of these are guaranteed to find the global minimum, but they may in many cases generate a local minimum which is close in energy to the global minimum (but not necessarily close in terms of structure). A very brief description of the ideas in these methods is given below. For simplicity we assume that the function optimization is an energy as a function of atomic coordinates, but it is of course equally valid for any function depending on a set of parameters.

14.7.1 *Stochastical and Monte Carlo Methods*

These methods starts from a given geometry, which typically is a (local) minimum, and new configurations are generated by adding a random "kick" to one or more atoms. In *Monte Carlo* (MC) methods the new geometry is accepted as a starting point for the next perturbing step if it is lower in energy than the current. Otherwise the Boltzmann factor $e^{-\Delta E/k_B T}$ is calculated and compared to a random number between 0 and 1. If $e^{-\Delta E/k_B T}$ is less than this number the new geometry is accepted, otherwise the next step is taken from the old geometry. This generates a sequence of configurations from which geometries may be selected for subsequent minimization. In order to have a reasonable acceptance ratio, however, the step size must be fairly small.

In stochastical methods the random kick is typically somewhat larger, and a standard minimization is carried out starting at the perturbed geometry. This may or may not produce a new minimum. A new perturbed geometry is then generated and minimized etc.[26] There are several variations on how this is done.

(1) The length of the perturbing step is important, a small kick essentially always returns the geometry to the starting minimum, a large kick may produce high energy structures, which minimize to high energy local minima.
(2) The perturbing step may be done directly in Cartesian coordinates, or in a selected set of internal coordinates, like torsional angles. The Cartesian procedure has the disadvantage that many of the perturbed geometries are high in energy, because two (or more) atoms are moved close together by the kick.
(3) The perturbing step may be taken either from the last minimum found, or from all the previous found minima, weighted by a probability factor which is such that low energy minima are used more often than high energy structures.

14.7.2 *Molecular Dynamics*

Molecular Dynamics (MD) methods solve Newton's equation of motion for atoms on an energy surface (see Section 16.2.1). The available energy for the molecule is distributed between potential and kinetic energy, and molecules are thus able to overcome barriers separating minima if the barrier height is less than the total energy minus the potential energy. Given a high enough energy, which is closely related to the simulation temperature, the dynamics will sample the whole surface, but this will also require an impractically long simulation time. Since quite small time steps must be used for integrating Newton's equation, the simulation time is short (pico- or nanoseconds). Combined with the use of "reasonable" temperatures (few hundreds or thousands of degrees), this means that only the local area around the starting point is sampled, and that only relatively small barriers (few kcal/mol) can be overcome. Different (local) minima may be generated by selecting configurations at suitable intervals during the simulation and subsequently minimizing these structures.

14.7.3 *Simulated Annealing*

Both MD and MC methods employ a temperature as a guiding parameter for generating new geometries. At sufficiently high temperature and long run time, all the conformational space is sampled. In *Simulated Annealing* (SA) techniques,[27] the initial temperature is chosen to be high, maybe 2000–3000 K. An MD or MC run is then initiated, during which the temperature is slowly reduced. Initially the molecule is allowed to move over a large area, but as the temperature is decreased, it becomes trapped in a minimum. If the cooling is done infinitely slowly (implying an infinite run time), the resulting minimum is the global minimum. In practice, however, an MD or MC run is so short that only the local area is sampled. The name, simulated annealing, comes from the analogy of growing crystals. If a melt is cooled slowly, large single crystals can be formed. Such a single crystal represents the global energy minimum for a solid state. A rapid cooling produces a glass (local minimum), a disordered solid.

14.7.4 Genetic Algorithms

Genetic Algorithms (GA)[28] take their concepts and terminology from biology. The idea is to have a "population" of structures, each characterized by a set of "genes". The "parent" structures are allowed to generate "children" having a mixture of the parent genes, allowing for "mutations" to occur in the process. The best species from a population are selected based on Darwin's principle, survival of the fittest, and carried on to the next "generation", while the less fit structures are discarded.

Consider for example a molecule having 10 torsional angles, which may have ~100 000 possible conformations. The species in an initial population of say 100 different conformations is characterized by their fitness, e.g. a low energy is equivalent to a good structure. These 100 structures are allowed to produce offspring with a probability depending on their fitness, i.e. low energy structures are more likely to contribute to the next generation than high energy conformations. Two child conformations can be generated by taking the first n torsional angles from one of the parents and the remaining $10 - n$ from the other ("single-point crossover"), with the second child being the complementary. A small amount of mutation is usually allowed in the process, i.e. angles are randomly changed to produce conformations outside the range contained in the current population. Having generated say 100 such children, their (minimized) energies are determined, and a suitable portion of the best parent and children structures are carried over to the next generation. The population is then allowed to evolve for perhaps a few hundred generations. There are several variations on the GA method: varying the size of the population, the mutation rate, the breading selection, the ratio of children to parent surviving to the next generation, single- or multi-point crossover, etc. Genetic algorithms have become popular in recent years as they are easy to implement and have proved to be robust for locating a point in parameter space close to the global minimum. Owing to the coding of the parameters into genes, the sampling is pointwise, and the final structures should therefore be refined using a standard gradient optimization.

14.7.5 Diffusion Methods

In *Diffusion Methods*[29] the energy function is changed so that it eventually contains only one minimum. The function may be changed for example by adding a contribution proportional to the local curvature of the function (second derivative). This means that minima are raised in energy, and saddle points (and maxima) are reduced in energy (negative curvature). Eventually only one minimum remains. Using the single minimum geometry of the modified potential, the process can be reversed, ending up with a minimum on the original surface which often (but not necessarily) is the global minimum. The mathematical description of this process turns out to be identical to the differential equation describing diffusion.

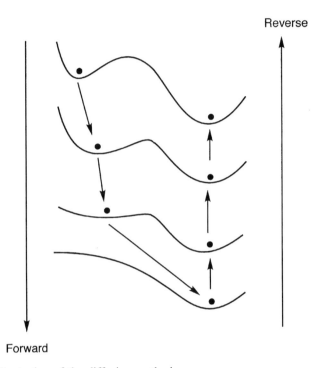

Figure 14.13 Illustration of the diffusion method

14.7.6 *Distance Geometry Methods*

The idea in *Distance Geometry* methods[30] is that trial geometries can be generated from a set of lower and upper bounds on distances between all pairs of atoms. The method was originally developed for generating possible geometries based on experimental information like NMR NOE effects, which place certain constraints on the distance between protons that may be far from each other in terms of bonding. The bonding information itself, however, also places restrictions on distances between all pairs of atoms. Having a set of upper and lower bounds for all distances, many different trial sets of distances may be generated by assigning random numbers between these limits. Such a random distance matrix can then be translated into a three-dimensional structure, this is known as *embedding*. Distance geometry can thus be used for generating trial conformations which can be optimized using conventional methods.

From the above it may be clear that MD, MC and stochastical methods tend to primarily sample the local area, generating a relatively large number of local minima in the process. The use of a larger step size in stochastical methods normally means that they are more efficient than MC or MD. Simulated annealing, genetic algorithms and diffusion methods, on the other hand, are primarily geared to locating the global minimum, and will in general only produce one final structure, this being the best estimate of the global minimum. Distance geometry methods are more or less random searches, where "impossible" structures are excluded. Simulated annealing normally explores a significantly smaller space than genetic algorithms, but there is currently no clear consensus on which method is the better for locating the global minimum. It is most likely that the best method depends on the problem at hand.

14.8 Intrinsic Reaction Coordinate Methods

The optimization methods described in Sections 14.1–14.5 concentrate on locating stationary points on an energy surface. The important points for discussing chemical reactions are minima, corresponding to reactant(s) and product(s), and saddle points, corresponding to transition structures. Once a TS has been located, it should be verified that it indeed connects the desired minima. At the TS the vibrational normal coordinate associated with the imaginary frequency is the reaction coordinate (Section 13.1), and an inspection of the corresponding atomic motions may be a strong indication that it is the "correct" TS. A rigorous proof, however, requires a determination of the *Minimum Energy Path* (MEP) from the TS to the connecting minima. If the MEP is located in mass-weighted coordinates, it is called the *Intrinsic Reaction Coordinate* (IRC).[31] The IRC path is of special importance in connection with studies of reaction dynamics, since the nuclei usually will stay close to the IRC, and a model for the reaction surface may be constructed by expanding the energy to second order for example, around points on the IRC (Section 16.2.4).

The IRC path is defined by the differential equation.

$$\frac{d\mathbf{x}}{ds} = -\frac{\mathbf{g}}{|\mathbf{g}|} = \mathbf{v} \tag{14.23}$$

Here $\mathbf{x}$ is the (mass-weighted) coordinates, s is the path length and $\mathbf{v}$ is the (negative) normalized gradient. Determining the IRC requires solving eq. (14.23) starting from a geometry slightly displaced from the TS along the normal coordinate for the imaginary frequency.

The simplest method for integrating eq. (14.23) is the *Euler* method. A series of steps are taken in the direction opposite to the normalized gradient.

$$\mathbf{x}_{n+1} = \mathbf{x}_n + \Delta s \mathbf{v}(\mathbf{x}_n) \tag{14.24}$$

This corresponds to a steepest descent minimization with a fixed step Δs. As discussed in Section 14.1, such an approach tends to oscillate around the true path, and consequently requires a small step size for following the IRC accurately.

A more advanced method is the *Runge–Kutte* (RK). The idea here is to generate some intermediate steps which allow a better and more stable estimate of the next geometry for a given step size. The *second-order Runge–Kutte* (RK2) method first calculates the gradient at a point corresponding to an Euler step with half the step size. The gradient at the half-way point is then used for taking the full step.

$$\mathbf{k}_1 = \Delta s \mathbf{v}\,(\mathbf{x}_n)$$
$$\mathbf{k}_2 = \Delta s \mathbf{v}\,(\mathbf{x}_n + \tfrac{1}{2}\mathbf{k}_1) \tag{14.25}$$
$$\mathbf{x}_{n+1} = \mathbf{x}_n + \mathbf{k}_2$$

The *fourth-order Runge–Kutte* (RK4) method generates four intermediate gradients, and combines the steps as follows.

$$\mathbf{k}_1 = \Delta s \mathbf{v}(\mathbf{x}_n)$$
$$\mathbf{k}_2 = \Delta s \mathbf{v}(\mathbf{x}_n + \tfrac{1}{2}\mathbf{k}_1)$$
$$\mathbf{k}_3 = \Delta s \mathbf{v}(\mathbf{x}_n + \tfrac{1}{2}\mathbf{k}_2) \tag{14.26}$$
$$\mathbf{k}_4 = \Delta s \mathbf{v}(\mathbf{x}_n + \mathbf{k}_3)$$
$$\mathbf{x}_{n+1} = \mathbf{x}_n + \tfrac{1}{6}\mathbf{k}_1 + \tfrac{1}{3}\mathbf{k}_2 + \tfrac{1}{3}\mathbf{k}_3 + \tfrac{1}{6}\mathbf{k}_4$$

Another method for following the IRC, which does not rely on integration of the differential equation (14.23), has been developed by Gonzales and Schlegel (GS)[32]. The idea is to generate points on the IRC by means of a series of constrained optimizations, as illustrated in Figure 14.14. An expansion point is generated by taking a step along the current direction with a step size of $1/2\ \Delta s$. The energy is then minimized on a hypersphere with radius $1/2\ \Delta s$, located at the expansion point. This is an example of a constrained optimization which can be handled by means of a Lagrange multiplier, Section 14.6. The GS procedure ensures that the tangent to the IRC path is correct at each point.

Although it is clear that RK4 is more stable and accurate than the Euler method for a given step size, this does not necessarily mean that it is the most efficient method. Since the RK4 method requires four gradient calculations for each step, the simple Euler can employ a step size four times as small for the same computational cost. Similarly, although the Gonzales–Schlegel method appears to be quite tolerant of large step sizes, each constrained optimization may take a significant number of gradient calculations to converge, which could also be used to advance the Euler algorithm at a slower pace. Nevertheless, the Gonzales–Schlegel method appears at present to be one of better methods for accurately following the IRC path. Which algorithm is the optimum will depend on the system at hand and the required accuracy of the IRC path. If only the

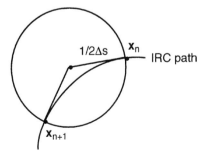

Figure 14.14   Illustration of the Gonzales–Schlegel constrained optimization method for following an IRC

nature of the two minima on each side of the TS is required, a crude IRC is sufficient, and a simple Euler algorithm may be the most cost-efficient. For use in connection with reaction path methods (Sections 12.3 and 16.2), however, the IRC needs to be located very accurately, and a sophisticated method[32] and a small step size may be required.

References

1. T. Schlick, *Rev. Comput. Chem.*, **3** (1992), 1; H. B. Schlegel, *Adv. Chem. Phys.*, **67** (1987), 249; H. B. Schlegel, *Modern Electronic Structure Theory*, Part I, ed. D. Yarkony, World Scientific, 1995, p. 459–500; R. Fletcher *Practical Methods of Optimizations*, Wiley, 1980; M. L. McKee and M. Page, *Rev. Comput. Chem.*, **4** (1993), 35.
2. A. Banerjee, N. Adams, J. Simons and R. Shepard, *J. Phys. Chem.*, **89** (1985), 52.
3. P. Culot, G. Dive, V. H. Nguyen and J. M. Ghuysen, *Theor. Chim. Acta*, **82** (1992), 189.
4. H. B. Schlegel, *Theor. Chim. Acta*, **66** (1984), 333.
5. G. Fogarasi, X. Zhou, P. W. Taylor and P. Pulay, *J. Am. Chem. Soc.*, **114** (1992), 8191.
6. C. Peng, P. Y. Ayala, H. B. Schlegel and M. J. Frisch, *J. Comput. Chem.*, **17** (1996), 49.
7. J. Baker, A. Kessi and B. Delley, *J. Chem. Phys.*, **105** (1996), 192.
8. F. Eckert, P. Pulay and H.-J. Werner, *J. Comput. Chem.*, **18** (1997), 1473.
9. T. A. Halgren and W. N. Lipscomb, *Chem. Phys. Lett.*, **49** (1977), 225.
10. S. Bell and J.S. Crighton, *J. Chem. Phys.*, **80** (1984), 2464.
11. C. Peng and H. B. Schlegel, *Isr. J. Chem.*, **33** (1993), 449.
12. M. J. S. Dewar, E. F. Healy and J. J. P. Stewart, *J. Chem. Soc., Faraday Trans. 2*, **80** (1984), 227.
13. C. Cardenas-Lailhacar and M. C. Zerner, *Int. J. Quantum. Chem.*, **55** (1995), 429.
14. D. A. Liotard, *Int. J. Quantum. Chem.*, **44** (1992), 723.
15. C. Choi and R. Elber, *J. Chem. Phys.*, **194** (1991), 751.
16. T. Fischer and M. Karplus, *Chem. Phys. Lett.*, **194** (1992), 252.
17. R. Czerminski and R. Elber, *Int. J. Quantum. Chem. Symp.*, **24** (1990), 167.
18. P. Y. Ayala and H. B. Schlegel, *J. Chem. Phys.*, **107** (1997), 375.
19. Y. Abashkin and N. Russo, *J. Chem. Phys.*, **100** (1994), 4477.
20. T. Helgaker, *Chem. Phys. Lett.*, **182** (1991), 503.
21. P. Csaszar and P. Pulay, *J. Mol. Struct.*, **114** (1984), 31.
22. F. Eckert, P. Pulay and H.-J. Werner, *J. Comput. Chem.*, **18** (1997), 1473.
23. K. Bondensgård and F. Jensen, *J. Chem. Phys.*, **104** (1996), 8025.
24. A. R. Leach, *Rev. Comput. Chem.*, **2** (1991), 1; A. E. Howard and P. A. Kollman, *J. Med. Chem.*, **31** (1988), 1669; C. D. Maranus and C. A. Floudas, *J. Chem. Phys.*, **100** (1994), 1247.
25. M. Saunders, K. N. Houk, Y.-D. Wu, W. C. Still, M. Lipton, G. Chang and W. C. Guida, *J. Am. Chem. Soc.*, **112** (1990), 1419.
26. G. Chang, W. C. Guida and W.C. Still, *J. Am. Chem. Soc.*, **111** (1989), 4379.
27. S. Kirkpatrick, C. D. Gelatt Jr and M. P. Vecchi, *Science*, **220** (1983), 671; S. R. Wilson and W. Cui, *Biopolymers*, **29** (1990), 225.
28. R. S. Judson, *Rev. Comput. Chem.*, **10** (1997), 1.
29. J. Kostrowicki and H. A. Scheraga, *J. Phys. Chem.*, **96** (1992), 7442.
30. J. M. Blaney and J. S. Dixon, *Rev. Comput. Chem.*, **5** (1994), 299.
31. K. Fukui, *Acc. Chem. Res.*, **14** (1981), 363.
32. C. Gonzales and H. B. Schlegel, *J. Chem. Phys.*, **95** (1991), 5853.

15 Qualitative Theories

Although sophisticated electronic structure methods may be able to accurately predict a molecular structure or the outcome of a chemical reaction, the results are often hard to rationalize. It therefore becomes difficult to apply the findings to other similar systems. Qualitative theories, on the other hand, are unable to provide accurate results, but they may be useful for gaining insight, e.g. why a certain reaction is favoured over another. They also provide a link to many concepts used by experimentalists.

15.1 Frontier Molecular Orbital Theory

Frontier Molecular Orbital (FMO)[1] theory attempts to predict relative reactivities based on properties of the reactants. It is commonly formulated in terms of perturbation theory, where the energy change in the initial stage of a reaction is estimated, and "extrapolated" to the transition state. For a reaction where two different modes of reaction are possible, this may be illustrated as shown in Figure 15.1. The reaction mode which involves the least energy change in the initial stage is assumed also to have the lowest activation energy. FMO theory uses low-order perturbation expansion with the reactants as the unperturbed reference, and it is clear that such a treatment can only be used to follow the reaction a short part of the whole reaction pathway.

From second-order perturbation theory (Section 4.8) the following equation for the change in energy can be derived.[2]

$$\Delta E = -\sum_{A,B}^{\text{atoms}} (\rho_A + \rho_B)\langle \chi_A|\mathbf{V}|\chi_B\rangle\langle \chi_A|\chi_B\rangle + \sum_{A,B}^{\text{atoms}} \frac{Q_A Q_B}{R_{AB}}$$
$$+ \left(\sum_{i\in A}^{\substack{\text{occ.}\\ \text{MO}}} \sum_{a\in B}^{\substack{\text{vir.}\\ \text{MO}}} + \sum_{i\in B}^{\substack{\text{occ.}\\ \text{MO}}} \sum_{a\in A}^{\substack{\text{vir.}\\ \text{MO}}} \right) \frac{2\left(\sum_\alpha^{AO} c_{i\alpha} c_{a\alpha}\langle \chi_{i\alpha}|\mathbf{V}|\chi_{a\alpha}\rangle\right)^2}{\varepsilon_i - \varepsilon_a} \tag{15.1}$$

Here A and B denote atoms in each of the two interacting molecules. The $\mathbf{V}$ operator contains all the potential energy operators from both molecules, and the $\langle \chi_A|\mathbf{V}|\chi_B\rangle$ integral is basically a "resonance" type integral between two atomic orbitals, one from each molecule. The ρ_A is the electron density on atom A, and the first term in (15.1)

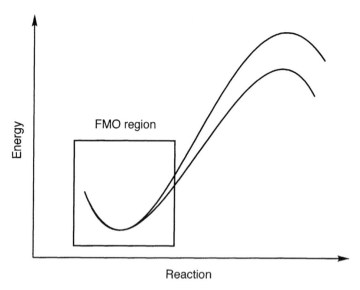

Figure 15.1 FMO region of a reaction profile

represents a repulsion ($\langle \chi_A | \mathbf{V} | \chi_B \rangle$ is a negative quantity) between occupied MOs (steric repulsion). This will usually lead to a net energy barrier for a reaction. The second term represents an attraction or repulsion between charged parts of the molecules, Q_A being the (net) charge on atom A. The last term is a stabilizing interaction ($\varepsilon_i - \varepsilon_a < 0$) due to mixing of occupied MOs on one molecule with unoccupied MOs on the other, $c_{i\alpha}/c_{a\alpha}$ being MO coefficients and $\varepsilon_i/\varepsilon_a$ MO energies. The summation is over all pairs of occupied/unoccupied MOs.

If we are comparing reactions which have approximatively the same steric requirements, the first term is roughly constant. If the species are very polar the second term will dominate, and the reaction is *charge controlled*. This means for example that an electrophilic attack is likely to occur at the most negative atom, or in a more general sense, along a path where the electrostatic potential is most negative. If the molecules are non-polar, the third term in (15.1) will dominate, and the reaction is *orbital controlled*.

All things being equal, the largest contribution to the double summation over orbital pairs in the third term will arise when the denominator is smallest. This corresponds to the *Highest Occupied Molecular Orbital* (HOMO) and *Lowest Unoccupied Molecular Orbital* (LUMO) pair of orbitals. FMO theory considers only this one contribution in the whole summation. From a purely numerical consideration this is certainly not a good approximation, the contributions from all the other pairs are much larger than the single HOMO–LUMO term. Nevertheless, it is possible to rationalize many trends in terms of FMO theory, thus the result justifies the means. If we furthermore consider a matrix element $\langle \chi_{i\alpha} | \mathbf{V} | \chi_{a\alpha} \rangle$ to be non-zero only between atoms where new bonds are being formed (where it furthermore is assumed to be roughly constant), the deciding factor becomes a sum over products of MO coefficients from the HOMO on one fragment with LUMO coefficients on the other. A few examples should help clarify this.

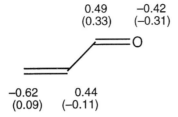

0.49   −0.42
(0.33)   (−0.31)

−0.62   0.44
(0.09)   (−0.11)

Figure 15.2 AM1 LUMO coefficients for acrolein with net charges in parenthesis

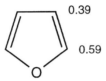

0.39

0.59

Figure 15.3 AM1 HOMO coefficients for furan

The reaction of a nucleophile involves addition of electrons to the reactant, i.e. interaction of the HOMO of the nucleophile with the LUMO of the reactant. If there is more than one possible centre of attack, the preferred reaction mode is predicted to occur on the atom having the largest LUMO coefficient. Figure 15.2 shows that the orbital component should prefer addition to the 4-position of acrolein (as a model for unsaturated carbonyl compounds in general), with the second most reactive position being C_2. The net charges, however, prefer position 2, as it is the most positive carbon. Experimentally it is found that attack at the 4-position is usually favoured (especially with "soft" nucleophiles, like organocuprates), but addition at the 2-position is also observed (and may dominate with "hard" nucleophiles, like organolithium compounds).[3] This is consistent with the reaction switching from being orbital controlled to charge controlled as the nucleophile becomes more ionic.

Similarly, the reaction of an electrophile will involve the HOMO of the reactant, i.e. the reaction should occur preferentially on the atom having the largest HOMO coefficient. The coefficients for furan shown in Figure 15.3 indicate that electrophilic substitution preferentially should occur at the 2-position, again in agreement with experimental results.[4]

Consider now the reaction between butadiene and ethylene, where both $2+2$ and $4+2$ reaction modes are possible. The qualitative appearances of the butadiene HOMO and ethylene LUMO are given in Figure 15.4. The MO coefficients are given as a, b and c, where $a > b > c$.

For the 2 + 2 pathway the FMO sum becomes $(ab - ac)^2 = a^2(b-c)^2$ while for the 4 + 2 reaction it is $(ab + ab)^2 = a^2(2b)^2$. As $(2b)^2 > (b-c)^2$, it is clear that the 4 + 2 reaction has the largest stabilization, and therefore increases <u>least</u> in energy in the initial stages of the reaction (eq. (15.1), remembering that the steric repulsion will cause a net increase in energy). Consequently the 4 + 2 reaction should have the lowest activation energy, and therefore occur easier than the 2 + 2. This is indeed what is observed, the Diels–Alder reaction occurs readily, but cyclobutane formation is not observed between non-polar dienes and dieneophiles.

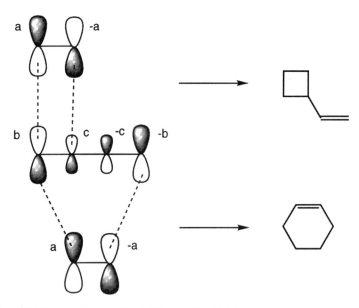

Figure 15.4 FMO theory favours the $4 + 2$ over the $2 + 2$ reaction

The appearance of the difference in MO energies in the denominator in eq. (15.1) suggests that a smaller gap between the diene HOMO and dieneophile LUMO in a Diels–Alder reaction should lower the activation energy. If the diene is made more electron rich (electron donating substituents), or the dieneophile more electron deficient (electron withdrawing substituents), the reaction should proceed faster. This is indeed the observed trend. For the reaction between cyclopentadiene and cyanoethylenes (mono-, di-, tri- and tetra-substituted), the correlation is reasonably quantitative, as shown in Figure 15.5.[5]

FMO theory can also be used for explaining the stereochemistry of the Diels–Alder reaction. Consider the reaction between 2-methyl butadiene and cyanoethylene. These may react to give two different products, the "para" and/or "meta" isomer. The MO coefficients for the p-orbitals on the isoprene HOMO and cyanoethylene LUMO (taken from AM1 calculations) are given in Figure 15.6. The FMO sum for the "para" isomer is $(0.594 \cdot 0.682 + 0.517 \cdot 0.552)^2 = 0.477$, while the sum for the "meta" isomer is $(0.594 \cdot 0.552 + 0.517 \cdot 0.682)^2 = 0.463$. FMO theory thus predicts that the "para" isomer should dominate, as is indeed observed (experimental ratio $70 : 30$). If cyanoethylene is replaced by 1,1-dicyanoethylene, the LUMO coefficients change to 0.708 and -0.511. The corresponding "para" and "meta" FMO sums change to 0.469 and 0.448, i.e. a larger difference between the two isomers. This is again reflected in the experimental data, the ratio is now $91 : 9$. The regio chemistry is thus determined by matching the two largest sets of coefficients and the two smaller sets, rather than making two sets of large/small.

FMO theory was developed at a time when detailed calculations of reaction paths were infeasible. As many sophisticated computational models, and methods for actually locating the TS, have become widespread, the use of FMO arguments for predicting reactivity has declined. The primary goal of computational chemistry, however, is not to

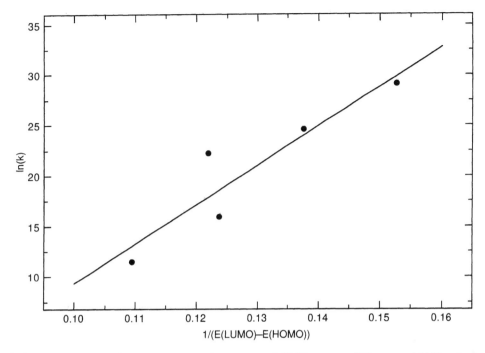

Figure 15.5   Correlation between reaction rates and FMO energy differences (AM1 calculations)

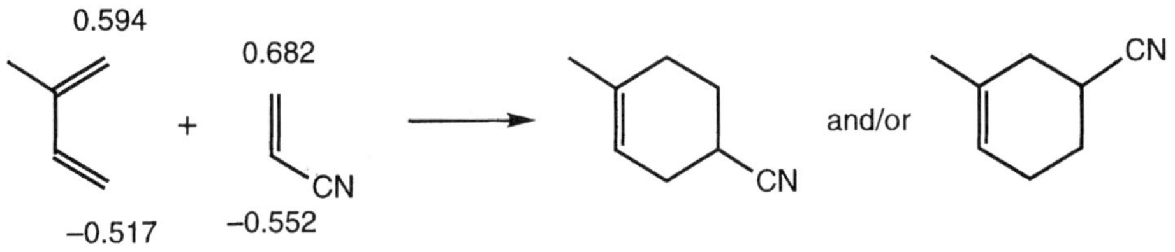

Figure 15.6   FMO rationalizes the stereochemistry of substituted Diels–Alder reactions

provide <u>numbers</u>, but to provide <u>understanding</u>. As such FMO theory still forms a conceptual model which can be used for rationalizing trends without having to perform time-consuming calculations.

15.2   Concepts from Density Functional Theory

The success of FMO theory is not because the neglected terms in the second-order perturbation expansion (eq. (15.1)) are especially small; an actual calculation will reveal that they completely swamp the HOMO–LUMO contribution. The deeper reason is that the shapes of the HOMO and LUMO resemble features in the total electron density, which determines the reactivity.

A reaction will in general involve a change in the electron density, which may be quantified in terms of the *Fukui function*.[6]

$$f(\mathbf{r}) = \frac{\partial \rho(\mathbf{r})}{\partial N} \tag{15.2}$$

The Fukui function indicates the change in the electron density at a given position when the number of electrons is changed. We may define two finite difference versions of the function, corresponding to addition or removal of an electron.

$$\begin{aligned} f_+(\mathbf{r}) &= \rho_{N+1}(\mathbf{r}) - \rho_N(\mathbf{r}) \\ f_-(\mathbf{r}) &= \rho_N(\mathbf{r}) - \rho_{N-1}(\mathbf{r}) \end{aligned} \tag{15.3}$$

The f_+ function is expected to reflect the initial part of a nucleophilic reaction, and the f_- function an electrophilic reaction, i.e. the reaction will typically occur where the $f_\pm$ function is large.[7] For radical reactions the appropriate function is an average of f_+ and f_-.

$$f_0(\mathbf{r}) = \tfrac{1}{2}(f_+(\mathbf{r}) + f_-(\mathbf{r})) = \tfrac{1}{2}(\rho_{N+1}(\mathbf{r}) - \rho_{N-1}(\mathbf{r})) \tag{15.4}$$

The $f_\pm$ functions may be written in terms of orbital contributions.

$$\begin{aligned} f_+(\mathbf{r}) &= \phi_{\mathrm{LUMO}}^2(\mathbf{r}) + \sum_{i=1}^{N} \frac{\partial \phi_i^2(\mathbf{r})}{\partial n_i} \\ f_-(\mathbf{r}) &= \phi_{\mathrm{HOMO}}^2(\mathbf{r}) + \sum_{i=1}^{N-1} \frac{\partial \phi_i^2(\mathbf{r})}{\partial n_i} \end{aligned} \tag{15.5}$$

In the frozen MO approximation the last terms are zero and the Fukui functions are given directly by the contributions from the HOMO and LUMO. The preferred site of attack is therefore at the atom(s) with the largest MO coefficients in the HOMO/LUMO, in exact agreement with FMO theory. The Fukui function(s) may be considered as the equivalent (or generalization) of FMO methods within Density Functional Theory (Chapter 6).

In the Atoms In Molecules approach (Section 9.3), the Laplacian ∇^2 (trace of the second derivative matrix with respect to the coordinates) of the electron density measures the local increase or decrease of electrons. Specifically, if $\nabla^2\rho$ is negative, it marks an area where the electron density is locally concentrated, and therefore susceptible to attack by an electrophile. Similarly, if $\nabla^2\rho$ is positive, it marks an area where the electron density is locally depleted, and therefore susceptible to attack by a nucleophile. It has in general been found that such maps of negative values of $\nabla^2\rho$ resemble the shape of the HOMO, and a map of positive values of $\nabla^2\rho$ resemble the shape of the LUMO.

The fact that features in the total electron density are closely related to the shapes of the HOMO and LUMO provides a much better rationale of why FMO theory works as well as it does, than does the perturbation derivation. It should be noted, however, that improvements in the wave function do not necessarily lead to a better performance of the FMO method. Indeed the use of MOs from semi-empirical methods usually works better than data from *ab initio* wave functions. Furthermore it should be kept in mind that only the HOMO orbital converges to a specific shape and energy as the basis set is improved in an *ab initio* calculation, the LUMO is normally determined by the most diffuse functions in the basis.

Besides the already mentioned Fukui function, there are a couple of other commonly used concepts which can be connected with Density Functional Theory (Chapter 6).[8] The *electronic chemical potential* μ is given as the first derivative of the energy with respect to the number of electrons, which in a finite difference version is given as half the sum of the ionization potential and the electron affinity. Except for a difference in sign, this is exactly the Mulliken definition of *electronegativity*.[9]

$$\mu = \frac{\partial E}{\partial N} \simeq -\frac{I+A}{2} \tag{15.6}$$

The second derivative of the energy with respect to the number of electrons is the *hardness* η (the inverse quantity η^{-1} is called the *softness*), which again may be approximated in term of the ionization potential and electron affinity.

$$\eta = \frac{1}{2}\frac{\partial^2 E}{\partial N^2} \simeq \frac{I-A}{2} \tag{15.7}$$

These concepts play an important role in the *Hard and Soft Acid and Base* (HSAB) principle, which states that hard acids prefer to react with hard bases, and vice versa.[10] By means of Koopmann's theorem (Section 3.4) the hardness is related to the HOMO–LUMO energy difference, i.e. a small gap indicates a "soft" molecule. From second-order perturbation theory it also follows that a small gap between occupied and unoccupied orbitals will give a large contribution to the polarizability (Section 10.6), i.e. softness is a measure of how easily the electron density can be distorted by external fields, for example those generated by another molecule. In terms of the perturbation equation (15.1), a hard–hard interaction is primarily charge controlled, while a soft–soft interaction is orbital controlled. Both FMO and HSAB theories may be considered as being limiting cases of chemical reactivity described by the Fukui function.[11]

15.3 Qualitative Molecular Orbital Theory

Frontier Molecular Orbital theory is closely related to various schemes of qualitative orbital theory where interactions between fragment MOs are considered.[12] Ligand field theory, as commonly used in systems involving coordination to metal atoms, can be considered as a special case where only the d-orbitals on the metal and selected orbitals of the ligands are considered.

Two interacting orbitals will in general produce two new orbitals, with lower and higher energies than the non-interacting orbitals. The magnitude of the changes is determined by the initial orbital energies ε_a and ε_b and the overlap S_{ab}. The overlap depends on the symmetries of the orbitals (orbitals of different symmetry have zero overlap), and the distance between them (the shorter the distance, the larger the overlap). The energies of the new orbitals can be calculated from the variational principle, and the qualitative result is shown in Figure 15.7:

$$\Delta \propto \frac{|S_{ab}|}{|\varepsilon_a - \varepsilon_b|} \tag{15.8}$$

There are two important features. The change in orbital energies is dependent on the

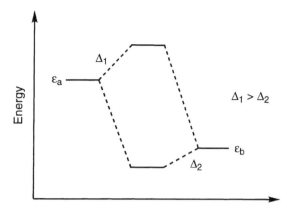

Figure 15.7 Linear combination of two orbitals leads to two new orbitals with different energies

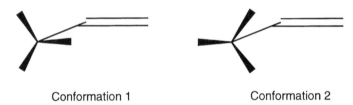

Conformation 1   Conformation 2

Figure 15.8 Possible propene conformations

magnitude of the overlap, $|S_{ab}|$, and inversely proportional to the energy difference of the original orbitals, $|\varepsilon_a - \varepsilon_b|$. Furthermore, the effect is largest for the highest energy orbital (anti-bonding combination), i.e. $\Delta_1 > \Delta_2$.

If the two initial orbitals contain one, two or three electrons, the interaction will lead to a lower energy, with the stabilization being largest for the case of two electrons (e.g. a filled orbital interacting with an empty orbital). If both initial orbitals are fully occupied, the interaction will be destabilizing, i.e. a steric type repulsion. By adapting a set of HOMO and LUMO orbitals for atomic or molecular fragments, the favourable interactions may be identified based on overlap and energy considerations. Qualitative MO theory may thus be considered as an intramolecular form of FMO theory, with suitable chosen fragments. Consider for example the two conformations for propene shown in Figure 15.8: which should be the more stable?

By "chemical intuition", the most important interaction is likely to be between the (filled) hydrogen s-orbitals and the (empty) π-orbital. The CH_3 group as a fragment has C_{3v} symmetry, and the three proper (symmetry adapted) linear combinations of the s-orbitals, together with the anti-bonding π-orbital, are given in Figure 15.9. The ϕ_1 orbital is lowest in energy, while the ϕ_2 and ϕ_3 orbitals are degenerate in perfect C_{3v} symmetry. The ϕ_1 and ϕ_2 orbitals have a different symmetry than the π^*-orbital and can consequently not interact ($S = 0$). The interaction of the ϕ_3 orbital with the π^*-system in the two conformations is shown in Figure 15.10.

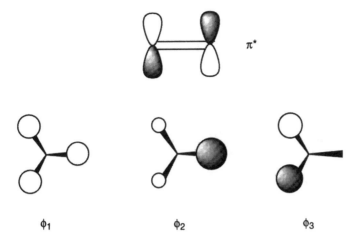

Figure 15.9 Fragment orbitals for propene

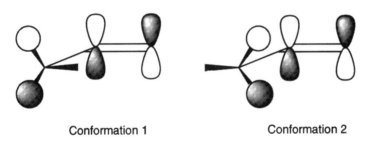

Conformation 1   Conformation 2

Figure 15.10 Fragment orbital interaction

The overlap between the nearest carbon p-orbital and ϕ_3 is the largest contribution, but it is the same in the two conformations. The overlap with the distant carbon p-orbital is of opposite sign, and largest in conformation 2 (the distance is shorter). The total overlap between the ϕ_3 and π^*-orbitals is thus largest for conformation 1, which implies a larger stabilizing interaction, and it should consequently be lowest in energy. Indeed, conformation 2 is a TS for interconverting equivalent conformations corresponding to 1.

It is important to realize that whenever qualitative or frontier molecular orbital theory is invoked, the description is within the orbital (Hartree–Fock or Density Functional) model for the electronic wave function. In other words, rationalizing a trend in computational results by qualitative MO theory is only valid if the effect is present at the HF or DFT level. If the majority of the variation is due to electron correlation, an explanation in terms of interacting orbitals is not appropriate.

15.4 Woodward–Hoffmann Rules

The *Woodward–Hoffmann* (WH) rules[13] are qualitative statements regarding relative activation energies for two possible modes of reaction, which may have different stereochemical outcomes. For simple systems the rules may be derived from a

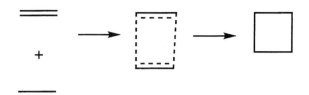

Figure 15.11   Reaction of two ethylenes to form cyclobutane under C_{2v} symmetry

conservation of orbital symmetry, but they may also be generalized by an FMO treatment with conservation of bonding. Let us illustrate the Woodward–Hoffmann rules by a couple of examples, the preference of the 4 + 2 over the 2 + 2 product for the reaction of butadiene with ethylene, and the ring-closure of butadiene to cyclobutene.

A face-to-face reaction of two π-orbitals to form a cyclobutane involves formation of two new C–C σ-bonds. The reaction may be imagined to occur under the preservation of symmetry, in this case C_{2v}, i.e. *concerted* (one-step, no intermediates) and *synchronous* (both bonds are formed at the same rate). Both the reactant and product orbitals may be classified according to their behaviour with respect to the two mirror planes present, being either symmetric (no change of sign) or anti-symmetric (change of sign). The energetic ordering of the orbitals follows from a straightforward consideration of the bonding/anti-bonding properties. Since orbitals of different symmetries cannot mix, conservation of orbital symmetry establishes the correlation between the reactant and product sides.

The *orbital correlation diagram* shown in Figure 15.12 indicates that an initial electron configuration of $(\pi_1 + \pi_2)^2(\pi_1 - \pi_2)^2$ (ground state for the reactant) will end up as a doubly excited configuration $(\sigma_1 + \sigma_2)^2(\sigma_1^* + \sigma_2^*)^2$ for the cyclobutane product. This by itself indicates that the reaction should be substantially uphill in terms of energy. It may be put on a more sound theoretical footing by looking at the *state correlation diagram* in Figure 15.13. The ground-state wave function for the whole system (all four active + remaining core and valence electrons) is symmetric with respect to both mirror planes, while the first excited state is anti-symmetric. The intended correlation is indicated with dashed lines, the lowest energy configuration for the reactant correlates with a doubly excited configuration of the product, and vice versa. Since these configurations have the same symmetry (SS), an avoided crossing is introduced, leading to a significant barrier for the reaction. The presence of a reaction barrier due to symmetry conservation for the orbitals makes this a Woodward–Hoffmann *forbidden* reaction. The reaction for the excited state, however, does not encounter a barrier, and is therefore denoted an *allowed* reaction.

The same conclusion may be reached directly from a consideration of the frontier orbitals. Formation of two new σ-bonds requires interaction of the HOMO of one fragment with the LUMO on the other. When the interaction is between orbital lobes on the same side (*suprafacial*) of each fragment (2s + 2s), this leads to the picture shown in Figure 15.14. It is clearly seen that the HOMO–LUMO interaction leads to formation of one bonding and one anti-bonding orbital, i.e. this is not a favourable interaction. The FMO approach also suggests that the 2 + 2 reaction might be possible if it could occur with bond formation on opposite sides (*antarafacial*) for one of the fragments as shown

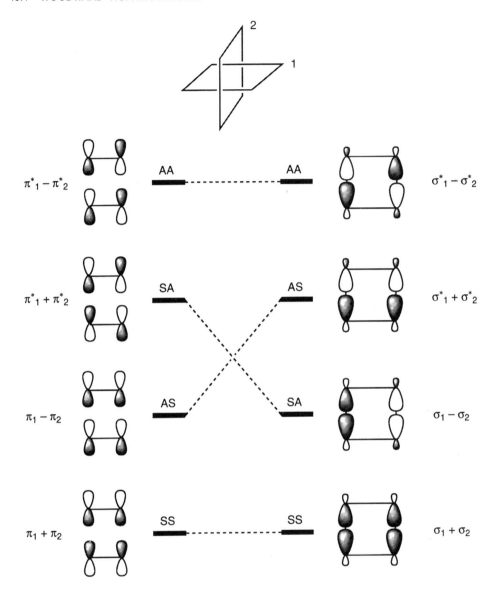

Figure 15.12   Orbital correlation diagram for cyclobutane formation

in Figure 15.15. Although the 2s + 2a reaction is Woodward–Hoffmann allowed, it is sterically so hindered that thermal 2 + 2 reactions in general are not observed. Photochemical 2 + 2 reactions, however, are well known.[14]

The 4s + 2s reaction of a diene with a double bond can in a concerted and synchronous reaction be envisioned to occur with the preservation of C_s symmetry, Figure 15.16. The corresponding orbital correlation diagram is shown in Figure 15.17. In this case the orbital correlation diagram shows that the lowest energy electron configuration in the reactant, $(\pi_1)^2(\pi_2)^2(\pi_3)^2$, correlates directly with the lowest energy electron configuration in the product, $(\sigma_1)^2(\sigma_2)^2(\pi_1)^2$. This is also shown by

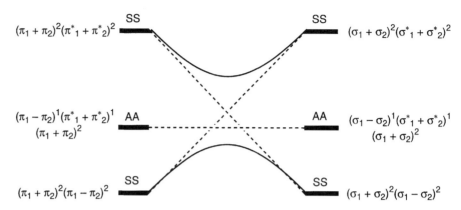

Figure 15.13   State correlation diagram for cyclobutane formation

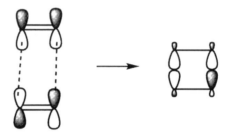

Figure 15.14   2s + 2s HOMO–LUMO interaction leading to two new σ-bonds

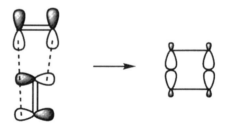

Figure 15.15   2s + 2a HOMO–LUMO interaction leading to two new σ-bonds

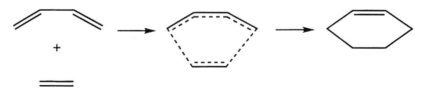

Figure 15.16   Reaction of butadiene and ethylene to form cyclohexene under C_s symmetry

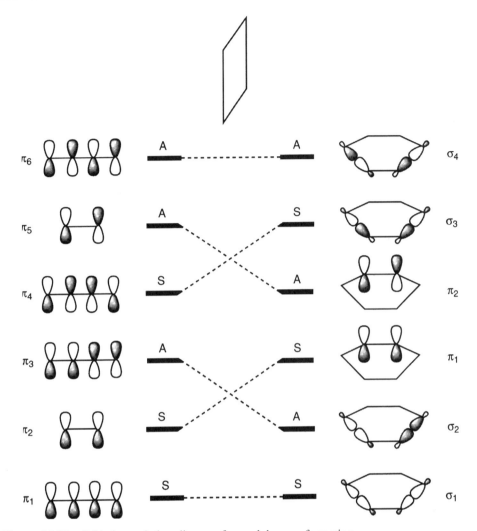

Figure 15.17 Orbital correlation diagram for cyclohexene formation

the corresponding state correlation diagram, Figure 15.18. In this case there is no energetic barrier due to unfavourable orbital correlation, although other factors lead to an activation energy larger than zero. The direct correlation of ground-state configurations for the reactant and product indicates a (relatively) easy reaction, and it is an allowed reaction. The lowest excited state for the reactant, however, does not correlate with the lowest excited product state, and the photochemical reaction is consequently forbidden.

The FMO approach shown in Figure 15.19 again indicates that the 4s + 2s interaction should lead directly to formation of two new bonding σ-bonds, i.e. this is an allowed reaction. The preference for a concerted 4s + 2s reaction is experimentally supported by observations which show that the stereochemistry of the diene and dienophile is carried over to the product, e.g. a *trans,trans*-1,4-disubstituted diene

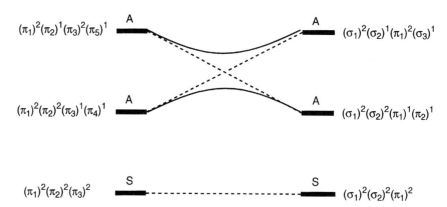

Figure 15.18   State correlation diagram for cyclohexene formation

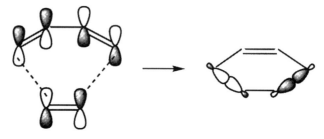

Figure 15.19   4s + 2s HOMO-LUMO interaction leading to two new σ-bonds

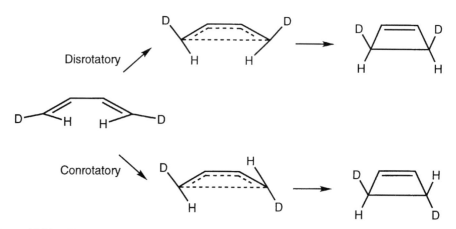

Figure 15.20   Two possible modes of closing a diene to cyclobutene

results in the two substituents ending up in a *cis* configuration in the cyclohexene product.[15]

The ring closure of a diene to a cyclobutene can occur with rotation of the two termini in the same (*conrotatory*) or opposite (*disrotatory*) directions. For suitable substituted compounds, these two reaction modes lead to products with different stereochemistry.

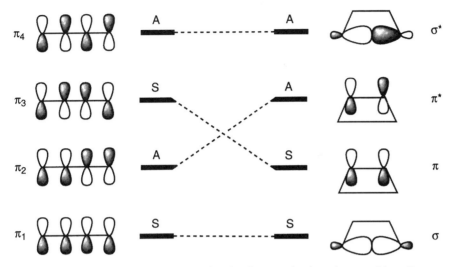

Figure 15.21 Orbital correlation diagram for the disrotatoric ring closure of butadiene

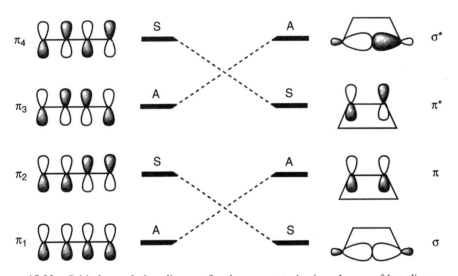

Figure 15.22 Orbital correlation diagram for the conrotatoric ring closure of butadiene

The disrotatory path has C_s symmetry during the whole reaction, while the conrotatory mode preserves C_2 symmetry. The orbital correlation diagrams for the two possible paths are shown in Figures 15.21 and 15.22. It is seen that only the conrotatory path directly connects the reactant and product ground-state configurations. Taking into account also the excited states leads to the state correlation diagram in Figure 15.23. The conrotatory path is Woodward–Hoffmann allowed for a thermal reaction, while the corresponding photochemical reaction is predicted to occur in a disrotatory fashion.

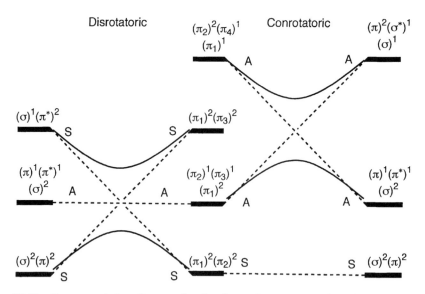

Figure 15.23   State correlation diagram for the dis- and conrotatory ring-closure of butadiene

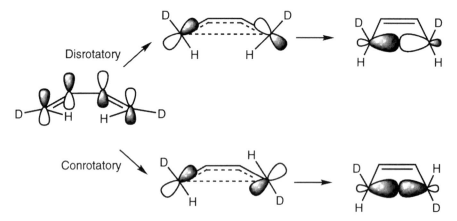

Figure 15.24   HOMO orbital for the ring closure of butadiene

The same conclusion may again be reached by considering only the HOMO orbital, Figure 15.24. For the conrotatory path the orbital interaction leads directly to a bonding orbital, while the orbital phases for the disrotatory motion lead to an anti-bonding orbital.

While the orbital and state diagrams can only be rigorously justified in the simple parent system, where symmetry is present, the addition of substituents normally only alters the shape of the relevant orbitals slightly. The nodal structure of the orbitals is preserved for a large range of substituted systems, and the "preservation of bonding" displayed by the FMO type diagrams consequently has a substantially wider predictive range. It may be used for analysing reactions where there is no symmetry element present in the whole reaction, as for example in the [1,5]-hydrogen shift in 1,3-pentadiene.

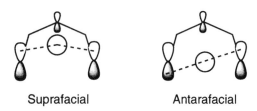

Figure 15.25 FMO interactions for the [1,5]-hydrogen shift in 1,3-pentadiene

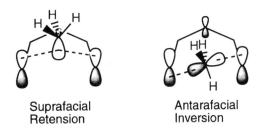

Figure 15.26 FMO interactions for allowed modes of [1,5]-methyl shift in 1,3-hexadiene

In the suprafacial migration the interaction of the pentadienyl radical singly occupied orbital with the hydrogen s-orbital is seen to involve breaking and making bonds where the orbital phases match, Figure 15.25. For the antarafacial path, however, the orbital in the product ends up being anti-bonding, i.e. a [1,5]-hydrogen migration is predicted to occur suprafacially, in agreement with experiments.[16]

In the general case the transferring group may migrate with either *retension* or *inversion* of its stereochemistry. A [1,5]-CH_3 migration, for example, is thermally allowed if it occurs suprafacially with retension of the CH_3 configuration, or if it occurs antarafacially with inversion of the methyl group, as shown in Figure 15.26.

The Woodward–Hoffmann allowed reactions can be classified according to how many electrons are involved, and whether the reaction occurs thermally or photochemically, as shown in Table 15.1.

The state correlation diagrams give an indication of the minimum theoretical level necessary for describing a reaction. For allowed reactions, the reactant configuration

Table 15.1 Woodward–Hoffmann allowed reactions

Reaction type	Number of electrons	Thermally allowed	Photochemically allowed
Ringclosure	$4n$	Conrotatory	Disrotatory
	$4n+2$	Disrotatory	Conrotatory
Cycloadditions	$4n$	Supra–Antara or Antara–Supra	Supra–Supra or Antara–Antara
	$4n+2$	Supra–Supra or Antara–Antara	Supra–Antara or Antara–Supra
Migrations	$4n$	Antara–Retension or Supra–Inversion	Supra–Retension or Antara–Inversion
	$4n+2$	Supra–Retension or Antara–Inversion	Antara–Retension or Supra–Inversion

smoothly transforms into the product configuration by a continuous change of the orbitals, and they are consequently reasonably described by a single determinant wave function along the whole reaction path. Forbidden reactions, on the other hand, necessarily involve at least two configurations, since there is no continuous orbital transformation which connects the reactant and product ground states. Such reactions therefore require MCSCF type wave functions for a qualitatively correct description.

While the state correlation diagram for the 2s + 2s reaction (Figure 15.13) indicates that the photochemical reaction should be allowed (and cyclobutanes are indeed observed as one of the products from such reactions), the implication that the product ends up in an excited state is not correct. Although the reaction starts out on the excited surface, it will at some point along the reaction path return to the lowest energy surface, and the product is formed in its ground state. The transition from the upper to the lower energy surface will normally occur at a geometry where the two surfaces "touch" each other, i.e. they have the same energy; this is known as a *conical intersection*.[17] Achieving the proper geometry for a transition between the two surfaces is often the dynamical bottleneck, and a conical intersection may be considered as the equivalent of a TS for a photochemical reaction. As conical intersections involve two energy surfaces, MCSCF based methods are required, and non-adiabatic coupling elements (Section 3.1) are important. The actual location of a geometry corresponding to a conical intersection for a multi-dimensional system may be determined using constrained optimization techniques (Section 14.6).

15.5 The Bell–Evans–Polanyi Principle/Hammond Postulate/ Marcus Theory

The simpler the idea, the more names, could be the theme of this section.[18] The overriding idea is simple. For similar reactions the more exothermic (endothermic)

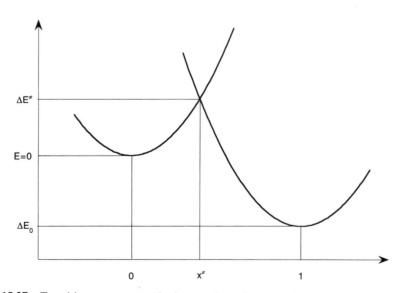

Figure 15.27   Transition structure as the intersection of two parabolas

reaction will have the lower (higher) activation energy. This was formulated independently by *Bell*, *Evans* and *Polanyi* (BEP) in the thirties,[19] and is commonly known as the *BEP principle*. The *Hammond postulate* relates the position of the transition structure to the exothermicity: for similar reactions the more exothermic (endothermic) reaction will have the earlier (later) TS.[20] Compared to FMO theory, which tries to estimate relative activation energies from the <u>reactant</u> properties, the BEP principle tries to estimate relative activation energies from <u>product</u> properties (reaction energies). The above qualitative statements have been put on a more quantitative footing by the *Marcus equation*. The Marcus equation was originally derived for electron transfer reactions,[21] but it has since been shown that the same equation can be derived from a number of different assumptions, three of which will be illustrated below.

Let us assume a reaction coordinate x running from 0 (reactant) to 1 (product). The energy of the reactant as a function of x is taken as a simple parabola with a "force constant" of a. The energy of the product is also taken as a parabola with the same force constant, but offset by the reaction energy ΔE_0. The position of the TS ($x^{\neq}$) is taken as the point where the two parabola intersect, as shown in Figure 15.27. The TS position is calculated by equating the two energy expressions.

$$
\begin{aligned}
E_{\text{reactant}} &= a(x)^2 \\
E_{\text{product}} &= a(x-1)^2 + \Delta E_0 \\
a(x^{\neq})^2 &= a(x^{\neq}-1)^2 + \Delta E_0 \\
x^{\neq} &= \frac{1}{2} + \frac{\Delta E_0}{2a}
\end{aligned}
\tag{15.9}
$$

For a thermoneutral reaction ($\Delta E_0 = 0$) the TS is exactly half-way between the reactant and product (as expected), while it becomes earlier and earlier as the reaction becomes more and more exothermic (ΔE_0 negative). The activation energy is given as

$$
\begin{aligned}
E(x^{\neq}) &= a\left(\frac{1}{2} + \frac{\Delta E_0}{2a}\right)^2 \\
E(x^{\neq}) &= \frac{a}{4} + \frac{\Delta E_0}{2} + \frac{\Delta E_0^2}{4a}
\end{aligned}
\tag{15.10}
$$

Let us define the activation energy for a (possible hypothetical) thermoneutral reaction as the *intrinsic activation energy*, $\Delta E_0^{\neq}$. As seen from eq. (15.10), $a = 4\Delta E_0^{\neq}$. The TS position and activation energy now become

$$
\begin{aligned}
x^{\neq} &= \frac{1}{2} + \frac{\Delta E_0}{8\Delta E_0^{\neq}} \\
\Delta E^{\neq} &= \Delta E_0^{\neq} + \frac{\Delta E_0}{2} + \frac{\Delta E_0^2}{16\Delta E_0^{\neq}}
\end{aligned}
\tag{15.11}
$$

The latter is, except for a couple of terms related to solvent reorganization, the Marcus equation. The central idea is that the activation energy can be decomposed into a component characteristic of the reaction type, the intrinsic activation energy, and a correction due to the reaction energy being different from zero. Similar reactions should have similar intrinsic activation energies, and the Marcus equation obeys both the BEP principle and the Hammond postulate. Except for very exo- or endothermic reactions (or a very small $\Delta E_0^{\neq}$), the last term in the Marcus equation is small, and it is seen that roughly half the reaction energy enters the activation energy. Note, however, that the activation energy is a parabolic function of the reaction energy. Thus for sufficiently exothermic reactions the equation predicts that the activation energy should <u>increase</u> as the reaction becomes <u>more</u> exothermic. The turnover occurs when $\Delta E_0 = -4\Delta E_0^{\neq}$. Much research has gone into proving such an "inverted" region, however, experiments with very exothermic reactions are difficult to perform.[22]

An alternative way of deriving the Marcus equation is again to assume a reaction coordinate running from 0 to 1. The intrinsic activation energy is taken as a parabola centred at $x = 1/2$. The reaction energy is taken as progressing linearly along the reaction coordinate, Figure 15.28. Adding these two contributions, and evaluating the position of the TS and the activation energy in terms of ΔE_0 and $\Delta E_0^{\neq}$, again leads to the Marcus equation.

Actually the assumptions can be made even more general. The energy as a function of the reaction coordinate can always be decomposed into an "intrinsic" term, which is symmetric with respect to $x = 1/2$, and a "thermodynamic" contribution, which is anti-symmetric. Denoting these two energy functions h_2 and h_1, it can be shown that the Marcus equation can be derived from the "square" condition, $h_2 = h_1^2$.[23] The intrinsic and thermodynamic parts do not have to be parabolas and linear functions, as in Figure 15.28; they can be <u>any</u> type of function. As long as the intrinsic part is the square of the thermodynamic part, the Marcus equation is recovered. The idea can be taken one step further. The h_2 function can always be expanded in a power series of even powers of h_1, i.e. $h_2 = c_2 h_1^2 + c_4 h_1^4 + \ldots$. The exact values of the c-coefficients only influence the appearance of the last term in the resulting Marcus-like equation (eq. (15.11)). As already mentioned, this is usually a small correction anyway. For reactions where the reaction energy is less than or similar to the activation energy, there is thus a quite general theoretical background for the following statement. <u>For similar reactions the</u>

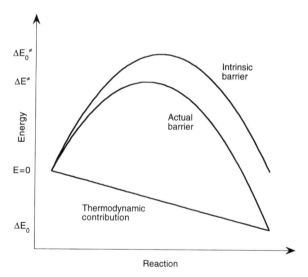

Figure 15.28 Decomposition of a reaction barrier into a parabola and a linear term

difference in activation energy is roughly half the difference in reaction energy. The trouble here is the word "similar". How similar should reactions be in order for the intrinsic activation energy to be constant? And how do we calculate or estimate the intrinsic activation energy? We will return to the latter question shortly.

The Marcus equation provides a nice conceptual tool for understanding trends in reactivity.[24] Consider for example the degenerate Cope rearrangement of 1,5-hexadiene and the ring-opening of Dewar benzene (bicyclo-[2,2,0]hexa-2,5-diene) to benzene, Figure 15.29. The experimentally observed activation energies are 34 kcal/mol and 23 kcal/mol, respectively.[25] The Cope reaction is an example of a Woodward–Hoffmann allowed reaction ([3,3]-sigmatropic shift) while the ring-opening of Dewar benzene is a Woodward–Hoffmann forbidden reaction (the cyclobutene ring-opening must necessarily be disrotatory, otherwise the benzene product ends up with a *trans* double bond). How come a forbidden reaction has a lower activation energy than an allowed reaction? This is readily explained by the Marcus equation. The Cope reaction is thermoneutral (reactant and product are identical) and the activation energy is purely intrinsic, while the ring-opening is exothermic by 71 kcal/mol, and therefore has an intrinsic barrier of 52 kcal/mol. The "forbidden" reaction occurs only because it has a huge driving force in terms of a much more stable product, while the allowed reaction occurs even without a net energy gain.

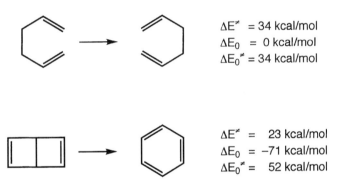

$$\Delta E^{\neq} = 34 \text{ kcal/mol}$$
$$\Delta E_0 = 0 \text{ kcal/mol}$$
$$\Delta E_0^{\neq} = 34 \text{ kcal/mol}$$

$$\Delta E^{\neq} = 23 \text{ kcal/mol}$$
$$\Delta E_0 = -71 \text{ kcal/mol}$$
$$\Delta E_0^{\neq} = 52 \text{ kcal/mol}$$

Figure 15.29   The Cope rearrangement and Dewar benzene ring-opening reaction

The goal of understanding chemical reactivity is to be able to predict how the activation energy depends on properties of the reactant and product. Decomposing the activation energy into two terms, an intrinsic and a thermodynamic contribution, does not solve the problem. The reaction energy is relatively easy to obtain, from experiments, various theoretical methods or estimates based on additivity. But how does one estimate the intrinsic activation energy? It is purely a theoretical concept, the activation energy for a thermoneutral reaction. But most reactions are <u>not</u> thermoneutral, and there is no way of measuring such an intrinsic activation energy. For a series of "closely related" reactions it may be assumed to be constant, but the question then becomes: how closely related should reactions be? Alternatively it may be assumed that the intrinsic component can be taken as an average of the two corresponding *identity* reactions. Consider for example the S_N2 reaction of OH^- with CH_3Cl. The two identity reactions are $OH^- + CH_3OH$ and $Cl^- + CH_3Cl$. These two reactions are thermoneutral and their activation energies, which are purely intrinsic, can in principle be measured by

Table 15.2 Comparing experimental activation barriers with those calculated by the Marcus equation

	$\Delta G^{\neq}$ (identity)	ΔG_0 (exp.)	$\Delta G^{\neq}$ (Marcus)	$\Delta G^{\neq}$ (exp.)
OH^-	40.5			
F^-	31.8	-22.5	25.8	26.1
Cl^-	26.5	-22.0	23.4	24.6
Br^-	23.7	-23.4	21.5	22.7
I^-	23.2	-21.3	22.1	23.2

isotopic substitution (for example $^{35}Cl^- + CH_3{}^{37}Cl \rightarrow CH_3{}^{35}Cl + {}^{37}Cl^-$). From the reaction energy for the $OH^- + CH_3Cl$ reaction, and the assumption that the intrinsic barrier is the average of the two identity reactions, the activation energy can be calculated. An example of the accuracy of this procedure for the series of S_N2 reactions $OH^- + CH_3X$ is given in Table 15.2.

Again this averaging procedure can only be expected to work when the reactions are sufficiently "similar". This is difficult to quantify *a priori*. The Marcus equation is therefore more a conceptual tool for explaining trends, than for deriving quantitative results.

15.6 More O'Ferrall–Jenks Diagrams

The BEP/Hammond/Marcus treatment only considers changes due to energy differences between the reactant and product, i.e. changes in the TS position along the reaction coordinate. It is often useful also to include changes that may occur in a direction perpendicular to the reaction coordinate. Such two-dimensional diagrams are associated with the names of More O'Ferrall and Jenks *(MOJ diagrams)*.[26]

Consider for example the Cope rearrangement of 1,5-hexadiene. Since the reaction is degenerate the TS will have D_{2h} symmetry (the lowest energy TS has a conformation resembling a chair-like cyclohexane). It is, however, not clear how strong the forming and breaking C–C bonds are at the TS. If they both are essentially full C–C bonds, the reaction may be described as bond formation followed by bond breaking. The TS therefore has character of being a 1,4-biradical, as illustrated by path **B** in Figure 15.30. Alternatively, the C–C bonds may be very weak at the TS; this corresponds to a situation where the TS can be described as two weakly interacting allyl radicals (path **C**). The intermediate situation, where both bonds are roughly half formed/broken can be described as having a delocalized structure similar to benzene, i.e. an "aromatic" type TS (path **A**).

In such MOJ diagrams the *x*- and *y*-coordinates are normally taken to be bond orders (Section 9.1) or $1 -$ bond order for the breaking and forming bonds, so that the coordinate runs from 0 to 1. A third axis corresponding to the energy is implied, but rarely drawn.

At the TS the energy along the reaction path is a maximum, but it is a minimum in the perpendicular direction(s). A one-dimensional cut through the (0,0) and (1,1) corners for path **A** in Figure 15.30 thus corresponds to Figure 15.28. A similar cut through the (0,1) and (1,0) corners will display a normal (as opposed to inverted) parabolic behaviour,

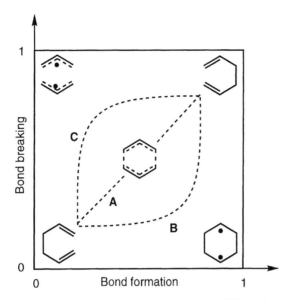

Figure 15.30   MOJ diagram for the Cope rearrangement of 1,5-hexadiene

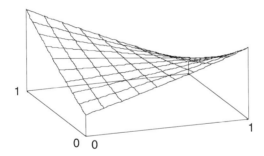

Figure 15.31   MOJ diagram corresponding to Figure 15.30 with the energy as the vertical axis

with the TS being at the minimum on the curve. The whole energy surface corresponding to Figure 15.29 will have the qualitative appearance shown in Figure 15.31.

There is good evidence that the Cope reaction in the parent 1,5-hexadiene has an "aromatic" type TS, corresponding to path **A** in Figure 15.30, i.e. a "central" or "diagonal" reaction path. The importance of MOJ diagrams is that they allow a qualitative prediction of changes in the TS structure for a series of similar reactions. Addition of substituents which stabilize the product relative to the reactant corresponds to a lowering of the (1,1) corner, thereby moving the TS closer to the (0,0) corner, i.e. towards the reactant. The one-dimensional BEP/Hammond/Marcus treatment thus corresponds to changes along the (0,0)-(1,1) diagonal.

Substituents which do not change the overall reaction energy may still have an influence on the TS geometry. Consider for example 2,5-diphenyl-1,5-hexadiene. The reaction is still thermoneutral, but the phenyl groups will preferentially stabilize the 1,4-biradical structure, i.e. lower the energy of the (1,0) corner. From Figure 15.31 it is clear that this will lead to a TS which is shifted towards this corner, i.e. moving the reaction from path **A** towards **B** in Figure 15.30. Similarly, substituents which preferentially stabilize the bis-allyl radical structure (like 1,4-diphenyl-1,5-hexadiene) will perturb the reaction towards path **C**, since the (0,1) corner is lowered in energy relative to the other corners.

From such MOJ diagrams it can be inferred that changes in the system which alter the relative energy <u>along</u> the reaction diagonal (lower-left to upper-right) imply changes in the TS in the <u>opposite</u> direction. Changes which alter the relative energy <u>perpendicular</u> to the reaction diagonal (upper-left to lower-right) imply changes in the TS in the <u>same</u> direction as the perturbation.

The structures in the (1,0) and (0,1) corners are not necessarily stable species, they may correspond to hypothetical structures. In the Cope rearrangement it appears that the reaction only involves a single TS, independently of the number and nature of substituents. The reaction path may change from **B** → **A** → **C** depending on the system, but there are no intermediates along the reaction coordinate.

In other cases one or both of the perpendicular corners may correspond to a minimum on the potential energy surface, and the reaction mechanism can change from being a one-step reaction to a two-step. An example of this would be elimination reactions. The x-axis in this case corresponds to the breaking bond between carbon and hydrogen, while the y-axis is the breaking bond between the other carbon and the leaving group. An E2 type reaction has simultaneous breaking of the C–H and C–L bonds while forming the B–H bond, and corresponds to the diagonal path **A** in Figure 15.32. Path **C** involves initial loss of the leaving group to form a carbocation (upper-left corner), followed by loss of H^+ (which is picked up by the base), i.e. this corresponds to an E1 type mechanism involving two TSs and an intermediate. Path **B**, on the other hand, involves formation of a carbanion, followed by elimination of the leaving group in a second step, i.e. an E1$_{cb}$ mechanism. Substituents which stabilize the carbocation thus

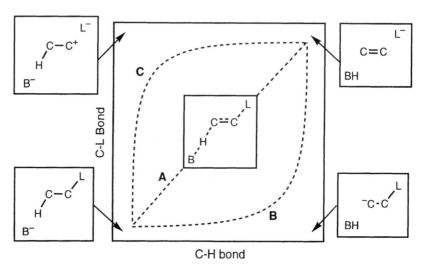

Figure 15.32 MOJ diagram for elimination reactions

shift the reaction from an E2 to an E1 type mechanism, while anionic stabilizing substituents will shift the reaction towards an E1$_{cb}$ path.

In principle MOJ diagrams can be extended to more dimensions, for example by also including the B–H bond order in the above elimination reaction, but this is rarely done, not least because of the problems of illustrating more than two dimensions.

References

1. I. Fleming, *Frontier Orbitals and Organic Chemical Reactions* Wiley, 1976.
2. A. Devaquet, *Mol. Phys.*, **18** (1970), 233.
3. A. Alexakis, C. Chuit, M. Commercon-Bourgain, J. P. Foulon, N. Jabri, P. Mangeney and J. F. Normant, *Pure Appl. Chem.*, **56** (1984), 91.
4. G. Marino, *Adv. Heterocycl. Chem.*, **13** (1971), 235.
5. J. Sauer, *Angew. Chem. Int. Ed.*, **6** (1967), 16.
6. R. G. Parr and W. Yang, *J. Am. Chem. Soc.*, **106** (1984), 4049.
7. Y. Li and J. N. S. Evans, *J. Am. Chem. Soc.*, **117** (1995), 7756.
8. F. D. Proft, S. Liu and R. G. Parr, *J. Chem. Phys.*, **107** (1997), 3000.
9. R. S. Mulliken, *J. Chem. Phys.*, **2** (1934), 782.
10. R. G. Pearson, *J. Am. Chem. Soc.*, **85** (1963), 3533.
11. Y. Li and J. N. S. Evans, *J. Am. Chem. Soc.*, **117** (1995), 7756.
12. A. Rauk, *Orbital Interaction Theory of Organic Chemistry*, Wiley, 1994; T. A. Albright, J. K. Burdett and M.-H. Whangbo, *Orbital Interactions in Chemistry*, Wiley, 1985.
13. R. B. Woodward and R. Hoffmann, *The Conservation of Orbital Symmetry* Academic Press, 1970; R. B. Woodward and R. Hoffmann, *Angew. Chem. Int. Ed.*, **8** (1969), 781.
14. N. Turro, *Modern Molecular Photochemistry*, The Benjamin/Cummings Publishing Co., 1978.
15. J. G. Martin and R. K. Hill, *Chem. Rev.*, **61** (1961), 537.
16. W. R. Roth, J. König and K. Stein, *Chem. Ber.*, **103** (1970), 426.
17. F. Bernardi, M. Olivucci and M. A. Robb, *Chem. Soc. Rev.*, **25** (1996), 321.
18. W. P. Jenks, *Chem. Rev.*, **85** (1985), 511.
19. R. P. Bell, *Proc. R. Soc. London, Ser. A*, **154** (1936), 414; M. G. Evans and M. Polanyi, *J. Chem. Soc., Faraday Trans.*, **32** (1936), 1340.
20. G. S. Hammond, *J. Am. Chem. Soc.*, **77** (1955), 334.
21. R. A. Marcus, *J. Phys. Chem.*, **72** (1968), 891.
22. D. M. Guldi and K.-D. Asmus, *J. Am. Chem. Soc.*, **119** (1997), 5744.
23. J. Donnella and J. R. Murdoch, *J. Am. Chem. Soc.*, **106** (1984), 4724.
24. See for example V. Aviyente, H. Y. Yoo, K. N. Houk, *J. Org. Chem.*, **62** (1997), 6121.
25. W. von Doering, V. G. Toscano, G. H. Beasley, *Tetrahedron*, **27** (1971), 5299; M. J. Goldstein and R. S. Leight, *J. Am. Chem. Soc.*, **99** (1977), 8112.
26. R. A. More O'Ferrall, *J. Chem. Soc. B* (1970), 274; W. P. Jenks, *Chem. Rev.*, **72** (1972), 705; S. S. Shaik, H. B. Schlegel and S. Wolfe, *Theoretical Aspects of Physical Organic Chemistry. The S$_N$2 Mechanism*, Wiley, 1992.

16 Simulations, Time-dependent Methods and Solvation Models

Electronic structure methods are aimed at solving the Schrödinger equation for a single or a few molecules, infinitely removed from all other molecules. Physically this corresponds to the situation occurring in the gas phase under low pressure (vacuum). Experimentally, however, the majority of chemical reactions are carried out in solution. Biologically relevant processes also occur in solution, aqueous systems with rather specific pH and ionic conditions. Most reactions are both qualitatively and quantitatively different under gas and solution phase conditions, especially those involving ions or polar species. Molecular properties are also sensitive to the environment.

A detailed description of methods for studying dynamic (i.e. time-dependent) phenomena and condensed phases is outside the scope of this book. The common feature for all these methods, however, is the need for an energy surface upon which the dynamics can take place. The generation of such a surface normally relies at least partly on results from calculations of the types discussed in Chapters 2–6, and it may therefore be of interest to briefly discuss the fundamentals.

There are various methods of treating solvation, ranging from a detailed description at the molecular level to reaction field models where the solvent is modelled as a continuous medium.[1] The molecular description involves a simulation where many configurations of the model solution is generated, and properly averaged to give macroscopic quantities, which may be compared with experiments. The configurations in a simulation are often generated by following the time evolution of a molecular system. Not all solvation models involve simulations, and simulations may also be used to study things other than solutions. Furthermore, configurations can also be generated by time-independent methods, and time-dependent methods are used for many purposes other than generating configurations in a simulation. Nevertheless, solvation is a common element for the topics covered in this chapter.

16.1 Simulation Methods

Simulation refer to models involving a <u>statistical</u> component, i.e. the results are obtained with an uncertainty arising from the finite length of the simulation.[2] A typical application is simulation of a liquid phase, i.e. a detailed description of a solvent or solution.

A macroscopic quantity X, such as the enthalpy or entropy, can be calculated from the partition function Q, as discussed in Chapter 12. For easy reference we reproduce eqs. (12.6–12.8) here.

$$U = k_{\rm B} T^2 \left(\frac{\partial \ln Q}{\partial T} \right)_V$$

$$P = k_{\rm B} T \left(\frac{\partial \ln Q}{\partial V} \right)_T$$

$$H = k_{\rm B} T^2 \left(\frac{\partial \ln Q}{\partial T} \right)_V + k_{\rm B} T V \left(\frac{\partial \ln Q}{\partial V} \right)_T$$

$$C_V = \left(\frac{\partial U}{\partial T} \right)_V \qquad (16.1)$$

$$A = -k_{\rm B} T \ln Q$$

$$S = k_{\rm B} T \left(\frac{\partial \ln Q}{\partial T} \right)_V + k_{\rm B} \ln Q$$

$$G = k_{\rm B} T V \left(\frac{\partial \ln Q}{\partial V} \right)_T - k_{\rm B} T \ln Q$$

The properties have intentionally been separated in to two groups, those involving derivatives of Q and those which depend directly on Q.

For an isolated molecule in the rigid rotor, harmonic oscillator approximation, the (quantum) energy states are sufficiently regular to allow an explicit construction of the partition function, as discussed in Chapter 12. For a collection of many particles the relevant energy states are those describing the vibrations, and translation and rotation of molecules relative to each other. Owing to the very closely spaced energy levels we can neglect quantum effects and treat the state distribution as continuous, replacing the discrete sum over energies by an integral over all coordinates ($\mathbf{r}$) and momenta ($\mathbf{p}$), called the *phase space*.

$$Q = \sum_i^{\text{all states}} e^{-E_i/k_{\rm B} T} = \int e^{-E(\mathbf{r},\mathbf{p})/k_{\rm B} T} \, \mathbf{dr dp} \qquad (16.2)$$

The partition function Q here describes the whole system consisting of N interacting particles, and the energy states E_i are consequently for all the particles (in Section 12.2 we considered N non-interacting molecules, where the total partition function could be written in terms of the partition function for one molecule, $Q = q^N/N!$). More correctly the partition function in eq. (16.2) should be written in terms of the Hamiltonian for the system, i.e. replacing E with $\mathbf{H}$. The kinetic and potential energy components, however, can be separated ($\mathbf{H} = \mathbf{T} + \mathbf{V}$). In the absence of potential energy, the Hamiltonian is purely the kinetic energy and the system is an ideal gas. The interesting component is therefore the potential energy part of the partition function, which we denote by E. In the large majority of cases the energy E is of the force field type described in Chapter 2.

The energy states associated with intermolecular translation and rotation are not only numerous, but also so irregularly spaced that it is impossible to derive them directly from molecular quantities. It is consequently not possible to construct the partition function explicitly. Nevertheless, we may derive formal expressions for U and A from eqs. (16.1) and (16.2).

Using that $\partial \ln Q / \partial T = Q^{-1} \partial Q / \partial T$ we get

$$U = \frac{k_B T^2}{Q} \frac{\partial Q}{\partial T} = \sum_i^{\text{all states}} E_i Q^{-1} e^{-E_i/k_B T}$$

$$A = -k_B T \ln Q = -k_B T \ln \left(\sum_i^{\text{all states}} e^{-E_i/k_B T} \right)$$

$$(16.3)$$

The Boltzman probability distribution function P may be written either in a discrete energy representation or in a continuous phase space formulation.

$$P(E_i) = Q^{-1} e^{-E_i/k_B T}$$

$$P(\mathbf{r}, \mathbf{p}) = Q^{-1} e^{-E(\mathbf{r}, \mathbf{p})/k_B T}$$

$$(16.4)$$

where Q^{-1} ensures normalization. This allows eq. (16.3) to be written as

$$U = \sum_i^{\text{all states}} E_i P(E_i)$$

$$U = \int E(\mathbf{r}, \mathbf{p}) P(\mathbf{r}, \mathbf{p}) \, d\mathbf{r} \, d\mathbf{p}$$

$$(16.5)$$

Eq. (16.5) shows that U is simply a sum of energies weighted by the probability of being in that state, i.e. U is the average (potential) energy of the system.

A similar expression for A may be derived by substituting 1 by $e^{E/k_B T} e^{-E/k_B T}$ in eq. (16.3).

$$A = -k_B T \ln \left(\sum_i^{\text{all states}} e^{-E_i/k_B T} \right) = k_B T \ln \left(\frac{1}{\sum_i^{\text{all states}} e^{-E_i/k_B T}} \right)$$

$$A = k_B T \ln \left(\frac{\sum_i^{\text{all states}} e^{E_i/k_B T} e^{-E_i/k_B T}}{\sum_i^{\text{all states}} e^{-E_i/k_B T}} \right)$$

$$A = k_B T \ln \left(\sum_i^{\text{all states}} e^{E_i/k_B T} P(E_i) \right)$$

$$(16.6)$$

Alternatively it may be written as an integral over phase space.

$$A = k_B T \ln \left(\int e^{E(\mathbf{r}, \mathbf{p})/k_B T} P(\mathbf{r}, \mathbf{p}) \, d\mathbf{r} d\mathbf{p} \right)$$

$$(16.7)$$

In general a macroscopic observable can be calculated as an average over a corresponding microscopic quantity weighted by the Boltzman probability distribution as in eq. (16.5).

It is not possible to carry out the summation over all states, or equivalently integrate over all phase space in eqs. (16.5–16.7). If, however, a representative collection of configurations can be generated, the sum over all states can be approximated by an average over a finite set of configurations. Representative here means that the number of configurations with a given energy is proportional to that given by the Boltzman distribution, and that all "important" parts of the phase space are sampled. We can denote such points either by their energies or by their positions and momenta, where M is the number of points.

$$\langle X \rangle_M = \frac{1}{M} \sum_i^M X(\varepsilon_i) = \frac{1}{M} \sum_i^M X(\mathbf{r}_i, \mathbf{p}_i) \tag{16.8}$$

A collection of configurations is called an *ensemble*, and (16.8) is called an ensemble average. The subscript indicates what is being averaged over. If the configurations are generated by following the time evolution of the system (as in Molecular Dynamics methods, Section 16.2.1), the average is formally a time average. By the *ergodic hypothesis* (which can be proven rigorously only for a hard-sphere gas) this is assumed to yield the same as an average over an ensemble.

$$\langle X \rangle = \lim_{\tau \to \infty} \frac{1}{\tau} \int_0^\tau X(t)\, dt = \lim_{M \to \infty} \frac{1}{M} \sum_i^M X_i \tag{16.9}$$

The ergodic hypothesis essentially states that no matter where a system is started, it is possible to get to any other point in phase space. For U and A this leads to the following expressions.

$$\langle U \rangle_M = \frac{1}{M} \sum_i^M E_i = \langle E \rangle_M$$

$$\langle A \rangle_M = k_B T \ln \left(\frac{1}{M} \sum_i^M e^{E_i/k_B T} \right) = k_B T \ln \langle e^{E/k_B T} \rangle_M \tag{16.10}$$

The average value of X calculated from (16.9) has a statistical uncertainty $\sigma(X)$ which is inversely proportional to the square root to the number of sampling points M.

$$\sigma(X) \propto \frac{1}{\sqrt{M}} \tag{16.11}$$

The statistical error can thus be reduced by averaging over a larger ensemble. How well the calculated average (from eq. (16.9)) resembles the "true" value, however, depends on whether the ensemble is representative. If a large number of points is collected from a small part of the phase space, the property may be calculated with a small statistical error, but a large systematic error (i.e. the value may be precise, but inaccurate). As it is difficult to establish that the phase space is adequately sampled, this can be a very misleading situation, i.e. the property appears to have been calculated accurately but may in fact be significantly in error.

The reason for dividing the properties into two sets in eq. (16.1) is now apparent. As seen from eqs. (16.5)–(16.7), a calculation of U involves an average over E, while the expression for A involves an average over $e^{E/k_B T}$. The Helmholt free energy therefore depends primarily on high energy states, which occur with a very low probability.

Computationally it is therefore difficult to achieve a reasonable statistical error for entropic quantities such A, S and G.

The difference can also be seen directly from eq. (16.1). Properties which depend on derivatives of Q are independent of the actual value of Q, only its variation with an external variable (T or V) matters. Thermodynamical functions such as A, S and G, however, depend on the actual value of Q, i.e. the whole volume of phase space.

A random selection of points for evaluating the average in eqs. (16.5–16.6) will suffer from an extremely slow convergence as the large majority of points will have high energies, and consequently contribute with a very small probability. The "magic" in simulations is in generating an ensemble that yields a good representation of the "important" phase space for the given property. Different parts of the phase space, however, may be important for different properties. An ensemble that gives an accurate value for one property may not necessarily be suitable for another property.

There are two major techniques for generating an ensemble: Monte Carlo and Molecular Dynamics. In *Monte Carlo* (MC) methods[3] a sequence of points in phase space is generated from an initial geometry by adding a random "kick" to the coordinates of a randomly chosen particle (atom or molecule). The new geometry is accepted as a starting point for the next perturbing step if it is lower in energy than the current. If the new geometry is higher in energy, the Boltzmann factor $e^{-\Delta E/k_B T}$ is calculated and compared to a random number between 0 and 1. If $e^{-\Delta E/k_B T}$ is less than this number the new geometry is accepted, otherwise the next step is taken from the old geometry. This *Metropolis* procedure[4] ensures that the configurations in the ensemble obey a Boltzman distribution, and the possibility of accepting higher energy configurations allows MC methods to climb uphill and escape from a local minimum. In order to have a reasonable acceptance ratio, however, the step size must be fairly small. This effectively means that even a few million MC steps (typical computational limit) only explore the local region around the starting geometry.

Molecular Dynamic (MD) methods generate a series of time-correlated points in phase space (a *trajectory*) by propagating a suitable starting set of coordinates and velocities according to Newton's second equation by a series of finite time steps, as described in more detail in Section 16.2.1. A typical time step is $\sim 10^{-15}$ s and a simulation involving 10^6 steps thus "only" covers $\sim 10^{-9}$ s. This is substantially shorter than many important phenomena, and MD methods, in analogy with MC, tend to sample only the region in phase space close to the starting condition.

A necessary (but not sufficient) requirement for producing a representative sampling is that the system is in equilibrium. The starting configuration may be generated by completely random positions (and velocities for MD), but is more often taken either from a previous simulation or by placing the particles at or near the lattice points of a suitable crystal. The system is then equilibrated by running perhaps 10^4–10^5 MC or MD steps, followed by perhaps 10^5–10^6 production steps. The averaging in eq. (16.8) should be over configurations which are uncorrelated, which is not the case for nearby points in an MD trajectory or sequence of MC steps. The whole set of points is therefore divided into blocks with a length that is sufficiently long to make equivalent points in two neighbouring blocks uncorrelated. A suitable block size can be determined by calculating the variance (eq. (16.11)) for increasingly larger block sizes (e.g. 1, 2, 4, 8, 16 etc.). The variance will increase with block size, but will level off and reach a plateau once the blocks become sufficiently large that the points between blocks are

Table 16.1 Constants in different ensembles, and corresponding equilibrium states

N	P	V	T	E	μ	Equilibrium	Name
×		×	×			A has minimum	Canonical
×		×		×		S has maximum	Micro-canonical
×	×		×			G has minimum	Isothermal–isobaric
		×	×		×	(PV) has maximum	Grand canonical

N, Number of particles; P, Pressure; V, Volume; T, Temperature; E, Energy; μ, Chemical potential; A, Helmholtz free energy; S, Entropy; G, Gibbs free energy.

uncorrelated. In MD methods the distance between (uncorrelated) blocks has the dimension of time, and is called the correlation time. A set of uncorrelated points may be extracted from the sample by taking for example the midpoint in each block (*systematic stratified sampling*). Alternatively, a random element may be taken from each block (*random stratified sampling*), or an average value may be calculated from all the points in the block and used subsequently for averaging over blocks (*coarse-grained or two-stage sampling*). A simulation consisting of 10^6 steps thus generates significantly fewer uncorrelated points (perhaps 10^3–10^5), which can be used for averaging. The error estimate in eq. (16.11) refers to the number of uncorrelated points. It is important to recognize that different properties may have different correlation times, and thus require a different sampling of the whole set of points.

An ensemble may be characterized by parameters which are fixed, and those which can be derived from the simulation data, as shown in Table 16.1. Ensembles generated by MC techniques are naturally of the constant NVT type, while MD methods naturally generate a constant NVE ensemble. Both MC and MD methods, however, may be modified to simulate other ensembles, as described in Section 16.2. Of special importance is the constant NPT condition, which relates directly to most experimental conditions. The primary advantage of MD methods is that time appears explicit, i.e. such methods are natural for simulating time-dependent properties, such as correlation functions, and for calculating properties which depend on particle velocities. Furthermore, if the relaxation time for a given process is (approximately) known, the required simulation time can be estimated beforehand (i.e. it must be at least several multiples of the relaxation time). The disadvantage of MD is that forces (energy derivatives) are required, while only energies are necessary for MC. For parameterized energy functions, such as those used in force field methods, this is not a limitation as forces can be calculated just as easily as the energy.

In order to reduce the statistical error, the averaging in eq. (16.8) is typically performed on 10^3–10^5 points in phase space. The requirement of calculating this many points and associated energies for a model consisting of several hundred particles means that the use of *ab initio* methods is extremely demanding, even for small systems and simple wave functions. For small molecules semi-empirical electronic structure methods may be used, implicitly accepting the low accuracy of these methods, but the large number of calculations necessary still makes this computationally intensive. The large majority of simulations is therefore carried out with an energy surface generated by a parameterized function of the force field type.

The expressions derived from statistical mechanics are often rewritten into a computationally more suitable form, which may be evaluated from the basic descriptors: positions $\mathbf{r}$, velocities $\mathbf{v}$ or momenta $\mathbf{p}$ and energies E.

The temperature is related to the average kinetic energy.

$$\frac{1}{2}(3N - N_c)k_B T = \frac{1}{2}\left\langle \sum_i^N m_i v_i^2 \right\rangle_M \tag{16.12}$$

Here $(3N - N_c)$ is the number of degrees of freedom, equal to three times the number of particles minus the number of constraints, N_c, which typically will be 3 (corresponding to conservation of linear momentum). In a standard MC simulation the temperature is fixed (NVT conditions), while it is a derived quantity in a standard MD simulation (NVE conditions).

The pressure is related to the product of positions and forces (for pairwise potentials). Here the first part is for an ideal gas.

$$PV = Nk_B T + \frac{1}{3}\left\langle \sum_{i<j}^N r_{ij} f_{ij} \right\rangle_M \tag{16.13}$$

The internal (potential) energy is a direct sum of energies, which is normally given as a sum over pairwise interactions (i.e. van der Waals and electrostatic contributions in a force field description).

$$U = \sum_i^N E_i = \sum_{i>j}^N E_{ij} \tag{16.14}$$

The internal energy will fluctuate around a mean value which may be calculated by averaging over the number of configurations, $\langle U \rangle_M$.

The heat capacity at constant volume is the derivative of the energy with respect to temperature at constant volume (eq. (16.1). There are several ways of calculating such response properties. The most accurate is to perform a series of simulations under NVT conditions, and thereby determine the behaviour of $\langle U \rangle_M$ as a function of T (for example by fitting to a suitable function). Subsequently this function may be differentiated to give the heat capacity. This approach has the disadvantage that several simulations at different temperatures are required. Alternatively, the heat capacity can be calculated from the *fluctuation* of the energy around its mean value.

$$C_V = \frac{1}{k_B T^2}\left\langle \left(U - \langle U \rangle_M \right)^2 \right\rangle_M = \frac{1}{k_B T^2}\left(\langle U^2 \rangle_M - \langle U \rangle_M^2 \right) \tag{16.15}$$

This approach requires only a single simulation. Since the fluctuation has a longer relaxation time than the energy itself, the ensemble average in eq. (16.15) must be over a larger number of points than for $\langle U \rangle_M$ to achieve a similar statistical error, i.e. the efficiency obtained by avoiding multiple simulations is partly lost owing to the longer simulation time required. Another disadvantage is that eq. (16.15) involves taking differences between large numbers, which is susceptible to round-off errors.

Distribution functions measure the (average) value of a property as a function of an independent variable. A typical example is the radial distribution function $g(r)$ which measures the probability of finding a particle as a function of distance from a "typical"

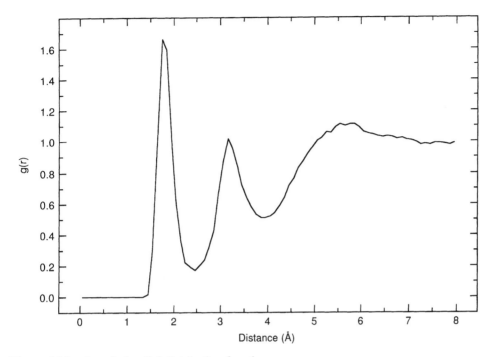

Figure 16.1 A typical radial distribution function

particle, relative to that expected from a completely uniform distribution (i.e. an ideal gas with density N/V). The radial distribution function is defined as

$$g(r, \Delta r) = \frac{V}{N^2} \frac{\langle N(r, \Delta r) \rangle_M}{2\pi r^2 \Delta r} \tag{16.16}$$

where $N(r, \Delta r)$ is the number of particles between r and $r + \Delta r$ from another particle, and $4\pi r^2 \Delta r$ is the volume of a spherical shell with thickness Δr.

For a solution the radial distribution function will typically have a structure as shown in Figure 16.1.[5] The figure displays the radial distribution function of hydrogen relative to the centre of mass for a benzene radical anion in an aqueous solution. At short distances the probability is zero owing to van der Waals repulsion. The distribution function then rises sharply to a value of $\sim$1.7 for a distance of $\sim$1.8 Å, indicating that it is 1.7 times more likely to find particles with this separation than in an ideal gas. This corresponds to water molecules which are located above or below the molecular plane. A second peak occurs at $\sim$3.2 Å, which corresponds to water molecules located around the edge of the benzene molecule. The integral under a peak gives the number of solvent molecules of a given type. At long range the distribution function levels off to a value of 1, i.e. the particles no longer sense each other and behave as in an ideal gas.

For molecules the radial distribution function can be extended with orientational degrees of freedom to characterize the angular distribution.

Correlation functions measure the relationship between two variables, x and y. A common definition is

$$C_{xy} = \frac{\langle (x - \langle x \rangle_M)(y - \langle y \rangle_M) \rangle_M}{\sqrt{\langle (x - \langle x \rangle_M)^2 \rangle_M \langle (y - \langle y \rangle_M)^2 \rangle_M}} \tag{16.17}$$

The correlation function is a number between -1 and 1, where 1 indicates that the two quantities are completely correlated, -1 that they are (completely) anti-correlated and 0 means that they are independent (uncorrelated).

Often such correlation functions are time dependent, and measure how the correlation between two quantities changes over time. They may be normalized by the corresponding static (i.e. $t = t_0$) limit.

$$C_{xy}(t) = \frac{\langle x(t_0)y(t) \rangle_{N, t_0}}{\langle x(t_0)y(t_0) \rangle_{N, t_0}} \tag{16.18}$$

Notice that the averaging is done over the number of particles N and t_0, but not the number of configurations M. Since an MD simulation produces a set of time-connected configurations, the number of a given configuration is directly related to the simulation time.

In the case where x and y are the same, $C_{xx}(t)$ is called an *autocorrelation* function, if they are different, it is called a *cross-correlation* function. For an autocorrelation function, the initial value at $t = t_0$ is 1, and it approaches 0 as $t \rightarrow \infty$. How fast it approaches 0 is measured by the *relaxation* time. The Fourier transforms of such correlation functions are often related to experimentally observed spectra, the far infrared spectrum of a solvent, for example, is the Fourier transform of the dipole autocorrelation function.[6]

16.1.1 Free Energy Methods

As noted above, it is very difficult to calculate entropic quantities with any reasonable accuracy within a finite simulation time. It is, however, possible to calculate <u>differences</u> in such quantities.[7] Of special importance is the Gibbs free energy, as it is the natural thermodynamical quantity under normal experimental conditions (constant temperature and pressure, Table 16.1), but we will illustrate the principle with the Helmholtz free energy instead. As indicated in eq. (16.1) the fundamental problem is the same. There are two commonly used methods for calculating differences in free energy: *Thermodynamic Perturbation* and *Thermodynamic Integration*.

16.1.2 Thermodynamic Perturbation Methods

Let us consider two systems **A** and **B** described by two different energy functions E_A and E_B. The difference in Helzholtz free energy is given by eq. (16.1).

$$A_A - A_B = -k_B T \ln Q_A + k_B T \ln Q_B = -k_B T \ln \frac{Q_A}{Q_B}$$

$$A_A - A_B = -k_B T \ln \left(\sum e^{-(E_A - E_B)/k_B T} \right) \tag{16.19}$$

The last equation follows from the definition of the partition function, eq. (16.2). Analogously to eq. (16.10) the free energy difference can be evaluated as an ensemble average.

$$A_A - A_B = k_B T \ln \left\langle e^{(E_A - E_B)/k_B T} \right\rangle_M \qquad (16.20)$$

The important difference is that the exponential now involves an energy difference. As long as this is sufficiently small compared with $k_B T$, the ensemble average can yield a good estimate of the difference in free energy within a reasonable simulation time. If the energy difference is large compared with $k_B T$, we may introduce intermediates states between A and B, which can be described in term of a coupling parameter λ ($0 \leq \lambda \leq 1$). The simplest approach involves a linear interpolation, but more complicated connections may also be used.[8]

$$E_\lambda = \lambda E_A + (1 - \lambda) E_B \qquad (16.21)$$

It should be noted that the transformation of **A** into **B** by the λ variable may or may not correspond to a physically realizable transformation. The change in free energy between two neighbouring points is then given analogously to (16.21), and the whole change is a sum over such terms.

$$A_A - A_B = k_B T \sum_\lambda \ln \left\langle e^{(\Delta E_\lambda)/k_B T} \right\rangle_M \qquad (16.22)$$

The number of intermediate states is chosen such that each ensemble average is performed over energy changes comparable to $k_B T$. To test the quality of the averaging, the perturbation is usually run in both directions (i.e. **A** $\rightarrow$ **B** and **B** $\rightarrow$ **A**), and the difference is a measure of how well ΔA is statistically converged. It should be noted that (too) short simulation times may lead to forward and backward calculated values which are in very good agreement, without the energy difference being calculated accurately. Calculation of free energy differences by means of eq. (16.22) is often called *Thermodynamic Perturbation*[9] or *Free Energy Perturbation* (FEP).

Instead of performing a series of simulations with a fixed energy function as in eq. (16.21), it is also possible to change it continuously during a single simulation by changing λ slightly in each time step. This is called the *Slow Growth* method, and requires that the increase in λ is slow enough for the system to remain essentially at equilibrium at all times. This is difficult to ensure in practice, and the slow growth method is therefore less commonly used.

16.1.3 Thermodynamic Integration Methods

Given an energy function as in eq. (16.21), the partition function and thereby also the free energy are functions of λ.

$$A(\lambda) = -k_B T \ln Q(\lambda) \qquad (16.23)$$

Differentiating this expressions yields

$$\frac{\partial A(\lambda)}{\partial \lambda} = -\frac{k_B T}{Q} \frac{\partial Q(\lambda)}{\partial \lambda} = -\frac{k_B T}{Q} \sum_i \left(-\frac{1}{k_B T} \right) e^{-E_i(\lambda)/k_B T} \frac{\partial E_i(\lambda)}{\partial \lambda}$$

$$\frac{\partial A(\lambda)}{\partial \lambda} = \sum_i \frac{\partial E_i(\lambda)}{\partial \lambda} P(\lambda) \qquad (16.24)$$

where the definition of Q and the Boltzman probability distribution (eq. (16.4)) have been used. Replacing the right-hand side by an ensemble average and integrating over λ gives

$$A(1) - A(0) = \int_0^1 \left\langle \frac{\partial E(\lambda)}{\partial \lambda} \right\rangle_M d\lambda \qquad (16.25)$$

The left-hand side is the desired free energy difference, and the right-hand side may be approximated by a discrete sum.

$$A_A - A_B = \sum_i \left\langle \frac{\partial E(\lambda)}{\partial \lambda} \right\rangle_M \Delta\lambda_i \qquad (16.26)$$

The use of eq. (16.26) for calculating ΔA is normally called *Thermodynamic Integration.*[10] The difference between eqs. (16.22) and (16.26) is that the former averages over finite differences in energy functions, while the latter averages over a <u>differentiated</u> energy function. For parameterized energy functions it is fairly easy to form the energy derivative with respect to the coupling parameter analytically, and the averaging in (16.26) is therefore no more complicated than averaging over energy differences as in eq. (16.22). Furthermore, it should be noted that the computational cost of performing the averaging is negligible compared to the cost of generating the ensemble, and the same ensemble can therefore be used to calculate the free energy difference by either eq. (16.22) or eq. (16.26). This allows a measure of the reliability of the calculated value to be obtained.

Free energy calculations are often combined with *thermodynamic cycles* to calculate properties which otherwise would require impossibly long simulation times.[11] A direct calculation of for example solvating acetone in water would require simulating the transfer of an acetone molecule from the gas phase (vacuum) to an aqueous phase, followed by solvent reorganization. If we wish to calculate the solvation energy of acetone relative to propane, this would require a second (long) simulation of transferring a propane molecule into an aqueous phase. Alternatively, the <u>difference</u> in solvation may be calculated by means of the thermodynamical cycle in Figure 16.2. The difference in solvation energy, $\Delta G_{solv,A} - \Delta G_{solv,B}$, which is difficult to calculate, may instead be obtained as $\Delta\Delta G_{solv} - \Delta\Delta G_{gas}$. If **A** and **B** are different molecules, such as acetone and propane, the $\Delta\Delta G$ values correspond to non-physical transformations. Theoreti-

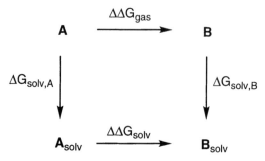

Figure 16.2 An example of a thermodynamic cycle for calculating differences in solvation energies

cally, however, it is quite easy to transform an oxygen atom into two hydrogens. The $\Delta\Delta G_{gas}$ value corresponds to differences in the internal (translational, rotational and vibrational) degrees of freedom, which can be calculated as discussed in Sections 12.2.1–12.2.3. This difference also is part of $\Delta\Delta G_{solv}$, but if the internal energy levels are assumed to be independent of the solvent, the solvent part of $\Delta\Delta G_{solv}$ is directly the difference in solvation.

In the acetone/propane example, the **A** to **B** change means that the oxygen atom gradually disappears, and two hydrogens gradually appear at the appropriate positions. In the force field energy expression, this corresponds to reducing or increasing van der Waals parameters and atomic charges, as well as changing all other parameters which are affected by the change in atom type. For $\lambda = 0.5$, the **A**/**B** "molecule" thus consists of a $CH_3–C–CH_3$ framework, with the central carbon having "half" a carbonyl oxygen and two "half" hydrogens attached. Absolute values of solvation energies may be calculated by transforming a solvent molecule to the solute, but if they are structurally very different, long simulation times may be required to ensure that equilibrium is attained.

The technique of thermodynamic cycles may be used for calculating relative free energies for a variety of other cases. Differential binding of two ligands to an enzyme, for example, requires transforming one ligand into the other in a pure solution, and when bound to the enzyme. The strength of free energy methods is that differences in free energies may be obtained with a statistical accuracy of a few tenths of a kcal/mol, at quite reasonable computational costs. Whether the calculated value agrees with experimental results depends on the quality of the force field; however, there are models for many solvents which are capable of providing an accuracy of better than 1 kcal/mol in terms of absolute values.

16.2   Time-dependent Methods

Dynamic methods may be classified as either classical, involving solution of Newton's equation, or quantal, involving solution of the (nuclear) Schrödinger equation (eq. (1.6)). Both of these are differential equations involving time, and can be solved by propagating an initial state through a series of small finite time steps.

16.2.1   Classical Methods

Newton's second law can be written as

$$-\frac{dV}{d\mathbf{r}} = m\frac{d^2\mathbf{r}}{dt^2} \tag{16.27}$$

$$\mathbf{F} = m\mathbf{a}$$

where V is the potential energy at position $\mathbf{r}$. The vector $\mathbf{r}$ contains the coordinates for all the particles, i.e. in Cartesian coordinates it is a vector of length $3N$. The left-hand side is the negative of the energy gradient, also called the force ($\mathbf{F}$) on the particle(s).

Given a set of particles with positions $\mathbf{r}_i$, the positions a small time step Δt later are given by a Taylor expansion.

$$\mathbf{r}_{i+1} = \mathbf{r}_i + \mathbf{v}_i\Delta t + \tfrac{1}{2}\mathbf{a}_i(\Delta t)^2 + \tfrac{1}{6}\mathbf{b}_i(\Delta t)^3 + \dots \tag{16.28}$$

The velocities $\mathbf{v}_i$ are the first derivatives of the positions with respect to time $(\mathrm{d}\mathbf{r}/\mathrm{d}t)$ at time t_i, the accelerations $\mathbf{a}_i$ are the second derivatives $(\mathrm{d}^2\mathbf{r}/\mathrm{d}t^2)$ at time t_i, $\mathbf{b}_i$ are the third derivatives etc.

The positions a small time step Δt <u>earlier</u> are derived from eq. (16.28) by substituting Δt with $-\Delta t$.

$$\mathbf{r}_{i-1} = \mathbf{r}_i - \mathbf{v}_i\Delta t + \tfrac{1}{2}\mathbf{a}_i(\Delta t)^2 - \tfrac{1}{6}\mathbf{b}_i(\Delta t)^3 + \ldots \qquad (16.29)$$

Addition of eqs. (16.28) and (16.29) gives

$$\mathbf{r}_{i+1} = (2\mathbf{r}_i - \mathbf{r}_{i-1}) + \mathbf{a}_i(\Delta t)^2 + \ldots \qquad (16.30)$$

This is the *Verlet* algorithm[12] for solving Newton's equation numerically. Notice that the term involving the change in acceleration (**b**) disappears, i.e. the equation is correct to third order in Δt. At the initial point the previous positions are not available, but may be estimated from a first-order approximation of eq. (16.29).

$$\mathbf{r}_{i-1} = \mathbf{r}_0 - \mathbf{v}_0\Delta t \qquad (16.31)$$

At each time step the acceleration must be evaluated from the forces, eq. (16.27), which then allows the atomic positions to be propagated in time and thus generate a trajectory. As the step size Δt is decreased, the trajectory becomes a better and better approximation to the "true" trajectory, until the practical problems of finite numerical accuracy arise (e.g. the forces cannot be calculated with infinite precision). A small time step, however, means that more steps are necessary for propagating the system a given total time, i.e. the computational effort increases inversely with the size of the time step.

The Verlet algorithm has the numerical disadvantage that the new positions are obtained by adding a term proportional to Δt^2 to a difference in positions $(2\mathbf{r}_i - \mathbf{r}_{i-1})$. Since Δt is a small number and $(2\mathbf{r}_i - \mathbf{r}_{i-1})$ is a difference between two large numbers, this may lead to truncation errors due to finite precision. The Verlet furthermore has the disadvantage that velocities do not appear explicitly, which is a problem in connection with generating ensembles with constant temperature, as discussed below.

The numerical aspect, and the lack of explicit velocities, in the Verlet algorithm can be remedied by the *leap-frog* algorithm. Performing expansions analogous to eqs. (16.28) and (16.29) with half a time step followed by subtraction gives

$$\mathbf{r}_{i+1} = \mathbf{r}_i + \mathbf{v}_{i+\frac{1}{2}}\Delta t \qquad (16.32)$$

The velocity is obtained by analogous expansions to give

$$\mathbf{v}_{i+\frac{1}{2}} = \mathbf{v}_{i-\frac{1}{2}} + \mathbf{a}_i\Delta t \qquad (16.33)$$

Eqs. (16.32) and (16.33) define the leap-frog algorithm, and it is seen that the positions and velocities updates are out of phase by half a time step. In terms of theoretical accuracy it is also of third order, as is the Verlet algorithm; however, the numerical accuracy is better. Furthermore, the velocities appear directly, which facilitates coupling to an external heat bath. The disadvantage is that the positions and velocities are not known at the same time, they are always out of phase by half a time step. Verlet and leap-frog algorithms are preferred over for example Runge–Kutte methods (Section 14.8) in MD simulations because they are time-reversible, which tends to improve energy conservation.[13]

The total energy of a system with a given set of positions and velocities is given as the sum of the kinetic and potential energy.

$$E_{\text{tot}} = \sum_{i=1}^{N} \frac{1}{2} m_i \mathbf{v}_i^2 + V(\mathbf{r}) \qquad (16.34)$$

The temperature of the system is proportional to the average kinetic energy (eq. (16.12), and therefore determines which parts of the energy surface the particles can exploit. Owing to the finite precision by which the atomic forces are evaluated, and the finite time step used, the total energy in practice is <u>not</u> constant (preservation of the energy to within a given threshold may be used to define the maximum permissible time step).

The maximum time step that can be taken is determined by the rate of the <u>fastest</u> process in the system (i.e. typically a factor of 10 smaller than the fastest process). Molecular motions (rotations and vibrations) typically occur with frequencies in the range 10^{11}–10^{14} s^{-1} (corresponding to wave numbers of 3–3300 cm^{-1}), and time steps of the order of femtoseconds (10^{-15} s) or less are required to model such motions with sufficient accuracy. This means that a total simulation time of one nanosecond (10^{-9}) requires $\sim 10^6$ time steps, and one microsecond (10^{-6}) requires $\sim 10^9$ time steps. A million time steps is already a significant computational effort, and typical simulation times are ~ 10 ps. Unfortunately there are relatively few process that occur on a picosecond time-scale, and many interesting phenomena, like protein folding or chemical reactions, have substantially longer time spans, of the order of seconds. Furthermore, a single trajectory may not be adequate for representing the dynamics; many runs must be carried out with different starting conditions (positions and velocities) and properly averaged.

For molecules the fastest processes are the stretching vibrations, especially those involving hydrogens. These degrees of freedom, however, have relatively little influence on many properties. It is therefore advantageous to freeze all bond lengths, which allows longer time steps to be taken, and consequently longer simulation times can be obtained for the same computational cost. As all atoms move individually according to Newton's equation, constraints must be applied to keep bond lengths fixed. This is normally done by the *SHAKE* algorithm,[14] where the atoms are first allowed to move under the influence of the forces, and subsequently forced to obey the constraints by an iterative procedure. This typically allows the time step to be increased by a factor of 2 or 3. Angles may also be frozen by adding a distance constraint on atoms 1,3 relative to each other. Angle bendings, however, affect calculated properties more than bond stretching, and fixing them may introduce unacceptable errors. Angle constraints are therefore used less frequently. A simulation can also be performed using fixed molecular geometries, i.e. only the positions and relative orientations of individual molecules are allowed to change. In such cases the natural variables to propagate in time are the centre of mass position and the three Euler angles of each molecule.

A standard MD simulation generates an *NVE* ensemble, i.e. the temperature and pressure will fluctuate. It is possible also to generate *NVT* or *NPT* ensembles by MD techniques by modifying the velocities or positions in each time step. As indicated in eq. (16.12) the instant value of the temperature is given by the average of the kinetic energy over the number of particles. If this is different from the desired temperature, all velocities may be scaled by a factor of $(T_{\text{desired}}/T_{\text{actual}})^{1/2}$ in each time step to achieve the desired temperature. Alternatively, the system may be coupled to a *"heat bath"* which gradually adds or removes energy to/from the system.[15] The kinetic energy of the system is again modified by scaling the velocities, but the rate of heat transfer is controlled by a coupling parameter τ.

$$\frac{dT}{dt} = \frac{1}{\tau}(T_{\text{desired}} - T_{\text{actual}})$$

$$\Updownarrow \qquad\qquad (16.35)$$

$$\text{scale factor} = \sqrt{1 + \frac{\Delta t}{\tau}\left(\frac{T_{\text{desired}}}{T_{\text{actual}}} - 1\right)}$$

In *Nosé–Hoover* methods[16] the heat bath is considered an integral part of the system, and enters the simulation on an equal footing with the other variables.

The pressure can similarly be held (approximately) constant by coupling to a "pressure bath". Instead of changing the velocities of the particles, the volume of the system is changed by scaling all coordinates by a factor closely related to that shown in (16.35).

$$\frac{dP}{dt} = \frac{1}{\tau}(P_{\text{desired}} - P_{\text{actual}})$$

$$\Updownarrow \qquad\qquad (16.36)$$

$$\text{scale factor} = \sqrt[3]{1 - \kappa\frac{\Delta t}{\tau}(P_{\text{desired}} - P_{\text{actual}})}$$

The constant κ is the compressibility of the system. Alternatively the pressure may be maintained by a Nosé–Hoover approach.

A realistic model of a solution requires at least several hundred solvent molecules. To prevent the outer solvent molecules from boiling off into space, and minimizing surface effects, *periodic boundary conditions* are normally employed. The solvent molecules are placed in a suitable box, often (but not necessarily) having a cubic geometry (it has been shown that simulation results using any of the five types of space filling polyhedra are equivalent[17]). This box is then duplicated in all directions, i.e. the central box is surrounded by 26 identical cubes, which again is surrounded by 98 boxes etc. If a

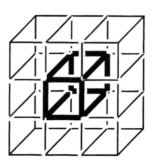

Figure 16.3   Periodic boundary condition

solvent molecule leaves the central box through the right wall, its image will enter the box through the left wall from the neighbouring box. This means that the resulting solvent model becomes quasi-periodic, with a periodicity equal to the dimensions of the box.

As mentioned in Section 2.5, the electrostatic interaction is long-ranged and will usually extend beyond the boundary of a box. Truncating the interaction by using a cut-off distance of say 10 Å has some rather unfortunate consequences for the distribution of the solvent molecules, producing "hot" and "cold" spots. Methods have therefore been developed which treat the electrostatic interaction "exactly" (to within a numerical threshold), but without having to perform the N^2 summation over all atoms. The *Ewald sum* method, which is only strictly valid for true periodic systems (like crystals) but may also be applied to quasi-periodic models arising from applying periodic boundary conditions, splits the interaction into a "near"- and "far"-field contribution.[18] The near-field is evaluated directly, while the far-field is calculated in the reciprocal space, which reduces the scaling from N^2 to $N^{3/2}$. A related method is the *Particle Mesh Ewald* (PME) method[19] which scales (only) as $N \ln(N)$. The *Fast Multipole Moment* (FMM) method[20] similarly splits the contribution into a near- and far-field, and calculates the near-field exactly. The far-field energy is calculated by dividing the physical space into boxes, and the interaction between all molecules in one box with all molecules in another is approximated as interactions between multipoles located at the centres of the boxes. The further away from each other two boxes are, the larger the boxes can be for a given accuracy, thereby reducing the formal N^2 scaling into something which approaches linear scaling.[21] The pre-factor, however, is rather large, and properly implemented it appears that the crossover point, where FMM becomes faster than PME, is around 100 000 particles. A disadvantage of FMM is that the maximum error (relative to an exact calculation) is significantly larger than for Ewald methods, i.e. there are certain particle pairs for which the error is larger than the average error by perhaps a factor of 10. The original FMM has also been refined by adjusting the accuracy of the multipole expansion as a function of the distance between boxes, producing the very *Fast Multipole Moment* (vFMM) method.[22]

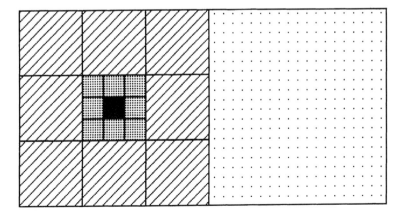

Figure 16.4 Illustration of the Fast Multiple Moment method

The major computational effort in classical simulations is the calculation of the forces on all particles at each time step. In principle any type of energy function can be used, force field, semi-empirical, *ab initio* electronic structure or DFT methods. The requirement of perhaps 10^6 force calculations, however, means that essentially only parameterized energy functions of the force field type have seen common use. For studying macromolecules and solvation, general force fields of the type discussed in Chapter 2 are normally used. Such functions are not suitable for studying chemical reactions where bond breaking/forming is taking place. To simulate chemical reactions a "global" energy surface may be constructed by fitting high-level *ab initio* results and experimental data to a suitable functional form.[23] For a sufficiently small time step, the result of a simulation is determined entirely by the quality of the energy surface. To obtain "converged" results for the dynamics, the energy surface must be accurate to perhaps 0.1 kcal/mol, over the whole surface which is accessible at the given energy (temperature). Constructing such high quality "global" energy surfaces is very demanding, and has only been done for a few systems. As mentioned in Chapter 1, the sheer dimensionality prevents an adequate sampling of a surface by point calculations for more than 3–4 atoms, and high-level dynamics have thus been limited to systems of this size.

Even for low-dimensional surfaces (3–5 atoms), it is often difficult to design a well-behaved fitting function. Alternatively, the fitting step may be bypassed and the dynamics performed "directly" by calculating the required forces in each step of an MD simulation. Traditionally such direct dynamics have involved determination of a completely converged wave function at each time step. This, however, is not necessary, as shown in an elegant breakthrough by *Car and Parrinello* (CP).[24] After having determined a converged wave function at the first point, the essence of the CP method is to let the orbitals evolve simultaneously with the changes in nuclear positions. This is achieved by including the orbital parameters as variables with fictive "masses" in the dynamics, analogous to the nuclear positions and masses. Except for the initial point, the nuclear forces are not correct, since the electronic wave function is not converged in the orbital parameter space. This error, however, can be controlled by suitable choices of the dynamic parameters. Such *ab initio* simulations can in principle be carried out with any type of wave function, but are still significantly more expensive computationally than traditional parameterized energy functions. For this reason only DFT methods are currently in use with CP methods. The great advantage over force field type functions is that electronic structure methods are able to describe bond breaking/formation, i.e. CP methods allow a direct simulation of chemical reactions.

16.2.2 Langevin Methods

MD methods generate detailed information about all the particles in the system, and are therefore well suited for calculating collective properties. In other cases the major interest is in the dynamics of a single molecule, in which case the surrounding molecules can be modelled by including only the average interactions. These average interactions are assumed to have a friction term (with a friction coefficient ζ) proportional to the atomic velocity, and a random component ($\mathbf{F}_{random}$) which averages to zero. These terms are in addition to the normal intramolecular forces ($\mathbf{F}_{intra}$) and possibly also external forces, for example from an electric field. The random force is associated with a temperature, and adds energy to the system, while the friction term removes energy. Typically the random force is taken to have a Gaussian distribution with a mean value of zero. The relevant dynamic equation is called the *Langevin* equation of motion, also called *stochastical* or *Brownian dynamics*.[25]

$$m\frac{d^2\mathbf{r}}{dt^2} = -\zeta\frac{d\mathbf{r}}{dt} + \mathbf{F}_{intra} + \mathbf{F}_{random} \qquad (16.37)$$

16.2.3 Quantum Methods

The nuclear Schrödinger equation is given as

$$\mathbf{H}\Psi = (\mathbf{T} + \mathbf{V})\Psi = i\frac{\partial\Psi}{\partial t} \qquad (16.38)$$

where $\mathbf{T}$ is the kinetic energy operator and $\mathbf{V}$ is the potential energy. The square of the wave function is the probability of finding a particle at a given position. Heisenberg's uncertainty principle means that a quantum description of a nucleus must be a continuous function, not a single specific position as in classical mechanics.[26] Such a continuous function is often denoted a *wave package*, and may be modelled by Gaussian functions (semi-classical methods) or numerically (quantum methods). Analogously to classical dynamics, the wave function may be propagated through a series of small, but finite, time steps.

$$\Psi_{i+1} = -i\mathbf{H}\Psi_i\Delta t = -i(\mathbf{T} + \mathbf{V})\Psi_i\Delta t \qquad (16.39)$$

Each time step thus involves a calculation of the effect of the Hamilton operator acting on the wave function. In fully quantum methods the wave function is often represented on a grid of points, these being the equivalent of basis functions for an electronic wave function. The effect of the potential energy operator is easy to evaluate, as it just involves a multiplication of the potential at each point with the value of the wave function. The kinetic energy operator, however, involves the derivative of the wave function, and a direct evaluation would require a very dense set of grid points for an accurate representation.

The kinetic energy operator is proportional to the square of the momentum, $\mathbf{T} = \mathbf{p}^2/2m$. In a momentum representation (i.e. using the particle momenta instead of positions as variables), $\mathbf{T}$ is a simple multiplication operator, analogous to $\mathbf{V}$ in position space. The transformation from position to momentum space can be achieved by a Fourier transformation. A numerical solution of the time-dependent Schrödinger equation can thus be done by switching back and forth between a position and momentum representation of the wave function, evaluating the effect of $\mathbf{V}$ in position, and the effect of $\mathbf{T}$ in momentum, space. Analogously to the leap-frog algorithm for the classical case (eqs. (16.32) and (16.33)), the update of the wave function by the potential and kinetic energy operators may be chosen to be out of phase by half a time step to improve the accuracy. The key to the popularity of this approach is the presence of highly efficient computer routines for performing Fourier transformations.

The requirement of an accurate global energy surface is even more important for a quantum mechanical treatment than for the classical case, since the wave function depends on a finite part of the surface, not just a single point. The updating of the positions and velocities is computationally inexpensive in the classical case, once the forces are available, but the requirement of two Fourier transformations in each time step makes the quantum propagation a significant computational issue. Furthermore, the representation of the wave function on a grid effectively limits the dimensionality to a maximum of three, i.e. di- and triatomic systems (one and three internal coordinates, respectively). Larger systems necessitate freezing some of the coordinates, or treating them classically.

16.2.4 Reaction Path Methods

The main problem in dynamical studies is the requirement of a <u>continuous</u> energy surface over a wide range of geometries. A simulation will normally be done with specification of an energy (or a temperature), and a surface must thus be available for all nuclear configurations which have an energy lower than the chosen simulation value. For quantum methods, the surface must also be available at higher energies, as the wave function has a tail which penetrates into classically "forbidden" areas.

Traditionally such "global" energy surfaces have been constructed by fitting a suitable functional form to energies (and possibly also first and second derivatives) calculated by *ab initio* methods at a large number (perhaps a few hundred or thousand) of geometries.[27] The function may be further refined by including experimental data (such as vibrational frequencies and geometries) in the fitting. For "large" systems (i.e. more than 3–4 atoms) a generation of an adequate number of fitting points is prohibitively expensive. In order to treat large systems, it is necessarily to concentrate the computational effort on the "chemically important" part of the potential energy surface.

In the simplest description, a chemical reaction takes place along the lowest energy path connecting the reactant and product, passing over the transition structure (Section 12.1) as the highest point. This is the *Minimum Energy Path* (MEP), which in mass-weighted coordinates is called the *Intrinsic Reaction Coordinate* (IRC) (Section 14.8). The idea in *Reaction Path* (RP) methods[28] is to consider only the energy surface in the immediate vicinity of a suitable one-dimensional reaction path, which usually (but not necessarily) is taken as the IRC. Typically the potential is expanded to second order along the reaction path, corresponding to modelling the perpendicular degrees of freedom as harmonic vibrational frequencies. The reaction path potential may be generated by a series of frequency calculations at points along the IRC, and the pointwise potential made continuous by interpolation. The potential may be generated prior to the reaction path calculation, or generated "on the fly" in a "direct" fashion.[29] The reaction path method may be generalized by having two "reaction coordinates" (a *reaction surface*) treated explicitly, and the remaining degrees of freedom treated approximately, or by having three "reaction coordinates" (a *reaction volume*).[30]

Inclusion of dynamical effects allows calculation of corrections to simple Transition State Theory, often described by a transmission coefficient κ to be multiplied with the TST rate constant (Section 12.1), or used in connection with variational TST (Section 12.3). Classical dynamics allow corrections due to recrossing to be calculated, while a quantum treatment is necessary for including tunnelling effects. Owing to the stringent requirement of a highly accurate global energy surface, there are only a few systems which have been subjected to a rigorous analysis.

The tunnel effect is sometimes approximated by inclusion of a semi-classical correction based on tunnelling through the barrier along the minimum energy path (i.e. the IRC). The *Bell* correction[31] is based on the assumption that the (one-dimensional) energy curve near the transition state can be approximated by a parabola. This yields a correction factor which depends only on the activation energy $\Delta E^{\neq}$ and the magnitude of the imaginary frequency ν_i, i.e. the curvature of the potential energy surface at the TS.

$$\kappa_{\text{Bell}} = \frac{\frac{1}{2}u^{\neq}}{\sin\frac{1}{2}u^{\neq}} - u^{\neq}e^{-\Delta E^{\neq}/k_B T}\left(\frac{e^{-\beta}}{2\pi - u^{\neq}} - \frac{e^{-2\beta}}{4\pi - u^{\neq}} + \frac{e^{-3\beta}}{6\pi - u^{\neq}} - \cdots\right)$$

$$u^{\neq} = \frac{h\nu_i}{k_B T} \qquad \beta = \frac{2\pi\Delta E^{\neq}}{h\nu_i}$$

(16.40)

Except for reactions with low barriers (i.e. <10 kcal/mol at $T = 300\,\text{K}$), or at high temperatures, the quantity $\Delta E^{\neq}/k_B T$ is large, and the last series can be neglected. The tunnel correction is then given completely in terms of the magnitude of the imaginary frequency. For small values of $u^{\neq}$ the first term may be Taylor expanded to give

$$\kappa_{\text{Bell}} = 1 + \frac{(u^{\neq})^2}{24} + \cdots$$

(16.41)

The first order term is known as the *Wigner* correction.[32]

It is possible to derive tunnel corrections for functional forms of the energy barrier other than an inverted parabola, but these cannot be expressed in analytical form. Since any barrier can be approximated by a parabola near the TS, and since tunnelling is most important for energies just below the top, they tend to give results in qualitative agreement with the Bell formula.

The main approximation of such one-dimensional corrections is that the tunnelling is assumed to occur along the MEP. This may be a reasonable assumption for reactions having either early or late (close to either reactant or product) transition states. For reactions where both bond breaking and formation are significant at the TS (as is usually the case), the dominant tunnel effect is "corner cutting" (Figure 16.5), i.e. the favoured

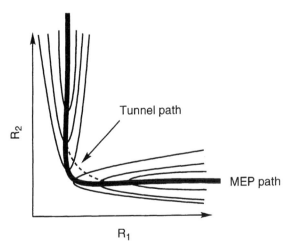

Figure 16.5   A contour plot illustration of the "corner cutting" tunneling path

tunnel path is not along the MEP. Although the energy increases away from the MEP, the barrier also becomes narrower on the concave side of the reaction path, which favours the tunnelling probability.

Truhlar and co-workers have developed different approximate schemes for including tunnelling in multi-dimensional systems.[33] In the *Minimum Energy Path Semiclassical Adiabatic Ground-state* (MEPSAC) approximation the tunnelling is assumed to occur along the MEP, analogously to the Bell approach, but for an arbitrary shape of the energy surface. The *Small Curvature Semiclassical Adiabatic Ground-state* (SCSAC) approximation allows tunnelling to occur within one vibrational half-amplitude perpendicular to the reaction path, while the *Large Curvature Ground-state* (LCG) approximation allows tunnelling to occur outside this region. The SCSAC method requires a knowledge of the (generalized) frequencies along the IRC (Section 13.1), which can be obtained by calculating the force constant matrix at suitable intervals and interpolating the results. The LCG methods require additional calculations away from the IRC.

16.3   Continuum Solvation Models

Methods for evaluating the effect of a solvent may broadly be divided into two types: those describing the individual solvent molecules, as discussed in Section 16.1, and those which treat the solvent as a continuous medium.[34] Combinations are also possible, for example by explicitly considering the first solvation sphere and treating the rest by a continuum model. Each of these may be subdivided according to whether they use a classical or quantum mechanical description.

A straightforward approach to modelling the effect of say water as a solvent, is to add water molecules one at a time until the property of interest no longer changes. Unfortunately an adequate description of the non-bonded interactions between water molecules, and between the solvent and solute, requires quite sophisticated methods, some type of *ab initio* wave function which includes electron correlation. Combined with the fact that a realistic model of a solution involves several hundred water molecules, this is clearly out of reach. Semi-empirical electronic structure and DFT methods are in general not sufficiently accurate for calculating intermolecular potentials. The only realistic methods are therefore based on parameterized energy functions for representing the intermolecular potential. Hybrid methods, where the solute–solvent interaction is calculated by electronic structure methods, and the solvent–solvent interaction by force field potentials, have been used in some cases (Section 2.10). Furthermore, the effect of a solvent is not determined by a single configuration, as discussed in Section 16.1. This for example means that the concept of a reaction path (Section 14.8) looses it significance as the reaction no longer follows a single trajectory.

The effect of a solvent may be divided into two major parts: specific solvation or "short-range" effects, for example hydrogen bonding or a preferential orientation of solvent molecules near an ion, and "macroscopic" or "long-range" effects, involving screening of charges (solvent polarization). The long-range part is responsible for generating a (macroscopic) dielectric constant different from 1. The specific solvation effect is mainly concentrated in the first solvation sphere, but the long-range effect requires a large number of solvent molecules. There are several methods for modelling

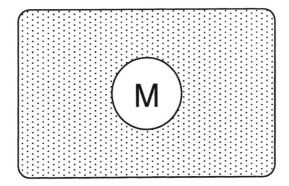

Figure 16.6 Reaction field model

the latter without explicitly considering the individual molecules, usually employing the concept of a "reaction field" in some way or another.

The *Self-Consistent Reaction Field* (SCRF) model considers the solvent as a uniform polarizable medium with a dielectric constant of ε, with the solute M placed in a suitable shaped hole in the medium.[35] Creation of a cavity in the medium costs energy, i.e. this is a destabilization, while dispersion interactions between the solvent and solute add a stabilization (this is roughly the van der Waals energy between solvent and solute). The electric charge distribution of M will furthermore polarize the medium (induce charge moments), which in turn acts back on the molecule, thereby producing an electrostatic stabilization. The solvation (free) energy may thus be written as

$$\Delta G_{\text{solvation}} = \Delta G_{\text{cavity}} + \Delta G_{\text{dispersion}} + \Delta G_{\text{electrostatic}} \qquad (16.42)$$

Reaction field models differ in five aspects.

(1) How the size and shape of the cavity is defined.
(2) How the dispersion contributions are calculated.
(3) How the charge distribution of M is represented.
(4) How the solute M is described, either classically (force field) or quantally (semi-empirical or *ab initio*).
(5) How the dielectric medium is described.

The simplest shape for the cavity is a sphere or possibly an ellipsoid. This has the advantage that the electrostatic interaction between M and the dielectric medium may be calculated analytically. More realistic models employ molecular shaped cavities, generated for example by interlocking spheres located on each nuclei. Taking the atomic radius as a suitable factor (typical value is 1.2) times a van der Waals radius defines a *van der Waals surface*. Such a surface may have small "pockets" where no solvent molecules can enter, and a more appropriate descriptor may be defined as the surface traced out by a spherical particle of a given radius rolling on the van der Waals surface. This is denoted the *Solvent Accessible Surface* (SAS) and illustrated in Figure 16.7.

Since an SAS is computationally more expensive to generate than a van der Waals surface, and since the difference is often small, a van der Waals surface is often used in practice. Alternatively, the cavity may be calculated directly from the wave function, for example by taking a surface corresponding to an electron density of 0.001.[36] It is

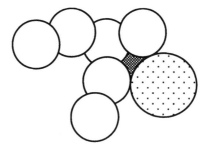

Figure 16.7 On a surface generated by overlapping van der Waals spheres there will be areas (hatch) which are inaccessible to a solvent molecule (dotted sphere)

generally found that the shape of the hole is important, and that molecular shaped cavities are necessary to be able to obtain good agreement with experimental data (such as solvation energies). It should be emphasized, however, that reaction field models are incapable of modelling specific (short-range) solvation effects, i.e. those occuring in the first solvation sphere.

The energy required to create the cavity (entropy factors and loss of solvent–solvent van der Waals interactions), and the stabilization due to van der Waals interactions between the solute and solvent (which may contain also a small repulsive component), is usually assumed to be proportional to the surface area. The corresponding energy terms may be taken simply as being proportional to the total surface area (a single proportionality constant), or parameterized by having a constant ξ specific for each atom type (analogous to van der Waals parameters in force field methods), with the ξ parameters being determined by fitting to experimental solvation data.

$$\Delta G_{\text{cavity}} + \Delta G_{\text{dispersion}} = \sum_i^{\text{atoms}} \xi_i S_i \qquad (16.43)$$

For solvent models where the cavity/dispersion interaction is parameterized by fitting to experimental solvation energies, the use of a few explicit solvent molecules for the first solvation sphere is <u>not</u> recommended, as the parameterization represents a best fit to experimental data <u>without</u> any explicit solvent present.

The charge distribution of the molecule can be represented either as atom centred charges or as a multipole expansion. For a neutral molecule, the lowest-order approximation considers only the dipole moment. This may be quite a poor approximation, and fails completely for symmetric molecules which do not have a dipole moment. For obtaining converged results it is often necessarily to extend the expansion up to order 6 or more, i.e. including dipole, quadrupole, octupole, etc. moments.

A classical description of M can for example be a standard force field with (partial) atomic charges, while a quantum description involves calculation of the electronic wave function. The latter may be either a semi-empirical model, such as AM1 or PM3, or any of the *ab initio* methods, i.e. HF, MCSCF, CISD, MP2 etc. Although the electrostatic potential can be derived directly from the electronic wave function, it is usually fitted to a set of atomic charges or multipoles, as discussed in Section 9.2, which then are used in the actual solvent model.

The dielectric medium is normally taken to have a constant value of ε, but may for some purposes also be taken to depend for example on the distance from M. For dynamical phenomena it can also be allowed to be frequency dependent[37] i.e. the response of the solvent is different for a "fast" reaction, such as an electronic transition, and a "slow" reaction, such as a molecular reorientation.

The charge density on the surface of the hole, $\sigma(\mathbf{r}_s)$, is given by standard electrostatics in terms of the dielectric constant, ε, and the electric field perpendicular to the surface, $\mathbf{F}$, generated by the charge distribution within the cavity.

$$4\pi\varepsilon\sigma(\mathbf{r}_s) = (\varepsilon - 1)\mathbf{F}(\mathbf{r}_s) \tag{16.44}$$

For spherical or ellipsoidal cavities, eq. (16.44) can be solved analytically, but for molecular shaped surfaces, it must be done numerically, typically by breaking it into smaller fractions which are assumed to have a constant σ. Once $\sigma(\mathbf{r}_s)$ is determined, the associated potential is added as an extra term to the Hamilton operator.

$$\mathbf{H} = \mathbf{H}_0 + \mathbf{V}_\sigma$$
$$\mathbf{V}_\sigma(\mathbf{r}) = \int \frac{\sigma(\mathbf{r}_s)}{|\mathbf{r} - \mathbf{r}_s|} \mathrm{d}\mathbf{r}_s \tag{16.45}$$

The potential from the surface charge $\mathbf{V}_\sigma$ is given by the molecular charge distribution (eq. (16.44)), but also enters the Hamiltonian and thus influences the molecular wave function. The procedure is therefore iterative.

The simplest reaction field model is a spherical cavity, where only the net charge and dipole moment of the molecule are taken into account, and cavity/dispersion effects are neglected. For a net charge q in a cavity of radius a, the difference in energy between vacuum and a medium with a dielectric constant of ε is given by the *Born* model.[38]

$$\Delta G_{el}(q) = -\left(1 - \frac{1}{\varepsilon}\right)\frac{q^2}{2a} \tag{16.46}$$

Using a set of (partial) atomic charges is often called the *generalized Born* model. It can be noted that the Born model predicts equal solvation for positive and negative ions of the same size, which is not the observed behaviour in solvents like H_2O.

The spherical cavity, dipole only, SCRF model is known as the *Onsager* model.[39] The *Kirkwood* model[40] refers to a general multipole expansion, if the cavity is ellipsoidal the *Kirkwood–Westheimer* model arise.[41] A <u>fixed</u> dipole moment of μ in the Onsager model gives rise to an energy stabilization.

$$\Delta G_{el}(\mu) = -\frac{\varepsilon - 1}{2\varepsilon + 1}\frac{\mu^2}{a^3} \tag{16.47}$$

This is not an SCRF model, as the dipole moment and stabilization are not calculated in a self-consistent way. When the back-polarization of the medium is taken into account, the dipole moment changes, depending on how polarizable the molecule is. Taking only the first-order effect into account, the stabilization becomes (α is the molecular polarizability, the first-order change in the dipole moment with respect to an electric field, Section 10.1.1).

$$\Delta G_{el}(\mu) = -\frac{\varepsilon - 1}{2\varepsilon + 1}\frac{\mu^2}{a^3}\left[1 - \frac{\varepsilon - 1}{2\varepsilon + 1}\frac{2\alpha}{a^3}\right]^{-1} \tag{16.48}$$

In the SCRF model the full polarization is taking into account, i.e. the initial dipole moment generates a polarization of the medium, which changes the dipole moment, which in turn generates a slightly different polarization etc.

The effect of induced dipoles in the medium adds an extra term to the molecular Hamilton operator.

$$\mathbf{V}_\sigma = -\mathbf{r} \cdot \mathbf{R} \tag{16.49}$$

where $\mathbf{r}$ is the dipole moment operator (i.e. the position vector). $\mathbf{R}$ is proportional to the molecular dipole moment, with the proportional constant depending on the radius of the cavity and the dielectric constant.

$$\mathbf{R} = g\mathbf{\mu}$$
$$g = \frac{2(\varepsilon - 1)}{(2\varepsilon + 1)a^3} \tag{16.50}$$

At the HF level of theory, the $\mathbf{V}_\sigma$ operator corresponds to addition of an extra term to the Fock matrix elements (Section 3.5).

$$\mathbf{F}_{\alpha\beta} = \langle \chi_\alpha | \mathbf{F}_i | \chi_\beta \rangle - g\mathbf{\mu} \langle \chi_\alpha | \mathbf{r} | \chi_\beta \rangle \tag{16.51}$$

The additional integrals are just expectation values of x, y and z, and their inclusion requires very little additional computational effort. Generalization to higher-order multipoles is straightforward.

The cavity size in the Born/Onsager/Kirkwood models strongly influences the calculated stabilization. Unfortunately there is no consensus on how to choose the cavity radius. In some cases the molecular volume is calculated from the experimental density of the solvent, and the cavity radius is defined by equating the cavity volume to the molecular volume. Alternatively, the molecular volume may be calculated directly from the electronic wave function, for example by using a contour surface corresponding to an electron density of 0.001, and multiplying the obtained radius with a suitable scale factor. Or the cavity size may be based on van der Waals radii of the atoms in the molecule. More recently it has been shown that for a given theoretical method there is only one consistent cavity size, which can be calculated from the (experimental) dielectric constant and the calculated dipole moment and polarizability.[42] In any case, the underlying assumption of these models is that the molecule is roughly spherical or ellipsoid, which is only generally true for small compact molecules.

In connection with electronic structure methods (i.e. a quantal description of M), the term SCRF is quite generic, and it does not by itself indicate a specific model. Typically, however, the term is used for models where the cavity is either spherical or ellipsoidal, the charge distribution is represented as a multipole expansion, often terminated at quite low orders (for example only including the charge and dipole terms), and the cavity/ dispersion contributions are neglected. Such a treatment can only be used for a qualitative estimate of the solvent effect, although relative values may be reasonably accurate if the molecules are fairly polar (dominance of the dipole electrostatic term) and sufficiently similar in size and shape (cancellation of the cavity/dispersion terms).

The *Polarizable Continuum Model* (PCM)[43] employs a van der Waals surface type cavity, a detailed description of the electrostatic potential, and parameterizes the cavity/ dispersion contributions based on the surface area. The *COnductor-like Screening*

MOdel (COSMO) and *Solvation Models* (SM$x, x = 1$–5) also employ molecular shaped cavities, and represent the electrostatic potential by partial atomic charges. COSMO was originally implemented for semi-empirical methods, but has recently also been used in connection with *ab initio* methods.[44] Two other widely available method, the AM1-SMx and PM3-SMx models[45] have atomic parameters for fitting the cavity/dispersion energy (eq. (16.43)), and are specifically parameterized in connection with AM1 and PM3 (Section 3.10.2). The generalized Born model has also been used in connection with force field methods in the *Generalized Born/Surface Area* (GB/SA) model.[46] In this case the Coulomb interactions between the partial charges (eq. (2.19)) are combined with the generalized Born formula (16.46) by means of a function f_{ij} depending on the internuclear distance and the Born radii for each of the two atoms, a_i and a_j.

$$
G_{el}(Q_iQ_j) = -\left(1 - \frac{1}{\varepsilon}\right)\frac{Q_iQ_j}{f_{ij}}
$$
$$
f_{ij} = \sqrt{r_{ij}^2 + a_{ij}^2 e^{-D}}
$$
$$
a_{ij}^2 = \sqrt{a_i a_j} \tag{16.52}
$$
$$
D = \frac{r_{ij}^2}{(2a_{ij})^2}
$$

The cavity/dispersion terms are parameterized according to the solvent accessible surface, as in eq. (16.43).

The "mixed" solvent models, where the first solvation sphere is accounted for by including a number of solvent molecules, implicitly include the solute–solvent cavity/dispersion terms, although the corresponding terms between the solvent molecules and the continuum are usually neglected. Once discrete solvent molecules are included, however, the problem of configuration sampling arises. Nevertheless, in many cases the first solvation shell is by far the most important, and mixed models may yield substantially better results than pure continuum models, at the price of an increase in computational cost.

Given the diversity of different SCRF models, and the fact that solvation energies in water may range from a few kcal/mol for say ethane to perhaps 100 kcal/mol for an ion, it is difficult to evaluate just how accurately continuum methods may in principle be able to represent solvation. It seems clear, however, that molecular shaped cavities must be employed, the electrostatic polarization needs a description either in terms of atomic charges or quite high-order multipoles, and cavity and dispersion terms must be included. Properly parameterized, such models appear to be able to give absolute values with an accuracy of a few kcal/mol.[47] Molecular properties are in many cases also sensitive to the environment, but a detailed discussion of this is outside the scope of this book.[25]

References

1. P. E. Smith and B. M. Pettitt, *J. Phys. Chem.*, **98** (1994), 9700.
2. J. M. Haile, *Molecular Dynamics Simulation*, Wiley, 1991; M. P. Allan and D. J. Tildesley, *Computer Simulations of Liquids*, Clarendon Press, 1987; W. F. van Gunsteren and H. J. C. Berendsen, *Angew. Chem. Int. Ed.*, **29** (1990), 992; A. R. Leach, *Molecular Modelling*.

Principles and Applications, Longman, 1996; D. Frenkel and B. Smith, *Understanding Molecular Simulations*, Academic Press, 1996.

3. W. L. Jorgensen, *Adv. Chem. Phys.*, **70** (1988), 469.

4. N. A. Metropolis, A. W. Rosenbluth, A. H. Teller and E. Teller, *J. Chem. Phys.*, **21** (1953), 1087.

5. K. V. Mikkelsen, P. Linse, P.-O. Åstrand and K. Karlström, *J. Phys. Chem.*, **98** (1994), 8209.

6. B. Guillot, *J. Chem. Phys.*, **95** (1991), 1543.

7. T. P. Straatsma, *Rev. Comput. Chem.*, **9** (1996), 81.

8. D. Frenkel and B. Smith, *Understanding Molecular Simulations*, Academica Press, 1996.

9. R. W. Zwanzig, *J. Chem. Phys.*, **22** (1954), 1420.

10. J. G. Kirkwood, *J. Chem. Phys.*, **3** (1935), 300.

11. P. A. Kollman and K. M. Merz Jr, *Acc. Chem. Res.*, **23** (1990), 246; P. Kollman, *Chem. Rev.*, **93** (1993), 2395.

12. L. Verlet, *Phys. Rev.*, **159** (1967), 98.

13. S. Toxværd, *Phys. Rev. E*, **47** (1993), 343.

14. J. P. Ryckaert, G. Ciccotti and H. J. C. Berendsen, *J. Comput. Phys.*, **23** (1977), 327; D. J. Tobias and C. L. Brooks III, *J. Chem. Phys.*, **89** (1988), 5115.

15. H. J. C. Berendsen, J. P. M. Postma, W. F. van Gunsteren, A. DiNola and J. R. Haak, *J. Chem. Phys.*, **81** (1984), 3684.

16. S. Nosé, *Mol. Phys.*, **52** (1984), 255; W. G. Hoover, *Phys. Rev. A*, **31** (1985), 1695.

17. H. Bekker, *J. Comput. Chem.*, **18** (1997), 1930.

18. P. P. Ewald, *Ann. Phys.*, **64** (1921), 253; V. Natoli and P. M. Ceperley, *J. Comput. Phys.*, **117** (1995), 171; Z.-M. Chen, T. Cagin and W. A. Goddard III, *J. Comput. Chem.*, **18** (1997), 1365.

19. H. G. Petersen, *J. Chem. Phys.*, **103** (1995). 3668.

20. L. Greengard and V. Rokhlin, *J. Comput. Phys.*, **73** (1987), 325.

21. J. M. Perez-Jorda and W. Yang, *Chem. Phys. Lett.,* **282** (1998), 71.

22. H. G. Petersen, D. Soelvason, J. W. Perram and E. R. Smith, *J. Chem. Phys.*, **101** (1994), 8870.

23. D. G. Truhlar, R. Steckler and M. S. Gordon, *Chem. Rev.*, **87** (1987), 217.

24. R. Car and M. Parrinello, *Phys. Rev. Lett.*, **55** (1985), 2471; D. K. Remler and P. A. Madden, *Mol. Phys.*, **70** (1990), 921.

25. S. He and H. A. Scheraga, *J. Chem. Phys.*, **108** (1998), 271, 278.

26. G. D. Billing and K. V. Mikkelsen, *Introduction to Molecular Dynamics and Chemical Kinetics*, Wiley, 1996; G. D. Billing and K. V. Mikkelsen, *Advanced Molecular Dynamics and Chemical Kinetics*, Wiley, 1997.

27. D. G. Truhlar, R. Steckler and M. S. Gordon, *Chem. Rev.*, **87** (1987), 217; T.-S. Ho, T. Hollebeek, H. Rabitz, L. B. Harding and G. C. Schatz, *J. Chem. Phys.*, **105** (1996), 10462.

28. M. A. Collins, *Adv. Chem. Phys.*, **93** (1996), 389.

29. Y.-P. Liu, D.-H. Lu, A. Gonzales-Lafont, D. G. Truhlar and B. C. Garrett, *J. Am. Chem. Soc.*, **115** (1993), 7806.

30. G. D. Billing, *Mol. Phys.*, **89** (1996), 355.

31. R. P. Bell, *The Tunnel Effect in Chemistry*, Chapman and Hall, 1980.

32. E. Wigner, *Z. Phys. Chem. B*, **19** (1932), 203.

33. Y.-P. Liu, D.-H. Lu, A. Gonzales-Lafont, D. G. Truhlar and B. C. Garrett, *J. Am. Chem. Soc.*, **115** (1993), 7806; J. C. Corchado, J. Espinosa-Garcia, W.-P. Hu, I. Rossi and D. G. Truhlar, *J. Phys. Chem.*, **99** (1997), 687.

34. K. V. Mikkelsen and H. Ågren, *J. Mol. Struct. (Theochem.)*, **234** (1991), 425.

35. J. Tomasi and M. Persico, *Chem. Rev.*, **94** (1994), 2027.

36. J. B. Foresman, T. A. Keith, K. B. Wiberg, J. Snoonian and M. J. Frisch, *J. Phys. Chem.*, **100** (1996), 16098.

37. K. V. Mikkelsen and K. O. Sylvester-Hvid, *J. Phys. Chem.*, **100** (1996), 9116.

38. M. Born, *Z. Physik*, **1** (1920), 45.

39. L. Onsager, *J. Am. Chem. Soc.*, **58** (1936), 1486.
40. J. G. Kirkwood, *J. Chem. Phys.*, **2** (1934), 351.
41. J. G. Kirkwood, F. H. Westheimer, *J. Chem. Phys.*, **6** (1936), 506.
42. Y. Luo, H. Ågren and K. V. Mikkelsen, *Chem. Phys. Lett.*, **275** (1997), 145.
43. M. Cossi, V. Barone, R. Cammi and J. Tomasi, *Chem. Phys. Lett.*, **255** (1996), 327.
44. A. Klamt, *J. Phys. Chem.*, **99** (1995), 2224; T. Truong and E. V. Stefanovich, *Chem. Phys. Lett.*, **240** (1995), 253.
45. C. J. Cramer and D. G. Truhlar, *Rev. Comput. Chem.*, **6** (1995), 1.
46. W. C. Still, A. Tempczyrk, R. C. Hawley and T. Hensrickson, *J. Am. Chem. Soc.*, **112** (1990), 6127; M. R. Reddy, M. D. Erion, A. Agarwal, V. N. Viswanadhan, D. Q. McDonald and W. C. Still, *J. Com. Chem.,* **19** (1998), 769.
47. V. Barone, M. Cossi and J. Tomasi, *J. Chem. Phys.*, **107** (1997), 3210.

17 Concluding Remarks

The real world is very complex. A complete description is therefore also very complicated. Only by imposing a series of often quite stringent limitations and approximations can a problem be reduced in complexity so that it may be analysed in some detail, for example by calculations. A chemical reaction in the laboratory may involve perhaps 10^{20} molecules surrounded by 10^{24} solvent molecules, in contact with a glass surface and interacting with gasses (N_2, O_2, CO_2, H_2O, etc.) in the atmosphere. The whole system will be exposed to a flux of photons of different frequency (light) and a magnetic field (from the earth), and possibly also a temperature gradient from external heating. The dynamics of all the particles (nuclei and electrons) is determined by relativistic quantum mechanics, and the interaction between particles is governed by quantum electrodynamics. In principle the gravitational and strong (nuclear) forces should also be considered. For chemical reactions in biological systems, the number of different chemical components will be large, involving different ions and assemblies of molecules behaving intermediately between solution and solid state (e.g. lipids in cell walls).

Except for a couple of rather extreme areas (like the combination of general relativity and quantum mechanics, or the unification of the strong and gravitational forces with the electroweak interaction), we believe that all the fundamental physics is known. The "only" problem is that the real world contains so many (different) components interacting by complicated potentials that a detailed description is impossible.

As this book hopefully has given some idea of, the key is to know what to neglect or approximate when trying to obtain answers to specific questions in predefined systems. For chemical problems only the electrostatic force needs to be considered; the gravitational interaction is a factor of 10^{39} weaker, and can be completely neglected. Similarly, the strong nuclear force is so short-ranged that it has no effect on chemical phenomena (although the brief claims regarding "cold" fusion for a period seemed to contradict this). The weak interaction, which is responsible for radioactive decay by the $n \rightarrow p + e$ process, is also much smaller than the electrostatic, although there have been various estimates of whether it might give rise to a symmetry breaking (i.e. preference of one enantiomer over its mirror image) which possibly could be detected experimentally. Similarly, the earth's magnetic field is so tiny that only under very special circumstances can it have any detectable influence on the outcome of a chemical reaction.

Other approximations used in computational chemistry are more severe, and affect the accuracy of the results. Some examples are:

(1) Neglect of relativistic effects, by using the Schrödinger instead of the Dirac equation. This is reasonably justified in the upper part of the periodic table, but not in the lower half. For some phenomena, like spin–orbit coupling, there is no classical counterpart, and only a relativistic treatment can provide an understanding.

(2) Quantum mechanics being replaced (wholly or partly) by classical mechanics. For electrons such an approximation would lead to disastrous results, but for nuclei (atoms) the quantum effects are sufficiently small that in most cases they can be neglected.

(3) Assumption of a rigorous separation of nuclear and electronic motions (Born–Oppenheimer approximation). In most cases this is a quite good approximation, and there is a good understanding of when it will fail. There are, however, very few general techniques for going beyond the Born–Oppenheimer approximation.

(4) Approximating the intermolecular interactions to only include two-body effects, e.g. electrostatic forces are only calculated between pairs of fixed atomic charges in force field techniques. Or the discrete interactions between molecules may be treated only in an average fashion, by using Langevin dynamics instead of molecular dynamics.

(5) Calculating ensemble or time averages over a relatively small number of points (perhaps a few million) and a limited number of particles (perhaps a few hundred), instead of something which approaches a macroscopic sample of perhaps 10^{20} molecules/configurations.

(6) Finite temperature being reduced to zero Kelvin, i.e. the use of static structures to represent molecules, rather than treating them as an ensemble of molecules in a distribution of states (translational, rotational and vibrational) corresponding to a (macroscopic) temperature.

(7) Approximating a many-electron wave function by a finite sum of Slater determinants, e.g. truncating the CI, CC or MBPT wave function to include only certain excitation types.

(8) Approximating a one-electron wave function (orbital) by an expansion in a finite basis set.

(9) Making approximations in the Hamiltonian describing the system, e.g. semi-empirical electronic structure methods.

(10) Approximating any external fields (electric or magnetic) by considering only their linear components. For normal conditions, this will be quite a good approximation, however, this is not the case for intense laser fields for example.

Most of these approximations are mainly of a computational nature; there are well-defined methods available for going beyond the approximations, but they are computationally too demanding. The key is therefore to be able to evaluate what level of theory (i.e. which approximations are appropriate) is required to obtain results which are sufficiently accurate to provide useful information about the question at hand.

Appendix A

Notation

Bold quantities are operators, vectors, matrices or tensors. Plain symbols are scalars.

$\boldsymbol{\alpha}$	Polarizability
α, β	Spin functions
$\boldsymbol{\alpha}$, $\boldsymbol{\beta}$	Dirac 4×4 spin matrices
$\alpha\beta\gamma\delta$	Summation indices for basis functions
α, β, γ, δ, ζ	Basis function exponents
α_A, β_{AB}	Hückel parameters for atom A, and between atoms A and B
a	Born radius for solvation cavities
$abcd$	Summation indices for virtual MOs
a_n, a_i, b_i, c_j..	Expansion coefficients
A	Electron affinity
A	Helmholtz free energy
$\mathbf{A}$	Antisymmetrizing operator
$\mathbf{A}$	Vector potential
$\boldsymbol{\beta}$	First hyperpolarizability
β_μ	Resonance parameter in semi-empirical theory
$\mathbf{B}$	Magnetic field (magnetic induction)
χ, μ, ν, λ, σ	Basis functions (atomic orbitals), *ab initio* or semi-empirical methods
$\mathbf{X}$	Gauge including basis function
χ^B	Out-of-plane angle for atom B
$\boldsymbol{\chi}$	Magnetic susceptibility
c	Speed of light
$c_{\alpha i}$	MO expansion coefficients
δ	An infinitesimal variation or quantity
δ_{ij}	Kroeniker delta ($\delta_{ij} = 1$ for $i = j$, $\delta_{ij} = 0$ for $i \neq j$)
$\delta(\mathbf{r})$	Dirac delta function ($\delta(\mathbf{r}) = 0$ for $\mathbf{r} \neq \mathbf{0}$)
Δ	An finite difference or quantity
d	Distance
$\mathbf{d}$	De-excitation operator
D	Dissociation energy

D	Density matrix
$D_{\alpha\beta}$	Density matrix element in AO basis
ε	Matrix eigenvalue
ε	van der Waals parameter
ε	Dielectric constant
ε	Energy, for one electron or as an individual term in a sum
$\boldsymbol{\varepsilon}$	Energy matrix
e	Excitation operator
E	Energy, many particles or terms
E_e	Electronic energy
$E[\rho]$	Energy functional
ϕ	Molecular orbital
Φ	Slater determinant or similar approximate wave function
f_i	Gradient component along a Hessian eigenvector
F	Electric field
F	Force constant matrix
F	Fock operator or Fock matrix
$F_{ij}, F_{\alpha\beta}$	Fock matrix element in MO and AO basis
γ	Second hyperpolarizability
γ_k	Density matrix of order k
g_e	Electronic g-factor
g_A	Nuclear g-factor
g	Two-electron repulsion operator
g	Hyperfine coupling constant
g	Gradient (first derivative)
G	Gibbs free energy
G_{xy}	Coulomb type matrix elements in semi-empirical theory ($x,y = $s,p,d)
G	Matrix containing square root of inverse atomic masses
η	An infinitesimal scalar
η	Absolute hardness
h	Planck's constant
$\hbar$	$h/2\pi$
h	Core or other effective one-electron operator
$h_{\alpha\beta}$	Matrix element of a one-electron operator in AO basis
$h_{\mu\nu}$	Matrix element of a one-electron operator in semi-empirical theory
$\mathbf{h}, \mathbf{h}_1, \mathbf{h}_2, \ldots$	Excitation and de-excitation operators
H	Enthalpy
H	Hessian (second derivative matrix of a function)
$\mathbf{H}, \mathbf{H}_e, \mathbf{H}_n$	Hamilton operator or Hamilton matrix (general, electronic, nuclear)
H_{ij}	Matrix element of a Hamilton operator between Slater determinants
H_{xy}	Exchange type matrix elements in semi-empirical theory ($x,y = $s,p,d)
$ijkl$	Summation indices for occupied MOs
I	Moment of inertia
I	Ionization potential
I	Nuclear magnetic moment or spin
J	Spin–spin coupling constant
J	Coulomb operator

J_{ij}	Coulomb integral
$J[\rho]$	Coulomb functional
κ	Lagrange multiplier
κ	Transmission coefficient
κ	Compressibility constant
k_B	Boltzmann constant
$k, k^{AB...}$	Force constant (for atoms A, B, . . .)
$\mathbf{K}$	Anharmonic constants (third derivative)
$\mathbf{K}$	Exchange operator
K_{ij}	Exchange integral
$K[\rho]$	Exchange functional
λ	Lagrange multiplier
λ	General perturbation strength
λ	Hessian shift parameter
L	Angular momentum quantum number
L	Lagrange function
$\mathbf{L}$	Angular momentum operator
μ	Mulliken electronegativity, chemical potential
μ	Reduced mass
$\boldsymbol{\mu}$	Dipole moment
μ_0	Vacuum permeability
μ_B	Bohr magneton
μ_N	Nuclear magneton
m	Mass, general or electron mass
M_A	Nuclear mass
M	Number of basis functions or configurations
ν	Vibrational frequency
n_i	Orbital occupation number
N	Number of nuclei, atoms, electrons or molecules
N_A	Avogadro's number
$\mathbf{O, P, Q, R, S}$	General operators
π	Generalized momentum operator
Π	Product of diagonal elements in a Slater determinant
$\mathbf{p}$	Momentum operator
$\mathbf{p}$	Momentum vector(s)
$\mathbf{P}, \mathbf{P}_{ij}$	Permutation operators (permuting indices i and j)
$\mathbf{P}_1, \mathbf{P}_2$	Perturbation operators (one- and two-electron)
q	Charge on a particle, integer
Q	Atomic charge (can be fractional), fitted or from population analysis
q	Partition function (one particle)
Q	Partition function (many particles)
$\mathbf{q}$	Normal coordinate
$\mathbf{Q}$	Quadrupole moment
ρ	Electron density
$\mathbf{r}$	Position vector(s), general or electronic
r_{ij}	Distance between electrons i and j
R	Trust radius

R	Gas constant
$\mathbf{R}$	Position vector, nuclear
R_{ij}, R_{AB}, R^{AB}	Distance between atoms or nuclei, i and j or A and B
r, θ, ϕ	Polar coordinates
σ	Order of rotational subgroup
σ	Charge density
$\boldsymbol{\sigma}_{x,y,z}$	Pauli 2×2 spin matrices
$\mathbf{s}$	Electron spin operator
S	Entropy
$\mathbf{S}^2$	Spin squared operator
$S_{\alpha\beta}$	Overlap matrix element in AO basis
$\theta(t)$	Heaviside step function ($\theta(t) = 0$ for $t < 0$, $\theta(t) = 1$ for $t > 0$)
θ^{ABC}	Angle between atoms A, B and C
Δt	Small (finite) timestep
t	Time
τ	Heat or pressure bath coupling parameter
t_i, t_{ij}	Cluster amplitudes
T	Temperature
$\mathbf{T}$, $\mathbf{T}_1$, $\mathbf{T}_2$, ...	Cluster operator (general, single, double, ... excitations)
$\mathbf{T}$, $\mathbf{T}_e$, $\mathbf{T}_n$	Kinetic energy operator (general, electronic, nuclear)
$T[\rho]$	Kinetic energy functional, exact
$T_S[\rho]$	Kinetic energy functional, calculated from a Slater determinant
U	Internal energy
$\mathbf{U}$	Unitary matrix
U_i	Matrix element of a semi-empirical one-electron operator, usually parameterized
v	Velocity
V	Volume
V, V^{AB}	Potential energy (between atoms A and B)
V_{ij}	Coulomb potential between particles i and j
$\mathbf{V}$, $\mathbf{V}_{ee}$, $\mathbf{V}_{ne}$, $\mathbf{V}_{nn}$	Potential (Coulomb) energy operator (general, electron–electron, nuclear–electron, nuclear–nuclear)
$V[\rho]$	Potential energy functional
ω	Frequency associated with an electric or magnetic field
ω	Harmonic vibrational frequency
ω^{ABCD}	Torsional angle between atoms A, B, C and D
$\boldsymbol{\Omega}$	Two-electron operator
W	Energy of an approximate wave function
W_i	Perturbation energy correction at order i
$\mathbf{W}$	Energy-weighted density matrix
$W_{\alpha\beta}$	Energy-weighted density matrix element in AO basis
$W^*_{\sigma\beta}$	Weighting factor in pseudospectral methods
ξ	Magnetizability
x_i, y_i, z_i	Cartesian coordinates for particle i
Δx_i	Component in a vector
Ψ, Ψ_e, Ψ_n	Exact or multi-determinant wave function (general, electronic, nuclear)

ζ	Spin polarization
ζ	Friction coefficient
ζ	Molecular surface parameter for calculating solvation energies
Z	Nuclear charge, exact
Z'	Nuclear charge, reduced by the number of core electrons
$\langle n|$	Bra
$|n\rangle$	Ket
$\langle n|\mathbf{O}|m\rangle$	Bracket (matrix element) of operator $\mathbf{O}$ between functions n and m
$\langle \mathbf{O} \rangle$	Average value of $\mathbf{O}$
$|\mathbf{O}|$	Norm of $\mathbf{O}$
$\langle\langle \mathbf{P};\mathbf{Q} \rangle\rangle$	Propagator of $\mathbf{P}$ and $\mathbf{Q}$
$[\mathbf{P},\mathbf{Q}]$	Commutator of $\mathbf{P}$ and $\mathbf{Q}$ ($=\mathbf{PQ}-\mathbf{QP}$)
∇	Gradient operator
∇^2	Laplace operator
$\cdot$	Dot product
$\times$	Cross product
$\nabla\cdot$	Divergence operator
$\nabla\times$	Curl operator
t	Vector transposition
$\dagger$	Complex conjugate

Appendix B

The Variational Principle

The *Variational Principle* states that an approximate wave function has an energy which is above or equal to the exact energy. The equality holds only if the wave function is exact. The proof is as follows.

Assume that we know the exact solutions to the Schrödinger equation.

$$\mathbf{H}\Psi_i = E_i\Psi_i, \quad i = 0, 1, 2, \ldots, \infty \tag{B.1}$$

There are infinitely many solutions and we assume that they are labelled according to their energies, E_0 being the lowest. Since the $\mathbf{H}$ operator is Hermitic, the solutions form a complete basis. We may furthermore chose the solutions to be orthogonal and normalized.

$$\langle \Psi_i | \Psi_j \rangle = \delta_{ij} \tag{B.2}$$

An approximate wave function can be expanded in the exact solutions, since they form a complete set.

$$\Phi = \sum_{i=0}^{\infty} a_i \Psi_i \tag{B.3}$$

The energy of an approximate wave function is calculated as

$$W = \frac{\langle \Phi | \mathbf{H} | \Phi \rangle}{\langle \Phi | \Phi \rangle} \tag{B.4}$$

Inserting the expansion (B.3) we obtain

$$W = \frac{\sum_{i=0}^{\infty} \sum_{j=0}^{\infty} a_i a_j \langle \Psi_i | \mathbf{H} | \Psi_j \rangle}{\sum_{i=0}^{\infty} \sum_{j=0}^{\infty} a_i a_j \langle \Psi_i | \Psi_j \rangle} \tag{B.5}$$

Using that $\mathbf{H}\Psi_i = E_i\Psi_i$ and the orthonormality of the Ψ_i (eqs. (B.1) and (B.2)), we get

$$W = \frac{\sum_{i=0}^{\infty} a_i^2 E_i}{\sum_{i=0}^{\infty} a_i^2} \tag{B.6}$$

The variational principle states that $W \geq E_0$ or equivalently, $W - E_0 \geq 0$.

$$W - E_0 = \frac{\sum_{i=0}^{\infty} a_i^2 E_i}{\sum_{i=0}^{\infty} a_i^2} - E_0 = \frac{\sum_{i=0}^{\infty} a_i^2 (E_i - E_0)}{\sum_{i=0}^{\infty} a_i^2} \geq 0 \qquad (B.7)$$

Since a_i^2 always is positive or zero, and $E_i - E_0$ always is positive or zero (E_0 is by definition the lowest energy), this completes the proof. Furthermore, in order for the equal sign to hold, all $a_{i \neq 0} = 0$ since $E_{i \neq 0} - E_0$ is non-zero (neglecting degenerate ground states). This in turns means that $a_0 = 1$, owing to the normalization of Φ, and consequently the wave function is the exact solution.

This proof shows that any approximate wave function will have an energy above or equal to the exact ground-state energy. There is a related theorem, known as *MacDonald's Theorem*, which states that the nth root of a set of secular equations (e.g. a CI matrix) is an upper limit to the $(n-1)$th excited exact state, within the given symmetry subclass.[1] In other words, the lowest root obtained by diagonalizing a CI matrix is an upper limit to the lowest exact wave functions, the 2nd root is an upper limit to the exact energy of the first excited state, the 3rd root is an upper limit to the exact second excited state and so on.

The Hohenberg–Kohn Theorems

In wave mechanics the electron density is given by the square of the wave function integrated over $N - 1$ electron coordinates, and the wave function is determined by solving the Schrödinger equation. For a system of M nuclei and N electrons, the electronic Hamilton operator contains the following terms.

$$\mathbf{H}_e = -\sum_{i=1}^{N} \frac{1}{2} \mathbf{V}_i^2 - \sum_{i=1}^{N} \sum_{A=1}^{M} \frac{Z_A}{|\mathbf{R}_A - \mathbf{r}_i|} + \sum_{i=1}^{N} \sum_{j>1}^{N} \frac{1}{|\mathbf{r}_i - \mathbf{r}_j|} + \sum_{A=1}^{M} \sum_{B=A}^{M} \frac{Z_A Z_B}{|\mathbf{R}_A - \mathbf{R}_B|}$$

$$(B.8)$$

Within the Born–Oppenheimer approximation, the last term is a constant. It is seen that the Hamilton operator is uniquely determined by the number of electrons and the potential created by the nuclei, $\mathbf{V}_{ne}$, i.e. the nuclear charges and positions. This means that the ground-state wave function (and thereby the electron density) and ground state energy are also given uniquely by these quantities.

Assume now that two different external potentials (which may be from nuclei), $\mathbf{V}_{ext}$ and $\mathbf{V}'_{ext}$, result in the same electron density, ρ. Two different potentials imply that the two Hamilton operators are different, $\mathbf{H}$ and $\mathbf{H}'$, and the corresponding lowest energy wave functions are different, Ψ and Ψ'. Taking Ψ' as an approximate wave function for H and using the variational principle yields

$$\langle \Psi' | \mathbf{H} | \Psi' \rangle > E_0$$
$$\langle \Psi' | \mathbf{H}' | \Psi' \rangle + \langle \Psi' | \mathbf{H} - \mathbf{H}' | \Psi' \rangle > E_0$$
$$E_0' + \langle \Psi' | \mathbf{V}_{ext} - \mathbf{V}'_{ext} | \Psi' \rangle > E_0 \qquad (B.9)$$
$$E_0' + \int \rho(\mathbf{r})(\mathbf{V}_{ext} - \mathbf{V}'_{ext}) d\mathbf{r} > E_0'$$

Similarly, taking Ψ as an approximate wave function for H' yields

$$E_0 - \int \rho(\mathbf{r})(\mathbf{V}_{ext} - \mathbf{V}'_{ext})d\mathbf{r} > E'_0 \tag{B.10}$$

Addition of these two inequalities gives $E'_0 + E_0 > E'_0 + E_0$, showing that the assumption was wrong. In other words, for the ground state there is a one-to-one correspondence between the electron density and the nuclear potential, and thereby also with the Hamilton operator and the energy. In the language of Density Functional Theory, the energy is a unique functional of the electron density, $E[\rho]$.

Using the electron density as a parameter, there is a variational principle analogous to that in wave mechanics. Given an approximate electron density ρ' (assumed to be positive definite everywhere) which integrates to the number of electrons, N, the energy given by this density is an upper bound to the exact ground-state energy, provided that the exact functional is used.

$$\int \rho'(\mathbf{r})d\mathbf{r} = N$$
$$E_0[\rho'] \geq E_0[\rho] \tag{B.11}$$

The Adiabatic Connection Formula

The Hellmann–Feynman theorem (eq. (10.25)) states that

$$\frac{\partial}{\partial\lambda}\langle\Psi_\lambda|\mathbf{H}_\lambda|\Psi_\lambda\rangle = \left\langle\Psi_\lambda\left|\frac{\partial\mathbf{H}_\lambda}{\partial\lambda}\right|\Psi_\lambda\right\rangle \tag{B.12}$$

With the Hamiltonian in eq. (6.3), this gives

$$\mathbf{H}_\lambda = \mathbf{T} + \mathbf{V}_{ext}(\lambda) + \lambda\mathbf{V}_{ee}$$
$$\frac{\partial}{\partial\lambda}\langle\Psi_\lambda|\mathbf{H}_\lambda|\Psi_\lambda\rangle = \left\langle\Psi_\lambda\left|\frac{\partial\mathbf{V}_{ext}(\lambda)}{\partial\lambda} + \mathbf{V}_{ee}\right|\Psi_\lambda\right\rangle \tag{B.13}$$

Integrating over λ between the limits 0 and 1 corresponds to smoothly transforming the non-interacting reference to the real system. This integration is done under the assumption that the density remains constant.

$$\int_0^1 \frac{\partial}{\partial\lambda}\langle\Psi_\lambda|\mathbf{H}_\lambda|\Psi_\lambda\rangle d\lambda = \int_0^1\left\langle\Psi_\lambda\left|\frac{\partial\mathbf{V}_{ext}(\lambda)}{\partial\lambda} + \mathbf{V}_{ee}\right|\Psi_\lambda\right\rangle d\lambda$$
$$\langle\Psi_1|\mathbf{H}_1|\Psi_1\rangle - \langle\Psi_0|\mathbf{H}_0|\Psi_0\rangle = \langle\Psi_1|\mathbf{V}_{ext}(1)|\Psi_1\rangle - \langle\Psi_0|\mathbf{V}_{ext}(0)|\Psi_0\rangle$$
$$+ \int_0^1\langle\Psi_\lambda|\mathbf{V}_{ee}|\Psi_\lambda\rangle d\lambda$$
$$E_1 - E_0 = \int\rho(\mathbf{r})\mathbf{V}_{ext}(1)d\mathbf{r} - \int\rho(\mathbf{r})\mathbf{V}_{ext}(0)d\mathbf{r} + \int_0^1\langle\lambda|\mathbf{V}_{ee}|\lambda\rangle d\lambda \tag{B.14}$$

Note that it is the <u>same</u> density that appears in the two integrals over $\mathbf{V}_{ext}(0)$ and $\mathbf{V}_{ext}(1)$, owing to the assumption that the density derived from Ψ_0 and Ψ_1 is constant. The energy of the non-interacting system is given by

$$E_0 = \langle \Psi_0 | \mathbf{T} | \Psi_0 \rangle + \int \rho(\mathbf{r}) \mathbf{V}_{\text{ext}}(0) d\mathbf{r} \tag{B.15}$$

and thereby

$$E_1 = \langle \Psi_0 | \mathbf{T} | \Psi_0 \rangle + \int \rho(\mathbf{r}) \mathbf{V}_{\text{ext}}(1) d\mathbf{r} + \int_0^1 \langle \lambda | \mathbf{V}_{\text{ee}} | \lambda \rangle d\lambda \tag{B.16}$$

Using that $\mathbf{V}_{\text{ext}}(1) = \mathbf{V}_{\text{ne}}$, separating out the Coulomb part of $\mathbf{V}_{\text{ee}}$ and using the definition of E_1 (eq. (6.7)), gives the *adiabatic connection formula* (eq. (6.32)).

$$E_{\text{xc}} = \int_0^1 \langle \Psi_\lambda | \mathbf{V}_{\text{xc}}(\lambda) | \Psi_\lambda \rangle d\lambda \tag{B.17}$$

Reference

1. J. K. L. MacDonald, *Phys. Rev.*, **43** (1933), 830.

Appendix C

First and Second Quantization

The notation used in this book is in terms of *first quantization*. The electronic Hamilton operator, for example, is written as (eq. (3.23))

$$\mathbf{H}_e = -\sum_{i=1}^{N} \frac{1}{2}\mathbf{V}_i^2 - \sum_{i=1}^{N}\sum_{A=1}^{M} \frac{Z_A}{|\mathbf{R}_A - \mathbf{r}_i|} + \sum_{i=1}^{N}\sum_{j>1}^{N} \frac{1}{|\mathbf{r}_i - \mathbf{r}_j|} + \sum_{A=1}^{M}\sum_{B>A}^{M} \frac{Z_A Z_B}{|\mathbf{R}_A - \mathbf{R}_B|}$$

(C.1)

and a single determinant wave function is given as an antisymmetrized product of orbitals (eq. (3.21)).

$$\Phi = \mathbf{A}[\phi_1 \phi_2 \dots \phi_N]$$

(C.2)

A matrix element of the Hamilton operator over the wave function in eq. (C.2) may be written as a sum over one- and two-electron integrals (eq. (3.33)).

$$\langle \Phi | \mathbf{H}_e | \Phi \rangle = \sum_i^N \langle \phi_i | \mathbf{h}_i | \phi_i \rangle + \frac{1}{2}\sum_{ij}^N (\langle \phi_i \phi_j | \mathbf{g} | \phi_i \phi_j \rangle - \langle \phi_i \phi_j | \mathbf{g} | \phi_j \phi_i \rangle) + V_{nn}$$

(C.3)

There is another commonly used notation known as *second quantization*.[1] In this language the wave function is written as a series of *creation operators* acting on the *vacuum state*. A creation operator $\mathbf{a}_i^\dagger$ working on the vacuum generates an (occupied) molecular orbital i.

$$\phi_i = \mathbf{a}_i^\dagger |0\rangle$$
$$\Phi = \mathbf{A}[\phi_1 \phi_2 \dots \phi_N] \Longleftrightarrow \Phi = \mathbf{a}_1^\dagger \mathbf{a}_2^\dagger \dots \mathbf{a}_N^\dagger |0\rangle$$

(C.4)

The opposite of a creation operator is an *annihilation* operator $\mathbf{a}_i$ which removes orbital i from the wave function it is acting on. The $\mathbf{a}_i^\dagger \mathbf{a}_j$ product of operators removes orbital j and creates orbital i, i.e. replaces the occupied orbital j with an unoccupied orbital i. The antisymmetry of the wave function is built into the operators as they obey the following anti-commutation relationships.

$$\mathbf{a}_i^\dagger \mathbf{a}_j^\dagger = -\mathbf{a}_j^\dagger \mathbf{a}_i^\dagger$$
$$\mathbf{a}_i \mathbf{a}_j = -\mathbf{a}_j \mathbf{a}_i \qquad\qquad (C.5)$$
$$\mathbf{a}_i^\dagger \mathbf{a}_j + \mathbf{a}_j \mathbf{a}_i^\dagger = \delta_{ij}$$

This shows that $\mathbf{a}_i^\dagger \mathbf{a}_i^\dagger = \mathbf{a}_i \mathbf{a}_i = 0$, i.e. attempting to create an already existing orbital, or trying to delete a non-existing orbital, generates a zero wave function.

The Hamilton operator (eq. (C.1)) in second quantization is given as (note that the summation now is over basis functions)

$$\mathbf{H}_e = \sum_i^M \langle \phi_i | \mathbf{h}_i | \phi_i \rangle \mathbf{a}_i^\dagger \mathbf{a}_i + \frac{1}{2} \sum_{ij}^M (\langle \phi_i \phi_j | \mathbf{g} | \phi_i \phi_j \rangle - \langle \phi_i \phi_j | \mathbf{g} | \phi_j \phi_i \rangle) \mathbf{a}_i^\dagger \mathbf{a}_j \mathbf{a}_j^\dagger \mathbf{a}_i + V_{nn}$$

$$(C.6)$$

The $\mathbf{a}_i^\dagger \mathbf{a}_i$ operator tests whether orbital i exists in the wave function, if that is the case, a one-electron orbital matrix element is generated, and similarly for the two-electron terms. Using the Hamiltonian in eq. (C.6) with the wave function in eq. (C.4) generates the first quantized operator in eq. (C.3).

There are several advantage of second quantization over first quantization.

(1) The number of particles (electrons) is transferred from the operator to the wave function, i.e. the Hamilton operator looks the same, independent of the size of the system.
(2) The Hamiltonian in the second quantization directly includes the approximations due to a finite basis set, i.e. only in a complete basis are the operators in eqs. (C.1) and (C.6) the same.
(3) The somewhat awkward antisymmetrizing operator necessarily in first quantization is replaced by formal rules for manipulating creation and annihilation operators.
(4) Expressions in general are easier to manipulate with formal rules, the same derivations in first quantization often require many explicit summation indices.
(5) The Slater–Condon rules (Section 4.2.1) are incorporated in the operators.
(6) There is a clear one-to-one correspondence between the theoretical expressions and the computational implementation in terms of one- and two-electron matrix elements. Implementations of the expressions are therefore facilitated.

Reference

1. J. Olsen, in *Lecture Notes in Quantum Chemistry*, ed. B. O. Roos, Springer-Verlag, 1992.

Appendix D

Atomic Units

In electronic structure calculations it is convenient to work in the *atomic unit* (a.u.) system, which is defined by setting $m_e = e = \hbar = 1$. From this follows related quantities, as shown in Table D.1.

Table D.1. The atomic unit system

Symbol	Quantity	Value in a.u.	Value in SI units
m_e	Electron mass	1	9.110×10^{-31} kg
e	Electron charge	1	1.602×10^{-19} C
$\hbar$	$h/2\pi$ (atomic momentum unit)	1	1.055×10^{-34} Js
h	Planck's constant	2π	6.626×10^{-34} Js
a_0	Bohr radius (atomic distance unit)	1	5.292×10^{-11} m
E_H	Hartree (atomic energy unit)	1	4.360×10^{-18} J
c	Speed of light	137.036	2.998×10^8 m/s
μ_B	Bohr magneton $(= e\hbar/2m_e)$	1/2	9.274×10^{-24} J/T
μ_N	Nuclear magneton	2.723×10^{-4}	5.051×10^{-27} J/T
$4\pi\varepsilon_0$	Vacuum permittivity	1	1.113×10^{-10} C^2/J m
μ_0	Vacuum permeability $(4\pi/c^2)$	6.692×10^{-4}	1.257×10^{-6} N s^2/C^2

Other commonly used energy units are kcal/mol and KJ/mol. The conversion factors are: 1 a.u. $= 627.51$ kcal/mol, and 1 kcal/mol $= 4.184$ KJ/mol.

Appendix E

Z-matrix Construction

All calculations need as input a molecular geometry. This is commonly given by one of the following three methods.

(1) Cartesian coordinates.
(2) Internal coordinates.
(3) Via a graphical interface.

Generating Cartesian coordinates by hand is only realistic for small molecules. If, however, the geometry is taken from outside sources, like for example an X-ray structure, Cartesian coordinates are often the natural choice. Similarly, a graphical interface produces a set of Cartesian coordinates for the underlying program, which carries out the actual calculation.

Generating internal coordinates, like bond lengths and angles, by hand is relatively simple, even for quite large molecules. One widely used method is the *Z-matrix* where each atom is specified in terms of a distance, angle and torsional angle relative to other atoms. It should be noted that internal coordinates are <u>not</u> necessarily related to the actual bonding, they are only a convenient method for specifying the geometry. The internal coordinates are usually converted to Cartesian coordinates before any calculations are carried out. Geometry optimizations, however, are often done in internal coordinates in order to remove the six (five) translational and rotational degrees of freedom.

Construction of a Z-matrix begins with a drawing of the molecule, and a suitable numbering of the atoms. <u>Any</u> numbering will result in a valid Z-matrix, although assignment of numerical values to the parameters is greatly facilitated if the bonding and symmetry of the molecule is considered (see the examples below). The Z-matrix specifies the position of each atom in terms of a distance, an angle and a torsional angle relative to other atoms. The first three atoms, however, are slightly different. The first atom is always positioned at the origin of the coordinate system. The second atom is specified as having a distance to the first atom, and is placed along one of the Cartesian axes (usually *x* or *z*). The third atom is specified by a distance to either atom 1 or 2, and an angle to the other atom. All subsequent atoms need a distance, an angle and a torsional angle to uniquely specify their position. The atoms are normally identified either by the chemical symbol or by their atomic number.

If the molecular geometry is optimized by the program, only a rough estimate of the parameters is necessary. In term of internal coordinates, this is fairly easy. Some typical bond lengths (Å) and angles are given below.

A–H: A=C: 1.10, A=O, N: 1.00, A=S, P: 1.40
A–B: A,B=C,O,N: 1.40–1.50
A=B: A,B=C,O,N: 1.20–1.30
A≡B: A,B=C,N: 1.20
A–B: A=C, B=S, P: 1.80

Angles around sp^3 hybridized atoms: $110°$
Angles around sp^2 hybridized atoms: $120°$
Angles around sp hybridized atoms: $180°$

Torsional angles around sp^3 hybridized atoms: separated by $120°$.
Torsional angles around sp^2 hybridized atoms: separated by $180°$.

Such estimates allow specification of molecules with up to $\sim 50–100$ atoms fairly easy. For larger molecules, however, this becomes cumbersome. In such cases the molecule is often built from preoptimized fragments. This is typically done by means of a graphical interface, i.e. the molecule is pieced together by selecting fragments (like amino acids) and assigning the bonding between the fragments.

Below are some examples of how to construct Z-matrices. Figure E.1 shows acetaldehyde.

C1	0	0.00	0	0.0	0	0.0
O2	1	1.20	0	0.0	0	0.0
H3	1	1.10	2	120.0	0	0.0
C4	1	1.50	2	120.0	3	180.0
H5	4	1.10	1	110.0	2	0.0
H6	4	1.10	1	110.0	2	120.0
H7	4	1.10	1	110.0	2	– 120.0

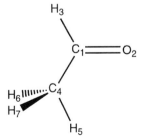

Figure E.1 Atom numbering for acetaldehyde

The definitions of the torsional angles are illustrated in Figure 2.7. To emphasize the symmetry (C_s) of the above conformation, the Z-matrix may also be given in terms of *symbolic* variables, where variables which are equivalent by symmetry have identical names.

C1							
O2	1	R1					
H3	1	R2	2	A1			
C4	1	R3	2	A2	3	D1	
H5	4	R4	1	A3	2	D2	
H6	4	R5	1	A4	2	D3	
H7	4	R5	1	A4	2	$-$D3	

R1 = 1.20
R2 = 1.10
R3 = 1.50
R4 = 1.10
R5 = 1.10
A1 = 120.0
A2 = 120.0
A3 = 110.0
A4 = 110.0
D1 = 180.0
D2 = 0.0
D3 = 120.0

Some important things to notice.

(1) Each atom must be specified in terms of atoms already defined, i.e. relative to atoms above.
(2) Each specification atom can only be used once in each line.
(3) The specification in terms of distance, angle and torsional angle has nothing to do with the bonding in the molecule, e.g. the torsional angle for C_4 in acetaldehyde is given to H_3, but there is no bond between O_2 and H_3. A Z-matrix, however, is usually constructed so that the distances, angles and torsional angles follow the bonding. This makes it much easier to estimate reasonable values for the parameters.
(4) Distances should always be positive, and angles always in the range $0-180°$. Torsional angles may be taken in the range $-180°-180°$, or $0°-360°$.
(5) The symbolic variables show explicit which parameters are constrained to have the same values, i.e. H_6 and H_7 are symmetry equivalent and must therefore have the same distances and angles, and a sign difference in the torsional angle. Although the R4 and R5 (and A3 and A4) parameters have the same values initially, they will be different in the final optimized structure.

The limitation that the angles must be between $0°$ and $180°$ introduces a slight complication for linear arrays of atoms, like for example a cyano group.

H3
\
C1———C2≡≡≡N6
/
H4
H5

Figure E.2 Atom numbering for methyl cyanide

Specification of the nitrogen in term of a distance to C_2 and an angle to C_1 does not allow a unique assignment of a torsional angle (it becomes undefined). There are two methods for solving this problem, one specifies N_6 relative to C_1 with a long distance:

C1						
C2	1	R1				
H3	1	R2	2	A1		
H4	1	R2	2	A1	3	D1
H5	1	R2	2	A1	3	−D1
N6	1	R3	3	A2	2	D2

R1 = 1.50
R2 = 1.10
R3 = 2.70
A1 = 110.0
A2 = 110.0
D1 = 120.0
D2 = 0.0

Note that the variables imply that the molecule has C_{3v} symmetry. Alternatively, a *Dummy Atom* (X) may be introduced:

C1						
C2	1	R1				
H3	1	R2	2	A1		
H4	1	R2	2	A1	3	D1
H5	1	R2	2	A1	3	−D1
X6	2	R3	1	A2	3	D2
N7	2	R4	6	A3	1	D3

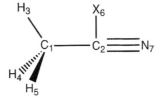

Figure E.3 Atom numbering for methyl cyanide including a dummy atom

R1 = 1.50
R2 = 1.10
R3 = 1.00
R4 = 1.20
A1 = 110.0
A2 = 90.0
A3 = 90.0
D1 = 120.0
D2 = 0.0
D3 = 180.0

A dummy atom is just a point in space and has no significance in the actual calculation. The above two Z-matrices give identical Cartesian coordinates. The R3 variable has arbitrarily been given a distance of 1.00, and the D2 torsional angle of 0.0 is also arbitrary, any other values may be substituted without affecting the coordinates of the real atoms. Similarly, the A2 and A3 angles should just add up to 180°, their individual values are not significant. The function of a dummy atom in this case is to break up the problematic 180° angle into two 90° angles. It should be noted that introduction of dummy atoms does not increase the number of (non-redundant) parameters, although there are formally three more variables for each dummy atom. The dummy variables may be identified by excluding them from the symbolic variable list, or by explicitly forcing them to be non-optimizable parameters.

When a molecule is symmetric, it is often convenient to start the numbering with atoms lying on a rotation axis or in a symmetry plane. If there are no real atoms on a rotation axis or in a mirror plane, dummy atoms can be useful for defining the symmetry element. Consider for example the cyclopropenyl system which has D_{3h} symmetry. Without dummy atoms one of the C–C bond lengths will be given in terms of the two other C–C distances and the C–C–C angle, and it will be complicated to force the three C–C bonds to be identical. By introducing two dummy atoms to define the C_3 axis, this becomes easy.

X1						
X2	1	1.00				
C3	1	R1	2	90.0		
C4	1	R1	2	90.0	3	120.0
C5	1	R1	2	90.0	3	– 120.0
H6	1	R2	2	90.0	3	0.0
H7	1	R2	2	90.0	3	120.0
H8	1	R2	2	90.0	3	– 120.0

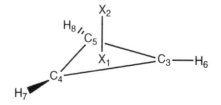

Figure E.4 Atom numbering for the cyclopropyl system

R1 = 0.80
R2 = 1.90

In this case there are only two genuine variables, the others are fixed by symmetry.

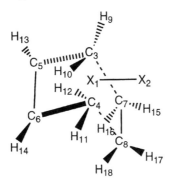

Figure E.5 Atom numbering for the transition structure of the Diels–Alder reaction of butadiene and ethylene

Let us finally consider two Z-matrices for optimization to transition structures, the Diels–Alder reaction of butadiene and ethylene, and the [1,5]-hydrogen shift in Z-1,3-pentadiene. To enforce the symmetries of the TSs (C_s in both cases) it is again advantageous to use dummy atoms.

X1							
X2	1	1.00					
C3	1	R1	2	90.0			
C4	1	R1	2	90.0	3	180.0	
C5	3	R2	1	A1	2	180.0	
C6	4	R2	1	A1	2	180.0	
C7	3	R3	1	A2	2	D1	
C8	4	R3	1	A2	2	−D1	
H9	3	R4	5	A3	6	D2	
H10	3	R5	5	A4	6	−D3	
H11	4	R4	6	A3	5	−D2	
H12	4	R5	6	A4	5	D3	
H13	5	R6	3	A5	1	−D4	
H14	6	R6	4	A5	1	D4	
H15	7	R7	8	A6	4	D5	
H16	7	R8	8	A7	4	−D6	
H17	8	R7	7	A6	3	−D5	
H18	8	R8	7	A7	3	D6	

R1 = 1.40
R2 = 1.40
R3 = 2.20
R4 = 1.10
R5 = 1.10

R6 = 1.10
R7 = 1.10
R8 = 1.10
A1 = 60.0
A2 = 70.0
A3 = 120.0
A4 = 120.0
A5 = 120.0
A6 = 120.0
A7 = 120.0
D1 = 60.0
D2 = 170.0
D3 = 30.0
D4 = 170.0
D5 = 100.0
D6 = 100.0

The mirror plane is defined by the two dummy atoms and the fixing of the angles and torsional angle of the first two carbons. The torsional angles for atoms C_5 and C_6 are dummy variables as they only define the orientation of the plane of the first four carbon atoms relative to the dummy atoms, and may consequently be fixed at 180°. Note that the C_5–C_6 and C_7–C_8 bond distances are given implicitly in terms of the R2/A1 and R3/ A2 variables. The presence of such "indirect" variables means that some experimentation is necessarily to assign proper values to the "direct" variables. The forming C–C bond is given directly as one of the Z-matrix variables, R3, which facilitates a search for a suitable start geometry for the TS optimization, for example by running a series of constrained optimizations with fixed R3 distances.

The [1,5]-hydrogen shift in Z-1,3-pentadiene is an example of a "narcissistic" reaction, with the reactant and product being identical. The TS is therefore located exactly at half-way, and has a symmetry different from either the reactant or product. By suitable constraints on the geometry the TS may therefore be located by a <u>minimization</u> within a symmetry constrained geometry.

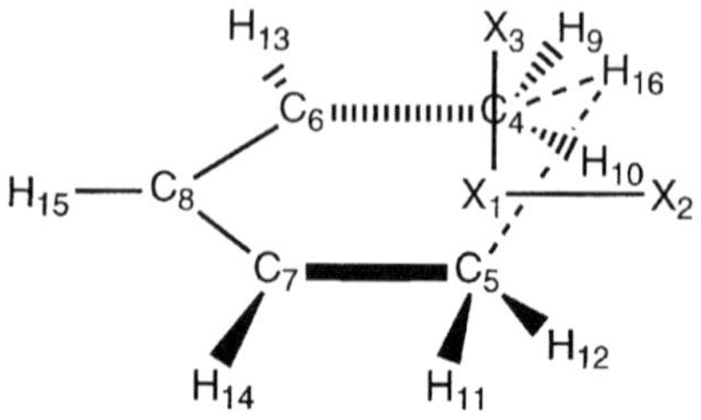

Figure E.6 Atom numbering for the transition structure for the [1,5]-hydrogen shift in 1,3-pentadiene

X1						
X2	1	1.00				
X3	1	1.00	2	90.0		
C4	1	R1	2	90.0	3	90.0
C5	1	R1	2	90.0	3	− 90.0
C6	4	R2	1	A1	2	180.0
C7	5	R2	1	A1	2	180.0
C8	1	R3	3	A2	2	180.0
H9	4	R4	6	A3	8	− D1
H10	4	R5	6	A4	8	D2
H11	5	R4	7	A3	8	D1
H12	5	R5	7	A4	8	− D2
H13	6	R6	4	A5	1	D3
H14	7	R6	5	A5	1	− D3
H15	1	R7	3	A6	2	180.0
H16	1	R8	2	A7	3	0.0

$R1 = 1.30$
$R2 = 1.40$
$R3 = 2.10$
$R4 = 1.10$
$R5 = 1.10$
$R6 = 1.10$
$R7 = 3.20$
$R8 = 0.70$
$A1 = 80.0$
$A2 = 90.0$
$A3 = 120.0$
$A4 = 120.0$
$A5 = 120.0$
$A6 = 90.0$
$A7 = 60.0$
$D1 = 160.0$
$D2 = 60.0$
$D3 = 160.0$

The mirror plane is defined by the dummy atoms. The migrating hydrogen H_{16} is not allowed to move out of the plane of symmetry, and must consequently have the same distance to C_4 and C_5. A minimization will locate the lowest energy structure within the given C_s symmetry, and a subsequent frequency calculation will reveal that the optimized structure is a TS, with the imaginary frequency belonging to the a'' representation (breaking the symmetry).

Index